THE

BUDGETT MEMORIAL VOLUME

THE WORK OF

JOHN SAMUEL BUDGETT

BALFOUR STUDENT OF THE UNIVERSITY OF CAMBRIDGE

BEING A COLLECTION OF HIS ZOOLOGICAL PAPERS, TOGETHER WITH A BIOGRAPHICAL SKETCH BY A. E. SHIPLEY, F.R.S., AND CONTRIBUTIONS BY RICHARD ASSHETON, EDWARD J. BLES, EDWARD T. BROWNE, J. HERBERT BUDGETT AND J. GRAHAM KERR

EDITED BY

J. GRAHAM KERR

CAMBRIDGE:
at the University Press
1907

CAMBRIDGE
UNIVERSITY PRESS

University Printing House, Cambridge CB2 8BS, United Kingdom

Published in the United States of America by Cambridge University Press, New York

Cambridge University Press is part of the University of Cambridge.

It furthers the University's mission by disseminating knowledge in the pursuit of education, learning and research at the highest international levels of excellence.

www.cambridge.org
Information on this title: www.cambridge.org/9781107424029

First published 1907
First paperback edition 2014

A catalogue record for this publication is available from the British Library

ISBN 978-1-107-42402-9 Paperback

PREFACE.

AT a meeting of friends of the late John Samuel Budgett held in Cambridge on February 8, 1904 it was decided to perpetuate his memory by the publication of a Memorial Volume which should contain reprints of his various Zoological papers together with descriptions of the more important material brought back by him on his various expeditions.

The Syndics of the Cambridge University Press were so good as to undertake the responsibilities of publication, the necessarily heavy expenses of illustration being met by a fund subscribed by Budgett's friends. Mr A. E. Shipley of Christ's College has kindly acted as Honorary Treasurer of this fund; he has also contributed a biographical sketch and he has helped in many ways towards the bringing out of the volume.

The preparation of the volume has taken a considerable time, particularly the working through of the extensive embryological material of *Gymnarchus* and *Polypterus* so as to make it possible to give a fairly complete sketch of the development of these forms.

It is right that I should say a word of thanks to those who have undertaken the working out of Budgett's material. To Mr Assheton special acknowledgement is due for the time and care which he has devoted to the working out of the development of *Gymnarchus*. At the same time it must be said—what must

be in the hearts of each one of those friends of Budgett who hav
contributed to the volume—that anything that each one of then
has done has been a labour of love and must seem to him bu
a small tribute indeed to the admiration and affection which h
bore to his friend.

The excellent portrait of Budgett which serves as a frontis
piece is from a photograph by Messrs Mason and Basèbé c
Cambridge: it has been admirably reproduced by Messr
T. & R. Annan & Sons of Glasgow.

Thanks are due to the Council of the Zoological Society o
London, the Council of the Cambridge Philosophical Society an
to the Editors of the *Quarterly Journal of Microscopical Science*
The Ibis and *The Field* for permission to reprint articles from thei
respective publications.

J. G. K.

University of Glasgow,
November 1, 1907.

TABLE OF CONTENTS.

Contents

PLATES.

LIST OF SUBSCRIBERS.

N. P. Adams, Esq.

Professor Sir T. Clifford Allbutt, K.C.B.

Dr H. K. Anderson

Anonymous

D. Appleby, Esq.

Mr & Mrs R. Assheton

Lieut.-Col. H. H. Godwin-Austen

P. H. Bahr, Esq.

H. C. Baker, Esq.

The Right Honourable. A. J. Balfour, M.P.

The Rev. Professor W. E. Barnes

W. Bateson, Esq.

W. A. J. Bell, Esq.

Professor A. A. Bevan

G. P. Bidder, Esq.

E. J. Bles, Esq.

C. L. Blew, Esq.

L. A. Borradaile, Esq.

G. A. Boulenger, Esq.

F. E. Bray, Esq.

Professor T. W. Bridge

H. H. Brindley, Esq.

C. N. F. Broad, Esq.

J. Y. Buchanan, Esq.

The late James Budgett, Esq.

R. S. Budgett, Esq.

R. P. Bullivant, Esq.

P. W. Bunting, Esq.

E. R. Burdon, Esq.

The Rev. Dr Butler, Master of Trinity College

The Rev. H. J. Buxton

P. M. S. Carmichael, Esq.

H. G. C. Carr-Ellison, Esq.

C. J. P. Cave, Esq.

P. A. Chasemore, Esq.

J. W. Clark, Esq., Registrary

W. H. Clayton-Greene, Esq.

Clifton College Library

The Rev. E. Cobham

G. G. Collet, Esq.

H. G. Comber, Esq.

Dr A. Cooke

C. Forster Cooper, Esq.

W. A. Cunnington, Esq.

Rev. Dr W. H. Dallinger

T. M. Davies, Esq.

L. Doncaster, Esq.

Captain C. Druce

Dr W. L. H. Duckworth

J. W. Eames, Esq.

H. J. Edwards, Esq.

A. H. Evans, Esq.

The Rev. T. C. Fitzpatrick, President of Queens' College

Dr W. M. Fletcher

The late Sir Michael Foster, K.C.B.

The Right Honourable Lewis Fry

E. G. Gallop, Esq.

J. Stanley Gardiner, Esq.

T. R. Glover, Esq.

J. G. Gordon, Esq.

P. B. Haigh, Esq.

Dr S. F. Harmer

K. J. Harper, Esq.

The Rev. F. W. Head

Professor W. A. Herdman

Dr A. Hill

A. W. Hill, Esq.

L. E. Hubbard

A. Hutchinson, Esq.

J. H. Jenkins, Esq.

Professor J. Graham Kerr

G. J. F. Knowles, Esq.

F. F. Laidlaw, Esq.

Professor J. N. Langley

Mrs Lidgett

Miss E. S. Lidgett

J. J. Lister, Esq.

F. Livesey, Esq.

H. M. Longridge, Esq.

Keith Lucas, Esq.

A. G. Lyttelton, Esq.

C. A. Mander, Esq.

G. Le M. Mander

Professor Howard Marsh, Master of Downing College

Dr F. H. A. Marshall

The Hon. E. S. Montagu, M.P.

Principal C. Lloyd Morgan

O. Mowatt, Esq.

R. S. Nairn, Esq.

A. F. Nicholson, Esq.

The late Professor Alfred Newton

Professor G. H. F. Nuttall

M. G. Ll. Palmer, Esq.

The Rev. R. St J. Parry

A. D. Pass, Esq.

Professor H. H. W. Pearson

J. B. W. Pennyman, Esq.

R. C. Punnett, Esq.

A. C. Radford, Esq.

G. A. Redman, Esq.

Professor S. H. Reynolds

Dr W. H. R. Rivers

A. W. Rogers, Esq.

The Honourable N. C. Rothschild

P. Russell, Esq.

J. T. Salusbury, Esq.

Dr P. L. Sclater

S. D. Scott, Esq.

Professor A. Sedgwick

Professor A. C. Seward

D. Sharp, Esq.

A. E. Shipley, Esq.

Captain R. B. Shipley

Mrs Sidgwick

R. W. Smith, Esq.

W. E. Taylor, Esq.

Dr H. W. Marett Tims

G. E. F. Torrey, Esq.

The University Natural Science Club

H. D. D. Walthall, Esq.

W. St A. Warde-Aldam, Esq.

W. W. Warde-Aldam, Esq.

L. J. Weatherley, Esq.

R. Wellesley, Esq.

C. A. Wells, Esq.

R. N. Willett, Esq.

Dr A. Willey

The Rev. J. V. Wilson

Professor G. Sims Woodhead

JOHN SAMUEL BUDGETT

WHEN John Samuel Budgett joined Trinity College in the autumn of 1894 he brought with him letters of introduction from Mr S. H. Reynolds of University College, Bristol, which at once introduced him to the circle of Zoologists at Cambridge. These letters dwelt on two matters, both of them eminently characteristic of our friend. One was his great modesty, a modesty which at times amounted almost to wholly unnecessary self-depreciation, the other was his quite extraordinary skill in making anatomical preparations.

J. S. Budgett was born at Redlands House, Bristol, on 16th June, 1872. Two years later his father moved to Stoke House, Stoke Bishop, and here the greater part of his life was spent. His first school was a kindergarten, from which he entered Clifton College, but he had to leave there when he was about fourteen years old, as he was suffering from a severe form of headache, the result of an accident. He continued his studies, however, with private tutors, and finally joined University College, Bristol.

As a boy he lived with his parents in the charming old stone house with stately carved portal in the pretty village of Stoke Bishop, at that time much more in the country than now, when the houses of Clifton threaten to draw near it. Here in the spacious grounds he was allowed to build aviaries and adapt outhouses for the shelter of his numerous pets. Here also he acquired some of his skill as a dissector, and a carpenter's shop in which the children had learned carpentry was gradually turned into a Laboratory and a Museum in which stuffed birds, skeletons of a cow, of a deer, and of the children's Shetland pony, and many wonderfully minute dissections were displayed. Budgett as a boy was fully abreast of modern methods, and used to make many preparations with natural surroundings, such as a stuffed swallow with its nest under a bit of tiled roof. He was a frequent visitor at the Clifton Zoological Gardens, and there he learned much. He always enquired attentively after any sick animals, not, it has been suggested, with a view to prolonging their life. In the summer he used often to go for long walks at 3 a.m. to enjoy the sunrise and to watch the awakening bird and beast life.

Mr W. H. Budgett, our friend's father, was a keen microscopist and a member of the Bristol Microscopical Society. He was also on the Council of the Bristol Museum, and was keenly interested in many branches of Natural History, a keenness he was eager his children should share.

Whilst Budgett was a boy his father's house was visited by many men of science. Amongst these, Professor W. K. Parker, who for twenty-three years in succession spent a fortnight at Stoke House, undoubtedly exercised a strong influence on the boy. At this period the Professor was advanced in life, and to quote his son's admirable biography[1] of him, "His habits became more retired, and almost his only outing was an annual visit to his friend Mr W. H. Budgett, at Stoke House, Bristol"; still, to the last, Professor Parker was brimming over with enthusiasm, and he can hardly have failed to fascinate the young zoologist. Professor Parker was remarkable for his skill in the preparation of delicate skeletons, such as those of tadpoles and of the minutest birds, and I have often thought that Budgett's extreme cleverness in making all sorts of anatomical preparations owed

---

[1] *William Kitchen Parker*, a Biographical Sketch by T. Jeffery Parker, London, Macmillan & Co. 1893.

much to the example of his elderly friend. Another visitor, the Rev. Dr Dallinger, well known for his great gift of exposition, must often have stirred the interest of the keen and eager lad.

After leaving school he still continued to study Zoology, and received much help from the Principal of University College, Bristol, Dr Lloyd Morgan, and from Mr (now Professor) S. H. Reynolds, but he was in a great measure self-taught. His knowledge of bird-life, his acquaintance with the methods of mounting objects, and his skill in keeping animals in captivity healthy, could not have been acquired in the Lecture-room. It must have been about this time that he began to design a new microtome which he laid aside for some years. During the last few months of his life, after his return from the Niger, he again took it up and had hopes that it would prove an effective instrument. He also devised a scheme of his own for making models of structures by cutting them into sections, drawing the sections on cardboard, cutting the drawings out, and sticking them together, one behind the other, in series.

His thorough knowledge of Natural History made him many friends on his arrival in the University, both among the senior and the junior members, and he was elected to the Cambridge University Natural Science Club in his first year, a somewhat unusual honour.

In 1896, at the end of his second year at Cambridge, he took Part I of the Natural Sciences Tripos. His University career was interrupted in his third year by a voyage to South America.

In 1890, when on the Page Expedition to the Pilcomayo, Mr Graham Kerr, now Professor of Natural History at Glasgow, had from a soldier the account of a fish dug up from the mud and eaten, and from the description given he suspected it to be *Lepidosiren*. This was confirmed in 1894 by the German Naturalist Bohls who first obtained this fish from the Gran Chaco[1]. The expedition came to an untimely end, the leaders of it either died

[1] According to an article in *Die Thierwelt*, VI. Jahrg. 16 Sept. 1896, the Gran Chaco habitat of *Lepidosiren* was first proved by Ternetz, "vor 3 bis 4 Jahren."

or deserted. The river disappeared and left their small steamer stranded in a dried-up bed. Mr Graham Kerr, as he then was, with the remnant of the expedition succeeded in reaching the coast and, returning to England, he joined Christ's College in 1892 and took his degree four years later. But during the time he spent as a student at Cambridge the haunt of the *Lepidosiren* was ever before him. In the autumn of 1896 he was able to start for the Gran Chaco of Paraguay, and he took Budgett with him.

Budgett was always a delightful travelling companion, considerate, thoughtful of others, very clever with his fingers, resourceful, and a good sportsman.

His knowledge of the fauna, particularly of the bird-fauna, was considerable, and he made a special study of the Anurous Amphibia of the swamps of Central South America, a study which resulted in his "Notes on the Batrachians of the Paraguayan Chaco." He kept very complete diaries. These were never meant for publication, and were written in every conceivable circumstance of discomfort, but I have reproduced passages from them practically as they stand, with only a very few verbal alterations where the sense seemed obscure. Several passages from that of his South American journey are worth quoting. They all show an eye alert for natural phenomena of every kind. On the 7th September, 1896, steaming up the great La Plata river on their way to Asuncion, he writes:

"Sky overcast, lightning and thunder at daybreak. Steaming through the delta all day, mostly hugging the western bank where the edge of the pampas formed a cliff about 60 ft. high. The channel by which we went up was as a rule about one mile in width, though the true eastern bank of the river was at least 50 miles distant. Perfectly flat shifting islands filled up the whole of the river at this part: they were covered with low but rich vegetation, and teemed with all kinds of wild fowl. We passed two settlements, San Nicolas and Villa Constitucion; at the latter place the manager had a very large English house. Lucern seemed to be the chief product here.

Birds: the most abundant *Milvago chimango*, perching in trees and on the ground. *Phalacrocorax brazilianus*, differing only from the British species in its yellow beak. *Euxenura maguari*, flying with legs straight out behind, wings dipping only slightly below the horizontal; six beats and then floating on the air for some distance, and then six beats again and so on. *Aechmophorus major*, somewhat larger than the British great crested grebe, mostly of a chestnut colour, with white secondaries. *Ardea cocoi*, tall herons, standing with neck outstretched, but at angle of 15° from the vertical. *Dendrocycna fulva* (the tree-duck), one flock; *Mareca sibilatrix* (a widgeon), one flock. *Polyborus tharus* (the Carancho or Caracara), two perching on distant tree.

Ran aground 5.30 p.m., trying to get to pier at Rosario, water very low. 100 tons of cargo to go to Rosario, hope she will then float."

Again:

"*September 9th*, 1896. Dull morning and close. 1 p.m. reached La Paz. The river is rather uninteresting to-day. About 3 p.m. regular Pampero thunderstorm, with violent wind from south-west, and rain in torrents. At first there was a black bank of cloud lying near the horizon, from which flashes of lightning passed to the earth, five or six in rapid succession. These series of flashes came at almost regular intervals. Other flashes extending almost right across the sky came more irregularly. Then the sky immediately above the horizon became brilliant yellow and streaked with clouds of sand; suddenly the storm burst upon us, lashing up the water and driving off the tops of the waves in clouds of spray. Then down came the rain in heavy torrents."

Near Las Palmas he notes that:

"The ground was bright with patches of red, blue, and pink verbena and innumerable other flowers. The most striking thing was the noise the frogs made; a regular chorus of little bells[1] was continually going on, while here and there we heard a noise like two cats mewing[2]; we did not, however, discover what species made these. Many very brightly coloured birds were seen, among which were the Carancho, some Picui Doves, and a pretty Snipe. I heard for the first time the oft-repeated call of the Martineta. A rabbit, or Brazilian hare more likely, bolted from beneath my feet. I noticed a nest of the Termite and also of the Oven bird, which it somewhat resembles."

Budgett had a keen eye for beauty, and in unadorned language was often able to reproduce the scene to those who perforce must remain at home. On 27th September, 1896, at Concepcion, he writes:

"'Laguna' was covered around the edges with several beautiful kinds of

---

[1] Probably *Bufo granulosus*, Spix. [2] Probably *Paludicola fuscomaculata*, Steind.

Camaloté[1] in fine bloom. The foliage and flowers were most luxuriant; beautifully plumaged birds flew here and there, gorgeous butterflies, and flimsy, delicate dragon-flies made the scene charming. A shower of rain refreshed the flowers and caused the gaily coloured Kingfishers to renew their angling efforts. We brought back our little bottle of spoil and did our first piece of scientific investigation with the microscope in South America. Soon crowds of natives shut the light out from our windows, among whom were some Lengua Indians. These are always to be seen wandering about the town or paddling across to the Chaco bank opposite."

And three days later:

"Started in the canoe about 7.30 a.m., intending to paddle up stream two leagues and down the Chaco side of the island which lies opposite Concepcion. We found, however, that the stream and wind were too much for us, so turned back and floated some distance down stream. At a rather likely place for snipe we went ashore with a few No. 7, and after walking inwards until we came to a lagoon, we saw on a little island a flock of about fifty Whistling Ducks. After a very cautious stalk I got within forty or fifty yards of them, trying if possible to get near enough to shoot a brace through the head, before they rose, but it was not to be. I was discovered too soon and up rose the flock, after which I sent two or three ineffectual shots. On the same little bit of water we bagged four nice Sand-snipe with the remaining cartridges; and then of course we had around us numbers of Snipe, Jaçanas, a large flock of Ibises and other birds. How could they know we had shot our last for that morning?"

His interest in the Frogs is shown by his diary for 13th October, 1896:

"Started out for walk at 6 a.m. with gun. Shot several birds and found some queer insects. Came back at 10 a.m. On the way passed two or three hundred vultures sitting on the palings of a slaughter house. Read newspapers till breakfast. After breakfast skinned three birds. Then went down to pool by river and discovered a beautiful little green frog[2] on the Camaloté. On almost every piece of Camaloté there was one. The males keeping constantly calling to one another. On the way back I found by much searching two new species of frogs. The Mosquitoes have been worse to-day than ever before.

---

[1] Budgett writes this word Camelota. Mr S. A. Skan has kindly suggested to me that it is probably the Camelote mentioned by Hieronymus in his *Plantae Diaphoricae Argentinae* and by Bettfreund in his *Flora Argentina* as the native name for the *Eichhornia azurea*, Keenth.

[2] *Pseudis limellum*, Cope.

Frogs. No. 1[1] and No. 2 were male and female of a frog[2] which seems to be more common than any at Concepcion. The male has a characteristic call, beginning on a rather low note, and ending on a higher one. It is very common and at sunset is everywhere in one's path. Its back is striped with parallel ridges of the skin. It is marked with black blotches on a greenish grey ground; the most characteristic mark is the triangular black patch between the eyes. The external ear is a very conspicuous tympanum.

No. 3 is a male. It makes a call similar to No. 1, but lower and softer. If it is chased, contrary to the usual rule, it still continues to call loudly, and thus does not easily escape the collector. This frog has a clumsy appearance, and its general shape is toad-like. The markings are distinct, chiefly black and green and a little yellow; the eyes are prominent. Very difficult to find in the grass, but with each croak it reveals itself by causing the yellow skin of the chin to become visible at the sides of the head.

No. 4. A very small brown frog[3], almost impossible to see against a background of earth. It has a shrill, sharp call, kept up constantly unless approached, when it immediately ceases. It is very agile and extremely shy. It is marked with small black spots on a yellowish-brown, dark ground. Pale yellow underneath. When many of these frogs are croaking the sound produced is nearly continuous. The tympanum is evident. Found in damp waste ground outside Concepcion.

No. 5. A species of *Rana*, with dark brown markings on a greyish brown ground. A good deal of yellowish marking about it. Tympanum evident. Broad part of back without markings. Female.

No. 6. Female[4]. Abundant on the Camaloté leaves at Concepcion. Capable of changing its colour greatly, from bright green to dull brown. Underneath silvery. Two white streaks run backwards from the eyes. The call is a succession of sharp croaks or vibrations resembling the sound made by castanets and caused by the inflation of the throat. Until one is accustomed to the sight they are difficult to see, so well does their colour and small size protect them, but once seen, they are everywhere. They are very lively and hop quickly *on the surface* of the water by means of their hind feet, which are webbed right up to the tips of the toes. I found a small Gasteropod partially digested in the stomach of one of them.

No. 7. Male of same."

And again two days later:

"After coffee went out for a walk in the camp north of Concepcion. Found all the country under water from last night's storm. Soon heard a pretty note.

---

[1] These numbers refer to the jars or bottles in which the specimens were preserved. The quotation shows the accurate and minute care which Budgett took in collecting specimens.

[2] *Leptodactylus ocellatus*, L.   [3] *Leptodactylus bufonius*, Blgr.   [4] *Pseudis limellum*, Cope.

I could not tell at first whether it were bird or frog. The continual repetition from the same quarter told me that it was the call of a frog (No. 8[1]) which I had not heard before. I stalked him for a considerable time, and at last saw, climbing over the blades of grass which were left sticking out of the water, a minute black frog with yellow and red spots. I stalked some more, but did not succeed in finding any but the two I had at first heard. On the way back I shot a fine brilliantly-coloured lizard. In the afternoon we walked across the island to the Mission Station. Cleaned guns and had tea. As we were passing homewards we saw the Indians catch two large Skates or Sting-rays[2], which were among the Camaloté. While they were lying on the bank speared through the spinal cord the large one gave birth to fifteen minute skates, with external gills, which were long and filamentous. We secured them and then started off round the island with the canoe, arriving at 8 p.m. by moonlight. On my way up from the boat we caught another frog hopping across the street by moonlight.
    Frogs. No. 8. Found in street after dark."

Collecting in the Chaco was not all fun; there was a good deal of discomfort, wet and storm; and beautiful and interesting as many of the creatures were, many of them were the reverse of attractive. On 27th October, 1896, he writes:

"We had been walking all the afternoon in water and long grass, and when we got within sight of the camp a violent storm of wind, lightning and rain came on; we had just time to shove all our things under cover when down came the rain, and huddling round the fire we ate our rice and got thoroughly soaked. The storm soon cleared off, and after a good warming I curled up under my mackintosh sheet and should have slept soundly but for a new pest, the *polvorinos*[3].

I was getting accustomed to mosquitoes and not to mind them so much, but these are too fearful for words, smaller than the midge at home and infinitely more painful. They attack one in thousands, and make one think that Dante might have got some useful hints for his *Inferno* by coming to the Paraguayan Chaco. Mosquitero is useless, one can only wrap one's head up in one's rug."

He took great interest in the natives, and recorded in the pages of his diary many observations on their habits and appearance. He also wrote down native vocabularies whenever he could get them. Writing about the Lenguas, a tribe who inhabit the Chaco, he says on 28th October, 1896:

---

[1] *Phryniscus nigricans*, Wigm.   [2] *Taeniura dumerilii*.   [3] *Simulium* sp.

"There seems to be a great scarcity of boys about 16 years old, though a good many of about 12. The Indians often kill their babes for no apparent reason. If one of the community die the Toldo[1] is removed. The dead are buried where they die with knees to chin. The doctors are great tyrants. If a man is ill food is kept from him; this together with the baby slaughter seems to effectually prevent their becoming a dominant race.

All the Lenguas wear round discs of wood in their ears. In the first place they are mere twigs, which are frequently replaced by larger ones.

The women are all small and fat and sometimes nice-looking. The race as a whole is very tame and do not fight, neither are they intelligent, but quite childlike, especially in their improvidence. The boys are very sharp and jolly, quite like nice English boys.

While we have meals or write there are always some of these Indians sitting round watching us. Their only garment is the woollen woven poncho; these are very well made, and will fetch in the market 20s.—30s.

They are of course very fond of beads, of which they only wear red and white. Their necklaces are made of various things, often pieces of *Bulimus* shell strung together or Cierbo[2] teeth. The men generally have the central part of their hair bound round in the shape of a stiff stick, which may lie forwards or backwards."

The life in the camp was very varied, and, owing to his keen interest in the anatomy and structure of all kinds of animals, Budgett spent much time in dissecting and in making many microscopic preparations. On 28th November, 1896, the following entry occurs:

"More bullock marking, continued dusty north wind and high temperature. The old bull gave a good deal of trouble, nearly hanging Mr Sibbet with the end of the lasso. Repaired camera, cleaned guns. Dissected young *Ceratophrys*, showing beautifully immature ovary and oviduct. Also the nervous system was exquisite, the sympathetic ganglia showing up bright orange. After tea went out with gun across the swamp and had a fearful time, rushes had grown high, some had fallen horizontally, so that at each step one's feet were sawn across by the fine teeth running along the under-surface of the midrib; socks were soon gone to shreds, and skin began to go. After half an hour's walk of this sort I reached the bird island, but was welcomed by such an attack from the garrison stationed there that I immediately turned and fled into the swamp

---

[1] More correctly "Toldoria." Toldo is a hut or shelter: toldoria is a collection of huts or an encampment.

[2] *Cariacus paludosus.*

again, fighting my way back to the Toldo, until I arrived pretty well exhausted, having only shot one small bird."

And two days later:

" Damp day, I walked round the Monte early. Found camera out of order; spent all morning and part of afternoon overhauling it. Mounted six slides in the afternoon. Mouth parts of *Tabanus*, Antennae of Dragon-fly, Mosquito male, Polvorino and larva of some kind of beetle. Had two *Ceratophrys* brought me, also a beautiful caterpillar with orange skin and black tufts of hair. Philip brought me a fine chrysalis. We discovered that the 'bicho' which has been heard splashing in the pool is the glorious green frog[1]. I must catch him.

1st December, 1896. At night before 6 p.m. the green Tree-frogs[2] caught by the pool laid white eggs with abundance of firm jelly. Spent the whole day preserving eggs in various stages of segmentation in Corrosive Acetic, Flemming's, Perenyi's, and Van Beneden's Fluids and in Formalin. In siesta time I photoed my laboratory and myself, developed in the evening.

2nd December, 1896. Preserved more eggs of green Tree-frog, but they appear to have died at the closing of the blastopore. Collected several Myriapods and began to make preparations of their external genital organs. After tea rode out on the black horse with Kerr, and had a fine bath, Tabanos very bad. At night a beautiful *Mantis* was brought in. Wind strong from north, it obliged me to put up awning. Beautiful specimens of leaf-mimicry in locusts came on to table at night.

3rd December, 1896. Wet morning, preserved more frog embryos, only a few apparently still alive, but in the afternoon discovered that I had still another batch tightly wrapped up in a leaf. These are all developing splendidly, and clear up the doubt as to how they are laid in nature, no water has had access to them, and yet they had developed far better than those to which I had allowed water and air free access. Apparently after the closure of the blastopore the epiblastic layer sinks downward, a milky fluid passing up through it. From this sunken area arises the blastoderm as a simple swelling, later stretching back to the swelling at the edge of the blastopore. Eggs are prominent and early developed. The neural canal closes over very gradually. The yolk sac became constricted off, the embryo curling over it, no tail is formed and rudiments of limbs are formed at once. Preserved one of the tree-frogs at night.

4th December, 1896. Preserved embryos of tree-frog in morning and dissected one also at 10 o'clock in the morning. External gills are well developed, and the circulation is going on, limbs not yet formed, tail now developes rapidly. Heart beating, optic involutions quite like chick. Auditory pits large; at 3 p.m. the first pair of gills greatly developed at expense of second pair. Blood red,

---

[1] *Pseudis paradoxa*, Linn.   [2] *Phyllomedusa hypochondrialis*, Cope.

tail much longer. A specimen of each in Formalin. Three fine frogs of a new species were brought in from swamp. Found another tree-frog on grass. At night fine frog caught by the pond, much like the ordinary fishing frog, and also much like the common frog found at Concepcion."

One more quotation from the South American diary must suffice. It was written on the 27th March, 1897, on the return of the expedition to Asuncion on their way home:

"At 6 a.m. we started on again through six leagues of Picada[1], passing under giant trees which almost shut out the light and kept us pleasantly cool, smaller trees of every description grew beneath these giants, while below these again was a very great variety of ferns, the tree-ferns standing sometimes eight or nine feet high. Sometimes we would have a long stretch of fairly level ground with nice soft sand where we could gallop; but more frequently we were clambering up and down steep water-courses with little rivulets running along the bottom of the gully. Sometimes even we had to lead our horses up and down these.

All the way along we were accompanied by scores of beautiful butterflies, often from one piece of dung we would disturb five or six great blue *Morphos*; these would then easily and gracefully flap their brilliant wings and circle round us as we rode.

But when we came to the little streamlets in the gullies, at first I could not believe my own eyes, for here and there were great flocks of butterflies sitting about over the rocks in masses of colour according to their species. There would be perhaps a patch two yards square literally covered with a large yellow butterfly, something like a Brimstone, only twice as large, the males being of a bright orange colour. Then a yard or two away there is another patch, perhaps more compact, composed of a fine species of a rich brown colour with a double bar of white and orange, stretching right across the four wings and the body. A little further there would be a patch less densely but uniformly populated with a variety of species, from a brilliant little fellow, red and blue above with concentric circles of sulphur below upon a black ground, to the huge *Morpho* with his blue wings measuring six inches from tip to tip. The large sulphur patches were perhaps more frequent. Then as one looked down into the gully, besides this Lepidopteran carpet of idleness covering the rocks, the air would be alive with more busily engaged individuals, partly of those already mentioned, but certainly also of the flashy energetic HELICONIIDAE with their brightly striped wings of black and red or orange; their mimics the DANAIDAE[2] were also there. But when I walked my horse into the stream, and dismounted to let him drink,

---

[1] A path cut through the forest.

[2] The South American Danaidae are now separated as Ithomiidae. The Ithomiidae have colour resemblances with both Heliconiidae and Pieridae.

the air seemed filled with the flapping of flimsy wings which gently fanned my face, and, as I waited a moment or so, they began to resume their basking, the social species collecting in their flocks, while the more varied assemblages settled everywhere. Twenty-five were perched upon me at one moment, the smallest measuring more than two inches across the wings, six were upon my gun barrels, my horse too was pretty well covered, while a great *Morpho* had alighted upon his forelock.

Truly one felt here, among the luxuriant fern foliage and clouds of coloured wings, almost overwhelmed with the abundance of life.

We came out about midday at the far end of the picada, and felt the heat of the sun in the last, shadeless, part of our ride very much. After another three hours' ride over some steep hills we arrived at Caa Guazú, the village of call for the Yerba[1] tramps. It is a most uninteresting place, and we felt glad that Caa Guazú itself was not what we went out for to see. The village consists of a solitary square upon a hill, with its four sides of attached thatched houses looking rather like farm buildings. The church (a thatched barn) stands in the centre of the square with its little belfry close by. This is all there is to say about the place. We were made fairly comfortable in a little store or public place, and started for Villa Rica again next morning at 7 a.m."

The expedition returned to England in the summer of 1897. It had been brilliantly successful. It brought back a large supply of fully grown specimens of *Lepidosiren paradoxa*, up to this date a fish so rare as to be seen only in a few Metropolitan Museums, and for which as much as £50 a piece had been paid but shortly before the expedition started. It also brought back a complete series of *Lepidosiren* eggs in all stages of development and a collection of the various phases which the larval *Lepidosiren* passes through. It was a remarkable feat. To go straight as an arrow to the place where this almost unknown fish lives, to arrive at about the time of the breeding season, quite unknown before, and to collect and preserve all the delicate and varying stages of development within some seven months, places the expedition in the first rank of Zoological exploration. But there is another factor which makes it even more remarkable, and that is the wonderful condition of the material when it arrived. Every zoologist knows the difficulty of preserving animals and eggs, which have not been worked at before, in the best way to render

---

[1] The so-called "Tea" or Maté of Southern South America.

permanent the minute details of microscopic structure. Even in the laboratory with every reagent and appliance around one, one often fails. The difficulty in improvised rooms, worried by all sorts of insects, by torrential rains and occasionally floods, by inquisitive and highly suspicious natives who have no glimmering as to what you are after, not forgetting the care of the horses and the need of provisioning the camp, must have been enormous, but it was overcome. The material brought back was preserved so that the finest histological details were revealed and this is true not only of the *Lepidosiren* but of the Amphibian collection which had been Budgett's special care. The success in this respect was certainly partly due to his skill in manipulation and his peculiar knowledge of the use of reagents.

On his return to Cambridge Budgett had to take up the task of reading for the Second Part of the Natural Sciences Tripos. He was never a good subject for examinations. His natural modesty made him distrust himself, and on such occasions he did not do himself justice. His reading had of course suffered during his visit to South America, although he took with him a number of standard zoological works, and diligently perused them in the Chaco; he also added much to his knowledge by constant dissection and careful observation. Still he had many drawbacks; and when the class-list came out in June, 1898, and his name was in the Second Class, his friends and he himself were pleased.

He almost immediately set to work upon his collection of South American frogs, some of the results of which appeared later in the year under the title "Notes on the Batrachians of the Paraguayan Chaco[1]," the memoir already named.

Budgett was a gifted draughtsman, and the figures which illustrated his various memoirs were both artistic and accurate. At times he made the most charming water-colour sketches, and

[1] The full title is "Notes on the Batrachians of the Paraguayan Chaco, with Observations upon their Breeding Habits and Development, especially with regard to *Phyllomedusa hypochondrialis*, Cope. Also a description of a new Genus." It appeared in the *Quarterly Journal of Microscopic Science*, Vol. 42, N.S. 1898.

two I possess of scenery on the Gambia show a great artistic power of depicting water and sky.

It was about this time that he made several improvements in an apparatus originally designed by Professor Graham Kerr for reconstructing solid figures from microscopic drawings. The principle of this apparatus consists in drawing each section on a plate of ground glass whose thickness bears a definite relation to the thickness of the microscopic section. Then the square plates of glass are placed one behind the other in a series, and the whole is immersed in a bath of oil which eliminates the opaqueness caused by the roughened glass, and allows a solid figure to appear which represents an enlarged view of the object cut into sections. Put thus shortly the matter seems simple enough but in practice there are many difficulties. Budgett showed his usual ingenuity in constructing devices to centre the glasses— a very important but a very difficult matter—and his exceptional knowledge of reagents in trying to find an oil which permits the maximum of light to pass through the bath. He also devised a Tropical Aquarium which has recently been figured and described by Mr E. J. Bles[1], of Glasgow University, who has used it with great success in the rearing of *Xenopus*.

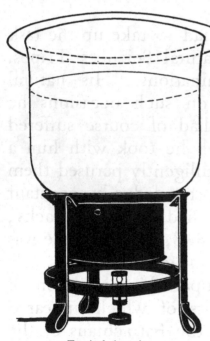

Tropical Aquarium.

But all this time, and indeed long before, when he was in the swamps of Paraguay, his thoughts had turned to one of the great unsolved problems of Zoology. There are a couple of remarkable fishes called *Polypterus* and *Calamichthys* found only in the rivers of Africa, the sole survivors of a vast group which flourished in the Palaeozoic and Mesozoic Periods. We knew little about them, little of their anatomy, nothing of their habits,

[1] *Transactions Roy. Soc. Edinburgh*, XLI. 1905, p. 794.

and nothing of their development which might throw much light upon the origin and upon the relationship of fishes. This problem Budgett determined to attack, and for the next five years, in spite of delay, in spite of every kind of discouragement, he followed his quest with dauntless courage, and in the end he succeeded. The measure of his success is written in this volume, but by another hand than his.

He at first thought of the Nile as the best river for his purpose, but at the suggestion of his friend, Mr P. L. Sclater, he chose the Gambia as the scene of his first exploration. He made careful preparations, and had a number of excellent figures of *Polypterus* prepared by our well-known Cambridge artist, Mr E. Wilson, which he afterwards used in explaining to the natives the fish he sought. On the 19th October, 1898, he sailed from Liverpool on the s.s. *Dahomey* for Bathurst at the mouth of the Gambia, which he reached on 3rd November. As usual he kept a diary, and though travelling alone, thousands of miles from anyone who sympathized with, or even understood, his aims, the diary contains no word of complaint, and hardly a mention of the dangers and difficulties he underwent. During his stay on the Gambia, which lasted over eight months, he made his headquarters at McCarthy Island some 150 miles from the mouth of the river. Here he met with much kindness from the officials, especially from the late Mr P. Wainewright, Travelling Commissioner, with whom he stopped at the Government House which he thus describes:

"The house itself lies back from the river and near it lie two or three ruins of the barracks of former days. All is now waste and overgrown with bushes and jungle. There are but four stone houses at McCarthy, the African huts lie further back."

As usual he paid great attention to the local fauna, and such entries as this are common:

"Walked out, passed the Cemetery, after early tea, and saw vast quantities of birds. Three sorts of Jays, one of which I shot, innumerable Finches. Shot one Pigeon for the pot on the way back. I noted the call of the English

Redstart, and there he was. I am not shooting birds for skins yet, as my supply of cartridges has not arrived.

The rest of the day I spent in unpacking boxes. In the evening I walked out with gun and shot a fine Spurwing Plover and a Pigeon. Missed right and left at Partridges. I walked along the edge of the swamp where the Mudfish was said to abound. Here two small boys promised to bring me the same of every size. A man begged the leg of my Plover of me, as I returned, to use as 'medicine.'

When having my bath I caught in it three small *Hylas*. Skinner, the B.T.C.[1] man, brought me in the evening a small *Gecko* caught in the house. This creature has the reputation of being very poisonous, and of coming to one's mouth while asleep in order to obtain moisture, thus poisoning the saliva. Skinner had charred the head in order to kill it, so that it is valueless as a specimen."

The first mention of finding the *Polypterus* is I think that of 15th November, 1898:

"When I got back I found that a very fine *Polypterus* had been brought for me, and was now in my aquarium. He was about eighteen inches in length, and spread his great pectoral fins on the bottom, using them like a seal's paddle. He snapped up greedily two beetles floating on the surface; but shortly after slowly died. The natural colour is a dirty straw colour with shadings of black above. While alive in my aquarium he was extremely lethargic." .

There are numerous other references to the " Sayo " as the natives call the *Polypterus*, but these will appear later in Professor Graham Kerr's article.

From time to time he accompanied Mr Wainewright on his tours through the district under his charge, and lived under all sorts of varying conditions. Thus on 24th November, 1898, he writes :

"The village of Nianimaru is the first African village that I have really lived in after the fashion of a native. The compounds are all placed close together, and the passages left between them constitute the streets of the town. The walls of the compound on either side of the street are six-foot wattles. Many of the huts look quite picturesque, being covered with the green leaves of climbing vegetable marrows."

At times things were dangerous :

"About 11 o'clock at night a man tried to get into the hut where we slept, but stumbled over the native boy who slept at the door.

[1] Bathurst Trading Company.

Then Wainewright heard two men at the other door, which was lightly barricaded, discussing the possibility of killing the white men. So we got the interpreter to sleep outside the hut, and we held our pistols in readiness, and passed the rest of the night in quietness."

On 12th December, 1898, when on one of these expeditions, he nearly got lost, as he records in his diary:

"Started away from McCarthy at daylight, and soon after leaving Larima Kota had the offer of horses to ride, which we readily accepted, and put on our own saddles. We passed through sparsely wooded country, now covered with canes about ten feet high, which later in the season die down and turn the country into magnificent game country. Passing through Fetu we had some fine cow's milk, and had to take on fresh carriers. About 11 a.m. we got to Dem Fai and were well received. Wainewright and I had a long talk after breakfast, and then I slept, and did my writing. At 5 o'clock I went out shooting, and on the way back lost my bearings, and almost made up my mind to spend the night in the bush. But listening I heard the sound of an ass, and made my way towards it; then I heard a cow low and made for this; and at last struck a road which led me to Dem Fai, but not the one I wanted. One of the people, however, led me to the right one, and I was much rejoiced to see Wainewright's lamp."

Whilst seeking for the breeding places of *Polypterus* Budgett had many opportunities of observing and collecting other fish. He brought back considerable collections, especially of the mudfish or *Protopterus*, called by the natives "Cambona," whose habits he carefully studied. On 12th January, 1899, he writes:

"Two Cambona cocoons which I had put into the pond were dissolved this morning and the Cambona are swimming about as lively as possible.

I tried to kill one to-day, but found it tremendously tenacious of life. The mark on the tail, where the mouth of the cocoon is sealed, remains for some time upon the side of the fish's tail. Twelve Cambonas have been brought me in all."

And again on the following day:

"More Cambonas brought me, about fifteen altogether. Put a small one into the *Polypterus* tank, but it scared them so that I had to remove it. For a long time they remain without moving, the limbs being glued to the sides by mucus. When freed they remain hanging vertically in the water with mouths at the surface. When first opened from the cocoon they will give a very sudden bite if handled roughly. I have turned three into my pond."

And later:

"14*th April*, 1899. Went out with men digging Cambonas. I learn that the largest Cambonas are found in the lower parts of the island where the ground is damp. Here they do not make a regular cocoon, but have the epidermis of the head only slightly dried, the tail does not even cover the head. At the beginning of the rains the holes will be found filled with water, while the surrounding ground is only moist.

From experiment I find that under these conditions the Cambona frequently glides up the hole to the surface of the water to breathe air. When the ground is flooded the Cambona quickly leaves its hole and swims freely in the water. If the water dries up it again makes a new hole. Shortly after the ground is well flooded the Cambona lays its eggs, not in a hole in the ground, but merely on the bottom of the swamp, having first made a shallow nest by clearing the grass.

The smaller Cambonas found in the rice swamps in dryer parts of the island make a regular cocoon by a secretion from the epidermis; this is very tough and air-tight, the only opening being into the mouth by a tube, formed from the cast of the mouth.

Where the Cambona has dried up very much, the tissues seem to undergo a regular degeneration. In some cases a large portion of the tail overlying the head and also the front part of the head are entirely absorbed. In this case when set free in water the tissues appear to be gradually regenerated."

And:

"14*th May*, 1899. Put two Cambonas in aquarium. Re-arranged latter. A Cambona which I have had alive in the pond for a long time, and which had had its limbs chopped off, has regenerated them all, and one which came out of its chrysalis with the fore-part of the head absorbed, and also a large slice of the tail, has completely regenerated them."

And on 1st June, 1899:

"Had more Cambonas brought, but cannot get more than one to live at a time in the tank; they always 'go for' one another, but when put in a place where there is plenty of mud it is all right. Had a tortoise brought me.

After the rain, all over the ground are countless winged termites and workers. As fast as they came up they were devoured by large *Scolopendras*. Find that the Cambonas which were put in water a fortnight ago had all developed eggs.

Labelled birds. Quite impossible to work by lamp in the evening because of the quantities of insects; termites, beetles, Homoptera.

Max. temp. 36° C. dry.

Min. temp. 26° C. dry."

In February, 1899, he came down to Bathurst, and was received, as he always was, in the kindest way, by the Governor, officers and other residents. On his return voyage to McCarthy Island on 15th February, 1899, he writes:

"At Nianimaru I found that my canoe had been recovered, and I sent the fisherman ashore in it. From here the river struck me as more beautiful than I had seen it before; the water-side shrubs seemed to look very fresh and green, while many of them, as well as the palm-trees, had assumed a most brilliant variety of tints.

At Sukatoa the contrasts of colour were very striking. Three vast pyramids of ground-nuts were piled in the clearing, up which the natives were clambering with their loads, clad in white, blue and red flowing pangs, tall monkey-bread trees formed a pinkish grey background, with here and there a towering cotton-tree; on either side of the clearing the banks were lined with laurel-like shrubs whose terminal groups of leaves were red as in the *Poinsettia*; behind, a row or so of a quaint branching palm of small growths, which bear a fruit that is said to cause a conflagration by spontaneous combustion. Behind them again the graceful Piaswa palm from which a valuable fibre is obtained, and, finally, a dense belt of forest and tangled undergrowth and creepers from which at intervals there rises up a waving majestic Rung palm.

Three smart cutters lay at anchor in front of the wharf, discharging their cargoes of nuts, and giving the finishing touch of brilliancy, by their characteristic painting of blue and white, to what to me was a fascinating picture. This place I have passed many times before, but have never been so struck.

Arrived at McCarthy I went ashore with Tuke of the "Magicienne." Showed him the ruins of the Barracks, and sent him over to Tamu Koti to see an ancient burial ground."

Towards the end of April he made an expedition up the Gambia as far as Netebulo, the highest navigable point, with a couple of English mineralogists, Mr Rey and Mr Picard, and Mons. Irimel, a French prospector, all of them seeking gold. One or two extracts from his diary at this time give some idea of the sport to be obtained on the Gambia:

"*22nd April*, 1899. Went ashore early for a little at Beru. Put my hammock up in the bows where it was beautifully cool and there was a fine view. Now and again we would pass through a range of hills where were high cliffs overhanging the river. In one such place we saw a large antelope of a very light fawn colour. We found a dead Dakoi or Roan Antelope[1] in the river, also some vast 'Hippos.'

We got to Yabutenda about 3 p.m. and stayed there until after dinner. I went ashore and got a couple of Sand Grouse. Then we started on in the lighter and left the "Madge."

A beautiful moonlight night, not a sound but the splash of the oars. We talked until late and then slept soundly, to wake up on

23rd April, 1899, at the wharf of Netebulo. We found that the town was four miles away, so decided to camp by the river side. This is the highest navigable point of the river. There is a very sharp bend with precipitous cliffs overhanging the river, and at the curve of the bend there is a grassy level. This is a famous place for shooting game as they come down to the river to drink.

Rey and Irimel went up to the town, Picard and I went out shooting. We saw Antelope in the distance and stalked them a long way; they turned out to be Roan Antelope, and just as we got within shooting distance they twigged our white helmets and started off.

We came back to breakfast. Just finished, and back came Rey and Irimel with the king's son. We were to wait until the next day for porters to carry up the loads.

All the baggage was put on shore and we spent the afternoon on the top of the cliff under shady trees.

Pitched tents by river side. After dinner we went out to try to get some game. Heard a lion coming towards us and slipt up trees till moon went in, saw nothing and came back.

24th April, 1899. Hot morning. The king came down and all the baggage was fetched up to the town. The town is large and compact, the king's house is made of very high mud walls. We were put into a comfortable compound and many delicacies were brought us.

The king and his son soon paid us a formal call, bringing fresh milk, etc., etc. It was most amusing. We returned his call, and he showed us all round his establishment and harem.

25th April, 1899. We were to have gone out early, but the huntsmen did not turn up until the sun was up. However, we soon came across a fine Senegambian Hartebeest[2], I had two shots at him, and then gave chase; as he passed quite close to us he fell down. I followed him up for some way, but he escaped wounded. Shortly afterwards we came across a large herd of the same, just returning to the hills, but they snuffed us and were off.

Then we saw two small deer, but the sun was too high, and we had to go back. Tracks everywhere of every kind of Antelope. I got a fine pair of horns of *Tragelaphus spekei* in the town.

---

[1] *Hippotragus equinus.*   [2] *Damaliscus corrigum.*

The king's son has been educated at St Louis, he talks French well and seems a very nice fellow. They both came in again to-day after breakfast and we exchanged signatures. He requested me to make known to the traders in the Gambia that he was most pleased for them to come over the border and trade, and that they would get every protection.

I intend starting back the day after to-morrow. The gold-seekers hope to start the next day, and have about a month's bad travelling.

In the evening we all went out shooting and saw several antelopes, but could not get near them. Came back to get a few hours' sleep.

Max. temp. 41° C.

26th April, 1899. Started off an hour before daylight with huntsmen. We soon came across a herd of Hartebeest, and stalking up to them I got a very good chance, but had a bad cartridge in. We then went on and soon came across three Tankong, waiting for them they came fairly close and I fired at one and the huntsman at the other, without stopping them.

I then saw a very small Antelope, but could not get near it. Again a solitary Senegambian Hartebeest. The sun was up now and I wended homewards. Hunted a flock of Guinea hen, but did not get them. Last night I picked up a fine skull and horns of the Senegambian Hartebeest.

The town of Netebulo is by the side of a series of springs forming in the dry season an underground stream.

I gave Sandian my red blanket for present. We changed films and took 'photos' of the castle.

Very pleasant evening. I am to carry many letters back.

Max. temp. 39° C."

As the wet season came on and the heat increased, the entries in the diary became shorter, and his work became more difficult to pursue. The frequent illness of his friend Mr Wainewright caused him constant anxiety.

On 27th May, 1899, he writes:

"Worked in the swamp between the showers of rain. Read and wrote. Wainewright better in the evening. Went out shooting first at a bush fowl flying overhead, and this immediately gave rise to a perfect pandemonium, for Jatto started off after it, and then thought I was after the dog-faced baboons, for we were, I saw then, surrounded by two or three hundred monkeys, some of them gigantic brutes. They began to go for the dog, and he then ran towards me, dropping his tail. I walked quietly on, the brutes following me a little way. I was very glad to leave them behind. Some rain again at night.

Max. temp. 26° C. dry.

Min. temp. 23° C."

During the month of June he dissected many specimens of *Polypterus*, and although their ovaries contained ripe eggs, and although in spite of rain and heat he sought diligently in the streams and backwaters, nowhere could he find the developing ova.

On 4th July, 1899, he tried to fertilize artificially the eggs of his fish, having some dozen Cambonas and twenty Sayos (*P. lapradii*) with ripe ova to work with, but on the following day he writes "Artificial fertilization has failed. Am a little bit seedy to-day. Not able to do much."

He was also trying to keep the fish in large wire cages or enclosures in the river and lagoons, but they failed to breed, and about this time he had the additional disappointment of mistaking the eggs of another fish, which subsequently turned out to be probably a species of *Hyperopisus*, for the eggs he sought. Toward the end of July he was again ill and confined to bed with fever; and on the 28th of that month he started for the coast, and having stayed about a week at the Government House at Bathurst he sailed for home.

He had not succeeded in his quest but he had acquired much useful experience and had made many valuable contributions to our knowledge of the fauna of the Gambia. During the autumn of 1899 and the spring of 1900 the following three Papers (i) "Observations on *Polypterus* and *Protopterus*[1]"; (ii) "General Account of an Expedition to the Gambia Colony and Protectorate in 1898–99[2]"; (iii) "List of the Fishes collected by Mr J. S. Budgett in the River Gambia," by G. A. Boulenger, F.R.S., F.Z.S., with notes by J. S. Budgett, F.Z.S.[3], were published on the results of his expedition.

Although Budgett had not succeeded in bringing back the material he sought, he had determined the time when the *Polypterus* breeds, and with great courage he again revisited the Gambia in the rainy season of 1900.

He reached McCarthy Island on 6th June and was astonished

---

[1] *Proceedings of the Cambridge Philosophical Society*, Vol. x. 1899.
[2] *Proceedings of the Zoological Society of London*, Nov. 1899.　　[3] *Ibid.* May, 1900.

to find " the country as dry as a bone." He set about constructing floating cages and wire enclosures, and very soon had them stocked with fish. The country was in a disturbed state, and he records a sad tragedy on 19th June, 1900:

"News has arrived that poor Sitwell and Silva have been ambushed and shot with their interpreters, cooks, Sergeant Cox and nine policemen: Details are not forthcoming, but plenty of rumours. There is some uneasiness here about the natives of Wuli and Sandugu.

In the early morning I made a fourth enclosure.

Mardeke called this morning to give the news. Iamei Ujorka called this morning, he seems straight.

I had two *P. lapradii* brought this evening.

Min. temp. 24° C.

Max. temp. 34° C.

20th June, 1900. A cutter has come up saying that the Commissioners were holding a palaver at Sankandi, when there was a sudden uproar and they were all set upon and slaughtered to a man, including the police force. The bodies were burnt with grass and the town destroyed by fire, the people all clearing over to the French country. All their throats were cut.

One Frenchman has died of yellow fever. Ingman is down with it.

The M.K.[1] has not come up. No mails and no news.

Dembo Dausa the messenger of Musomola says we are quite safe here."

For the next three months he worked incessantly, and in spite of the heavy rains and consequent floods he achieved much. He traced the development of *Gymnarchus niloticus* and made innumerable observations on other fish and on his favourite frogs, on mosquitos and on other insects. He added to his knowledge of *Protopterus* and collected material for the study of its development. On 1st August, 1900, he writes:

"Light rain in morning. Went over large specimens and made up new spirit. About midday Sory came to say he had found a Cambona nest. I went out immediately to the far swamp, and there, where water had entirely dried up, were several holes in the mud amongst the grass full of water. In one of them the tail of a Cambona was ceaselessly lashing to and fro, thus causing the surface water to penetrate the nest, which was a long hole coming to the surface again some feet off.

[1] The "Mansah Kilah," the river steamer.

On being startled, he fled off down the hole. The entrance to the hole was full of larvae identical with those of *Lepidosiren* only about half the size.

Four plumose pairs of external gills were placed just anterior to and above the budding limb. Pigment was beginning to appear about the head region.

The water in the hole was very muddy, and the larvae could only be seen occasionally, as they swam up to the surface. I collected a large number from three similar nests, all where there was no surrounding water. They were all in the same stage. The mudfish in the hole, tending the eggs, was a male.

The circumstances under which these nests were found were, I think, exceptional, since for nearly three weeks we have scarcely had any rain, and the water where the holes were found was a foot deep a week ago. I hope shortly to get eggs now that the nests have been seen.

The rain is now coming down in sheets, and the thunder is crashing. I preserved another batch of the "Suyo" larvae[1], the external gill filaments grow quite as much as the body. They are now a perfect sight, the yolk sac being an inch and a half long with an anterior and posterior efferent and afferent blood vessel, while the gill filaments are a quarter of an inch long. The body is about an inch and a quarter long, the tail absolutely diphycercal and showing no sign of the whip-like appendage of the adult. Pigment is abundant on the head, and the jaws may now be seen opening and shutting.

Max. temp. 31° C.

Min. temp. 27° C.

2nd August, 1900. After very heavy rain last night, light rain this morning. Cleaned out aquarium and put in rain water. The larvae do not like it.

Made drawing of larval *Protopterus*. Preserved six more *Protopteri*. Fed fish in evening, got caught in rain. Brought in twelve more larvae of the Suyo, they now begin to look quite like small *Gymnarchi*, though the external gills are still increasing in length. The swim-bladder has appeared. The larvae keep fairly well, and tend to cling temporarily to the side of the glass.

Max. temp. 31° C.

Min. temp. 26° C."

And again on 6th August:

"Went out to swamp with Sory; hunted for Cambona and Suyo eggs. Found two Cambona nests, one with advanced larvae, the other apparently not yet laid.

These nests were both in quite shallow water, not more than six inches. From the deeper water a way led up to the hole where the grass was brushed aside at the end of the way, a small hole running some distance underground and coming to surface again a yard off. One of them branched.

[1] These subsequently turned out to be larvae of *Gymnarchus niloticus*.

In the afternoon I brought away a large number of larvae. They do well in a jam jar half full of water, where they cling to the side of the glass. They hold their gills at right angles to the body, when at rest, very elegantly."

And two days later:

"The Cambona larvae when in glass vessel hang on to the sides with their suckers and keep their gills out at right angles to the body and bent slightly forwards."

The last entry in his journal on the trip is on 12th August, 1900, and reads as follows:

"Preserved Cambona eggs. Went out to the nest and got a fresh lot. The gill-folds are now becoming conspicuous, the eye and pronephros, also the folds of brain, are visible.

The fishermen came back, five of them not having caught one Suyo in two days.

This year though in all 127 *P. senegalus* and 36 *P. lapradii* have been got by me, yet by far the greater number were got by bailing little inlets from the river; a certain number in my trap, and a few in the basket. Last year, at one particular time, hundreds were caught in the last way.

This opportunity has not presented itself. It is now practically impossible to get Suyo. I do not think it is that they have not come up, but I think that they have spread away far more gradually this year, and having once got into the swamp cannot now be caught.

I had wanted to try artificial fertilization on freshly caught specimens. I shall now wait to see if my fish in the enclosures spawn, and as in last resort, shall try artificial fertilization on them.

Max. temp. 30° C.
Min. temp. 23·5° C."

On Budgett's first return from the Gambia to England in the autumn of 1899 he found the country on the eve of the South African war. He had always been very patriotic, and, as in so many cases, his love for his country was strengthened and deepened by what he had seen abroad. At the outbreak of hostilities he felt very strongly that he ought to go out to the front as a volunteer, and he had precisely the character and the aptitude essential to the making of a soldier. At the same time he was far from strong, and was liable to recurring attacks of malaria acquired in the American and African swamps, so that his friends felt greatly relieved when finally he abandoned the idea of volunteering for the front. But he felt the urgent necessity of training men for the defence of their country, and he joined the Mounted Infantry section of the Cambridge University Volunteers. For the next three years the care of the " M.I." or " Mounted Infants " as they are called, was that to which he devoted all his spare moments.

My friend, Lieutenant W. St A. Warde-Aldam of the Coldstream Guards, has kindly written the following account of his work with the section:

" Budgett was a very good example of the saying that the " busiest people can find time for everything. In the middle of " all his scientific work, he was one of the most proficient and " hard-working members of the University Volunteers. He joined " the Mounted Infantry section when it was first formed in 1900; " his character and experience at once showed his value; and in " July, 1901, he took over the command of the section, becoming " its first commissioned officer. This command he retained till " the time of his death. During these two and a half years he " developed the organisation and efficiency of the Mounted " Infantry in a very marked way: he helped to institute an " annual volunteer Gymkhana, in connexion with which he " started a Gymkhana Club; he organised funds for hiring and " insuring horses; he arranged fortnightly field-days, and annual " route marches; he trebled the number of members attending " camp; moreover he established the section on a permanent " basis as a University institution.

" In connexion with the ' M.I.' Budgett made many friends " among the Inns of Court Rifle Volunteers, to whose Mounted " Infantry Company he was attached for a week's camp in August, " 1901. In the following April he spent a month at the Yeo- " manry School at Aldershot with a class of some twenty Yeomanry " officers. Thus he lost no time in studying the professional duties " of his new commission.

"His character was perhaps more prominently brought out "by his objects and methods than by their results. His objects "are best given in his own words: '1. To foster and turn "to a useful purpose the love of riding in members of the "University. 2. To train those members, who are fond of "riding, to be efficient soldiers. 3. To induce a large number "of members of the University to join the Auxiliary forces "on "going down."'

"His method was the deliberate and gradual 'building up' "of a social system; he was just the man to be the heart of "such a system. His travels had enhanced the value of his "natural gifts of tact, method, broad-mindedness, capacity for "work, and capability to command; they had also given him a "large knowledge of men, and a practical experience of many "of a soldier's duties.

"Budgett applied all these gifts to his new interest with a "keenness balanced by forethought; it was characteristic of this "latter, as well as of his modesty, that he was continually trying "to avoid centering anything on himself; he often used to "explain how the section ought to be worked automatically, "and without his guidance being necessary.

"He was a personal friend of every member of the 'M.I.,' "and always welcomed them and their friends to his rooms; "he always endeavoured to keep up this friendship after they "had 'gone down'; on the day he sailed from Liverpool for "the last time, he wrote, 'Whether I am connected officially or "not in the future with the "M.I.," I shall do my utmost to get "together at least annually as many as possible of those excellent "fellows who have worked together so well.' It was this spirit "that gave so much force to his example, the reality and value "of which were so thoroughly brought home to his 'M.I.' friends "by his untimely death."

Budgett was always an enthusiastic member of the Cambridge University Natural Science Club, a somewhat exclusive body whose members are limited to twelve undergraduates and a few graduates.

He was elected in 1895, his first year, and was a constant attendant at their Saturday evening meetings, where immemorial custom limits the refreshments to coffee and anchovies on toast, which long-established tradition compels us to call whales. He was Secretary to the Club in the Michaelmas Term, 1897, Vice-President in the Easter Term, 1898, and President in the Michaelmas Term, 1899. He read the following four Papers before the Club:

"The Monotremata" ....on 8th Feb. 1896.
"A Zoological Journey in South America" ....27th Nov. 1897.
"Fishing in the Gambia" ....4th Nov. 1899.
"Natural History of some West African Fishes"...10th Nov. 1900.

He was also a member of the Cambridge Cruising Club, which he joined in December, 1897, and took keen interest in the periodical sailing matches which the Club holds from time to time on the Ouse, but I believe he never attended any of the Annual Meetings on the sea-coast. He was placed on the Committee, and elected Librarian in February, 1901. He read one Paper before this Club entitled "Life on a Gambia Cutter" (12th February, 1900) which was reproduced with illustrations in the "Yachtsman," 15th March, 1900. Latterly, too, he joined the Pitt Club, and after taking his M.A. Degree usually dined there. During his absence on the Gambia in December, 1898, he had been elected a Fellow of the Zoological Society and in February, 1901, he was elected a Fellow of the Cambridge Philosophical Society.

In May, 1901, he was appointed Assistant Curator in the Zoological Museum, and during the year that he held this post he made the wonderful series of preparations which add so greatly to the beauty of the University collections. To him an anatomical preparation was a work of art, and no trouble was for him too great if in the end he could make it tell its story simply and completely. He was constantly devising new methods, and experimenting with new reagents and materials, and would mount an object over and over again until it satisfied his high standard. During the following year (1902) he used to give two or three hours in each day to this task, while the rest of his time was spent in working up the valuable material he had collected on the Gambia. He published the following four Papers on his results, though the last of them did not appear until six months after he had left for Uganda.

(i) On some points in the Anatomy of *Polypterus*[1].
(ii) On the Ornithology of the Gambia[2].
(iii) On the Breeding-habits of some West African Fishes, with an Account of the External Features in Development of *Protopterus annectens*, and a Description of the larva of *Polypterus lapradii*[3].
(iv) On the Structure of the Larval *Polypterus*[4].

During the Lent Term, 1902, Budgett delivered a course of lectures on "The Geographical Distribution of Animals" for Professor Newton, who had recently given up lecturing. He had much first-hand knowledge of the Vertebrates of South America, Africa and his own country, and the course was greatly appreciated by the comparatively few students who study this subject.

About this time he became acquainted with the Rev. John Roscoe, of the Church Missionary Society, then resident in Uganda. Mr Roscoe, who was spending the winter in Cambridge, was unusually well-informed both as to the natives and the natural history of the country in which he had been so long a resident, and he was convinced, and succeeded in convincing Budgett, that his best chances of procuring the developing eggs of *Polypterus* was to visit the Albert Nyanza and the neighbouring streams.

In March, 1902, he was elected to the Balfour Studentship, which may be regarded as the Zoological blue ribbon of

---

[1] *The Transactions of the Zoological Society of London*, Vol. xv. April, 1901.
[2] *The Ibis*, July, 1901.
[3] *The Transactions of the Zoological Society of London*, Vol. xvi. August, 1901.
[4] *Ibid.* October, 1902.

Cambridge. The income of the Studentship, aided by a grant from the funds in the hands of the Managers, and by a further grant from the Zoological Society, made the projected trip possible, and after the most careful preparations he started at the end of May, sailing for Mombasa from Naples on the 6th June.

He reached Mombasa on 24th June, 1902, and at once "called on Sir Charles Eliot, who had not had instructions concerning me, but was very agreeable." His impressions of the seaport are interesting:

"The town of Mombasa is rather picturesque, though I think it lacks trees. What surprises me most is the apparent smallness of the trade here, just a few stores to fit out caravans for the interior and that is all. Bathurst on the Gambia is an imposing city compared to this. The number of officials' residences is great, and they are nicely situated overlooking the sea. The whole place is very gay with bunting for the 26th" [the date at first settled for the King's Coronation].

He travelled part of the way up to Victoria Nyanza with Sir Charles Eliot, the Commissioner of British East Africa, and crossed the lake to Entebbe with him, where the Commissioner was received with a salute of ten guns. The next nine days were occupied in arranging his "safari," a Swahili word which seems to serve both for a caravan and for a trek or march.

In addition to procuring the eggs of *Polypterus* the expedition had a second object, which was to see if anything could be done to procure a live *Okapi*, and for a time this greatly occupied Budgett's thoughts, but finally it had to be given up.

Whilst at Entebbe Mr Walter G. Doggett, a son of the well-known naturalist at Cambridge, who was shortly afterwards drowned in the Kagera river, called upon him, and they discussed the *Okapi*, Doggett being most anxious to accompany him on his expedition.

On the 11th July, 1902, the "safari"

"Started with forty porters, four askaris, two headmen and four boys. I started on the 'bike' an hour later by the bicycle road, and overtook the porters and went on, waiting for them where the two roads crossed. Passed through varied country, in places patches of luxuriant forest."

Next day an entry appears which is the first of those indicating that the search for the *Okapi* may perhaps be given up:

"The fact that from all accounts the *Okapi* does not live in British territory, that males and females have been got, and that *Polypterus* is apparently much more plentiful in the swamps of the Upper Nile, has induced me to make a trial of the north end of the lake first, then back to the Semliki."

On 14th July, 1902, the "safari" started on an eleven days tramp to Hoima, which lies east of Albert Nyanza, about half-way along the lake. Wherever he could he rode his bicycle, which attracted a good deal of notice from the natives, many of whom had never seen one.

"Started for Hoima 7.30 a.m. I started an hour and a half later. Curving round the Protestant Mission Hill the road passed the Tomb of King Mtesa and then went straight N.N.W. through almost endless elephant grass, in the lower parts Palm groves and Papyrus swamps. Here and there the road was almost unrideable on the 'bike,' being a very narrow track with deep ditches on either side, and often I had an exciting shave of going roly-poly into the ditch. Towards the end of the day's march the road was so steep that the 'bike' had to be carried up and down some of the hills. The crests of some of these hills are covered with a mass of loose and jagged boulders, evidently produced by denudation. We camped at a place where there are a few native huts and a mule shed. I have my tent pitched across the road, it being the only decently clean place.

Soon after I had had my lunch we had a terrific thunderstorm and a downpour of rain. It has rained every afternoon for the last week."

And on

"18th *July*, 1902. Started by 7 a.m. Arrived at the Maanja River about 8.30 a.m. The road all the way led through low meadows and swamps teeming with life. I noticed as I rode along numbers of *Vidua principalis*, the Serena and their sombre little wives; a large yellow-shouldered, black weaver bird; *Scopus umbretta*, the Hammer-head, and a black and white Shrike, taking the place in West Africa of *Corvinella corvina* in its habit of associating in groups and following or preceding the caravans. From the bridge over the Maanja I made a sketch of the view, which was certainly very pretty.

I then caught up my safari at a pleasant camp on top of rising ground; here we stop for the night. Very short march, in by 10 a.m.

We have had visitors all the way, and they have brought plenty of food. The chief usually pays his visit about sunset or just after tea, bringing ten to twenty loads of bananas or sweet potatoes, neatly tied up in banana leaves. These are laid out in a row in front of my tent, the more important men shake hands, and then leave the headman to sit down and have a chat. Generally I 'asked for more' in the way of milk and eggs for myself. Then we would talk of the game in the country. In Uganda the chiefs were always clothed, generally simply in white calico and a turban of the same, sometimes in quaint odds and ends of European clothing. Always they took a great interest in the 'bike,' many never having seen one before."

The following pages from his diary give a vivid idea of life "on a safari":

"It took the boys several days to get drilled into packing up camp, tent, etc., cooking breakfast, so that no one was kept waiting by anyone else.

Salim (small boy, very sharp, clean, and hard as nails) woke me at daylight or just before, put water in my basin, and set out my clothes. Men laid table outside for breakfast. Two whistles then brought Bafirawala to pack bedding and camp furniture, while the four askaris undid the tent ropes.

As soon as I was dressed, breakfast of porridge (poriki), eggs (magai), and potted meat (nyama kidogo), and tea (chai), was served.

Meantime tent packed by Bafirawala and the four askaris, all porters tied their own kit to their own particular load. Pipe lit, three whistles brought Abdulla with 'bike' well cleaned. Mackintosh, glasses, baccy, and sketching apparatus, camera and collecting gun adjusted.

Meantime kitchen-box packed. All loads placed in a row, porters behind each. Headman Simba called 'Tayari?'

'Tayari Owana.' Three-quarters of an hour.

On reaching camp, Salim unpacks table and chairs. Spot for tent chosen. If necessary twenty porters set to work to pull the grass and level and clean the ground. Four askaris and Bafirawala pitch the tent, five porters go for water, five for wood, ten gather good clean grass, which lines the tent and the verandah to tent.

Meanwhile Salim has set up the bed, mosquito-net and camp furniture, and got the bath ready.

At the same time Ali the cook and Abdulla have begun to get 'déjeuner' ready. After the bath Salim is whistled for, he searches for jiggers, empties bath, takes away clothes to dry and air, lays table for déjeuner.

Whiskey and bitters. Bacon, eggs. Curried fowl. Fried cod roes. Stewed peaches or rice pudding. Coffee (kawa), cigarette.

At night Simba comes in to ask for orders as to where we camp to-morrow."

And:

"Amused to find that Salim and Bafirawala have engaged a small 'Boyango' to do their cooking for them, for which he gets 1*d.* a month!"

On 25th July the "safari" reached Hoima which they found in rather a perturbed state:

"Four lions have been taking natives here every night, one was poisoned last night.

The headman of Butiaba, who was dressed in flowing white robes, and wearing a scarlet and silver mess-jacket, was here, and I had a talk with him. He accurately described *Polypterus*, and said they had eggs now.

Here note the great difficulty of getting accurate information from natives. Although one may be very careful and avoid putting leading questions, yet the interpreter, knowing what the information desired is, twists the question into a leading form, and the result is that the information obtained is surprisingly like that desired."

After a couple of days' rest Budgett started for Butiaba with thirty-nine porters of his own and fifteen borrowed, and on

"*29th July*, 1902. Early in morning a lion roared round camp and came within forty or fifty yards, but before I could get out he had cleared away.

To-day the road was through deep wooded valleys. Soon after starting, I saw that Elephants had just crossed the road. Soon we saw, moving parallel to the road, about two hundred, many with large tusks. I tried several snap-shots with V. 3, 4, 5, and tried to intercept them. They crossed the road just in front, and were screened from view. It was a wonderful sight to see the moving mass of dark brown bodies throwing trunks into air, flapping ears, and smashing down trees and branches. A large flock of Guinea-hen rose just after, but I missed them. Soon we reached the top of the hills and beheld a magnificent view of the Albert Nyanza. Immediately below lay a great strip of grass land, about two miles wide and dotted over with low bush between the foot of the hills and the lake. A little to the south a long neck of land ran out into the lake, very narrow and sandy and curving to the north, partially enclosing a large sheet of water as a natural harbour. On the point of this spit of land there was a fair-sized fishing village. Another village lay opposite on the mainland side of the bay, and here could be seen several small lagoons cut off from the lake.

Far away, seven or eight miles to the north, another large lagoon, apparently connected with the lake, was clearly the lagoon I had been advised to work at;

5

across the lake a vast blue shimmering wall appeared, the Blue Mountains. Near the edge of this wall there could distinctly be seen many a waterfall and torrent apparently plunging straight into the lake.

Butyaba is on a hill-top overlooking the lake. Camped in compound. Row with porters because on arriving my small boy Salim came excitedly to me to say that a lion had taken one of my men on the road. Lion in Swahili is Simba. Simba is the name of my headman, and I soon found that Simba the headman had struck a porter in the wind and he promptly fainted. Had him carried in hammock. Not much the matter.

In the afternoon · two sergeants from Gondokoro arrived by boat; we dined together; they are staying down at the lake-side (where I go to-morrow) until their porters arrive."

Almost immediately on reaching the shores of the lake he succeeded in procuring specimens of *Polypterus senegalus*:

"Was just starting to new camp by steel boat, my tent was packed, loads all tied up, and we had started towards the boat, when headman brought several fish, and among them *Polypterus senegalus*, seventeen inches in length, they said more common here than at the new camp, so am stopping a day or two. Female had eggs, but mostly shed.

When I got back had five more 'Intonto¹' brought, one female full of eggs, but evidently they have just laid. *The males do not seem to have milt.*"

The following extracts from the diary show the difficulties of shooting in the forest:

"Went out to shoot late, and, after a difficult stalk, shot a buck. He went down, and, as I went up to him, he suddenly sprang up and away into the thicket; we followed and tracked for half an hour and got torn to pieces. It is wonderful how easily apparently, and with how little noise a wounded antelope will clear a way for itself through the thickest tangle of aloes, acacias, euphorbias, etc., so that it is only possible to follow by practically cutting one's way, and even then getting the clothing torn off one's back. Shooting in the open, when the grass is as high as it was at this time, is a very disagreeable pastime. The seeds of the grass have hard, sharp, pointed bases, with hairs sloping backwards from the point, so that the seed lodging in one's shirt immediately makes inwards, and, protruding inside the shirt, prods one with its needle-like point with every movement. One cannot walk a quarter of a mile without having hundreds of these tormentors worrying one's already heated skin.

At last I felt done up and had to give it up. Sent back boys, who ran buck down and polished him off."

¹ The local name for *Polypterus*.

About this time it seems that he definitely made up his mind to abandon the search for the Okapi, and to press forward towards the north, where he heard that the *Polypterus* was in plenty and where its breeding season was later.

On 2nd August, 1902, he writes:

"In the afternoon went up to Boma to find details of sailing, etc., between here and Gondokoro. I have now made up my mind to try down the Nile. The Okapi is now no attraction. The birds of Toro are well known, and the prospects with *Polypterus* are better down the Nile than up here, for as far as I can see the season is earlier here than down the Nile.

Shot a Guinea-hen here this afternoon. Porters returned with twenty loads of food from Kajura."

And in the following words he concisely sums up his position:

"OBJECTS IN GOING TO TORO:
1. To work at *Polypterus* in Semliki.
2. To observe and catch Okapi.
3. To collect generally in Semliki valley.

POSITION AT PRESENT:
1. Season with *Polypterus* evidently advanced. Natives absolutely ignorant. Steel boats down Nile.
   About three weeks' journey to Semliki, where natives worse than here.
2. Okapi found in large numbers in Belgian territory; easy means of transport down the Congo.
3. Jackson's nephew, Archer, has shot all Toro.
4. Account of journey down the Congo, both as to difficulties and expense, hardly justifies attempt.
   Polypterus the main object; without a doubt the Nile the best for this (working homewards), especially as the season here is earlier. £100 can be saved by this way home and refunded to Zoological Society[1]."

After which he had a wretched night:

"After turning in, thunderstorm burst, brought down tent and drenched everything. All transferred to shed. Up to this time my tent had weathered out all the storms safely, and I was beginning to rely on it as proof against wind and rain. Down here, however, there was little holding power in the sand

[1] This was done.

5—2

on which my tent was pitched. I had turned in about half-an-hour, when thunder and lightning began, then a great wind came rushing down from the hills towards the lake, with heavy driving rain. Some instinct told me this was more than my tent would stand. I leapt up. Tried to light my lamp, and called to my askaris, who were supposed to be on guard, and then seized the tent-pole, but it was no good, the tent-pegs were drawn, and in a few seconds my tent was in ruins and half blown away, myself and all my belongings exposed to torrents of rain. Luckily there was an old shed near by, which, though by no means safe, was still intact. Thither I fled in my drenched night garments. At first my scoundrelly servants paid little attention to me, not having seen what happened, for they were looking after their own belongings, their tents having already gone before mine.

Very soon, however, the whole camp got to work. A roaring fire was lit inside the shed, the miserable-looking men brought in article after article from the pouring rain with a most woe-begone air. Suddenly the comic side of the thing struck me, and I laughed, at which they all roared with laughter; the tragedy became a joke, and before long I was settled down for the night again in my new quarters by the fireside and slept soundly till the morning. Never again did I regain confidence in my tent; at the approach of each thunderstorm I nervously got up, lit the lamp, and kept my boys in the verandah of the tent constantly going round the pegs until the storm was over.

After this I set my porters to build a temporary house for me. This was much interrupted by my having to send back porters to buy food; at length it was finished: the sides were made of wattle fences and stout poles, and a roof of rather a low pitch was made sun and rain proof by binding quantities of grass to it. It was just finished, and I was on the point of moving into it, when, with a crash, the roof collapsed inwards. The sand gives so little hold to the side-poles, that if the roof has not a very high pitch the weight of the grass on the roof will drive the walls asunder and make it fall inward.

These houses have to withstand not only rain and wind, but also earthquakes, which are by no means uncommon in these parts. Down on the lakeshore I experienced two, the first being of considerable violence."

On 16th August, 1902, the caravan started for Masindi on the way to Fajao, near the head of Albert Nyanza, where the Victoria Nile enters the lake; on their way they saw a large herd of elephants crossing their path.

On the 18th August the following interesting entry occurs:

"6.30 a.m. start. 'Bike' carried hung from tent-pole. Tall grass and thin forest at first. Noticed all along the path a four-leaved plant about eight inches across and lying quite flat on the soil. This plant was of the brightest green, and as it lay on the dark damp soil of the footpath it seemed almost to invite one to tread on it. It lay so flat on the soil that no injury was done to it by treading on it, and it certainly seemed as though specially evolved to live upon the pathway of human beings. More probably, however, it had taken refuge here as, being the only plant able to survive in such a situation, it was therefore out of the way of competition.

After an hour and a half the path became very hilly, and we entered real forest; at length coming to a deep ravine with pretty watercourse at bottom, lovely epiphytes and lianas everywhere, with the rubber-plant climbing up the trunks of the big trees; the going for the porters was very difficult because of the lianas, the rocks, and steep climbs. And yet it was wonderful to see the way in which these porters would glide along without hesitation, bending to this side and that with their 65 lb. load, often of unwieldy size, passing within a hair's breadth of some awkward obstacle, but very rarely touching it. These men seem to develope a sense of having their individuality extended into their loads, much as an omnibus-driver in London feels exactly the extent of the axles of his wheels. That this knowledge is something of great importance to the porter is seen by the fact of his clinging obstinately day after day to the load with which he started. He resents his load being changed, even for a lighter one. The first day of the march there is a terrific scramble for the light loads; once, however, he has had his load allotted, he will tie a little bit of rag or cord to it by which he may know it each day, and it is his to the end of the journey. I often used to recognise the porters and call for any particular one I wanted, more by the load he carried than by his own name. Human porterage of this kind is the best possible for delicate apparatus, as long as the path is good underfoot. Occasionally, however, on a bad stony road a porter will trip, try to save himself, and then fall headlong, his load coming with a crash to the ground, which is well calculated to destroy anything of a breakable nature. Then we came on to high ground where the path was more open and better, and, half an hour before reaching camp, to a high rock with a charming view of the Albert Nyanza and mouth of the Nile.

Camped at 10 a.m. at Kitoro.

In the afternoon had brought me a very large Puff-adder 4 ft. 5 in. long. I tried to dissect the head, but it was much broken. Apparently the poison fang on each side is in a pouch, loosely attached by flesh only!

Four days later they came in sight of the Victoria Nile and the Murchison Falls.

"*22nd August*, 1902. Started 6 a.m. and 'biked' most of the way, road much as yesterday. Saw golden Oriole of sp.? and a new bird, blue above, white below.

Eventually the road ran out along a spur, flat-topped and of red conglomerate pebble. Coming to the edge a grand view of the Victoria Nile winding away in the distance. At the end of the spur suddenly we came in view of Fajao nestling below and the Nile swirling along speckled with foam.

Four or five miles before we reached Fajao I heard the roar of the Murchison Falls, not a continuous sound, but a pulsating, thunder-like sound, but now within half a mile we cannot hear a sound.

There is a hill about a hundred feet high in front overlooking the river, and commanding the most enchanting view I ever remember to have seen. Away to the right, the gorge and Falls are seen with clouds of mist rising up; winding round the foot of the hill is the surging, swirling, Victoria Nile, which to the left winds away to the Albert Lake.

The water swarms with splashing fish, crocodiles, and 'hippos,' while the banks and valleys running riverwards are covered with the most beautiful foliage. In some degree I am reminded of Schaffhausen, but I have not had a close view of the Falls yet. It seems a pity that the Fort on the top of this hill has been abandoned and burnt.

The fish mostly caught are of two kinds, one an *Alestes* got at Butyaba, but more brilliant in colour (bright blue above, silver below), and a fish I do not know; it reminds me of a red mullet. It is golden above, and each of the lower scales are dashed with brilliant red, while the ventral and pectoral fins are also bright red, as also the eye and lower part of the head. Many of the fish caught here were of a much more brilliant colouration than the same species caught in the lake.

The natives say they catch the 'Intonto' of large size here."

At Fajao they stopped several days, Budgett eagerly fishing. All the female "Intontos," as the natives call *Polypterus*, were "either laying or having laid eggs." He is certain that the *Polypterus* fry must swarm in the floating "sudd[1]," but he failed to trap them, though he caught the fry of many other species of fish. And artificial fertilization again proved a failure. It is pathetic to read on 27th August, 1902:

"No Intonto caught. Many young fry. All yesterday eggs proved unfertilized. The gelatinous envelope in those eggs which decomposed last was swelled up almost like frogs' eggs.

[1] The sudd is the floating vegetation which sometimes almost blocks the upper reaches of the Nile. The plants that form it are described in Sir William Garstin's *Report on the Bahr el Jehel* published in 1901 and in "Some Notes on the 'Sudd'-Formation of the Upper Nile." By A. F. Brown, Director of Woods and Forests in the Soudan. *J. Linn. Soc.* XXXVII. 1905, p. 51.

Have made a number of small fry-traps of specimen tubes and placed in the sudd. In placing these traps I got an uncomfortable wetting through the sudd giving way.

The 'hippos' were fighting to-day while I was hunting the waterweeds for eggs, a splendid sight.

The eggs here can hardly be placed in the decayed vegetation on the bottom, for this is always shifting and fresh layers being deposited.

The enormous mass of vegetation makes it impossible to be very hopeful of success, and yet I have felt here that there must be abundance of material within a few yards of me."

At Fajao he had another instance of the power of the bicycle:

"Had seven Intontos brought, all *females*, either laying or having laid eggs. Caught on the hook in the 'sudd.'

Spent afternoon making fry-traps. Evening went down to spiller again, many hooks gone, three *Clarias* caught. Set four traps in 'sudd.' Very beautiful on the river at sunset. Native Chief called, wanted me to interfere in some palaver; exhibition of bicycle; I found the bicycle of great use and not only for locomotion. When a chief was in a perturbed state about some grievance in which he wanted me to interfere, instead of having to tell him that I was not a government official, and could do nothing, I had only to ask him if he had ever seen a white man fly, and then I rode round on the 'bike,' and at once his troubles were forgotten.

Polypterus fry must swarm in the 'sudd,' no flooded lands to go to. Water perfectly clear. It seems an ideal place for getting the *Polypterus* material, but cannot catch the large Intonto as thick as the arm which the natives talk of catching."

On the 29th August, 1902, Budgett gave it up, and again set his face north, crossing the Victoria Nile with all his "safari" in one hour and five minutes. On the last of the month he reached Wadelai, having bicycled a considerable part of the distance. Here it became necessary to pay off a number of his porters who refused to go further, though whilst witnessing the flogging of one of their number they became so cheerful and good-tempered that many of those who at first hung back were now ready "to go anywhere." Some of the porters were sent without loads north to Nimule at the junction of the Unyame with the Nile, whilst Budgett went with the baggage by boat. For the latter he had to wait till 8th September, when he started down the river in a boat with six oars and a sail, manned by a crew of the "Uganda Marines." The same day they passed Emin Pasha's old camp.

Rowing down the river, one day was much like another. The following is typical:

"10*th September*, 1902. All day gliding along through endless grass without variation. Shot a 'hippo' dead, but could not wait for him to rise. Rowing all day long, only reached our camp after sunset. Much difficulty in reaching the shore through the sudd. This sudd begins to be formed by the rapidly propagating *Pistia*. Seeds of a small floating rush then lodge in the roots and several plants of the *Pistia* become entangled. Next the tall water-grass lodges on the raft, which thus increases in size until it becomes a floating island. Very curious to watch these rafts as they glide past the fixed grasses. At times the sudd was so thick, we seemed to be stationary, as the whole of the water was choked with the vegetation which slid with us down the stream. Practically no bird life seen.

11*th Sept.*, 1902. Approaching and rowing along the south side of the Nimule mountains. Lovely scenery with lagoons of water-lilies, foreground of *Pistia* and background of bold wooded hills. These were the same hills we had seen nearly all the way from Wadelai, from the south appearing a small isolated range; as we approached however, one solitary peak reared itself up into the clouds, forming a landmark for nearly 100 miles in every direction, while the range extended to the north in a series of lower mountains, forming the western bank of the Nile. We now seemed to be steering straight for this mountain, and indeed the Nile does make directly for this range and plunges right into the Nimule gorge, which cuts off its eastern spur.

Passed a large rock amid-stream with several 'hippos' standing on it. They dashed into the water before we could get near enough for a 'photo.' On the south bank we passed a low hill about 10 a.m. covered with vast monoliths of basalt in curious attitudes, reminding one of Stonehenge."

On 11th September, 1902, they passed Dufile the Belgian post at three in the afternoon, and an hour and a half later reached Nimule, where the porters had just arrived. Two days later the caravan again started treking north. On the 19th they passed Kiri, a considerable native town, where they were pleased to find plenty of milk, an unknown article in the Unyoro country where the cattle had been swept off by cattle-plague. The following entry on 21st September, as they were approaching Legu, records an adventure with an elephant:

"Started 5.30 a.m., wound up into the Legu hills, pretty scenery with fine distant views of various ranges. Then down into the same country we left the other side. After seeing elephant tracks for three hours all the way, there was a sudden panic with the advanced guard and all fled at top speed as we had come upon about a dozen elephants. The grass was so long I could not see them, so going up the nearest tree I saw them apparently not much disturbed, but moving off slowly. Taking B. and an askari, I went after a solitary old tusker lagging behind the rest and getting on to some rising ground, had a good view of him at 100 yards, not daring to get nearer as I could not have seen him through the long grass, so had a shot, but was vexed to see him start, throw up his trunk, and move slowly away. I went after him, hoping to get a nearer shot, but lost track of him.

Another three hours brought us down into sandy country, and at length to the Nile at the junction of another river with it. Crossing this we camped on the other side.

Saw some more Gambian birds, a bee-eater, *Merops nubicus*; several birds I have not seen before and do not know. I meant to get specimens, but was tempted to go on an unsuccessful water-buck shoot.

Serious loss to-day, one of my porters fell with fishing box and smashed my last bottle of formalin."

Gondokoro was reached the following day. Here he dined with the entire white population, and next day paid off his "safari," re-sorted stores, and sold his superfluous gear. The four days Budgett spent in Gondokoro were as usual spent in fishing, but with no success. On 27th September at 8 a.m. he left by the steamer "Abuklea," and within ten minutes stuck upon a sandbank. They

"Reached Lado about 1 o'clock. Monsieur René came on board, having heard that I was coming to the Congo Free State; he said that all the stations had been directed to give me every assistance. He also said that Coquilhatville was the best place to stay for fishing in May, June, and July (dry)!"

The journey down the upper waters of the Nile was uneventful. Budgett reached Fashoda on 5th October, and went on immediately to Khartum, where he arrived on 10th October. In Khartum he inspected the fish-market and found there "almost all the fish" he had caught on Albert Nyanza, "also *Gymnarchus*." Whilst there he came across the servant of Mr Loate, who for some time before had been collecting the fishes of the Nile for the British Museum. This man told him that Mr Loate had taken "very great numbers of *Polypterus* at Fashoda." He immediately engaged this fisherman and some other servants, and with indomitable courage determined to retrace his steps up the Nile. He started back upon 17th October and reached Fashoda on 24th October, having lost a man overboard. On the last day of the month he reviews the whole situation in these words:

REVIEW OF POLYPTERUS QUESTION.

"BUTYABA, LAKE ALBERT, 31ST JULY—15TH AUGUST. Small numbers of *Polypterus* (20) caught in pool cut off from lake, in size up to 17 inches. *Females* mostly having laid eggs. *Males* without abundant milt. Not looking healthy. Netted two pools, lake and lagoon connected with lake, but got no fry. Fishermen very ignorant. Magungas only catch *Polypterus* in the pools.

FAJAO, VICTORIA HILL, 22ND AUGUST—29TH AUGUST. Small numbers of *Polypterus* caught on the hook in the sudd, laying or having laid eggs.

Here no flooded lands to go to, fry must swarm in the sudd, which near the falls is limited in amount. Here and in Lake Albert only *P. senegalus*.

Six *Polypterus senegalus* caught in one trap. Three females, three males. One female had about 50 eggs free in each oviduct, which were both much inflamed. Fertilization tried, but no really motile spermatozoa.

Sudd was then examined and hunted persistently for eggs or fry. Set many fry-traps in same place but took no fry. The river here formed a small bay, this was out of the force of the current and about 18 inches deep.

The eggs were coated with adhesive mucus, which swelled up and set with contact with the water.

The bottom of river here is soft, deep, vegetable débris, and it seems certain to my mind that the eggs are attached to the stems of the water weeds.

Fertilization unsuccessful, but with those eggs which decomposed last the jelly swelled up almost like frogs' spawn, that is 12 hours after laying.

The eggs are heavy, and if not attached would surely fall among the débris at the bottom.

Wadelai to Gondokoro, 31st August—12th September. From Wadelai to Nimule enquired all the way about *Polypterus*, but seldom caught any; current very swift and few landing-places in the sudd. Near Nimule the river opens into wide shallows, which seem more suitable for *Polypterus*. At Nimule *Polypterus* hardly known. Overland to Gondokoro here were steep banks and very swift current. At this time of year impossible to catch *Polypterus*. Gondokoro to Khartum *Polypterus* very common in the sudd.

Fashoda, 25th October—5th November. Here employed two Arab fishermen with casting net. Impossible to employ the natives, they are too wild; the government has little hold over them.

Several *Polypterus* caught with casting net. Females of *P. senegalus*, 18 inches long, almost all ova have been shed.

Male *P. endlicheri*, 26 inches long, abundant motile spermatozoa, with mulberry sperm-mother-cells, the first male *Polypterus* which I have ever seen in this condition. Caught about a dozen small specimens, 5—10 in., but no very small fry.

From what I have seen I have no hesitation in saying that there is no place above Khartum to compare with McCarthy Island on the Gambia, where the river rises and falls with the tide, and the *Polypterus*, apparently spawning, get stranded in pools at the side, and where the natives can help to catch specimens in large numbers.

Fajao the most favourable place, but here it is impossible to catch large numbers.

The main reason for going back to Fashoda is to try and trap the young fry, for here Loate (whose fisherman I have employed) caught large numbers of *Polypterus* in December, when the water had fallen in the channel nearest Fashoda.

Though a few are caught in my fry-traps, no great number, and no very small ones. Tried hard with tow nets, small trawl and lift-net; hundreds of fry of all kinds caught, but no *Polypterus* fry.

1. It must be possible to trap the young fry such as I caught with the limp net in the Gambia, if a sufficient number of traps are laid in the right place, *i.e.* shallows amongst grasses and out of the current.

2. Fertilization the best chance of getting the young stages.

Though breeding begins with the rains, it seems more probable that the fish of large size, the most abundant breeders, spawn at the height of the rains.

The ova accumulate in the oviducts before being laid up to about 100. The spermatozoa accumulate in the urino-genital sinus, and are found ripe and motile in a very small proportion of males caught at one time."

At the beginning of November he recognised that Fashoda, like McCarthy Island, the Albert Nyanza, Fajao, etc., was a failure as regards the obtaining of developing *Polypterus* ova, and on 6th November he took the boat to Khartum, where he arrived on the 10th. His visit coincided with that of Lord Kitchener, who was spending a few days at Khartum on his way to India.

His last entry in the Uganda diary is on the 19th November, when he visited Karnac. He was home by the end of the month.

It had always been Budgett's ambition to be associated with the Zoological Gardens in London. Even as a boy he had shown remarkable skill in keeping live animals and in keeping them healthy. When at home he was especially fond of spending his time in the Clifton Zoological Gardens, and thoroughly understood the management of animals in captivity.

Almost immediately after his return to England he was persuaded, rather against his habitual modesty, to apply for the post of Secretary to the Zoological Society, then vacated by his old friend Dr P. L. Sclater. He sent in a form of testimonial, with a list of his published Papers, and with testimonials from Professor Lloyd Morgan, Colonel the Hon. J. E. Lindley, late commanding the Imperial Yeomanry School, Colonel H. J. Edwards of the University Rifle Volunteers, who dwelt on Budgett's power of organization shown by his work with the Mounted Infantry, and from several of the Cambridge Zoologists. His application was carefully considered, and I have reason to believe his claims favourably impressed the Council, but in the end the Society not unnaturally preferred an older man.

Budgett, however, did not give up hope that he might in the end receive some post at the Gardens. During the year 1903 Mr de Winton was occupying the position of Resident Superintendent at the Gardens in Regent's Park, and it was understood that he did not propose to remain there permanently. If this post became vacant Budgett meant to apply for it. In pursuance of this wish he and I paid a visit to the Continent at Easter, 1903, and inspected the Jardin des Plantes and the Jardin d'Acclimatation at Paris, and the Zoological Gardens at Frankfort, Leipzig, Berlin, Hamburg, Hanover, Amsterdam and Antwerp. During this short holiday I was much impressed with my friend's extensive knowledge of vertebrate life and habits, and the keen practical insight he possessed as to their housing, the arrangement of their cages, their food, and so on. He was particularly anxious to devise a better system of labelling the various exhibits. He wanted to introduce into the labels more information than is conveyed by the recital of a generic and specific name. Labels, telling a story, after the fashion of those with which the British Museum in South Kensington has familiarised us, were his aim. The only way to effect this seemed to have cheaply-printed and easily replaceable cards enclosed in a weather-proof frame. The difficulty was to find the latter, and on his return to Cambridge he began experiments in this direction which were only interrupted by his death.

During the Lent Term and the early part of the May Term in 1903 Budgett continued to work at his fishes, but he could not give up the idea of having another try to procure the developing ova of *Polypterus*. Some short time before Dr Ansorge had brought back specimens of the young larvae of *Polypterus* from the delta of the Niger. There was no doubt that the fish lived and bred there, but, though this must have been equally true of the Gambia and the Nile basin, there was always the chance that the material would be more easy to get at. Budgett at any rate felt the chance worth trying, and, although he was well aware of the danger he ran, he declined to talk much about it or dwell upon it.

An agreeable incident in connexion with Budgett's start must be here recorded. His first idea was to get a passage in a King's ship as far as Sierra Leone, and he wrote to the Prime Minister on the subject. Mr Balfour—whose interest in the success of the Balfour Studentship is well known—did all he could, and when he found that the request could not be granted, most generously defrayed Budgett's passage from his own private purse.

He left Liverpool in the s.s. "Nigeria" on 27th June, 1903, and reached Sierra Leone on 7th July. After calling at Monrovia, Sekondi, Cape Coast Castle, Acra and Lagos on 14th July, they made the mouth of the Forcados River, the most westerly of the many streams into which the Niger splits. His plan was to go first to Assé, where Dr Ansorge obtained his larvae, and if the prospects were not good after a week or two, to go up the river to Lokoja and Dakmon, returning to Assé in September if unsuccessful at Dakmon.

He started first for Burutu, which he describes as "the most dreary spot I have seen, in the midst of mangrove swamps, consisting of a few tin huts, two rest-houses and the stores" of the Niger Company. It rained incessantly night and day. At Burutu he embarked on a stern-wheeler for Lokoja. Passing up-stream he was surprised at the great number of villages on both sides of the river, "the water swarming with canoes." He landed at Assé, but finding from the reports of the inhabitants that, although there were *Polypterus* in the river "as large as crocodiles," they could not catch them, he rapidly altered his plans and continued on the boat to Lokoja. At Onitsha, Mr Nelson, the Company's agent, told him that he had often seen scales an inch square, which Budgett thought "could be no other than those of *Polypterus*"; this seemed to corroborate the statements of the natives at Assé.

Between Ida and Lokoja they passed some fine scenery:

"19*th July*, 1903. At daylight in sight of the hills, soon the scenery became really very fine. High hills 500—700 feet, deep valleys, grassy slopes and rocky islands. Many of the hills topped with rugged crags. Sand-banks with fishing huts and canoes. Everywhere variety and change of interest. Very few water-birds. A few Egrets and Kingfishers.

Lokoja lies on a slope on the north bank of the Niger, opposite the Benue and under a long flat-topped hill. An island lies opposite, and here I am told *Polypterus* is caught. I have not yet interviewed the fishermen. It being

Sunday I had the old difficulty of getting porters to bring up my gear. Most of the necessary stuff is up now, however.

I rather fear there are too many white men here for me to get to know the fishing people much; if I do not find *Polypterus* plentiful here I think I shall go on to Dakmon, where I can get the whole village to work for me."

After inspecting the possible fishing places, and finding them not very promising, he determined to go further up the river to Dakmon, his reason being "that the river rises up there first, and later in Southern Nigeria" On 22nd July he notes:

"This country is much preferable to the Gambia and a collector's work here not a bad life.

The headman of the town came to court to-day in great style, all the horses covered with silver trappings and fine leather-work.

Watched the polo in the evening with Ryan, dined with Migeod, and played billiards all the evening."

On the 25th July he started up stream in the "Kampala," and was agreeably surprised to find that on the upper river they only had a few showers at night time, though it was the middle of the rainy season.

At Muriji, which he reached on the evening of 27th July, he determined to stop a few days, as he had heard through Major Burdon and Mr A. H. Cheke that the *Polypterus* was plentiful there. He at once began to fish, but with comparatively little success, and he could gather nothing about the breeding habits of *Polypterus* from the native fishermen. Although the fish were shy they caught a couple of Manatees, but Budgett had no time to dissect them. He continued fishing for some days longer, and collected some thirty or forty different species, but "there is no means here...of getting large numbers of *Polypterus*, and I doubt whether they can get any small fry."

On 5th August he decided to go on to Dakmon, and after a difficult passage arrived there two days later:

"Though not a very good fishing spot it is very well situated and one might work well with the people. At the present time the water is well below the bank, and there is not much chance of finding *Polypterus* spawning. The headman, a very intelligent Mahomedan, says that when the river overflows its banks many fish come up to spawn, including *Polypterus*, *Gymnarchus*, *Heterotis*, etc. He declares that he knows the egg of *Polypterus* but I fancy he thinks of toad's spawn. He knows well *Gymnarchus* and *Heterotis* nests.

A native brought me a nice specimen of *Xenopus*, which I am keeping alive, it looks somewhat different from those I have seen in England from the Cape.

I set the bolter with a few hooks, but only caught one fish, a *Synodontis*. In the afternoon I climbed the hill behind Dakmon and had a lovely view, there are large flats on the far side of the river partly flooded now, but the headman says it will be two months before the river will really overflow."

Dakmon also proved a failure, and he determined to return down stream to Assé, which he reached on 12th August. On the next day he writes:

"Early morning setting house in order. I have a fine verandah to work in, overlooking the Forcados River, an island in front, where are shallows and creeks of good prospect. A few light showers and sunshine in the afternoon. Took canoe up to small village, where I was much interested in the fish-traps and nets.

Very great ivory rings are worn as ornaments on the arms and legs. Apparently the natives on the river banks are Ijos, those in the bush Igabos. The latter are said to catch *Polypterus* in large numbers.

There is a large lagoon and village somewhere at the back, but it seems difficult to get to it at this time. I had some small specimens of *Ophiocephalus* brought in, and also a small fish with a head like a pipe-fish and long eel-like body, which I do not know.

The natives are of a very low type, but I think in a short time I shall get on with them all right.

The Niger Company have quite a good garden with oranges, mangoes, and plums. The river is nothing like up yet. Though I think the natives will probably bring me small specimens of *Polypterus*, I fancy I shall have to catch the adults myself, unless I can get to an Igabo village. *Hyperopisus* seems exceedingly common here, and I saw one *Gnathognemus* with brilliant yellow colouration. I hope to get small specimens alive.

It seems to be impenetrable forest all around, though there is open ground on the island.

On the whole I feel much more hopeful here than up river, chiefly I think because of the innumerable rain-water swamps here."

During his stay at Assé it seems to have rained all night and very nearly all day. In spite of the continuous downpour, he worked hard, both fishing and collecting and interviewing the native fishermen.

On 19th August he joyfully records the capture of " six *Polypterus senegalus*, between two and a half and three inches in length, with large external gills! Preserved in Bles' fluid."

On the 22nd August he visited a lagoon:

"Went down to see nest of *Gymnarchus*. Opposite Bari entered creek and then a small stream very winding, full of natives fishing and fish-traps. Took many *Polypterus* from canoes. After half a mile came to swamp choked with *Pistia* and floating grass. Here we found an old disused nest of *Gymnarchus*. Then crossing, entered another small stream, all the way through dense forest carpeted with ferns; after a quarter of a mile came to large lagoon free from sand. In a small creek from this, on the far side, we found a nest of *Gymnarchus* full of young ready to leave nest. Meanwhile I talked to the guide; he knows the eggs of *Polypterus*; has seen them spawning at the beginning and end of the rise of Niger, always at west end of lagoon, attaching eggs to stick under water in great quantity. He says they are hatched in two or three days. I must wait until water begins to go down. Have offered him £2. 0s. 0d. for eggs. In the evening had small *Polypterus senegalus* brought with huge gills."

On the 25th August he again tried artificial fertilization:

"In the afternoon I opened three female *P. senegalus*, each of which had the oviducts crammed. Two of them appeared to have been captured some days ago, and the ova had begun to decompose. The third one was quite healthy. I fertilized with teazed testes and spread on the bottom of small glass dishes to which they firmly adhered. Fine day with a few showers.

26th August. The ova appear to be segmenting, about 60 per cent. have decomposed. The pigment seems to be redistributed irregularly. One egg I examined, as it looked more normal than the rest. The upper pole was covered with fine cells of a light brown colour, the lower pole cells were white and about twice the size. The upper cells easily visible with a ×8 Leitz lens. Round one half of the egg there is a deep constriction between the brown cells and the white cells. In attempting to free it from the envelope the egg broke, but I preserved it in formalin.

Later (*mid-day*) another egg was examined, the constriction surrounded the whole egg about $\frac{1}{4}$ white $\frac{3}{4}$ brown. Preserved in formalin.

4 *p.m.* Another examined; the white area is now a mere plug, evidently the 'yolk plug of the blastopore.' The shape of the egg is oval, the gelatinous envelope does not fit the egg closely, and is pretty firm. Preserved in formalin.

Most of the eggs appear to be backward or not properly fertilized. They are clearly segmenting but irregularly, the pigment is scattered in patches and streaks, chiefly following the segmental furrows.

At five to-day a Bari fisherman brought me two female *P. senegalus*, one had the oviducts full, the other had also a large portion of the ova free in the body cavity. I had two good males and tried to thoroughly fertilize them, then, spreading them on hatching tray, set them in hatch box, in the river. I noticed that many of the ova from the oviducts appeared to have begun segmenting, the body cavity was full of fluid, and I am inclined to think that there is internal fertilization.

The vent of the female was swollen and protruding, suggesting that the female receives the milt from the anal fin of the male, together with a certain amount of water, though there must also be sufficient serum added to prevent coagulation of the gelatinous envelope.

The tubules of the testis of the male used were greatly distended, and the sperm was clear and not opaque as in the other males.

Hot sunshine in morning, downpour of rain in the afternoon. Temp. mid-day, 28° C. Temp. of water in river 28° C. 8 a.m.

Am not making drawings until material more abundant and normal.

27th August. About 70 per cent. of the ova fertilized yesterday are developing.

At 8 a.m. the blastopore was just closed while the embryonic plate extended in a pear-shaped manner from the latter.

A series during the day were put in formalin.

First series at 10 a.m. showed first appearance of the embryo.

Second series at 2.30 p.m. showed uprising of neural folds around the plate.

Third series at 8 p.m. showed closing of the folds over the plate. Brown pigment is irregularly scattered, chiefly on the upper pole of the eggs. Much pigment sinks into the neural groove. The head portion of the neural plate is last closed in, and then it is a little broader than the body portion.

The yolk plug does not seem to be included in the neural groove."

For the next five days he was watching one of the most wonderful sights this world affords us, the development of an animal from the egg. After years of patience, after three unsuccessful journeys into the heart of Africa, he had at last succeeded where all others had failed, and as he watched under the microscope the gradual unfolding of the ovum, the formation of the layers, the building up of the organs, he must have experienced a joy peculiar to men of science, and experienced by but few of them.

In a letter to Lieutenant Warde-Aldam, dated 9 September, 1903, he writes:—" Within a fortnight of getting here (Assé), I success-" fully brought off the artificial fertilization of about a thousand " eggs, and accomplished that which I have been trying to do for " the past four years. The interest to me of the following week, " day by day, and hour by hour, I fear you can hardly appreciate, " as I saw confirmed my views as to this extraordinary fish's re-" lations to other animals.... I was not to have it all my own way " however, for at the end of a week soon after hatching an infernal " fungus attacked them, and they all died. Had they continued to " live another week I should have had them packed up, and made " the best of my way home. You will be amused when I tell you " that I am bringing twenty little tubes that I would not part with " for £20 a piece, ... It rains almost continuously, everything " is mildew and rust. We are surrounded with dense tropical " forest, and move nowhere except by canoes. The natives' " respect for my 'ju-ju,' in that I made fishes' eggs live, is very " amusing. The landing is crowded every afternoon with canoes " with fish for me to examine.... I shall be really glad to turn " my face homewards once more, the depression of this vapour-" bath is almost unbearable, especially when Fortune closes her " hand to me."

The several periods into which he divided the stages of the developing ova are described later in this volume. He carefully preserved some specimens of each, and made elaborate drawings. He succeeded in rearing a few young larvae, but these were very delicate, and were liable to be killed by a fungus which infested his jars and tanks. About this time he suffered much from the effect of formalin on his hands, and the sores thus set up troubled him for some weeks after his return to England.

Through the whole of September he continued to artificially fertilize the ova, and greatly increased his supply of material.

On the 23rd September he writes:

"Of the three larvae left in hatching frame yesterday only one found to-day. This was five days old, and in exactly the stage to which I reared the eldest of the original batch, and this died. So that I now have a continuous series."

The diary closes on the last day of September.

When Budgett returned to England it was evident that he was not well. He suffered from successive attacks of malarial fever, and his hands caused him much inconvenience, but he came up to Cambridge and began to work out his material.

Towards the end of the year his friend, Mr de Winton[1], retired from the post of Resident Superintendent at the Zoological Gardens, a post Budgett would have dearly liked to have filled. But it fell out otherwise, and he bore the disappointment with his usual quiet courage, though it undoubtedly added to the depression which the malaria induced.

He spent a few days before Christmas with me at Englefield Green, and then went to Clifton to spend the remainder of his holidays with his mother, who wrote to me that he was in better spirits than she had seen him since his return from the Niger.

He returned to Cambridge early in January, and on the 9th had finished his drawings of the external features of the developing *Polypterus* ova. The same evening he was seized with blackwater fever, the symptoms of which he knew only too well. He was very ill, but for some days held his own. The last time I saw him he said, "I've pulled through. I'm going to get better," but although the blackwater fever had materially diminished, a bad attack of malaria supervened, and on Tuesday, 19th January, he died, the very day announced for his Paper at the Zoological Society on his recent success.

Robert Louis Stevenson has somewhere remarked that "to be wholly devoted to some intellectual exercise is to have

[1] *Vide* page 44.

succeeded in life." Never has been a student more "wholly devoted" to his subject than Budgett. The patient persistence of his quest for the eggs of *Polypterus* under crushing difficulties forms one of the most courageous episodes in the history of Zoology. He succeeded after years of toil, but in succeeding met his death.

It is difficult to describe anyone's character, almost impossible to describe a friend's. Budgett had all the features of the best type of Englishmen. He was courageous, courteous, long-suffering and absolutely loyal, patient when ill and cheerful under physical discomfort and suffering, very pertinacious, with a strong sense of duty and of personal honour, kind to a degree, but always considerately and quietly so. He hated advertisement, and always kept himself in the background. He was just the sort of man who makes the backbone of the governing classes of our Colonies and Dependencies, reliable, resourceful, the man whom we can trust. The Colonies of Great Britain are full of such men, "a people scattered by their wars and affairs over the whole earth, and home-sick to a man," as Emerson describes them; but each of them doing his duty. Whenever in his wide travels Budgett came across these makers of Empire, he was recognized as of their stamp, and everywhere he was welcomed and made at home.

Latterly he had a great influence with the Undergraduates, and he came across many in connexion with the Mounted Infantry. One of them writes to me that his influence was always welcomed and reciprocated, which is not always the case in the relations of an older man with youth. The same tact and courtesy which made him so successful a Commanding Officer in the "M.I." made him a successful leader of a "safari." The natives trusted him, they knew him to be just, and they knew also that he had a quick eye for shirking, and that he was a good judge of men. He further gained the respect of his porters by his sportsman-like abilities. He was a good shot, sat a horse well, and could sail a boat.

Budgett was a zoologist of the best type. He was a keen and accurate field-naturalist. He was no mere anatomist interested only in the structure of animals that had once been alive. He loved to watch their homes, their play and their habits. He had a wide knowledge; especially of fishes, frogs, and birds.

Mr G. A. Boulenger of the British Museum has kindly written for me the following paragraph upon Budgett's contributions to our knowledge of African fishes and frogs:

"The breeding habits of fishes living between the tropics "are among the secrets of Nature which are most difficult to "unravel, and which have most taxed the acumen and patience "of zoologists. Collectors of zoological specimens there are in "plenty, but they are seldom in a position to make observations "on the breeding habits of the lower vertebrates. Several attempts "had been made with the object of procuring the developmental "stages of the African fishes *Polypterus* and *Protopterus*, but "in vain. Budgett determined not to rest until he had attained "the long-sought prize; he succeeded, but for this success he "paid with his life.

"In pursuit of the breeding places of *Polypterus* and "*Protopterus* he visited the Gambia, the Victoria Nile, and the "Niger, and from each of these rivers he brought home not "only most valuable notes on the habits of the fishes he came "across, habits which were then totally unknown, and embryo-"logical material which has made his name famous, but very "important collections of the fishes themselves, which it has "been my privilege to name and describe. Several new species "were discovered by him in the Gambia and in the Niger: "*Marcusenius budgetti*, *Gnathonemus gilli*, *Clarias budgetti*, "*Synodontis ocellifer*. He also paid attention to frogs and their "larval stages, and a new species from the Gambia was described "by me as *Rana budgetti*.

"Frogs had been the first subject on which he published. "When accompanying Professor Graham Kerr to the Paraguayan "Chaco in 1896–7, he made a very important collection of

" Batrachians, of which he published a list, accompanied by
"interesting notes on the habits, and a description of the
"nesting habits and development of the quadrumanous tree-frog
"*Phyllomedusa hypochondrialis.* He also discovered on this
"occasion a new frog which could not be referred to any of
"the known genera, and for which he proposed the name
"*Lepidobatrachus.*"

I have dwelt upon Budgett's gift in the setting up of Museum
specimens; but he was equally skilful in the methods of the
Laboratory. His knowledge of reagents and materials, and a
certain mechanical turn which his mind possessed, led to endless
experiments, most of which marked a distinct advance in the
technique of the subject.

His original work is included in this volume. Much of
his material has been worked out by others, but, in the short
time that he had for research, he had already added substantially
to our knowledge of Vertebrate Anatomy. We owe to him the
first accurate account of the urino-genital organs of *Polypterus*,
and the demonstration that the crossopterygian fin is really a
uniserial archipterygium; besides a series of invaluable observations
upon the life-history and breeding habits of many tropical frogs
and fishes.

His work was characterized by an almost fastidious degree
of accuracy; he had a somewhat critical respect for the work
of others, and he always liked to satisfy himself by reinvestigation
that things were really as they were described.

As Cowley wrote about the greatest of English physiologists, he

> "sought for Truth in Truth's own Book
> The creatures, which by God himself was writ;
> And wisely thought 'twas fit
> Not to read comments only upon it,
> But on the Original itself to look."

A. E. SHIPLEY.

CHRIST'S COLLEGE,
CAMBRIDGE,
1906.

ZOOLOGICAL MEMOIRS BY J. S. BUDGETT.

I. NOTES ON THE BATRACHIANS OF THE PARAGUAYAN CHACO, WITH OBSERVATIONS UPON THEIR BREEDING HABITS AND DEVELOPMENT, ESPECIALLY WITH REGARD TO PHYLLOMEDUSA HYPOCHONDRIALIS, COPE. ALSO A DESCRIPTION OF A NEW GENUS[1].

With Plates I—IV.

LIST OF BATRACHIA COLLECTED BY J. S. BUDGETT IN THE PARAGUAYAN CHACO, 1897.

		Length
I.	LEPTODACTYLUS OCELLATUS, L.	120 mm.
II.	LEPTODACTYLUS TYPHONIUS, Daud.	45 mm.
III.	LEPTODACTYLUS BUFONIUS, Boul.	55 mm.
IV.	LEPTODACTYLUS PŒCILOCHILUS (Cope)	50 mm.
V.	PHRYNISCUS NIGRICANS, Wiegm.	33 mm.
VI.	PALUDICOLA FUSCOMACULATA, Steindachn. ...	40 mm.
VII.	PALUDICOLA SIGNIFERA, Boul.	25 mm.
VIII.	PALUDICOLA FALCIPES (Hensel)	15 mm.
IX.	ENGYSTOMA OVALE, Schn. ... ... ♀ 40 mm., ♂ 25 mm.
X.	ENGYSTOMA ALBOPUNCTATUM, Boul.	18 mm.
XI.	PSEUDIS PARADOXA, L.	50 mm.
XII.	PSEUDIS LIMELLUM (Cope)	20 mm.
XIII.	BUFO MARINUS, L.	150 mm.
XIV.	BUFO GRANULOSUS, Spix.	50 mm.
XV.	PHYLLOMEDUSA HYPOCHONDRIALIS, Cope ...	40 mm.
XVI.	PHYLLOMEDUSA SAUVAGII, Boul.	70 mm.
XVII.	HYLA SPEGAZINII, Boul.	80 mm.
XVIII.	HYLA VENULOSA, Laur.	70 mm.
XIX.	HYLA NANA, Boul.	22 mm.
XX.	HYLA PHRYNODERMA, Boul.	43 mm.
XXI.	HYLA NASICA, Cope	28 mm.
XXII.	CERATOPHRYS ORNATA, Bell.	120 mm.
XXIII.	LEPIDOBATRACHUS ASPER, n. sp.	80 mm.
XXIV.	LEPIDOBATRACHUS LÆVIS, n. sp.	80 mm.

I. LEPTODACTYLUS OCELLATUS, L.

An extremely common frog, frequently found in the streets of Concepcion at sunset and on both sides of the river Paraguay. At Concepcion, black markings on a greyish-green ground. At Waikthlatingmayalwa the ground is usually of a brighter green.

[1] From *Quart. Journ. Micr. Sci.* Vol. 42, N.S.

A triangular black spot at the back of the eyes is very constant.

The natives, who are Lengua Indians, name this frog *Nukkmikkting*, and use it largely for baiting their hooks. The largest measure 50 mm. from snout to vent.

The call is regularly repeated, beginning on a low note and ending on a high one, and is constantly heard in wet weather. There is, however, another call, which is heard immediately after rain; this is a drumming like that of a snipe.

A large variety found in the Chaco is called by the Lenguas *Yattnukk-mikkting*; these measure up to 120 mm., and are only found down in the swamps. I think this may be *L. bolivianus*.

II. LEPTODACTYLUS TYPHONIUS, Daud.

Not nearly so common as *L. ocellatus*; I procured only two specimens, though I saw a few others. These were all seen at Caraya Vuelta on the river bank. The general colour is lighter than *ocellatus*, the spots are more numerous and smaller, and there is a bright gold band on either side running from the eye to the hips. It appears to be about the same size as *ocellatus*.

No Lengua name was obtained for it.

III. LEPTODACTYLUS BUFONIUS, Boul.

Small brown frog with blackish spots above, beneath pale yellow. Most inconspicuous on a background of earth. It is very agile and extremely shy.

In damp waste places on the outskirts of Concepcion I found it in great numbers, but very difficult to capture.

The call is a shrill sharp "ping" kept up constantly until approached, when it immediately ceases. The croaking of so many of them at a time produces an almost continuous sound.

Though only one specimen was secured, it was frequently heard on both sides of the river. This is probably the young form of *L. bufonius*, which grows to about the same size as *L. ocellatus*. I never detected large individuals of this form calling, and I am convinced that during the continuous calling described above the individuals about were of the small form almost entirely. It would appear, then, that either young forms have the habit of calling to one another, or that there is a small and a large variety. Lengua name *Ukksa-liapertikk*. In Lengua Uksaelia means a coin or disc. The name refers to the spots.

IV. LEPTODACTYLUS PŒCILOCHILUS (Cope).

This frog is much less common than *L. ocellatus*. It is of a more slender build; the toes are thin and long, especially the second toe. The markings are all in the form of stripes rather than spots. These are dark brown on a greyish-brown ground. At the side yellowish. One broad dark stripe runs down the back on either side at the edges of the transverse processes of the vertebræ. One specimen was found at Concepcion and one at Waikthlatingmayalwa.

I do not know if it has a native name.

V. PHRYNISCUS NIGRICANS, Wiegm.

This is a brilliantly coloured frog of toad-like appearance. The ground colour is black, and is irregularly spotted with yellow, or sometimes with large yellow blotches on the upper surface. Beneath it is black, with scarlet blotches; the palms of the hands and the soles of the feet are scarlet.

The variety found at Concepcion had on the under surface scarlet blotches extending to the throat, while the variety found at Waikthlatingmayalwa had the scarlet confined to the lower part of the abdomen. This form, too, had yellow blotches irregularly arranged on the back, while the Paraguayan form had small yellow spots more regularly arranged. On the journey between these two regions I twice met with large numbers of small black frogs which seem to be of this species. They were characterised by their smallness, and by the absence of either yellow or red markings.

At the breeding season the males and females have a call which consists of two clear musical "pings," followed by a long descending "trill" like that of our British greenfinch. The eggs are laid in separate globules of jelly, which float freely on the surface of the water, and are heavily pigmented.

This frog, which at ordinary times is the slowest and most bold of frogs, is now active and excessively shy. Swimming rapidly between the blades of grass it climbs a tuft, and, dilating its throat, repeats its call, but if in the least disturbed it is suddenly gone. This change of habit is very remarkable.

The spawn is found in quite temporary pools in grassy ground; the development is excessively rapid. Segmentation beginning at 10 a.m., they were hatched and wriggling about by 7 a.m. the following day. They probably are washed down into deeper pools by the retreating waters, and for this purpose the manner in which the eggs are laid, i.e. in separate globules of jelly, seems especially suited.

The native Lengua name is *Pithpaya.*

The eggs and larvae do not seem to differ in any great degree from those of Rana. There is, however, a very large yolk-plug, which remains evident after the closure of the neural groove.

VI. PALUDICOLA FUSCOMACULATA, Steindachner.

This is the largest of the genus that I found in the Chaco. It is a short-limbed frog, with spreading slender toes and small head. The upper surface is marked with characteristic marblings, which vary, however, greatly in colour. The metatarsal tubercles are large, horny, shovel-shaped, and black.

The peculiar cry which is so constantly heard in the neighbourhood of shallow pools, and resembles that of a kitten, is produced by the alternate inflation of throat and abdomen. When fully inflated the frog appears to be the size of a golf ball, but, if startled, instantaneously shrinks to one fifth of that size, so that it seems to have vanished. It has also the power of ventri-loquising. In the spawning time it was found at night floating on the surface of pools in the distended condition, and crying to the females in a most mournful way. On coming to the surface it fills its lungs with a few gasps, greatly distending the walls of the abdomen, and then drives the air into the throat diverticula of the pharynx, causing them to become distended as the stomach collapses, and giving rise to a kitten-like cry.

The eggs are chiefly laid in January, and are found embedded in a frothy mass floating upon the surface of the water. The eggs themselves measure 1 mm. in diameter, and are without pigment and with extremely little yolk. They become free-swimming within from eighteen to twenty-four hours of the time of the first segmentation. When ready for hatching they wriggle their way through the froth to the water below, and hang into it from the floating froth.

In this rapidly hatching, free-swimming larva many of the processes of development are blurred, and as it were hurried over. The external gills never reach a high state of development. The cell layers are many cells deep and diffuse, and the involutions and evolutions are difficult to follow.

The natives call this frog *Zing Ye*, which of course applies to the genus generally, for the species differ very slightly.

In this species the testes are much pigmented and lobulated. Its food consists largely of water-beetles.

VII. Paludicola signifera, Boul.

This is considerably smaller than *fuscomaculata*, and is usually an olive-green on the back without conspicuous markings.

Its general habits seem to be the same as those of *P. fuscomaculata*, as also its cry. It is most agile. I put this species into a cage in which were many brightly coloured frogs, including *Phryniscus nigricans* and also *Phyllo-medusa hypochondrialis*. In this cage was also a small grass snake. Hitherto it had taken no interest at all in the gaudy frogs in its cage; but as soon as the little Paludicola made its first spring, it was caught in mid-air by the snake.

VIII. Paludicola falcipes (Hensel).

Only one specimen found at Concepcion by the river side. Its toes are even, long, and slender. Many of the specimens in the British Museum are marked with one broad light band running from nose to vent. But by no means all have this, neither does it depend on sex. In the specimen which I procured this stripe is very much marked.

IX. Engystoma ovale, Schn.

This frog has a small head and pointed nose. The eyes are set far forward, and there is an encircling fold just behind the eyes. The fore-limbs are very small, and the general shape of this frog proclaims it at once to be a burrower. The skin is perfectly smooth. It is greenish-brown above, yellow beneath, and a bright yellow band passes up the thighs and over the vent. In the male this band is bright red. The male is somewhat smaller. The natives call it *Po it*, being convinced that the cry which sounds to them thus proceeds from this frog. However, in each case that I tracked down the frog calling with this cry I found a *Leptodactylus ocellatus*. The cry was heard everywhere, but I only found one male and one female.

I think the native boys were here mistaken again; they pointed out to me holes in the ground beneath fallen tree trunks, of the size of a cricket ball and lined with a froth containing white eggs and also tailed larvae. The entrance to the hole was about a centimetre in diameter. This they said was the nest of the *Po it*.

I reared some of the eggs, and one as far as the four-legged stage, when the young frog bore a very strong resemblance to a Paludicola, but unfortunately escaped from my tank before it had lost its tail.

Though the information obtained from the natives generally turned out to be fairly accurate, yet I feel sure that in some instances they were quite wrong.

To whatever frog these nests belonged, it is certain that they were a most ingenious contrivance for collecting water and keeping the eggs and larvae at least moist, between the storms of the wet season. They were always found within the forest belts which lay on the highest ground.

I found with these larvae that they would exist for a very long time in a small quantity of water without increasing in size, but that when removed to a tank they grew enormously, and very soon left the water.

These eggs were somewhat larger than the minute eggs of Paludicola, $1\frac{1}{4}$ mm., and pigmentless. As far as my investigations have gone these eggs develop much as Paludicola, though they are rather more heavily yolked.

X. Engystoma albopunctatum, Boul.

About half the size of *E. ovale*, and found under a heap of decaying vegetation in the forest. Plum-coloured and very glossy above, and greyish with white spots below. One specimen found. Native name unknown.

(The specimens collected by Bohls, in Paraguay, are all brightly spotted above.)

XI. Pseudis paradoxa, L.

A water frog never seen on land, and extremely shy. Though often seen floating in a shallow pool, it was caught with great difficulty.

In life most beautifully coloured with bronze, bright green, and black markings above; underneath a satiny sulphur-yellow, with brown spots on the trunk and brown stripes on the thighs. On killing, all the brilliant colours of the back turned to a dull uniform brown in a few minutes.

Though there were a pair of these in a pool all through the early part of the wet season, yet the pool did not contain any of the well-known gigantic larvae with reference to which the frog is named.

No native name known to me.

XII. Pseudis limellum (Cope).

Small green frog abundant on the camelote leaves at Concepcion. Capable of changing its colour greatly from bright green to dull brown, underneath silvery. Two white streaks run backwards from the eyes. The call is a succession of sharp croaks or vibrations resembling the sound made by castanets. The throat is inflated for each series.

They hop quickly over the surface of the water, and perch on the camelote leaves and stems. They are enabled thus to hop on the surface of the water by reason of the very large webs of the hind feet. The tips of the toes also have dilated discs. Their food consists mainly of small fresh-water Gasteropods. Females larger than the males.

No native name known to me.

XIII. Bufo marinus, L.

The common toad of South America, up to 150 mm. in length from snout to vent. It feeds on all kinds of insects, and is very useful in helping to keep down the mosquitoes. One half-grown toad, sitting by one man's foot, picked off fifty-two mosquitoes in the space of one minute, flicking them up with his tongue as they settled.

This toad, which may be found in every shed or outhouse, is called by the natives *Pinnikk*. Its call consists of three bell-like notes, the middle one being the highest. The parotid glands are enormously developed, and, if the toad is roughly handled, are discharged like squirts. When wet weather comes it hops out from its hiding-place, and proceeds to sit in a puddle, with its head out.

XIV. BUFO GRANULOSUS, Spix.

A very common small toad, found in great numbers near water after rain. Dark above, with black, brown, and greenish blotches, and a light vertebral line. Skin much tuberculated. Calls with a continuous bell-like tinkle, the vocal sac being greatly distended. A great deal of variety in colour.

Native name *Kelaelik*.

This species forms the chief food of the two newly described species of Lepidobatrachus.

XV. PHYLLOMEDUSA HYPOCHONDRIALIS, Cope.

A brilliantly coloured grass frog, which I found breeding freely in the Paraguayan Chaco, about 120 miles due west of Concepcion (fig. 34). Above it is brilliant green, which may become brown, grey, or bluish at will; below granular, white. The flanks are scarlet with black transverse bars; and the plantar surfaces are a deep purplish-black.

The *Wollunnkukk*, as it is called by the Lengua Indians, from the call of both male and female at pairing time, is extremely slow in its movements, and is active only at night. At this time, if it is seen by the aid of a lantern as it slowly climbs over the low bushes and grass, it is very conspicuous, as shown in the figure. In the daytime, however, nothing is seen but the upper surfaces of the body as it lies on the green leaf or caraguata plant, and here it is most inconspicuous.

This small Hylid has a remarkable power of changing the colour of its skin to harmonise with its surroundings, and can effect a change from brightest green to a light chocolate in a few minutes. The skin is also directly sensitive to light; for if the frog is exposed to the sun while in a tuft of grass in such a way that shadows of blades of grass fall across it, on removal it will be found that dark shadows of the grasses remain on the skin, while the general colour has been raised to a lighter shade. Its food consists largely of young locusts. The ovaries on each side are divided into five distinct clusters. The rectum has a large saccular diverticulum, which is very heavily pigmented.

In the breeding season—December to February—this beautiful grass frog collects in considerable numbers in the neighbourhood of pools. During the night-time they call incessantly to one another, and produce a sound as of a dozen men breaking stones, well imitated by the native name "Wollunnkukk."

As regards the native names for frogs, most species had their separate names; for instance, two species so closely like one another as *Leptodactylus ocellatus* and *L. bufonius* had their names respectively "Nukkmikkting" and "Ukseliapertic," but with the tree frogs it was not so. I could get no name for any frog with dilated discs but "Wollunnkukk," whether they had a call resembling this name or not, and whatever their form, colour, and size. I may mention also that they had no general name for frog, though they had a general name for bird and fish.

Breeding Habits.—On November 30th, 1896, I caught six of these frogs at the edge of a shallow pool late at night, and put them with some leaves in a tin until the morning. Next morning I discovered batches of white eggs, in masses of firm jelly, lying about at the bottom of the tin. I put some of these in water, and some I kept damp. Those which I put in water died immediately; those which I kept merely moist I watched segmenting and developing until December 5th, and preserved several eggs of each stage, but on this day the last of the embryos died, and I tried hard to get some more, and to find out how they were laid in nature.

On December 31st I discovered a small leaf overhanging a pool of water, and containing a batch of the Wollunnkukk eggs. At this same pool I found within the next three weeks about twenty leaves enclosing batches of eggs, in no case more than two feet from the water.

On January 15th I had an opportunity of watching the process of egg-laying. About 11 p.m. I found a female carrying a male upon her back, wandering about apparently in search of a suitable leaf. At last the female, climbing up the stem of a plant near the water's edge, reached out and caught hold of the tip of an overhanging leaf, and climbed into it. With their hind legs both male and female held the edges of the leaf, near the tip, together, while the female poured her eggs into the funnel thus formed, the male fertilising them as they passed (fig. 35). The jelly in which the eggs were laid was of sufficient firmness to hold the edges of the leaf together. Then moving up a little further more eggs were laid in the same manner, the edges of the leaf being sealed together by the hind legs, and so on up the leaf until it was full.

As a rule two briar leaves were filled in this way, each containing about 100 eggs.

The male hurried away immediately the laying was over, and he did not embrace the female except during the act of laying eggs. The time occupied in filling one leaf was three quarters of an hour.

Life History.—Development proceeds very rapidly; within six days the embryo increases from 2 mm. (the diameter of the egg) to 9 or 10 mm., when it leaves the leaf as a transparent glass-like tadpole whose only conspicuous part is its eyes (fig. 30). These are very large and of a bright metallic green colour, so that when swimming in the water all that is seen are pairs of jewel-like eyes.

The newly hatched tadpole has also a bright metallic spot between the nostrils somewhat in front of the pineal spot. This is the point which touches the surface of the water when the tadpole is in its favourite position. Whether it is a protective coloration, or some mechanical arrangement for holding the surface, I cannot say.

The leaves containing the eggs are not always directly over water, and the newly hatched tadpole has often to make his way many inches to the water.

This migration to the water usually takes place during a shower of rain, when the larvae tend to be washed into the pool, but they also assist themselves by jumping several inches into the air. They are intensely sensitive to light and shock.

During the embryonic development the jelly surrounding the embryo becomes more and more dilated by the growth of the embryo, and also by the accumulation of fluid within. Towards the close of embryonic life the embryo comes to lie quite freely within a membranous capsule.

The eggs are very heavily yolked, and some yolk persists until the tadpoles are ready to leave the capsule.

On the third day external gills are well developed, and the red blood-corpuscles may be seen coursing through them, and the heart beating rapidly. These external gills reach their highest state of development about the fifth day, when they extend beyond the vent, and are of course bright red (fig. 27).

The tadpole is hatched without a trace of yolk, the external gills have completely disappeared, there is a median spiracle, and the lungs are already clearly visible shining through the transparent body-wall (fig. 30).

The day after the tadpoles are set free, pigment begins to be developed about the head and upper surface of the body. There is a conspicuous absence of pigment for some time over the pineal body (fig. 26).

Black pigment appears first, then green. At the end of about five more weeks the tadpole has begun to develop its hind limbs. During this period it has grown to a length of 8 cm. The upper surfaces are now a glossy green, beneath silver and rosy; the tail is still transparent, and the red blood-vessels give it a bright red colour (fig. 31).

At the time of the development of the hind limbs there is a very great accumulation of black pigment at the middle of the tail, especially below (fig. 31).

The tail is absorbed very rapidly up to this point; the final absorption of the proximal part of the tail is postponed for some days.

The young frog, having now grown both pairs of legs, leaves the water and betakes itself to the blades of grass close by (fig. 32).

Here it sits during the time of absorption of the remainder of the tail. When lying in the blade of grass, only the brilliant green upper surfaces are visible, and the tail helps to make the young frog still less noticeable by shading off the body, and causing it to become merged in the green of the grass blade.

The young frog at the close of its metamorphosis is two thirds the length of the adult frog, and at this time acquires the red flanks barred with black (fig. 33).

There is a certain stage in the life of this larva when it will not bear transferring from the pool to aquaria. If the larvae are transferred at the time when pigment in the tail is just beginning to accumulate, that is when they are 3 cm. in length, they invariably die, though both younger and older larvae stand the change quite well.

Development.

External Characters.—Segmentation is holoblastic, though not so regular as in Rana and most Batrachia (figs. 1—6). The blastopore is formed by a more general overgrowth of epiblast, and is from the first circular; it is only just before closure that it is possible to tell from an external view which is the anterior and which the posterior side of the pore (figs. 7, 8). It is quite during the last stages of gastrulation that the closing mouth of the gastrula swings up to the posterior edge of the egg. When the blastopore has reached this position it becomes pointed at the anterior end, and there can now be seen running forward from this point a groove showing unmistakably the line of fusion of the edges of the mouth of the gastrula (fig. 9).

Finally the yolk retreats, and a slit-like open blastopore remains at the posterior end of the pear-shaped neural plate (fig. 10).

While this fusion has been taking place the centre of gravity has been continually shifting; for along the line of fusion there is a greater accumulation of yolkless protoplasm, i.e. epiblast and mesoblast, than elsewhere. Yolk is heavy, therefore the fused edges of the mouth of the gastrula come to the upper side.

Finally the entire egg has rotated through 180°. The anterior end of the archenteron is now in the position that the posterior end occupied at the beginning of gastrulation. The area occupied by the neural plate has been formed chiefly by the downgrowth of the lateral and anterior edges of the blastoporic rim; however, the posterior edge here takes a greater share in gastrulation than in Rana, and in consequence the blastopore comes to lie not at the extreme posterior end of the archenteron, but further towards the middle, while the neural plate extends beyond the blastopore. The anus, however, makes its appearance at the extreme posterior end of the archenteron, far from the position now occupied by the blastopore (Plate II, Sections II, III, *an.*).

The centre of the neural plate becomes slightly depressed, and here the blastoporic scar is seen running forwards from the edge of the blastopore along the whole plate as the "primitive streak" (fig. 11).

The neural folds now begin to approach one another at the anterior and posterior ends of the groove, but there is no well-marked anterior transverse fold (fig. 12).

Posteriorly the folds enclose the remains of the blastopore, which then opens only into the neural canal formed by the complete fusion of the edges of the two folds. A tail fold develops of a crescentric form encircling the posterior end of the neural plate, on the posterior convex side of which the anus is formed (Section III, *an.*).

Between the blastoderm and the egg membrane there is now present a considerable space, filled with a milky fluid (fig. 12, *sp.*).

When the neural folds have completely met, i.e. fifty hours after laying, then the anterior end of the neural plate expands to form the optic vesicles, and an elevation extends forwards from them, homologous with the so-called *Sense-plate* of Morgan. Behind the optic vesicles extending laterally and anteriorly on either side is seen the gill-plate or branchial fold.

Later this grows to completely encircle the sense-plate, which now shows a depression at the anterior end, the rudiment of the stomodaeum (fig. 20, *Stom.*).

The right and left halves of the "sense-plate" thus divided are very conspicuous features at this stage in development, and for some time later. A little later they become formed into a regular pair of mandibular bars, which only just meet below the stomodaeum (fig. 21, *Mnd.*).

In section they appear quite like the succeeding hyoid arches, which are very slightly developed, and also the larger first and second pair of branchial arches.

There is a total absence of suckers such as are borne behind the mouth in most Batrachian larvae, and the embryo has now more the appearance of a young larva of Acipenser than of Rana.

The *gill-plate* in life appears as a single elevation on either side, but after fixing with appropriate reagents it may be seen almost from the first to consist of three branchial pouches of the pharynx; the two anterior of these alone persist. The first pouch is between the hyoid and the first branchial arch. The second pouch is between the first and second branchial arch. The third pouch is between the second and third branchial arch (fig. 14, *3rd br. f.*).

The *optic bulbs* early begin to bud out from the fore-brain, and now just behind the gill-plate is seen the first rudiment of the pronephros, a slight but defined elevation tapering posteriorly; mediad to this are seen four or five mesoblastic somites (fig. 14, *mes. som.*).

The *auditory vesicles* are not easily visible until after the appearance of the external gills.

Up to this time the embryo has lain almost flat upon the surface of the yolk, preserving in all a spherical form; now, however, it begins to rise from the surface of the yolk, both anteriorly and posteriorly, but the yolk is still nearly spherical (fig. 15).

The eye-bulbs increase greatly in size, and are exceedingly large in comparison with what is found in Rana at a corresponding stage. The ocular muscles are developed very early, and the eye may be seen to be rotated by them on the fourth day of development. A very conspicuous feature of this stage is the dilated condition of the double gill-pouch (fig. 15). Viewed from the dorsal surface, the head region has now in outline the form of a trefoil.

The tail now begins to grow back from the surface of the yolk, the dilation of the branchial folds ceases, and in proportion to the latter the head portion increases greatly (fig. 16).

The first pair of external gills now may be seen budding out from the first branchial arch. Below the cleft post-oral region, formed from the sense-plates, the rudiment of the heart is clearly visible (figs. 21, 24, *ht.*).

In a side view, more or less transparent as in life, there are to be seen the heart, first pair of external gills, well-formed eye with conspicuous choroid fissure, auditory vesicles, somites, caudal notochord, and extended cloaca. The yolk-sac still retains its spherical form (fig. 24).

The changes in external form which now take place mainly consist in the appearance of the second pair of external gills, which do not reach nearly so high a state of development as the first pair ; also in the rapid growth of the first pair of gills, so as to extend beyond the vent as blood-red filaments through which the corpuscles stream along, propelled by the now rapidly pulsating heart (figs. 25, 27, 17—19).

A dense plexus of vitelline veins ramifies over the surface of the yolk, while the dorsal aorta and cutaneous veins give to the elongated tail a copious supply of blood (fig. 29). Indeed, so noticeable is this, that I am quite inclined to agree with Mr Kerr's suggestion that the tail of this larva is an important organ of respiration. This view is further strengthened by my observation that in hatched larvae the tail often remains motionless as a whole, while the extremity of the tail is kept rapidly vibrating. As the larvae are not propelled by this motion through the water, I am tempted to think that the object of it is that a stream of water may be kept constantly running along the surface of the proximal part of the tail.

The *operculum* now grows down from the hyoid arch and encloses the gill arches. The external gills become rapidly absorbed (but I think that a study of the origin of the filaments of the internal gills shows them to be really of the same nature as the external gills), the stomatodaeal aperture breaks through, and the young frog has reached the end of its embryonic life (fig. 26).

Internal Characters.—On account of the short space of time at my disposal it seems advisable not to attempt a continuous account of the internal phenomena of development, but merely to figure and describe sections illustrating some of the more important points of interest, leaving the fuller account for a future occasion.

Section I[1] passes transversely through the blastopore before the formation of the neural groove. The main points to be noted are the smallness of the arch-enteron, the absence of a yolk-plug, the abundance of yolk, and the mesoderm extending only as far as the equatorial region of the egg.

Section II passes longitudinally through the blastopore (*bl.*), the walls of the neural groove being closed anteriorly, but not yet posteriorly. The points to be noted are the anterior position of the blastopore, the fusion of the embryonic layers before and behind the blastopore, and anteriorly the beginning of the first branchial pouch (*br. f.*).

Section III, a sagittal section of an embryo after the closure of the neural groove, showing the comparatively anterior position of the neurenteric canal (*n. en.*), the brain vesicles, the notochord not yet differentiated posteriorly, the archenteron, and anteriorly the branchial fold (*br. f.*) of the pharyngeal wall, which is continuous across the middle line; also posteriorly the depression which will later become the anus (*an.*).

Section IV is of the same series as the last, but further from the middle line, showing the large lateral cavity of the archenteron caused by the upraising of the branchial pouch (*br. f.*). The section also passes through the optic vesicle (*op. ves.*).

Section V is a transverse vertical section passing through the head region of an embryo in which the body of the embryo is just beginning to rise up off the yolk. To be noted are the regularly formed optic vesicles (*op. ves.*) and stalks, there being as yet no trace of the lens. Below is seen the pharyngeal region of the archenteron (*ph.*).

Section VI is of the same series as the last, passing between the optic and auditory region; it shows the single branchial pouch (*br. f.*) and the accompanying epidermal thickening.

Section VII of the same series, passing through the auditory region. There are seen in section the front end of the notochord (*N. ch.*), the auditory thickening of the epidermis (*Aud.*) and the rudiment of the pronephros being differentiated off from the general mesoderm (*P. n.*). There is seen also a faint indication of a neural crest (*N. cr.*).

Section VIII .of the same series, passing through the posterior end of the archenteron, where it is seen to be constricted into two portions, the upper being the opening of the neurenteric canal (*N. en.*), the lower the rudiment of the rectum (*Rect.*)

Section IX shows the fusion of the layers in the region of the neurenteric canal, and the separation of the latter from the rectum.

Section X of the same series, through the tail and vent, shows the opening of the neurenteric canal into the neural tube; also the fusion of the epidermis with the hypoblast at the anus (*An.*).

---

[1] The numbers here given correspond with those of the sections on Plates II and III.

Section XI, a transverse vertical section quite at the anterior end of the head of an embryo, in which the first pair of external gills are beginning to bud. The section passes through the bottom of the stomodaeum (*Stom.*), and obliquely through the mandibular arches at the point where they meet (*Mnd.*).

Section XII is of the same series, but further back, and on the right side in the figure passes through the centre of one of the eyes, showing the attenuation of the posterior wall of the optic cup (*Op. w. p.*) and the thickening of the anterior wall (*Op. w. an.*), the rudiment of the retina. The lens is also seen arising as a regular involution of the nervous layer of the epiblast, the epidermal layer (*l.*) remaining stretched across as a very thin membrane. The section also passes through the middle portion of the mandibular arches (*mnd.*). The pharynx and pericardium are also cut through (*Ph., P. c.*).

Section XIII of the same series passes through the centre of the opposite eye. The proximal parts of the mandibular arch are here cut through (*mnd.*). The formation of the pericardium and heart, with its mesodermal membranous lining, is well seen (*P. c., ht.*).

Section XIV of the same series passes a very little further back through the infundibulum, pharynx, and the two lateral extensions of the pericardium overlying the sinus venosus (*S. v.*).

Section XV is of a slightly older embryo, passing transversely through the eyes. The lens is now nearly completely constricted off, and the back wall of the lens has begun to thicken and fill up the hollow of the lens (*l.*).

Section XVI is a sagittal section through the pineal eye of an embryo about two days before hatching. It shows the pineal stalk, still allowing free communication between the pineal body and the brain-cavity; this passage is now distinctly ciliated (*cil. st.*). Blood-sinuses are seen in front and behind.

Section XVII is a transverse section of an embryo just before hatching. It passes through the root of the first external gill (*Ext. G.*), and shows the developing first and second internal gills (*Int. G.*).

From this and similar sections there certainly does not seem to me to be any very marked difference in the nature of the external and internal gills.

As regards the development of the *Pronephros* and its duct, my sections indicate that there is considerable variation. Though by the time the external gills are developed there are invariably three nephrostomes, as in Rana, the first and third being lateral, the second dorsal, yet previously to this stage I find often only two nephrostomes, and in some instances two on one side with three on the other, and in one case but one. This seems to me to indicate that the pronephric tubules do not arise in the way usually described for Rana, namely, by the primitive pronephric groove becoming a closed tube and remaining in open communication with the coelom at three points, but rather as a solid rod of mesoderm (Section VII), which later becomes hollow and acquires perforations into the coelom, at first one, later three points—the nephrostomes.

In comparing the development of *Phyllomedusa hypochondrialis* with that of Rana, Bombinator, Pelobates, and other Batrachians with free-swimming larvae, the first thing that strikes one as regards external characters is that, throughout, this embryo maintains a greater similarity to ichthyic forms, especially Ganoids, on the one hand, and to the Urodela on the other, than do the free-swimming larvae of other Batrachians.

Again, we find this difference in general development of the young larva intensified in such forms as undergo a still more abbreviated embryonic development; for instance, in *Paludicola fuscomaculata*, where the embryonic development is shortened to something between twelve and twenty-four hours. All the points in which Rana appears to be a more modified form of development than Phyllomedusa are intensified, and the external characters are ill-defined. However, a minute comparison cannot yet be made until I shall have had time to study more carefully the details of development in Paludicola. The study of the internal anatomy leads to the same conclusion, namely, that in this protracted development we do not find the course of development distorted and blurred, but on the contrary every organ, so far as I can find, develops as in the ordinary frog, only more clearly and more definitely, and at the same time more as we see it develop in other great groups, Elasmobranchs, Ganoids, and the higher Vertebrates.

Take for instance the *eye* of a free-swimming batrachian larva, and compare it with the eye of Phyllomedusa. The evolution of the optic cup and lens is hurried over and blurred in the former, so that they are often difficult to trace, while here in Phyllomedusa it is as regular and diagrammatic as in any Vertebrate there is. Contrary to what we find in most Batrachia, the lens develops as an involution of a single layer of nervous epiblast rather than a mere thickening of that layer. In the free-swimming form the eye has been required to become functional as rapidly as possible, while here it has been suffered to go through its normal course of development in peace.

Take again the *suckers* of the free-swimming forms. They are evidently new adaptations without phylogenetic significance. Through the presence of these structures the form of the mandibular arches has become quite obliterated, while here in Phyllomedusa these would compare favourably with those of an Elasmobranch, reptile, or bird.

The peculiarly symmetrical *gastrulation* that this egg exhibits must be supposed, I think, to be the effect of a large amount of food-yolk, as it can hardly be supposed that, at a stage previous to hatching in either mode of development, Phyllomedusa should be more primitive than the free-swimming forms.

I think the *median spiracle* may also be looked upon as a primitive feature.

The manner in which the *branchial fold* encircles the head reminds one strongly of Salensky's figure of Acipenser at a similar stage.

From the study of the development of Phyllomedusa, of which I have described the points of more general interest, I am distinctly inclined to think that we are not always warranted in attributing to alecithal free-swimming larvae a greater biological importance, as far as retaining ancestral characters is concerned, than to heavily yolked embryos.

I think, moreover, that this is what we should expect, for from the time that the larva is hatched onwards it is subjected to the iufluence of natural selection.

Indeed, in this particular case of Batrachian development it would seem rather that the shortening of the embryonic period may be a specialised and not a primitive condition.

The fact that the majority of frogs have a shorter embryonic life does not seem to me to prove that the minority are the specialised forms in this respect. This particular mode of development is not confined to this species.

Von Jhering has described the oviposition of *Phyllomedusa Jheringii*, which agrees very closely with that here described. The eggs were laid between two or more leaves instead of being rolled in one, as with *Phyllomedusa hypochondrialis*. Von Jhering did not, however, work out the development of this species; in all probability it would not differ from this one.

This year S. Ikeda, of Tokio, has published an account of the oviposition in a species of Rhacophorus; from what he mentions of the appearance of the embryos, which develop in a froth, much as is the case with Paludicola, I think the development of this form will be found to be quite like that of Phyllomedusa; indeed, Professor Mitsukuri, who has seen them both, assures me that this is so.

To a paper by Gasser published in the *Sitz. d. Kön. Ak. Marburg*, 1882, upon the development of the midwife toad, *Alytes obstetricans*, I have not yet been able to get access, but I feel quite prepared to find that it exhibits the same features that characterise the development of *Phyllomedusa hypochondrialis*.

XVI. PHYLLOMEDUSA SAUVAGII, Boul.

This handsome tree frog was brought to me in the Chaco, but I am not able to state anything about its habits.

XVII. HYLA SPEGAZINII, Boul.

This fine Hyla was fairly common; I often caught or saw young specimens swimming from stem to stem of the Papyrus grass as we travelled through the reed-choked swamps. The full-grown specimens, however, were always taken either from palm tops just felled or from the trees overhead.

When caught in the water by daylight they were a bright light yellow, but at night they turned to a darker shade, and became marbled on the upper surface with brown markings. The full-grown specimens did not in this way become dark at night.

The largest specimens taken measured 80 mm. The eggs in the cloaca appear to be quite like those of Rana in size and colour, and are probably laid and reared in the same way.

One full-grown specimen I obtained in Central Paraguay on the Tibicuari, the rest in the Paraguayan Chaco.

XVIII. HYLA VENULOSA, Laur.

In life the markings are olive-green or grey upon a whitish ground. When taken from amongst foliage the whitish ground colour is suffused with green. It is a powerful and energetic frog, the large toe-discs having a tenacious sucking power.

The skin glands are strongly developed, emitting a very sticky white slime.

XIX. HYLA NANA, Boul.

This small frog was abundant in the swamps, usually found by moonlight sitting on the broad-leaved plants of the swamp, and calling with a rather highly pitched scraping note.

The upper surfaces, as *H. spegazinii*, are light straw-colour by day, but brown by night. Flanks and underneath pigmentless.

XX. HYLA PHRYNODERMA, Boul.

A light golden colour, shaded with darker above. The discs fairly strongly dilated. The skin is warty and extremely delicate, and it is not easy to catch one uninjured.

They are not common, but make themselves known by their constant call, which is just like the quacking of a duck. All the specimens I obtained were about the palm fencing and sheds.

XXI. HYLA NASICA, Cope.

This is the most common Hyla in the Chaco. It is found everywhere, usually upon palm or palm fencing, where it is most inconspicuous.

Its call and habits are much like *H. phrynoderma*; the note is, however, somewhat lower. The colours are chiefly olive-green and brown, but the markings are variable. It is of more slender build than *H. phrynoderma*, the body being longer in proportion to the width of the head.

XXII. CERATOPHRYS ORNATA, Bell.

I obtained some half a dozen specimens of this curious and well-known South American frog, commonly known as the *Escuerso*.

Its ferociousness is its most striking characteristic. If it is approached to within two feet it will make a vicious spring at one with its gigantic mouth wide open. If it succeeds in seizing any part of its tormentor, it holds on like a bull-dog. The habit it has of distending its lungs to their fullest when teased has given rise to the idea amongst the Argentine people that if teased sufficiently it will burst.

It is perhaps needless to say that I was disappointed in my efforts to obtain this end. The Ceratophrys lives chiefly off other frogs and toads, but it is said also that it will seize and devour young chickens. The largest I saw was 120 mm. in length.

LEPIDOBATRACHUS, J. S. B. N. g.

Pupil horizontal. No vomerine teeth; transverse processes of sacral vertebræ not dilated. Large teeth in upper jaw; also two large teeth in dentaries of lower jaw. Tongue circular and free behind. Nostrils the most elevated portion of the head. Eyes close together, not more than the diameter of the eye apart. Fontanelles in the parietal region. Outer metatarsal tubercle very large. Great development of membrane bones in the head; width of jaw very great. Tympanum fairly distinct.

XXIII. LEPIDOBATRACHUS ASPER, J. S. B. N. sp.

Hind legs carried forward, toes reach barely to the eyes. Tips of toes horny. Skin of dorsal surface a dull leaden colour, much tuberculated and tough.

This frog lives continually in muddy pools. Its habit is to float with just the eyes and nostrils above the surface. If disturbed it slowly sinks to the bottom, leaving no ripple on the surface of the water. It feeds largely on *Bufo granulosus*.

XXIV. LEPIDOBATRACHUS LÆVIS, J. S. B. N. sp.

This may possibly be the same species as the last, but differs from it in the greater width of the skull, greater length of the hind legs, which carried forwards reach tip of snout, and in the skin being smooth, thin, and slimy, with the organs of the lateral line showing clearly upon it. Also the tympanum is larger and more evident. The tips of the toes do not bear horny caps as in the preceding species.

Below is a comparative list of measurements in millimetres of two specimens of XXIII and one specimen of XXIV.

		Total Length	Hind Legs	Width of Jaw	Eye to Eye	Eye to Ear	Ear to Ear
XXIII.	a.	80	62	34	4	8	24
	b.	70	60	33	$3\frac{1}{2}$	$7\frac{1}{2}$	23
XXIV.	c.	80	70	38	5	9	28

It is a source of great regret to me that I am obliged to abandon for the present my work in this direction. I have a considerable amount of material at my disposal of the developmental stages of several of the species of frogs, concerning which I have here merely stated the observations which I made a note of while yet in the Paraguayan Chaco. I sincerely hope that I may be able to return to this work at a future date.

Concerning the species *Phyllomedusa hypochondrialis* I should state that, although I have gone more fully into its development than others of my collection, here also my work has been cut short.

In concluding, I should like to say that I am very greatly indebted to Mr Graham Kerr for the opportunity he afforded me of obtaining my material, and also for much help and advice in my work.

EXPLANATION OF PLATE I.

The figures on this plate are all drawn under a magnification of eighteen diameters.

FIGS. 1—6. Illustrating the character of segmentation. Figs. 1—4 are views from above; Figs. 5 and 6 from the side.

FIGS. 7 and 8. Views of egg from below, showing diminution in size of the blastopore.

FIG. 9. Egg seen from below, at a time when the blastopore is much reduced in size, and has nearly reached the level of a horizontal plane passing through the centre of the egg. From the pointed anterior end of the blastopore there passes forwards a distinct groove, indicating the line of fusion of the gastrula lips.

FIG. 10. View of egg from above and behind, showing the continuation forwards of the slit-like blastopore as a faint groove along the axis of the medullary plate.

FIG. 11. View of egg from above, showing neural plate and early condition of neural folds.

FIG. 12. View of anterior end of embryo, with well-formed neural folds. *sp.* Space between embryo and egg membrane.

FIG. 13. Similar view where the neural folds are arching over towards one another.

FIG. 14. View of middle of trunk region of an embryo in which the pronephros has appeared on each side (*p.n.*). *mes. som.* Mesoblastic somites. *3rd br. f.* Position of third branchial pouch.

FIGS. 15—19. Figures illustrating the further development in general form of the embryo.

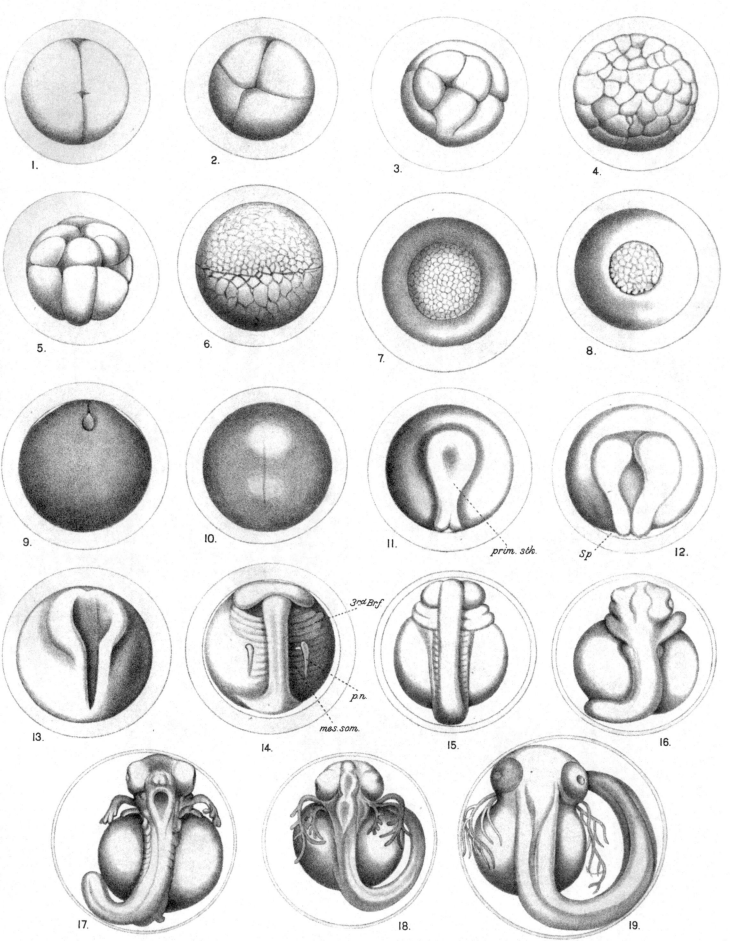

PLATE II.

EXPLANATION OF PLATE II.

FIG. 20.   View of anterior end of embryo, showing the first trace of stomodaeum (*Stom.*).

FIG. 21.   Oblique view of embryo, showing the mandibular bars (*Mnd.*) and rudiment of heart (*ht.*).

FIGS. 22 and 23.   Views of anterior end of head, showing the fusion of the mandibular bars (*mnd.*) in the mid-ventral line.

FIGS. 24—26.   Side views of larvae, showing the further changes in form up till the time of hatching. *ht.* in Fig. 24, rudiment of heart.   In Fig. 25 the external gills are at about their maximum.

FIG. I.   Transverse section through blastopore before formation of neural groove.   *bl.* Blastopore. *ep.* Epiblast.   *mes.* Mesoblast.   *hyp.* Hypoblast.

FIG. II.   Longitudinal vertical section of embryo with neural groove closed in anteriorly.   *bl.* Blastopore. *An.* Depression marking position where anus will appear.   *Arch.* Archenteron.   *Mes.* Mesoblast. *br. f.* Branchial outgrowth of archenteron.

FIG. III.   Longitudinal vertical section of an embryo after the closure of the neural groove. *n. en.* Neurenteric canal.   *an.* Anal depression.   *Notoch.* Notochord.   *hyp.* Hypoblast.   *br. f.* Branchial outgrowth of archenteric wall.   *Ves* [i], *Ves* [ii], *Ves* [iii].   Brain vesicles.

FIG. IV.   Section parallel to the last figured, but more lateral in position.   *op. ves.* Optic vesicle. *br. f.* Branchial outgrowth from archenteron.

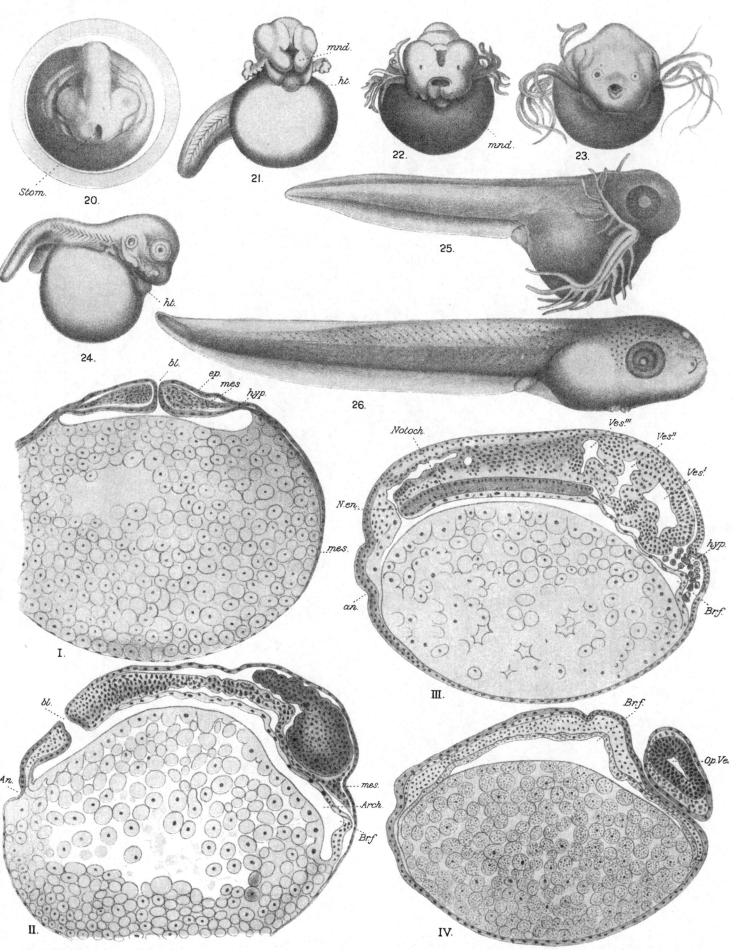

PLATE III.

EXPLANATION OF PLATE III.

FIG. V.   Transverse section through head of an embryo which was just beginning to be folded off the yolk. *op. ves.* Optic vesicle. *ph.* Pharynx. *mes.* Mesoblast.

FIG. VI.   Section of same series as that shown in Fig. V, through the single branchial pouch (*br. f.*) and the ectodermal thickening accompanying it.

FIG. VII.   Section of same series through auditory region. *Aud.* Commencing auditory invagination of ectoderm. *N. ch.* Notochord. *N. cr.* Neural crest. *P. n.* Rudiment of pronephros.

FIG. VIII.   Section of same series through posterior end of archenteron. *Rect.* Rectum. *N. en.* Neurenteric canal opening into this. *hyp.* Hypoblast. *N. ch.* Notochord.

FIG. IX.   Section of same series further back. *N. en.* Neurenteric canal. *Rect.* Rectum.

FIG. X.   Section of same series showing opening of neurenteric canal (*N. en.*) into neural canal; also anus (*An.*).

FIG. XI.   Transverse section through anterior end of an embryo, in which the first pair of external gills were beginning to develop. *Stom.* Cavity of stomodaeum. *Mnd.* Mandibular arch close to its junction with its fellow.

FIG. XII.   Section of same series passing through the rudiment of the eye. *Op. w. an.* Anterior layer of optic cup. *Op. w. p.* Posterior wall of ditto. *l.* Outer layer of epiblast continued over mouth of lens invagination. *Ph.* Pharynx. *P. c.* Pericardium. *mnd.* Mandibular arch.

FIG. XIII.   Section of the same series as that shown in Fig. XII. *mnd.* Mandibular arch. *ht.* Heart. *P. c.* Pericardium.

FIG. XIV.   Section from same series through infundibulum (*Inf.*). *P. c.* Pericardium. *S. v.* Sinus venosus.

FIG. XV.   Transverse section through head of slightly older embryo, showing later stage in the formation of the lens (*l.*).

FIG. XVI.   Sagittal section through pineal body (*Pin.*) of embryo about two days before hatching. *cil. st.* Pineal Stalk with ciliated lining. *Sin.* Blood-sinus.

FIG. XVII.   Portion of transverse section of embryo just before hatching, passing through the origin of the first external gill (*Ext. G.*). *Int. G.* Internal gill.

PLATE III.

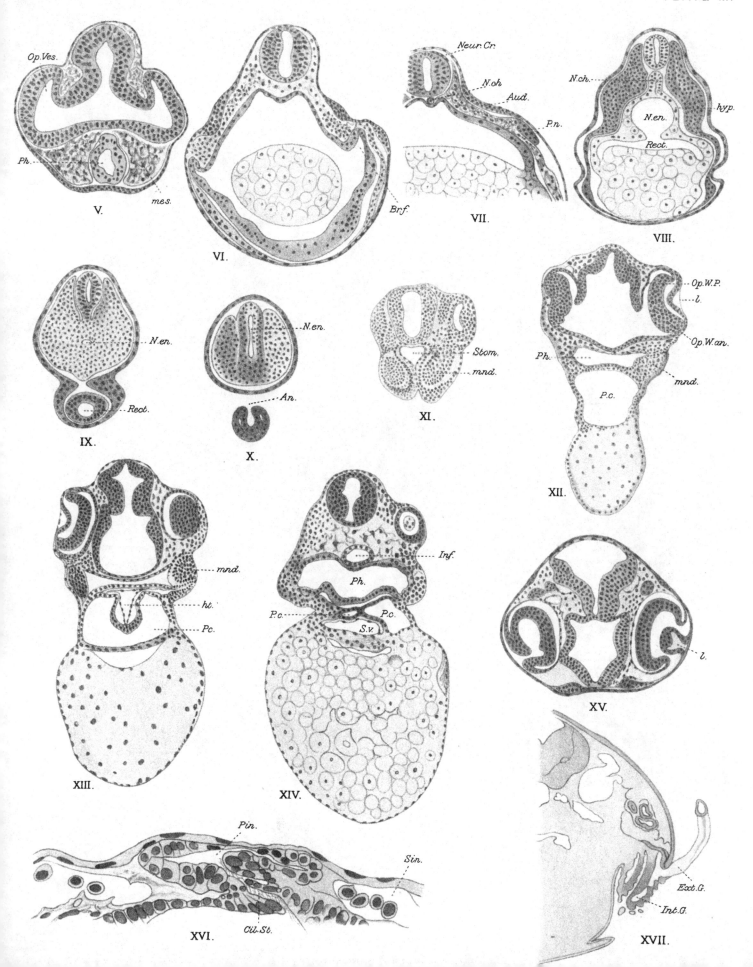

PLATE IV.

EXPLANATION OF PLATE IV.

FIGS. 27—30 are drawn under a magnification of eight diameters.

FIGS. 31—35. Natural size.

FIGS. 27--29. Side view of larvæ during the last day of intra-oval development.

FIG. 30. Larva just after hatching.

FIG. 31. Side view of larva at time of development of hind limbs, showing accumulation of pigment in the tail.

FIG. 32. Young frog after leaving water.

FIG. 33. Young frog after completed metamorphosis.

FIG. 34. Adult specimen. This figure by comparison with Fig. 35 illustrates the extent of reflex colour change.

FIG. 35. Pair of adults during the process of oviposition.

27.

28.

29.

30.

31.

32.

33.

34.

35.

E.Wilson, Cambridge.

DEVELOPEMENT OF PHYLLOMEDUSA

II. GENERAL ACCOUNT OF AN EXPEDITION TO THE GAMBIA COLONY AND PROTECTORATE IN 1898—99[1].

I propose to give a short general account of an expedition recently made by me, under instructions from the Council of this Society, to the river Gambia. This expedition had for its object the general study of the vertebrate fauna of the Gambia, and especially the investigation of the habits of *Protopterus* and *Polypterus*.

The river Gambia lies between the 13th and 14th parallels of north latitude. It flows due west through country which, lying about 100 miles to the north of the equatorial forest-region, is nowhere densely wooded but mostly covered with a somewhat sparse vegetation consisting largely of leguminous trees interspersed with gigantic baobabs (*Adansonia digitata*), the African mahogany (*Khaya senegalensis*), figs and sycamores.

Extensive open plains, which in the rainy season become flooded, border this river along the greater part of its course, while at a very variable distance from the river-bank low hills of dark red conglomerate rise, often abruptly, and occasionally in steep cliffs, to form level plateaux, which in the upper river may be 200 feet high.

The river-bank itself is clothed throughout the year with a rich luxuriant vegetation extending usually about 100 yards from the water's edge. Though here the trees and creepers remain green the year round, yet away from the river the trees lose their foliage in the dry season as completely almost as our own trees in winter.

From the mouth of this river to the country just below Nianimaru, the river is shut in by an almost impenetrable wall of mangroves, sometimes 30 feet in height. Above this point the river, though tidal, is perfectly fresh. The tides in the dry season make themselves felt for over 200 miles up the river, in fact to the end of navigable water, where there is about a foot rise.

The dry season extends from November to May. Tornadoes usually begin in June, while during July, August, and September there is a total rainfall of about 50 inches. During these months, though the tides make themselves felt, yet there is no change in the direction of the flow, while in August there runs a steady current of about 3 or 4 miles an hour.

In passing up the river the first place of interest is the old Fort James, which was formerly the port of export of the Gambia for the black-ivory trade. It is now being slowly washed away.

[1] From the *Proceedings of the Zoological Society of London*, November 28, 1899.

About 20 miles further up, the Vintang creek joins the Gambia, and at the junction of the two streams is the village of Vintang; it is seldom that a purely native village is seen at the water's edge, as they are usually on higher ground a mile or so from the river. If there are any tall trees in these villages, they are sure to be the nesting-places of Pelicans and Marabou-birds, which in the neighbourhood of the villages are strictly preserved. The vast flocks of these birds and also of the Balearic Crane are a great feature in the lower river, where there is little else to be seen but continual walls of mangroves, though now and again the monotony is broken by the passage of a native canoe or some trading cutter; but further up the variety of the vegetation is much greater.

Of particular interest to myself were patches of a Pandanus growing in the swampy ground at the river-side. The native name of this was Fang jani, which means "It burns itself." It certainly looked as though it deserved this name, for wherever it was seen a portion of every patch was charred with fire, and it was not easy to imagine how this could have been set alight by an external agency[1].

The great trading station on the Upper Gambia is McCarthy Island. To this place the trading cutters bring their cargoes of ground-nuts, the fruit of the plant *Arachis hypogæa*, to be shipped to Europe by the ocean steamers which make their way up to this island.

On McCarthy Island there are two trading establishments or "factories" as they are termed, and the remains of an ancient military settlement, consisting of Government House, Officers' quarters, and Barracks, formerly occupied by a detachment of the West India Regiment, which was withdrawn about 1870. The Government House alone of these buildings has been kept in repair; and here I established myself in company with Mr Wainewright, the Commissioner of the district, who, though usually travelling about the district, yet spends a considerable portion of his time here as Governor of the island. I stayed on McCarthy Island about one-third of my time. To the Governor of the Colony, Sir Robert Llewellyn, I am indebted for allowing me the free use of the Colonial steamer, "Mansah Kilah," and also for much hospitality. To Mr Wainewright, the Travelling Commissioner in the McCarthy Island district, I am greatly indebted for allowing me the use of a portion of the Government House at McCarthy Island, and also for the use of his huts in the main towns of his district. Very soon after my arrival at my headquarters, I made a tour through the district with the Commissioner to get some idea of the kind of country that surrounded me.

We started from Nianimaru, which was subsequently made my second headquarters, and where I spent even more time than at McCarthy. The chief interest in this tour lay in the people themselves, the country we travelled through not being of great interest from the point of view of its scenery.

Travelling was not difficult, as porters were plentiful, and were employed

[1 The charred appearance is actually caused by the presence of a peculiar fungus. Ed.]

from one village to the next at the rate of 3*d.* a man, if the distance was not more than 5 miles. At the important towns a court was held, and a stay was made of two days. The courts were held in the open, the chief, the head-man of the town, and the people all sitting round the Commissioner's chair.

There was plenty of time for shooting and no need to carry much in the way of provisions. The bag usually consisted of Bush-fowl and the Barbary Quail, *Pterocles quadricincta*, Guinea-hen, *Œdicnemus*, various Spur-winged Plovers, especially *Lobivanellus senegalus* and *Hoplopterus spinosus*, also Doves and Pigeons as many as were required. The finest of these, as game, was the Green Pigeon (*Treron calva*), which is never seen to approach the ground, being especially fond of the fruit of the fig-tree.

The commonest birds around us, which were not shot for the pot, were numbers of four species of *Coracias*, a *Centropus* known as the "foolish-bird" from its fearless habits and its call, which resembles a soft laugh, several species of *Bucerotidæ*, generally seen flying clumsily from tree to tree in small flocks; while overhead hovered large flocks of Bee-eaters (*Merops nubicus*), swallow-like in flight and song.

Other common birds everywhere seen in large flocks were the Metallic Starlings (*Lamprocolius auratus* and *L. caudatus*); Wood-Hoopoes (*Irrisor senegalensis*) seen in smaller flocks; while the commonest solitary birds were the Long-tailed Shrike (*Corvinella corvina*) and a species of Drongo (*Dicrurus assimilis*). The bushes of course swarmed with *Ploceidæ* and *Nectariniidæ*.

It being the beginning of the dry season, the grass was everywhere yet high, and it was out of the question to do any mammal-shooting; the only mammals visible were Climbing Squirrels and Monkeys. Burrows of *Orycteropus* were seen, though the animal does not appear to be very common in this region.

The towns visited during this tour were mostly far from the river and were taken in the order Nianimaru, Sukuta, Kaihai, Demfai, Tabanani, Sami, Koreantab, and back to McCarthy Island.

Near Kaihai there was news of a Giraffe having been seen, but they appear to be extremely rare in these parts. I heard indirectly that there were two in captivity at Kaies on the Senegal river.

Having returned to McCarthy Island on December 5, I devoted myself again to fishing and catching *Polypterus*. I found that all the specimens of *Polypterus lapradii* had already returned to the river from the swamps, where they come up to spawn in the wet season. However, large numbers of the young of *Polypterus senegalus* could still be caught by damming up the swamp-outlets.

This is a favourite way of fishing with the natives. They make dams across the creeks at short intervals, and then leave them in connection with each other for some days. Then damming up the connections, they bale out the water from the lowest compartment, collect the fishes, and proceed to the next compartment.

Very much more difficult is it to catch the *Polypteri* in the river. Nets which were very successful with other river fishes, failed utterly with *Polypterus*. The seine-net and trammel were given up, and the native cast-net was used with better success. The results of weeks of patient work were not encouraging however, and I gradually realised that the time to catch *Polypterus* was during the rainy season, when it had betaken itself to the flooded lands.

However, during these fishing days at Nianimaru, many interesting fishes were caught, and most of the common small Passerine birds were skinned. Moreover, this fishing was not without its dangers and excitement, as a look-out had ever to be kept for Hippopotami which swarmed in all the creeks. Moreover, frequently in the morning, when the trammel-net was examined, a Crocodile (*Crocodilus cataphractus*) or a Sawfish (*Pristis perotteti*) had to be slain. Several specimens of the latter were thus caught up as far as McCarthy Island, some of them measuring 9 feet in length.

Fly-fishing was tried without success. The line and hook were used more by the natives than myself. The trammel was found to be the best kind of net to use for the *Mormyridæ*, which were seldom caught in other ways. The Mormyrids apparently keep to the bottom of the river, and were seldom taken in the seine near shore.

It was noticed that a very large proportion of the fishes caught in this river were brilliantly coloured red in the ventral posterior portion of the body. Of fishes I believe 40 species were obtained, including 2 Selachians, *Protopterus annectens*, *Polypterus lapradii* and *P. senegalus*, 8 species of Siluroids and 7 Mormyridæ, and 18 others belonging to various groups. Most of the fish were tried as food, but there was only one that was really good eating: this was, I believe, a grey mullet and was taken far up the river.

Often the creeks in which the cast-net was thrown were very narrow, and the canoe slid silently amongst the most luxuriant vegetation abounding with Bee-eaters and Flycatchers. Altogether representatives of 108 species of birds were shot, measured, and described; but skins were made only of the smaller birds, of which examples of 52 species were obtained, belonging to 23 families.

With Dr Gadow's assistance, most of these have been identified. Of the *Upupidæ*, in addition to the gregarious *Irrisor* already mentioned, several specimens of *Scoptelus aterrimus* were seen and a skin of a male preserved.

On April 4 I took my two fishermen, my cook, and canoe up to a small village in the Kunchow creek called Alimaka, and there had some huts built. At this place again the trammel, the seine, and the cast-net were worked with hope of obtaining numbers of *Polypterus*. As a rule in the afternoon I went out to shoot, and found it a fair place for game.

During the fortnight thus spent at Alimaka, only six *Polypteri* were caught. There were caught also in this creek several specimens of *Gymnarchus niloticus* and some fine specimens of a freshwater Turtle, *Cyclanorbis senegalensis*. Lions were heard here frequently, and Leopards were seen, but at neither did I get a chance of a shot.

On April 20 two English gentlemen and a Frenchman arrived at McCarthy Island, on their way to some supposed gold-mines about 300 miles to the east of McCarthy Island. I accompanied them a short distance beyond the eastward British frontier to the town of Netebulu; the river is not navigable beyond that point.

Netebulu is an important native town, where a powerful chief named Sandian had his castle and harem. Here we stayed several days as the guests of the chief, and then I parted from the gold expedition, and made my way back overland to McCarthy Island, staying on the way a week at Koina.

About 50 miles above McCarthy Island the river-banks become high and precipitous, the country around being composed of high plateaux intersected by valleys. Frequently, however, the edges of the plateaux retreat from the river-bank a mile or so, surrounding wide plains, where one could be fairly certain of finding game.

Along the steep cliffs of the river-bank, vast numbers of Dogfaced Baboons (*Cynocephalus babuin*) might be seen wending their way. Sometimes the cliffs extended so far along the river-side that the Antelope were forced to come down to drink at certain places, and here the ground would be covered with their spoor.

April and May are the best months for big-game shooting. At Koina, large herds of Tankong (*Damaliscus korrigum*) were seen almost every day. Several were shot and a complete skin was made, which, however, suffered severely from the attacks of dogs and insects before it reached England. The herds were composed of males, females, and young of every age. The largest males seemed to lead the herd, though fine males mingled with the females and young as they daily made their way back in long procession from the river-banks to the higher lands.

Large herds were also seen of *Hippotragus equinus*, the Roan Antelope, or Dakoio as the natives call it, but this species was not so plentiful as the Tankong in these parts. A herd of Elands (*Oreas derbianus*) are believed to have been seen in the distance, and I was presented with a skull taken by Mr Wainewright from a carcase floating down the river.

In the open plains, where clumps of tall dead grass were shaded by a few trees, one might generally count on starting a Konkotong (*Cobus kob*), some Gazelles, or a Harnessed Antelope. The smaller solitary Antelopes were usually found in pairs. Enquiries were instituted everywhere as to the existence in this region of a Zebra, but I could hear nothing of it.

The horns either collected by me or from the natives included those of 9 species:—*Bubalis major, Damaliscus korrigum, Cobus unctuosus, Cobus kob, Cervicapra redunca, Hippotragus equinus, Tragelaphus scriptus*, a second species of *Tragelaphus* not yet determined, and *Oreas derbianus*.

12

Buffaloes were said to be common on Deer Island, but they were not seen by me, though horns of two forms were obtained from natives.

On the way back, a cutter was taken from Fatotenda to McCarthy Island, and after a few days spent at the Government House attending to my collections, and my living fishes and reptiles, I paid a final visit to Nianimaru. During this time, being the latter part of May, the rainy season began and the swampy places became filled with water. The Frogs began to spawn, and several series of stages in development of the different forms were preserved.

Here I first obtained free swimming *Protopterus* with ripe ovaries: examples of 8 Frogs, 3 Chelonians, 5 Lacertilia, and 9 Ophidia, including a *Typhlops*, were also collected about this time.

Returning to McCarthy Island, it was found that a number of *Polypterus lapradii* which had been kept in a pool connected with the river in the hope of getting them to spawn had been set free by the rising river. However, during the latter part of June and July a large number of *Polypterus* of both species were obtained, the females of which were crowded with ripe eggs. Artificial fertilization was tried with these, without success. Many were kept in confinement, and some, of which a pair are now exhibited, were successfully brought alive to England.

About the 10th of July, in the same swamp where these fishes were obtained, several nests of eggs were found. These eggs coincided in measurement exactly with the ovarian eggs of *Polypterus*. The young larvae possessed cement-organs on the front of the head so characteristic of Ganoid larvae; and other characters led me to assume that they were the young of *Polypterus*. None were reared beyond the larval state, and their identity could not well be established. However, having stayed on the Gambia three months longer than I had intended, and having a number of healthy *Polypteri* full of spawn, I decided to return home.

Just a day or so before leaving McCarthy Island I obtained eggs of *Protopterus*. These were watched through the early stages of segmentation, but the young could not be reared. On July 25 I left McCarthy Island and returned to England.

Several *Polypteri* and *Protopteri*, 12 young *Crocodilus cataphractus*, a Python, 3 *Cyclanorbis senegalensis*, 2 Hinged Tortoises, some Chameleons, and a Serval Cat were brought home alive.

Since my return to England, I have definitely decided that the eggs and larvae obtained are not those of *Polypterus*. I have, however, I believe, learned enough about the habits of *Polypterus* to encourage me to make a second attempt next year to obtain the developmental stages.

In conclusion, I wish to thank the Society for lending me influence and support, without which the little that has been done by this expedition could not have been accomplished.

III. OBSERVATIONS ON POLYPTERUS AND PROTOPTERUS[1].

The following observations on the Crossopterygian Polypterus and the Dipnoan Protopterus were made on the river Gambia 1898—99.

The Gambia flows due west, and lying between the 13th and 14th parallels of north latitude is tidal throughout its navigable waters, that is, to a distance of 260 miles from the sea.

a. Two species of *Polypterus* were obtained, *Polypterus lapradii*, Steind. and *Polypterus senegalus*, Steind. The former was taken 31 inches in length. The latter never more than $12\frac{1}{2}$ inches.

The two species are perfectly distinct. *P. lapradii* has not fewer than 13 free dorsal finlets. The head is flattened dorsoventrally and elongated. The body-surface is variegated with dark greyish-green markings on a yellow ground, taking the form of spots on the sides of the head and on the pectoral fin, but of blotches arranged in rows on the sides of the body elsewhere. The external gill is retained on the operculum until the young fish is 9 or 10 inches in length.

P. senegalus, on the other hand, has not more than 9 free dorsal finlets, the first arising on the 19th row of scales, whereas in *P. lapradii* the first arises on the 13th row. The head is shorter, round and deeper, while the body-surface is a uniform dark green above and bright yellow below. The external gill in this species is lost by the time the young fish is $3\frac{1}{2}$ inches in length.

In both species there is a marked difference between the *male and female.* The anal fin in the male is broad and fleshy with deep folds; in the female it is narrow, thin and pointed.

In the Gambia the rainy season begins in June and continues until October. In the early part of June *Polypterus* begins to migrate from the river to the flooded, low-lying plains; it spawns during August and September, returning to the river in October and November.

In the river *Polypterus* is one of the most difficult fishes to catch. *P. lapradii* however was caught from time to time in the river from January to May, all the specimens being about 18 or 19 inches in length and all without ova. Most of these were caught with the native cast-net, though a few were also caught with the seine-net.

In November and December when the swamps had nearly dried up large numbers were to be obtained from the mouths of the small creeks leading from

[1] From the *Proceedings of the Cambridge Philosophical Society*, vol. x. p. 236.

the swamps to the river. This was done by damming the creeks at intervals of about 30 yards but leaving a passage for the fish to pass from pool to pool. Each time the natives wished to collect the fish they stopped up these passages and then baled out the water. In this way I was able to obtain great numbers of young Polypteri on their way to the river. These all measured from 4 to 8 inches, and were invariably *P. senegalus*.

At the same time I had a few Lapradii brought me from 8 to 10 inches in length, but am not certain where they came from ; all these had external gills. When, however, these fish were kept in aquaria they soon lost their external gills, a small trace remaining for some time after the gills had been almost absorbed.

P. lapradii.	Nov.	Dec.	Jan.	Feb.	Mar.	Apr.	May	June	July	
♂	1	—	—	—	—	—	3	1	11	16
♀	1	—	2	—	—	—	9	3	12	27
Young	3	4	—	—	—	—	—	—	—	7
Undetermined	—	1	9	1	1	3	—	—	15	30
	5	5	11	1	1	3	12	4	38	80
P. senegalus.										
♂	—	1	—	—	—	—	—	17	—	18
♀	2	1	—	—	—	—	—	30	—	33
Young	15	50	—	—	—	—	—	—	—	65
Undetermined	—	—	—	—	—	3	—	—	4	7
	17	52	—	—	—	3	—	47	4	123
Total	22	57	11	1	1	6	12	51	42	203

From the table it will be seen that only 31 specimens of both species were obtained in the months from January to May, while during June and July 93 specimens were caught on their way to the swamps, and during November and December 79 young specimens were taken on their way to the rivers.

It is also seen that large numbers of *P. senegalus* were taken on their way from the river in June, while large numbers of *P. lapradii* were not taken until July. All the females taken in June and July were crowded with ripe ova, but *P. senegalus* was not taken above 13 inches, and yet no female less than 12 inches contained ripe ova. *P. lapradii*, however, was taken 31 inches in length, but females were crowded with ova when only 18 inches in length. It would seem therefore that *P. senegalus* does not produce young until it is practically full grown.

In June and July practically mature individuals were taken in numbers together and the sexes were noted. It appeared to be the case with both species that three-fifths of each lot taken together were females. This discrepancy can scarcely be accounted for by the greater activity of the males enabling them to elude capture, for they were caught in shallow pools connected with the river at high water. The water was baled from these pools when the tide was low.

When the fish are in the river they are always caught in the shade of the river-banks, up small creeks or under the bushes at the sides of the river. For this reason it is almost impossible to catch Polypterus in the trammel or seine-net, and with the cast-net it is very difficult, for the habit of the fish is to spend much time lying motionless on the bottom and it does not strike upwards into the cast-net thrown over it as do most fish, but slowly wriggles off, snakelike.

Polypterus was watched in the wild state and also in captivity. It lies for long periods in the mud at the river bottom with the fore part slightly raised and resting upon its pectoral fins like a seal upon its paddles. If the water is a little stale it may be seen to move slowly forwards by the action of its pectoral fins, which are worked very much as a lady uses her fan; the ventral fringing rays are deflected first, the more dorsally placed ones later, giving the action, in which the whole shaft of the fin is involved, a screw-like appearance.

As it nears the surface, however, the whole of the body and tail is brought into action, with a dart it strikes the surface, gulps in the air with its mouth, lets out the excess by opening its spiracles, and with lightning rapidity returns to the bottom.

The only time Polypterus was seen to feed in the wild state there was a small shoal of *P. senegalus* making their way slowly along the river-bank. One of their number seized a fresh-water crustacean, two others gave chase and, stirring up the mud, they all disappeared. When seizing young fry or tadpoles it proceeds stealthily after them, propelling itself by the flutters of its fan-like fins until within striking distance, and then with a sharp snap they are gulped down.

If the water is perfectly well aërated Polypterus may lie for a long time without breathing air. But a specimen which had been perfectly happy in tolerably fresh water for some days, when allowed to reach the surface, succumbed in a few hours when prevented from so doing. On the other hand one specimen lived for 24 hours in a landing net, with no more water than the moisture of the atmosphere, and finally had to be killed.

Polypterus may be watched for a considerable time, and give the impression that it is a sluggish and inactive fish. If, however, one is lucky enough to observe a male and female sporting together, it will be seen that they are capable of wonderful activity; executing the most lithe and supple movements, turning, twisting, darting and pausing in an extremely graceful manner, they thoroughly justify the native Mandingo name of Sayo or snake-fish.

The *ovary* of Polypterus is not hollow but originally a fold of the coelomic wall, it is early divided into compartments, both longitudinally and transversely. Each compartment contains from one to three ova, which are developed on the external wall of the ovary. Later, the external wall becomes much plicated, nevertheless at the internal and external summits of each fold the ova retain their original position, being attached to the wall of the ovary by the now pigmented pole. It thus comes about that, through the internal wall of the ovary, which is a smooth thin sheet, only the pigmentless poles of the ova are seen, while through the external much folded wall only the pigmented poles are seen.

The attachment of the ovary extends backwards obliquely across the short, thin walled and wide oviduct. The latter, about 3 inches in length in a 20 inch fish, opens to the exterior just behind the vent.

The spermatozoa in the male are extremely small; not more than $\frac{1}{6}$ the diameter of a red corpuscle in length.

It is seen then that both species of Polypterus found in the Gambia migrate to the flooded lands to spawn; that without doubt Polypterus uses its air bladder as an accessory organ of respiration and seldom as a hydrostatic organ, as a rule being unable to float, though it should be mentioned, that, preparatory to sporting near the surface, it was seen to take in several gulps of air in succession. The spiracle is used for the emission of air and not for the passage of water. The pectoral fins are important organs of propulsion and not mere balancers, as in almost all Teleostomes.

b. The dry season habits of *Protopterus* on the Gambia have long been known. Nothing however has been recorded hitherto of their habits in the rainy season.

Protopterus annectens emerges from its cocoon at the commencement of the rainy season in the early part of June, the eggs are usually ripe by the end of July.

There is no certain sexual difference, even at the height of the breeding season, except in old males, in which the head is stouter than in old females, and the pectoral limbs appear to be broader. During the breeding season, however, in both sexes the limbs and also the tip of the tail become much elongated and attenuated. In some cases the pectoral limbs extended beyond the vent and ended in extremely fine threads.

Though the nests of Protopterus were never actually seen in their wild state, eggs were obtained which proved to be fertilized.

The *ova* of *Protopterus annectens* measure 5 mm. in diameter, are enclosed in a thin horny capsule, and are light salmon colour above, and slightly tinted with greenish in the lower hemisphere.

Segmentation is complete, but very unequal. After the first three or four cleavages it is quite irregular, and results in an upper hemisphere of small cells, and a lower hemisphere of larger cells. The eggs were not reared beyond the beginning of gastrulation.

Larvae two inches in length were obtained from the swamps, presumably hatched in the preceding breeding season. These had three pairs of well developed external gills, arising however apparently at a common point of origin.

In one case these external gills were as long as the head, and provided with long vascular fringes. The *spermatozoa* of Protopterus are very large, twice the diameter of a red corpuscle in length.

As I was obliged to return to England early in the breeding season my observations on the life history of Polypterus and Protopterus are necessarily very incomplete. I hope however to return to their breeding grounds next season to complete the study.

IV. LIST OF THE FISHES COLLECTED BY MR J. S. BUDGETT IN THE RIVER GAMBIA[1]. By G. A. Boulenger, F.R.S., F.Z.S. With Notes by J. S. Budgett, F.Z.S.

The collection made by Mr Budgett fills a gap in our knowledge of the African river-fishes. Extraordinary as it may appear, the fishes of the Gambia have been little collected before, and it would have been impossible to draw up a list complete enough to compare with that of the Senegal. As might be expected, the fishes are very similar in the two rivers; I have nevertheless to describe as new two species, which are represented by several specimens in the collection. Mr Budgett intends to present examples of these to the British Museum.

CROSSOPTERYGII.

POLYPTERIDÆ.

1. POLYPTERUS LAPRADII Stdr.
2. POLYPTERUS SENEGALUS Cuv.

DIPNOI.

LEPIDOSIRENDÆ.

3. PROTOPTERUS ANNECTENS Ow.

[On this and the two preceding species, see Mr Budgett's notes in the preceding paper.]

TELEOSTEI.

ELOPIDÆ.

4. ELOPS LACERTA C. & V.

MORMYRIDÆ.

5. MORMYROPS DELICIOSUS Leach.—"Known as 'Suyi.' The natives are fond of all the Mormyridæ as food."

6. GNATHONEMUS SENEGALENSIS Stdr.—"Called by the natives 'Suyi-furu.' Brazen, blue and pink sheen."

7. MORMYRUS JUBELINI C. & V.—"Known as 'Suyi-nala.' All the Mormyridæ are easily caught with the trammel net, but with difficulty by other means."

8. HYPEROPISUS BEBE Lacép.

9. GYMNARCHUS NILOTICUS Cuv.—"Only caught in the Kunchow creek. Five specimens seen."

[1] From the *Proceedings of the Zoological Society of London*, May 8, 1900.

NOTOPTERIDÆ.

10. NOTOPTERUS AFER Gthr.—"This is not common. Its native name is 'Liffi lafo.' Two specimens only were taken near McCarthy Island."

OSTEOGLOSSIDÆ.

11. HETEROTIS NILOTICUS Cuv.—"Native name 'Fanntanng.' Breeds in the swamps; the fish occurs also in the river. I have seen no specimen over 20 inches."

CHARACINIDÆ.

12. SARCODACES ODOË Bl.—"Known by the natives as 'Saunko.' It was taken several times at Nianimaru, 130 miles from the sea."

13. HYDROCYON BREVIS Gthr.—"Native name 'Sokkoro.' Very common; a specimen was found with a large *Alestes* in its gullet. Ventral lobe of caudal fin brilliant red. Used by the natives for food."

14. ALESTES DENTEX Hasselq.

15. ALESTES SETHENTE C. & V.—"Known as 'Ballaunta.' Everywhere common, largely eaten and used as bait. The scales have a sky-blue tint, tail bright red."

16. ALESTES LEUCISCUS Gthr.

17. CITHARINUS GEOFFROYI Cuv.—"Known as 'Tara.' Very common, 12 inches in length. The ventral lobe of the caudal fin and the ventral fins bright red, the rest silver grey."

CYPRINIDÆ.

18. LABEO COUBIE Rüpp.—"Known as 'Kulinumma.' The whole fish has a fine rosy tint; it is esteemed as food, and fairly common; the largest seen being about 18 inches in length."

19. LABEO SELTI C. & V.—"Known as 'Jotto.' Fairly common at McCarthy Island. It is good eating. Silver white."

SILURIDÆ.

20. CLARIAS BUDGETTI, sp. n.

Vomerine teeth granular, forming a crescentic band which is as broad as or a little narrower than the praemaxillary band; the latter about 7 times as long as broad. Depth of body $6\frac{1}{2}$ or 7 times in total length, length of head (to extremity of occipital process) 3 or $3\frac{1}{4}$ times. Head $1\frac{1}{2}$ or $1\frac{1}{3}$ as long as broad, very feebly granulate; occipital process angular; frontal fontanelle 4 or 5 times as long as broad, its length about 4 times in length of head; occipital fontanelle small, in advance of occipital process; eye small, 3 or 4 times in length of snout, $5\frac{1}{2}$ to 7 times in interorbital width, which nearly equals width of mouth and is contained $2\frac{1}{3}$ or $2\frac{1}{2}$ in length of head; nasal barbel about $\frac{1}{2}$ length of head; maxillary barbel as long as or a little shorter than the head (a little longer in the young); outer mandibular barbel $1\frac{1}{4}$ or $1\frac{1}{3}$ as long as inner, which measures $\frac{1}{2}$ or $\frac{2}{3}$ length of head. Gill-rakers closely set, about 40 on first arch. Dorsal 68 to 73, its distance from the occipital process $\frac{1}{6}$ or $\frac{1}{7}$ length of head, its distance from the caudal fin greater than the diameter of the eye. Anal 46 to 50, narrowly separated from the caudal. Pectoral not quite $\frac{1}{2}$ length of head; the spine serrated on the outer border, about $\frac{2}{3}$ length of the fin. Ventrals midway between end of snout and caudal. Caudal $\frac{1}{2}$ length of head. Olive above, marbled with black, white beneath; anal with a light edge; a blackish streak from the angle of the mouth to the base of the pectoral.

Total length 330 millim.

Three specimens.

Very nearly allied to *C. senegalensis* C. & V., with which it may ultimately have to be united. The less rugose head and the narrower frontal fontanelle are the characters which induce me to regard it, provisionally, as distinct.

"Called 'Connoconno' by the natives. Often eaten by natives but not much esteemed. Lives chiefly in shallow swamps."

21. SCHILBE SENEGALENSIS C. & V.

22. ARIUS LATISCUTATUS Gthr.—"Known as 'Wollinyaba.' It is used as food. Seen two feet in length."

23. CHRYSICHTHYS CAMERONENSIS Gthr.

24. CHRYSICHTHYS NIGRODIGITATUS Lacép.

25. AUCHENOGLANIS BISCUTATUS Geoffr.

26. SYNODONTIS CLARIAS Hasselq.—"Not very common; taken several times at Nianimaru. Brilliant red tail and little red near the head. Native name 'Konnkrikonng.' Not used as food."

27. SYNODONTIS GAMBIENSIS Gthr.—"Called by the natives 'Kosso.' Was very abundant in the river. Frequently when taken from the water made a faint cry. It is seldom eaten by the natives. Dull grey."

28. SYNODONTIS OCELLIFER, sp. n.

Praemaxillary teeth in several irregular series, forming a broad band; mandibular teeth 23 to 30, hooked, simple, measuring barely one-sixth the diameter of the eye. Depth of body 3 times in total length, length of head $3\frac{1}{2}$ times. Head slightly longer than broad, convex on the occiput; snout obtusely conical, $\frac{1}{3}$ length of head; eye supero-lateral, its diameter 5 times in length of head, $1\frac{2}{3}$ to $1\frac{3}{4}$ in interorbital width; upper surface of head moderately granulate from between the eyes; frontal fontanelle narrow. Occipito-nuchal shield obtusely tectiform, longer than broad, granulate like the upper surface of the head, and terminating in two obtuse processes. Gill-cleft not extending below base of pectoral. Maxillary barbel without distinct fringe, longer than the head, reaching middle or posterior third of pectoral spine; mandibular barbels with long, slender, simple branches, inserted on a straight transverse line, outer nearly as long as the head, once and two-thirds as long as inner. Lips rather feebly developed. Humeral process granulate, acutely pointed, extending nearly as far as occipito-nuchal shield. Dorsal II. 7; spine strong, a little shorter than the head, curved, striated, with 12 or 13 feeble serrae behind in its upper half. Adipose fin 3 to $3\frac{1}{2}$ times as long as deep, a little longer than the head, 4 to 5 times as long as its distance from the dorsal. Anal IV. 7–8. Pectoral spine slightly longer than dorsal, striated, feebly serrated on the outer edge, with 18 to 21 strong antrorse serrae on the inner edge. Ventral not reaching anal. Caudal deeply forked, with pointed lobes, upper longest. Caudal peduncle as long as deep. Skin of body smooth. Grey-brown above and on the sides, white beneath; body and adipose fin with large black-and-white ocellar spots disposed with greater or less regularity at considerable intervals; caudal fin with numerous small round blackish spots.

Total length 490 millim.

Three specimens.

This species is most nearly allied to *S. nigrita* C. & V., from the Senegal, from which it differs, however, in the larger adipose fin and the longer and more slender branches of the mandibular barbels, as well as in the presence of the ocellar spots on the body.

"Never found in main river; chiefly taken in the Kunchow Creek. Eye-spots very bright black with white centre. Upper parts brown."

29. Malopterurus electricus Gm.—"Is known to the natives of Gambia by the name of 'Tingo.' It is common. The largest specimen seen was 14 inches in length. Two small specimens 3 inches in length were kept alive for several weeks in an aquarium; they appeared to browse on the algæ, and were capable when even so small of giving a very considerable shock."

Mugilidæ.

30. Mugil falcipinnis C. & V.—"Taken frequently in the Kunchow Creek. No fish in the river can compare with this as food, most of the river-fish being soft-fleshed and tasteless."

Polynemidæ.

31. Polynemus quadrifilis C. & V.

Sphyrænidæ.

32. Sphyræna guachancho C. & V.

Ophiocephalidæ.

33. Ophiocephalus obscurus Gthr.—"Called by the natives 'Pattukoma' or 'sleeping-fish,' from a curious habit it has when caught in shallow pools by draining the water: it does not attempt to get to the deeper parts of the pools but lies stranded as though dead. If placed in deep water it will suddenly dart away. In these pools it bears from above a striking resemblance to *Polypterus*, with which it is generally found."

GOBIIDÆ.

34. ELEOTRIS SENEGALENSIS Stdr.

CARANGIDÆ.

35. TRACHYNOTUS OVATUS L.

SCIÆNIDÆ.

36. CORVINA NIGRITA C. & V.

CICHLIDÆ.

37. HEMICHROMIS FASCIATUS Ptrs.
38. PELMATOCHROMIS JENTINKI Stdr.
39. TILAPIA GALILÆA Gm.
40. TILAPIA LATA Gthr.—"Known as 'Furu.' Much esteemed as food. Very common, the largest seen being 10 inches in length; great numbers taken with the seine-net. They are chiefly found in shallow water."

PLEURONECTIDÆ.

41. CYNOGLOSSUS SENEGALENSIS Kaup.—"Native name 'Juso' (heart). Taken from above McCarthy Island."

V. ON SOME POINTS IN THE ANATOMY OF POLYPTERUS[1].

With Plates V—VII. and text-figures 1—7.

CONTENTS.

I. INTRODUCTION.

Having obtained a large amount of material of both the species of *Polypterus* found in the Gambia, in the spring of 1899, I have thought it advisable to attempt to fill up some gaps in our knowledge of the anatomy of this most interesting of fishes.

A great deal of work has already been done in this direction by such eminent zoologists as Agassiz, Geoffroy St-Hilaire, Joh. Müller, Leydig, and Hyrtl, as well as by numerous other authors in recent years. The anatomy of the head of *Polypterus* has been described in detail by Pollard, the brain by Waldschmidt, the skull by Traquair and by Bridge, while Hyrtl has described the blood-supply of the external gill. The foundation of this work of recent years had been laid, however, by Joh. Müller in his *Bau und Grenzen der Ganoiden*.

In the present paper I have, I believe for the first time, described in detail the urinogenital system of the male and female *Polypterus*, together with the later stages in the development of these organs. I have also added some observations upon the vascular system, the external gill, the abdominal pores, the anal fin, and the skull.

[1] From *The Transactions of the Zoological Society of London*, Vol. XV. Part VII. April, 1901.

II. The Male Organs.

The following observations have been made on adult male specimens of *Polypterus senegalus* and *P. lapradei* taken at the commencement of the breeding-season. Series of the urinogenital organs of the young *P. senegalus*, 13 cm. and 9 cm. in length, have also been carefully studied.

In the adult male a ridge of testicular tubules extends the entire length of the body, lying parallel to and directly over the kidney on either side (Pl. V. figs. 2, 3, *t.r.*). Each testis-ridge is enlarged towards its anterior end, forming a conspicuous lobulated testis (fig. 3, *t.*): that on the left side is situated more anteriorly than that on the right.

The testis and testis-ridge are covered by peritoneum, the two folds of which are approximated at the base of the testis to form a mesorchium, in which spermatic veins pass to the cardinals (fig. 3, *sp.v.*).

The tubules of the testis and testis-ridge open by very numerous short ducts into a longitudinal canal extending the whole length of the gland lying in the mesorchium at the base of the testis, and further back between the testis-ridge and the kidney (Pl. V. fig. 3, *t.d.*; Pl. VI. figs. 10, 11, *t.d.*). Posteriorly this duct leaves the testis-ridge as the vas deferens (Pl. V. figs. 5, 6, *v.d.*), and passing backwards in the same sheath of connective-tissue as the ureter, opens upon a papilla into the narrow neck of the urinogenital sinus just before it opens to the exterior (Pl. V. fig. 5; Pl. VI. figs. 12, 14, *g.ap.*).

This duct is of even calibre throughout the greater part of its length, but is somewhat dilated in the region of the lobulated testis.

The tubules of the adult ripe testis are very numerous, and dilated with spermatozoa and what appear to be sperm mother-cells. The tubules are embedded in lymphoid tissue with deeply-staining nuclei (Pl. VI. fig. 10, *t.*). The spermatozoa are very small, about the diameter of a red corpuscle in length, thickened anteriorly and tapered posteriorly (Pl. VI. fig. 13).

The arrangement of the tubules of the testis-ridge is a simplification of that of the testis-tubules. Here three or four longitudinal tubules are connected by numerous transverse tubules with the testis-duct (Pl. VI. fig. 15, *t.r.tbs.*). The walls of these tubules are lined with columnar epithelium, and are not surrounded by lymphoid tissue as are the tubules of the testis, but by dense connective-tissue.

The tubules of this testis-ridge do not appear to be functional testis-tubules, although amongst them were found what appeared to be traces of spermatozoa. It seems possible that the tubules of this ridge assist in carrying away the sperm from the testis to the vas deferens, but do not themselves actually produce spermatozoa.

In the young Polypterus senegalus, 13 cm. in length, the tubules of the testis which are embedded in lymphoid tissue are lined by a single layer of large-celled columnar epithelium, while the lumina are small (Pl. VI. fig. 16, *t.tbs.*).   The tubules of the testis-ridge differ little from those of the adult male (Pl. VI. figs. 15, 17, *t.r.tbs.*); the tubules, however, are smaller as compared with the testis-duct in the young than in the adult.   It is noteworthy also that the vas deferens, which in the young is very thick-walled and has a larger lumen, is much larger as compared with the ureter in the young than in the mature individual (Pl. VI. figs. 14, 18, *v.d.*).

In the young *Polypterus* the duct of the testis runs forward a short distance anteriorly to the testis.   This portion of the duct could not be traced with certainty in the adult.   No opening in the young male was found, however, into the body-cavity.

In the very young male, 9 cm. in length, the tubules of the testis are fore-shadowed by the nuclei being arranged in double rows, but there are no lumina; the duct, however, has a wide lumen and is well formed (Pl. VII, fig. 19, *t.d.*). At this stage the tubules of the testis-ridge are not yet developed; the duct, however, is here well formed, as opposite the testis.   Posteriorly the vasa deferentia end blindly in the wall of the ureter (Pl. VII. fig. 20, *w.v.u.*).

III.   The Female Organs.

The following observations have been made on adult female specimens of *P. senegalus* and *P. lapradei* taken at all times of the year, and also upon specimens of the young female *P. senegalus* 9 cm. in length.

The funnel-like opening of the oviducts into the body-cavity (Pl. V. fig. 1, *p.ap.od.*) were mentioned and figured by Joh. Müller; the ducts were figured in more detail by Hyrtl, and the ovaries were described.   My own observations, however, do not in some respects agree with those of the latter author.   According to Hyrtl the two oviducts unite to form a urinogenital sinus, into which the two ureters open by a common mid-dorsal aperture.

I have carefully studied the adults of both species found in the Gambia, and I find that the ureters are dilated posteriorly, lying closely approximated to each other, but not communicating, except immediately before opening to the exterior (Pl. V. figs. 1, 4, *u.*; fig. 4, *s.u.g.s.*).   Shortly before the ureters open to the exterior the oviducts open into their lateral walls precisely as do the vasa deferentia in the male (Pl. V. fig. 4, *g.ap.*).   Further, in the young female 9 cm. in length the course of these ducts has exactly the same relation to the ureter as in the male, only that the oviducts are considerably more dilated; they lie immediately over the genital ridge, which anteriorly is developed into the ovary (Pl. VII. figs. 21, 22, 23, *mes.o.*).

The great difference between the sets of organs in the male and female is that in the male the genital gland discharges directly into the duct, whereas in the female the genital products are shed free into the body-cavity, and thence find their way to the mouth of the duct. Were the outer wall of the duct in the female carried forwards to enclose that side of the ovary from which the ova are shed, or were the testis-duct in the male open anteriorly to receive the products of the testis from the body-cavity, the arrangement would be precisely analogous in the male and female. The latter appears actually to be the case in *Polyodon folium*, where, according to Hyrtl, the duct of the testis, as well as the duct of the ovary, opens into the body-cavity by means of a peritoneal funnel.

At what stage the opening of the oviduct into the cœlom is acquired in *Polypterus* I cannot definitely say. It was, however, open in my youngest specimen 9 cm. in length (Pl. VII. fig. 22, *p.ap.od.*).

In young females 9 cm. in length the genital ducts have not a free opening into the ureter, but, as in the male, the ducts end blindly in the wall of the latter (Pl. VII. fig. 24, *w.o.u.*). The communication is complete, however, by the time the young *Polypterus* is 12 cm. in length.

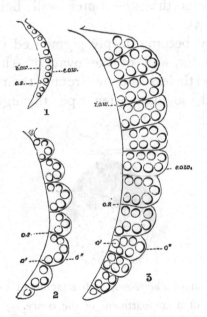

Figs. 1—3. Ovary of *Polypterus*.

Fig. 1. Diagrammatic representation of the developing ovary.
Fig. 2. The outer wall of the ovary between each septum beginning to bulge.
Fig. 3. The outer wall thrown into deep folds, as in the ripe ovary.

e.o.w., external wall of ovary; *i.o.w.*, internal median wall of ovary; *o'*, white side of ovum; *o''*, black side of ovum; *o.s.*, septum dividing ovary into compartments.

Though it would be unwise to form any definite opinion as to the significance of these ducts until their primary origin shall have been made clear, yet, seeing that very immature specimens have been examined, it seems worth while suggesting that the very high development of the genital ducts at an early stage in both sexes, and their similar arrangements, point to their being homologous with one another and also with the embryonic Müllerian duct. The discussion of the *à priori* objections and the consequences of this conclusion need not here be discussed, in view of the probability of the question being settled by a study of the early development of *Polypterus*.

The ovary in *Polypterus* develops as a genital ridge lying along the ventral surface of the kidney, and separated from it posteriorly by the genital duct (Pl. VII. figs. 21, 22, 23).

A large vein and artery traverse the median wall of the ovary (Pl. VII. fig. 21, *b.v.*). The ovary becomes divided into numerous compartments, on the outer wall of which the ova are developed (Pl. VII. fig. 21, *o.at.*, and text-fig. 1, p. 103).

With increased development of the ova in size and number, the outer wall becomes greatly enlarged in surface, while the median wall is not thus enlarged (text-fig. 2). It thus happens that the outer wall bulges out into great folds between the septa (text-fig. 3).

As the ova develop they become deeply pigmented upon the pole of the ovum which is attached to the outer wall of the ovary. Although the outer wall becomes so much folded, nevertheless the ova retain their original "orientation" at the extremities of each fold, so that, when ripe, through the median wall of the

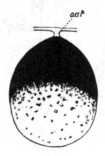

Fig. 4. The ovarian ovum of *Polypterus*, with attachment (*o.at.*) to the outer wall of a compartment of the ovary.

ovary, which is a thin smooth sheet, only the pigmentless poles are seen (text-fig. 3, *o'*), while through the external much-folded wall only the pigmented poles are seen (text-fig. 3, *o''*). It has been stated that the outer wall of the ovary of *Polypterus* is without a covering of peritoneum. It will be understood from the above-given description of the growth of the ovary, and from the examination of text-fig. 4, that this is not strictly correct.

IV. The Kidneys.

The kidneys are similar in male and female. They consist of two bands of excretory tubules, glomeruli, and embedding lymphoid tissue (Pl. VI. figs. 10, 11, 16) lying on either side of the vertebral column between the peritoneum and the muscles of the body-wall, and are constricted metamerically by the projecting myocommata. The kidneys of the two sides do not unite (Pl. V. figs. 2, 3, *k*.).

The minute structure of the kidney of the young *Calamoichthys* has been described by Lebedinsky. His description of the arrangement of the tubules in *Calamoichthys* will answer perfectly for that in *Polypterus*, except that in my youngest specimen 9 cm. in length I can find no trace of the nephrostomes opening into the cœlom described by him in "larvæ" 12 cm. in length. It seems, then, that *Calamoichthys*, which is so much smaller in the adult than *Polypterus*, is in a more larval condition at 12 cm. length than is the *Polypterus* at 9 cm.

It is possible that the nephrostomes never entirely close in *Calamoichthys*. However this may be, I have found in *Polypterus* no trace of nephrostomes opening into the cœlom.

In my youngest specimens the uriniferous tubules were arranged in distinct metameric masses, the metamerism disappearing in older specimens. The glomeruli did not appear to be thus arranged, there being a very indefinite number of these structures to each metamere (Pl. VI. figs. 10, 16, *m.cps.*). The openings of the tubules into the ureter are far more numerous and irregular in the adult than in the young.

V. The Ureters.

The ureters lie along the whole length of the kidney between the outer ventral edge of the latter and the body-wall (Pl. V. figs. 2, 3 ; Pl. VI. figs. 10, 11, *u.*) (text-figs. 5, 6). They receive the kidney-ductules. The ureter on passing ventralwards from the hind end of the kidney becomes dilated, and, in the male, joining its fellow of the opposite side, forms a large urinary sinus (Pl. V. figs. 2, 5, 6 ; Pl. VI. fig. 12, *u.s.*). The urinary sinus still passing ventralwards to a position just dorsal to the rectum becomes constricted to a narrow neck and, just before opening to the exterior, receives on either lateral wall the opening of the genital duct, then opens to the exterior in a depression just posterior to the anus (Pl. V. figs. 6, 7, 8, *u.g.ap.*).

In the female the ureters do not become confluent until immediately before they open by a slit-like aperture just posterior to the anus. Otherwise they resemble entirely those of the male (text-fig. 6, p. 106).

VI.   GENERAL CONSIDERATIONS ON THE STRUCTURE AND GROWTH OF THE
URINOGENITAL ORGANS.

It having been shown by Balfour and Parker and also by Semon that the
testis of *Lepidosteus*, and perhaps also of *Acipenser*, was connected with the kidney-
tubules, it was of extreme interest to see whether this was the case also in
Polypterus, in many respects the most archaic of recent Teleostomes.

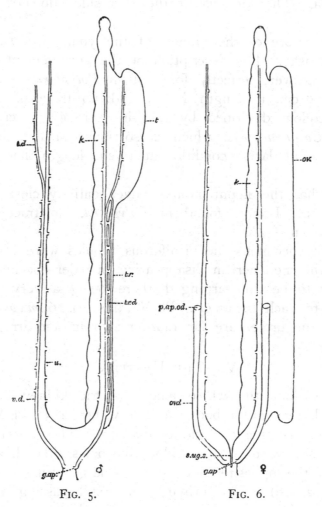

FIG. 5.   FIG. 6.

FIG. 5.   Diagram of the urinogenital organs in the male *Polypterus*.
FIG. 6.   Diagram of the urinogenital organs in the female *Polypterus*.

g.ap., genital aperture;   *k.*, kidney;   *ov.*, ovary;   *ovd.*, oviduct;   *p.ap.od.*, peritoneal aperture
of oviduct;   *s.ug.s.*, septum of urinogenital sinus;   *t.*, testis;   *t.d.*, testis-duct;   *t.r.*, testis-
ridge;   *t.r.d.*, duct of testis-ridge;   *u.*, ureter;   *v.d.*, vas deferens.

No such connections as a matter of fact exist. The products of the testis
pass out by a well-developed duct, which, running the same course as the ureter,
opens into the lateral wall of the latter close to its termination (text-fig. 5).

It is well known that the oviduct of the female *Polypterus* is short and is open anteriorly by a wide peritoneal funnel, the ova being shed through the external wall of the ovary into the body-cavity. This duct in the female runs a course similar to the genital duct in the male, opening into the ureter on its lateral wall shortly before its termination (text-fig. 6).

The resemblance of these ducts in the male and female is brought out still more strongly by a study of the quite young fish 9 cm. in length. Apart from the fact that the oviduct opens into the body-cavity, the arrangements in the two sexes are identical.

The question as to whether these ducts are both homologous with the Müllerian ducts of Elasmobranchs, Amphibia, and Amniota can of course only be settled by a knowledge of their actual origin.

The fact remains, however, that in this most primitive fish the arrangement of the genital ducts is nearly identical in the two sexes, and would seem to substantiate the view which was latterly held by Balfour that primitively the testis had a duct of its own, derived from a Müllerian duct like that of the female Elasmobranch, or a structure *sui generis*, as held by Howes and Max Weber, and that, on the other hand, the connection of the testis with the tubules of the kidney found in the male Elasmobranch, Amphibian, and Amniot is a secondary one.

There is, however, a difficulty with regard to the latter view in the fact that *Lepidosteus*, the Ganoid fish which, as regards its ovary and oviduct, most closely resembles the Teleostean arrangement, in having a closed ovary continuous with its duct, is the very one which shows this supposed secondary connection of testis with kidney.

It is possible, however, that this acquirement is confined to *Lepidosteus*, while other Ganoids retain the primitive condition; and it may be that it is a feature which has been frequently acquired independently. So that the Elasmobranch, Amphibian, and Amniot are not necessarily a separate evolutionary line from the Crossopterygian, Dipnoon, Ganoid, and Teleost, but the Amphibia may have acquired the Elasmobranch arrangement after they split off from the Dipnoi, which have not acquired it.

That the arrangement in *Lepidosteus* is not primitive seems probable from the fact that the testis-tubules open into a well-marked longitudinal collecting-duct, which lies along the ureter in the same position as the testis-duct in *Polypterus*, and it is from this longitudinal duct that the transverse tubules pass to the kidney.

That the arrangement in *Polypterus* is not secondary seems probable from the fact that not only is it the simpler method of conveying the testis-products outwards, but is, on the whole, closely similar to the arrangement in the female *Polypterus*, and we can hardly suppose that in the primitive vertebrate the ova and spermatozoa found exit by totally different means.

VII. Abdominal Pores.

In both male and female, abdominal pores are present in *Polypterus*. They have been correctly described in the adult as fine canals opening to the exterior on either side of the vent (Pl. V. figs. 7, 8). In the young female 9 cm. in length there is a very fine nucleated diaphragm cutting off the communication of the cœlom with the exterior. This is, however, extremely delicate (Pl. V. fig. 9, *ab.p.*).

In the young male 9 cm. in length the cœlom is completely shut off from the exterior, the abdominal pores not being yet formed.

VIII.   The Anal Fin.

Traquair has already noted in *Calamoichthys* that the males have an enlarged anal fin. In *Polypterus*, during the breeding-season at least, this difference is not merely one of size but also of shape and form, as shown in Pl. V. figs. 1 and 2.

The anal fin in the female is narrow and pointed, while in the male it is twice as deep as in the female, and its surface is thrown into deep folds between the successive fin-rays. The muscle of the anal fin is greatly enlarged in the male, protruding as a rounded mass into the cœlom. It is this mass which causes the ureters and genital ducts to turn so abruptly ventralwards in the male (Pl. V. figs. 2, 5, 6, *an.f.m.*).

I may mention that in an abnormal male specimen in which the anal fin was absent this muscle was completely absent, and the excretory and genital ducts ran backwards into an extension of the cœlom, then forwards ventrally to open in the normal position behind the vent.

This sexual character almost entirely disappears out of the breeding-season.

Leydig has suggested that there is internal fertilization in *Polypterus* from the fact that in the cœlom of a female *Polypterus* he found masses of filaments which he took for spermatozoa. That these filaments are not spermatozoa can at once be seen by comparing the figure he has given with my figure of spermatozoa from the ripe testis (Pl. VI. fig. 13).

When the arrangement of the oviducts is considered, it seems extremely improbable that the spermatozoa would find their way into a duct which opens into the urino-genital sinus upon a papilla. It remains to be seen to what use the male *Polypterus* puts this modified anal fin.

IX.   The Vascular System.

The blood-supply to the external gill has been worked out by Hyrtl, while the main roots of the arterial system were described by Joh. Müller. I would call attention, however, to a few additional details. Having injected a male specimen with salt-solution when killed, it was re-injected with a coloured gelatine in the laboratory. The specimen was adult and had no external gill. The details of the blood-supply to the external gill were made out on a young specimen in which it was possible to inject the hyoidean artery with a coloured fluid. The figure of the arterial system was made by a combination of these two dissections.

As Hyrtl has shown, the hyoidean artery arises at the anterior end of the ventral aorta immediately in front of the first afferent branchial artery, and passes to the base of the operculum, at the centre of which it meets the efferent hyoidean artery, to run with the latter to the posterior edge of the operculum and thence to the external gill.

The point I wish to call attention to is that the afferent and efferent arteries at the extremity of the gill are continuous one with another, forming a drawn-out loop. From the afferent limb branches run to the pinnæ, at the extremity of which they loop back to the main efferent limb; similar tertiary loops pass into the pinnules.

At the root of the external gill there is a dorsal and a ventral muscle; towards the extremity of the gill these break up into numerous isolated bundles (Pl. VII. figs. 25, 26, 27). The whole arrangement is quite similar to that of an Amphibian or Dipnoan external gill.

Hyrtl, from the arrangement of the main blood-supply to the external gill, argues that this must be homologous with the pseudobranch of *Acipenser*, which has the same structure as the succeeding gills.

The external gills of larval batrachians are borne upon the first two or three gill-arches, there being to each of these arches an external epidermal gill and an internal, probably endodermal, gill.

Kerr has shown that in larval Dipnoi which possess external gills there is to each arch an internal, probably endodermal, gill and an external epidermal gill, both being supplied by the same afferent artery.

It appears probable therefore that in the external gill of the hyoid arch in *Polypterus* we have not the homologue of the internal endodermal pseudobranch of *Acipenser*, but of the external epidermal gill of Dipnoi and Amphibia.

Moreover there is in *Polypterus* at the base of the operculum a stout branch from the efferent artery (Pl. VII. fig. 25, *eff.'*) which runs parallel with the afferent artery. The presence of this branch is suggestive, as indicating the position of the pseudobranch, corresponding to the pseudobranch of *Acipenser*, of which in *Polypterus* there is no further trace.

Pollard states that he could find no trace of a connection between the last efferent branchial artery and the dorsal aorta, the blood from the last hemibranch passing only to the air-bladder. Part of the blood from this gill does, however, pass to a branch joining the third efferent artery, which on the right side meets the cœliac artery (text-fig. 7, *br.IV.eff.*). The main part of the blood from the hemibranch of the IVth arch passes to the air-bladder on either side and is returned from them by veins passing to the hepatic veins, as shown by Joh. Müller. The vein on the right side is of great size, corresponding to the size of the right air-bladder, and posteriorly unites with the caudal vein.

It is difficult to see how these air-bladder veins in *Polypterus* could get converted into the pulmonary veins in Amphibia. It seems more probable that the great vein of the right air-bladder corresponds to the anterior abdominal vein of Amphibia, though it is notorious that veins frequently make secondary connections.

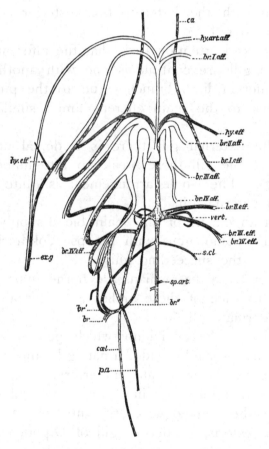

FIG. 7.   The arterial system of *Polypterus*.

br., brachial; *br.'*, branch of subclavian to shoulder-girdle; *br.''*, branch of subclavian which becomes the coronary artery; *br. I. II. III. IV. aff.*, branchial afferent I., II., III., IV.; *br. I. II. III. IV. eff.*, branchial efferent I., II., III., IV.; *ca.*, carotid; *cœl.*, cœliac; *ex.g.*, external gill; *hy. art. aff.*, hyoidean afferent; *hy. eff.*, hyoidean efferent; *hy. eff.'*, hyoidean efferent branch; *p.a.*, pulmonary; *s.cl.*, subclavian; *sp. art.*, spermatic.

The subclavian artery of *Polypterus* passes outwards dorsal to the vagus and the cœliac and pulmonary arteries, and curving ventralwards gives off, near the pericardium, a branch on either side (text-fig. 7, *br.''*). The posterior branch, the brachial, is a stout artery passing to the pectoral fin ; the anterior branch runs dorsally again parallel with the main subclavian, and divides to supply the muscles of the shoulder-girdle (text-fig. 7, *br.'*).   After giving off these two branches, the subclavian is continued as a small coronary artery along the sides of the pericardium, at the anterior end of which it passes to the walls of the conus and ventricle (text-fig. 7, *br.''*).

The blood from the conus and ventricle is returned to the ductus Cuvieri by a thick-walled, deeply-pigmented, coronary vein, which runs free in the pericardial chamber from the ventricle to the ductus Cuvieri.

From the dorsal side of the dorsal aorta just behind the junction with it of the second efferent branchial artery there passes outwards on either side a small vertebral artery, which curving dorsally round the spinal column appears to enter the spinal canal.

Pollard has pointed out the extremely primitive condition of the dorsal arterial system in *Polypterus*.   The single efferent vessel from each gill-arch uniting with a median dorsal aorta, resembles most the condition found in *Chlamydoselachus* and the embryos of other Selachians.

X.   THE CRANIUM.

In this paper I have no intention of dealing with the anatomy of the head and cranium of *Polypterus*, except that I wish to mention that the peculiar differences in the condition of the sphenoid bone which Bridge has described in two specimens which he examined, and called specimen A and specimen B, have been found to be specific in the two species from the Gambia.   The sphenoid bone of Bridge's specimen A, resembling that figured by Traquair, in which the lateral wings of the sphenoid did not curve inwards and fuse anteriorly in the frontal region, resembles exactly that of *Polypterus senegalus* Cuv.   The sphenoid (or "sphenethmoid") bone of Bridge's specimen B, which he suspected of being *Polypterus lapradei* Steind., resembles exactly that species[1].

XI.   CONCLUSION.

If in this paper I have occasionally reiterated facts which have long been known, I trust such has not been done without adding at least some details of interest.   In any case I hope I have always acknowledged the authority.

I feel that no apology is needed for having entered into minute details in some cases, for an accurate knowledge of the anatomy of a creature of such surpassing interest as the *Polypterus* seems to me in every way desirable.

In obtaining my material for this investigation I have been aided by a grant

---

[1] Since this paper was read, I have examined a young specimen of *P. lapradei*, 12 cm. in length, and find that at this age the sphenoid has not yet developed its specific character, but resembles that of the adult *P. senegalus*.

of £50 from the Balfour Fund of the University of Cambridge. The work has been done in Mr Adam Sedgwick's laboratory. To Mr Sedgwick, Mr Graham Kerr, and Prof. Howes my thanks are due for much help and advice in my work.

XII. Bibliography.

AYERS: "Morphology of the Carotids," in Bull. Mus. Comp. Zool. Harvard, vol. xvii., 1889.

BALFOUR: Comp. Embryology, vol. ii., p. 606.

BALFOUR AND PARKER: "The Structure and Development of *Lepidosteus*." Phil. Trans. Roy. Soc. Lond., part ii., 1882.

BRIDGE, T. W.: "Some points in the Cranial Anatomy of *Polypterus*," in Birm. Phil. Soc., vol. vi., part i., 1886.

GARMAN: "*Chlamydoselachus anguineus*," in Bull. Mus. Comp. Zool. Harvard, vol. xii., 1885.

HOWES: "On some Hermaphrodite Genitalia of the Codfish." Journ. Linn. Soc., Zool., vol. xxiii.

HUXLEY: "Classif. of Devonian Fishes," in Mem. Geol. Survey, Dec. 10, 1861.

    „   "On the Oviducts of *Osmerus*." Proc. Zool. Soc. Lond. 1883, p. 132.

HYRTL: "Ueber d. Pori abdominales, d. Kiemen-Arterien, u. d. Glandula thyroidea d. Ganoiden." SB. d. Wiener Akad., Bd. viii., 1852.

HYRTL: "Ueber den Zusammenhang der Geschlechts- und Harnwerkzeuge bei den Ganoiden." Denksch. d. k. Akad. d. Wiss., Bd. viii., Math.-naturw. Classe, Wien, 1854.

HYRTL: "Ueber die Blutgefässe der äusseren Kiemendeckel-Kieme von *Polypterus lapradei*, Steind." SB. d. k. Akad. d. Wiss. Wien, Math.-naturw. Classe, lx., Bd. i., Abth. 1869.

JUNGERSEN: "Beiträge zur Kenntniss der Geschlechts-Organe der Knochenfische." Arbeit. zool.-zoot. Inst. Würzburg, Bd. ix., 1890.

KERR, J. GRAHAM: "The External Features in the Development of *Lepidosiren paradoxa*, Fitz." Phil. Trans. Roy. Soc., vol. 192 B, 1900.

LEBEDINSKY: "Ueber die Embryonalniere von *Calamoichthys calabaricus*." Arch. mikr. Anat., Bd. xliv., p. 216.

LEYDIG: "Histologische Bemerkungen über den *Polypterus bichir*." Zeitschr. wiss. Zool., 1854.

MÜLLER, JOHANNES: Ueber den Bau und die Grenzen der Ganoiden. Berlin, 1846.

POLLARD, H. B.: "On the Anatomy and Phylogenetic Position of *Polypterus*." Zool. Jahrb. Anat. u. Ont., Bd. v., 1892.

SEMON, R.: "Zusammenhang der Harn- und Geschlechtsorgane bei den Ganoiden." Morph. Jahrb., Bd. xvii.

STEINDACHNER: "*Polypterus* aus dem Senegal." SB. d. k. Akad. d. Wiss. Wien, Math.-naturw. Classe, Bd. lx., 1869.

TRAQUAIR: "Cranial Osteology of *Polypterus*." Journ. Anat. and Phys., vol. v., 1871.

    „   "On the Anatomy of *Calamoichthys*." Proc. Roy. Soc. Edinburgh, vol. v., 1866.

WEBER, MAX: "Die Abdominalporen der Salmoniden, nebst Bemerkungen über die Geschlechts-organe der Fische." Morph. Jahrb., Bd. xii.

XIII. Explanation of the Lettering of the Figures.

ab.p.	Abdominal pore.		*ovd.ep.*	Epithelium of oviduct.
ab.p.c.	Abdominal pore-canal.		*p.ap.od.*	Peritoneal aperture of oviduct.
ab.v.	Abdominal vein.		*p.ep.*	Peritoneal epithelium.
aff.art.	Afferent artery.		*pl.*	Pinnule.
an.f.	Anal fin.		*pn.*	Pinna.
an.f.m.	Anal-fin muscle.		*r.*	Rectum.
ao.	Aorta.		*s.u.g.s.*	Septum of urinogenital sinus.
b.c.	Blood-corpuscle.		*sp.*	Spermatozoa.
b.v.	Blood-vessel.		*sp.art.*	Spermatic artery.
b.w.m.	Body-wall muscle.		*sp.ep.*	Spermatic epithelium.
c.v.	Cardinal vein.		*sp.v.*	Spermatic vein.
cd.v.	Caudal vein.		*sw.bl.*	Swim-bladder.
con.tis.	Connective-tissue.		*t.*	Testis.
d.m.	Dorsal muscle.		*t.d.*	Testis-duct.
e.ov.w.	External wall of ovary.		*t.d.ep.*	Epithelium of duct of testis.
eff.	Branch of hyoidean efferent artery.		*t.d.sh.*	Testis-duct sheath.
eff.art.	Efferent artery.		*t.r.*	Testis-ridge.
f.b.	Fat-body.		*t.r.tbs.*	Testis-ridge tubules.
g.ap.	Genital aperture.		*t.tbs.*	Testis-tubules.
g.p.	Genital papilla.		*u.*	Ureter.
k.	Kidney.		*u.ep.*	Epithelium of ureter.
k.v.	Kidney-vein.		*u.g.ap.*	Urinogenital aperture.
l.t.	Lymphoid tissue.		*u.g.s.*	Urinogenital sinus.
m.cps.	Malpighian corpuscles.		*u.s.*	Urinary sinus.
m.ov.w.	Median wall of ovary.		*u.sh.*	Sheath of ureter.
mes.o.	Mesoarium.		*u.tbs.*	Uriniferous tubules.
mes.t.	Mesorchium.		*v.d.*	Vas deferens.
n.o.	Nucleus of ovum.		*v.d.ep.*	Epithelium of vas deferens.
o.	Ovum.		*v.m.*	Ventral muscle.
o.at.	Attachment of ovum.		*w.o.u.*	Point where oviduct ends in wall of ureter.
op.m.	Opercular muscle.			
ov.	Ovary.		*w.v.u.*	Point where vas deferens ends in wall of ureter.
ov.d.	Oviduct.			

XIV. Explanation of the Plates.

EXPLANATION OF PLATE V.

Fig. 1. General side view of the posterior part of the viscera in a female *Polypterus lapradei* (p. 102), showing the course of the oviducts and the anal fin of a female in the breeding-season.

Fig. 2. General side view of the posterior part of the viscera in a male *Polypterus lapradei* (p. 101) in which the ureters have been injected blue, showing the testis-ridge and duct running along the ventral external wall of the ureter. The anal fin of a male in the breeding-season is also shown with the anal-fin muscle projecting into the posterior part of the cœlom.

Fig. 3. A median ventral view of the testis, the kidney, and their ducts of the same male specimen dissected out (p. 101). The ureter, as in fig. 2, is injected blue. Figs. 2 and 3 together give a complete view of the urinogenital organs in the male.

Fig. 4. Ventral view of the urinogenital sinus in the female (pp. 102, 108), showing the openings of the oviducts upon papillæ, the septum of the urinogenital sinus, the narrowness of the anal fin in the female, and the corresponding smallness in size of the anal-fin muscle. The rectum has been partially cut away and the ventral wall of the urinogenital sinus removed.

Fig. 5. A similar view of the urinogenital sinus in the male (pp. 101, 108), showing the genital aperture, the narrow neck of the urinogenital sinus, the great width of the base of the anal fin and the great development of its muscle.

Fig. 6. View from the right side of the urinogenital sinus in a large male (p. 101), the anal fin and its muscle, showing the course of the vas deferens; these organs, together with the rectum, having been dissected from the body.

Fig. 7. View from below of the anal region in a female (pp. 105, 108), showing the slit-like urinogenital aperture and the abdominal pores.

Fig. 8. A similar view in a male (pp. 105, 108), showing the wide urinogenital aperture.

Fig. 9. A vertical section of a very young female *Polypterus senegalus*, 9 cm. in length (p. 108), passing through the abdominal pores, showing a thin nucleated diaphragm closing the abdominal pores, at the same time showing an exudation of the cœlomic fluid. The caudal vein and abdominal vein are cut through just anterior to their junction.

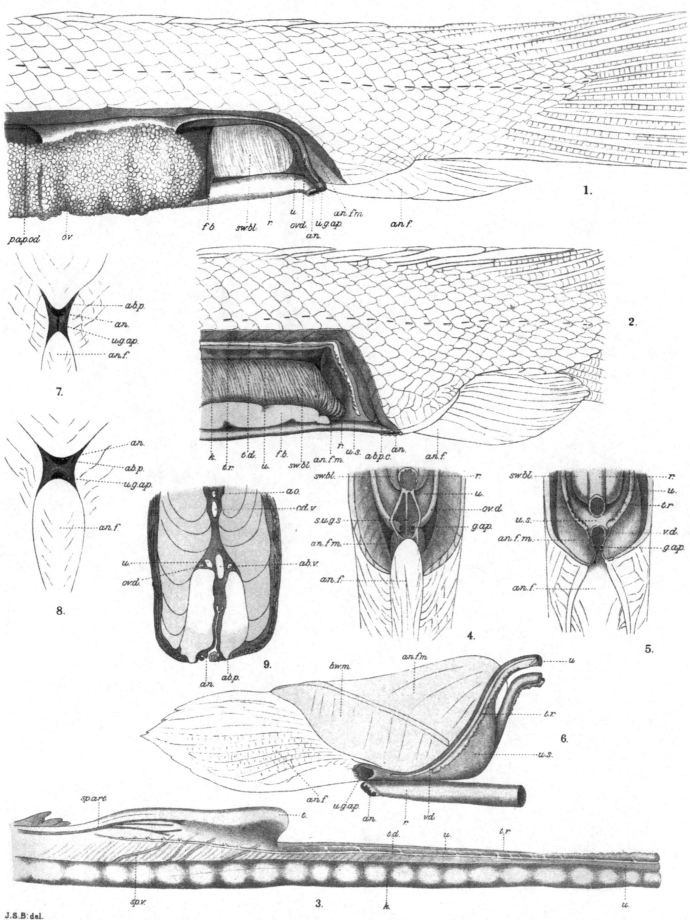

PLATE VI.

EXPLANATION OF PLATE VI.

FIG. 10. A vertical section through the testis and kidney of an adult male (pp. 101, 105). The ureter is injected blue. The testis-tubules are shown opening into the testis-duct, and a kidney-ductule opening into the ureter. The magnification is 6 diameters.

FIG. 11. A similar section behind the testis (pp. 101, 105), showing the relation of the testis-ridge and duct to the ureter in an adult male.

FIG. 12. A similar section in the region of the urinogenital sinus (pp. 101, 105), showing the duct of the testis opening upon a papilla into the urinogenital sinus.

FIG. 13. Spermatozoa compared with a blood-corpuscle (pp. 101, 108). Zeiss ocular 3, objective E.

FIG. 14. An enlarged drawing of the right half of fig. 12 (pp. 101, 102), showing the stout columnar epithelial lining of the vas deferens.

FIGS. 14 to 24 are all magnified about 70 diameters.

FIG. 15. An enlarged drawing of the testis ridge of fig. 11 (p. 101), showing the similar appearance of the ductules of the ridge and the main testis-duct. They are lined by a large-celled epithelium and embedded in connective-tissue with a few small blood-vessels.

FIG. 16. A vertical section through the kidney and testis of a young male *P. senegalus*, 13 cm. in length (pp. 102, 105), showing the minute structure of the testis. The tubules are lined by large glandular cells, and are embedded in lymphoid tissue similar in appearance to the lymphoid tissue of the kidney. The section shows also the similar appearance of the testis-duct and ureter. The uriniferous tubules of the kidney are embedded in a mass of lymphoid tissue.

FIG. 17. A similar section passing behind the testis (p. 102). When compared with fig. 15 the figure shows the relatively great development of the main duct of the testis-ridge, the small development of the ductules of the ridge in number and size, and the small relative size of the ureter.

FIG. 18. A similar section in the region of the urinogenital sinus (p. 102), showing the opening of the vas deferens into the latter and its lining of large columnar cells. The large relative size of the vas deferens is seen on comparing with fig. 14.

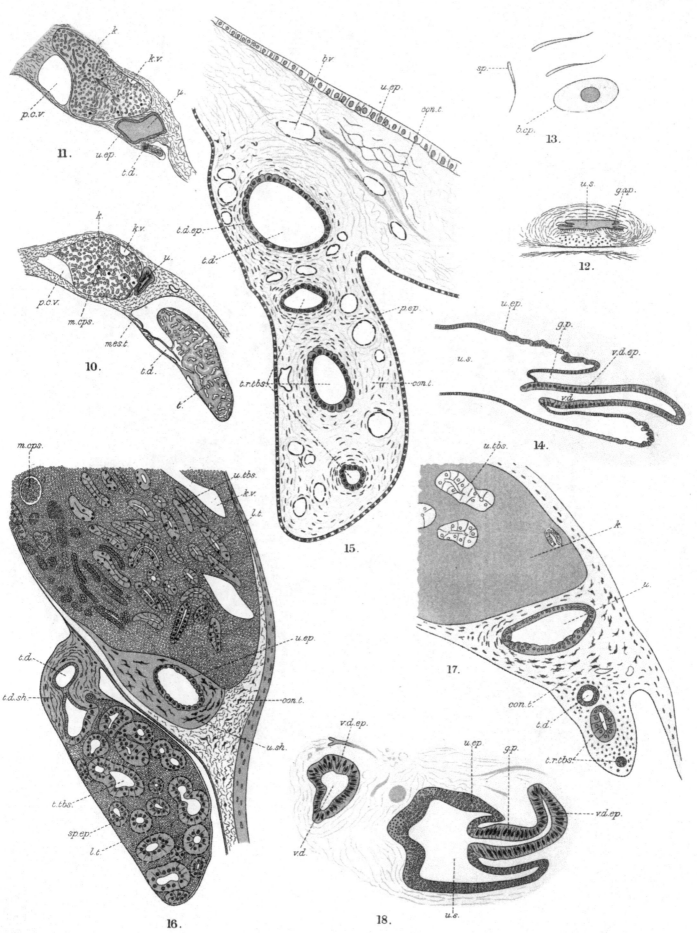

PLATE VII.

EXPLANATION OF PLATE VII.

FIG. 19. A vertical section through the testis and ureters of a very young male *P. senegalus*, 9 cm. in length (p. 102), showing the precocious development of the testis-duct, the tubules of the testis not being yet formed.

Fig. 20. A similar section through the posterior termination of the testis-ducts (p. 102), showing that they are only just acquiring their connection with the urinogenital sinus.

FIG. 21. A vertical section through the developing ovary and ureter of a very young female *P. senegalus*, 9 cm. in length (pp. 102, 104), showing the ova developing from the external wall of the ovary, the ovary divided into a number of loculi, and the artery and vein in the median wall.

FIG. 22. A similar section behind the ovary passing through the peritoneal opening of the oviduct (pp. 102, 103).

FIG. 23. A similar section behind the peritoneal opening of the oviduct (pp. 102, 104), showing the oviduct lying between the mesoarium or female genital ridge and the ureter. To be compared with fig. 17.

FIG. 24. A similar section through the posterior termination of the oviducts (p. 103), showing that they have not yet acquired their connection with the urinogenital sinus. To be compared with fig. 20.

FIG. 25. Left operculum bearing external gill of *Polypterus lapradei*, 30 cm. in length (p. 109), showing the blood-supply to the gill and the efferent artery giving off a branch which runs parallel with the afferent artery. The muscles of the operculum and the gill are also shown.

FIG. 26. An enlarged drawing of the tip of the external gill (p. 109), showing its minute structure and the continuation of the afferent into the efferent artery at the end of the gill.

FIG. 27. A pinnule of the external gill highly magnified (p. 109).

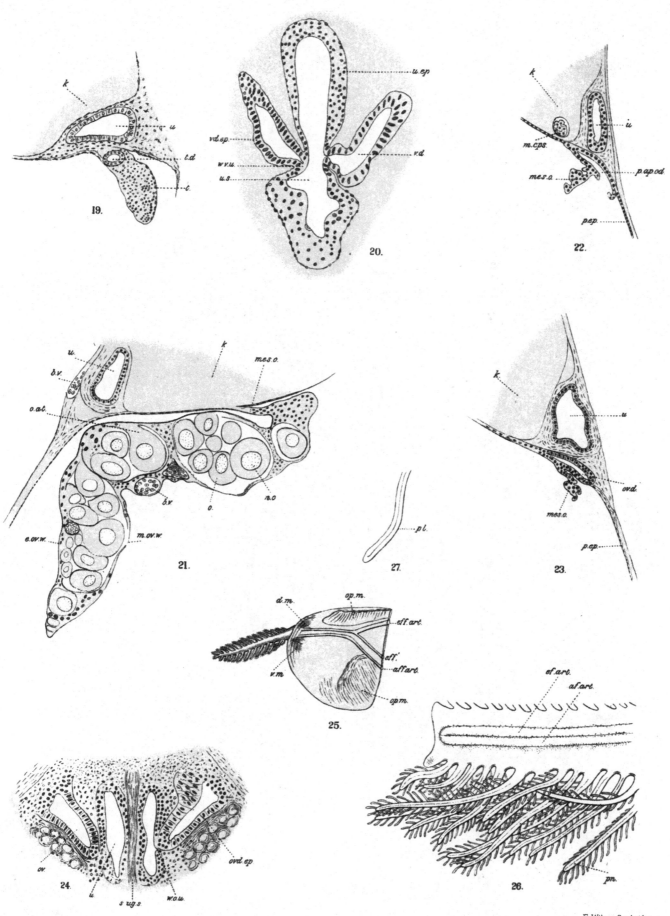

E. Wilson, Cambridge.

PLATE VI

VI. ON THE BREEDING-HABITS OF SOME WEST-AFRICAN FISHES, WITH AN ACCOUNT OF THE EXTERNAL FEATURES IN THE DEVELOPMENT OF PROTOPTERUS ANNECTENS, AND A DESCRIPTION OF THE LARVA OF POLYPTERUS LAPRADEI[1].

Received and read December 4, 1900.

With Plates VIII. and IX. and text-figures 8—12.

CONTENTS.

I. INTRODUCTION.

The months of June, July, and August of this year I spent on McCarthy Island, in the river Gambia, hunting for the eggs of *Polypterus*. But in this paper I have recorded not only the result of my work in this direction, but also my observations upon the breeding-habits of the African Lung-fish, *Protopterus annectens*, of the Teleosteans *Gymnarchus niloticus*, *Heterotis niloticus*, and *Sarcodaces odoë*, and also my observations upon the nests of another Teleostean, presumably *Hyperopisus bebe*. Towards the expenses of this expedition I received contributions from the Government Grant, from this Society, and from the Cambridge University Balfour Fund. I should here like to express my gratefulness for this assistance, which enabled me to undertake what turned out to be, so far as I was concerned, an intensely interesting collecting trip. When my results are completed, the expenditure will, I hope, be found to have been justified.

[1] From the *Transactions of the Zoological Society of London*, Vol. XVI. Part II. August, 1901.

The Island of McCarthy is six miles long and about one mile wide, and is situated 160 miles up the river Gambia. The whole of the island is low-lying, and the greater part, in the wet season, is completely under water. On the highest part of the island is the native town of Ginginberri, and the ruins of the old military settlement where I lived. The eastern half of the island is partly under culti-vation, and here the natives plant rice on the low land, and kuskus or millet on the high land. The western half of the island is little cultivated, and here was my hunting-ground.

The whole island is traversed by one main swamp, which has the appearance of having been at one time an old bed of the river, and which is seldom quite dry, even in the dry season. Parallel with this lie several shallower and more irregular swamps, all of which become perfectly dry in the dry season. These swamps are separated by belts of low forest, composed largely of leguminous trees, palms, and fig-trees. The swamps themselves are mostly choked with papyrus and other swamp-grasses ; while in the middle there is often a little open water covered with several kinds of lovely water-lilies.

The amount of open water depends largely on the rapidity of the oncoming of the rains. In the present season there was little open water, as the rains came on very slowly and gave the grass time to grow abundantly.

As the rainy season advances the swamps become filled with water, the river rises and soon becomes connected with the swamps by narrow creeks, up which great numbers of fishes pass to the flooded grasslands to spawn. Conspicuous amongst these are the two species of *Polypterus*—*P. lapradei* Steind. and *P. sene-galus* Cuv.

II.   Results of the Search for *Polypterus* Eggs.

From what I observed in my former expedition of its habits, I concluded that the eggs of *Polypterus* might be obtained by one of three methods :—

(1) Hunting the flooded grasslands for the eggs in a state of nature.

(2) Inducing *Polypterus* to breed in captivity.

(3) Artificially fertilizing the eggs of the female.

The first method was persistently tried without success, especially on the occasion of spawning females being brought in, when, taking a number of natives to the spot, we examined every foot of water within a hundred yards of the place where the female had been caught. Five spawning females were caught, but in no case did I find a single egg.

For the purpose of the second method I took out to the Gambia with me a large number of wire hurdles, with which I made four enclosures in the swamp. During the second half of June about fifty *Polypteri* were caught in pools by the river-side. As the rains had scarcely begun, and consequently there was no water as yet on the land where I had made my enclosures, I temporarily turned my fish into large floating cages in the river, and fed them regularly on minced meat. A few days after turning them in, I came one evening to feed them, and found that the two cages had been torn to pieces by crocodiles and the fishes had escaped. I then rigged up some temporary cages in pools by the river-side, into which I turned six pairs of *P. lapradei* and fifteen pairs of *P. senegalus*. By the 30th of July there was sufficient water in my swamp-enclosures to turn in the fishes, and there they seemed quite happy, as the grass was now two feet high and I fed them regularly every night. The enclosures were each eighteen feet long and six feet wide. As soon as I put the food into the water at one end of the enclosure the *P. senegali* came hurrying through the grass from all parts, and greedily devoured it without the least appearance of shyness. The *P. lapradei*, however, were considerably more shy, and continually damaged themselves against the wire netting. Thus the *Polypteri* remained until the 5th of September, and though I examined every inch of the enclosure every other day, I found no traces of eggs.

The third method was not tried so frequently as I could have wished, as I could not obtain anything like the number of *Polypteri* that I procured in the previous season. The natives at this time use a kind of basket for catching fishes, which is called the "wusungu." This they deftly drop over the fishes as they see them move in the grasses, and putting their hand into the basket from above, draw forth the captives. Whereas by this method I obtained an abundant supply of *Polypterus* last year at the price of sixpence apiece, this year I raised the price to two shillings apiece and was yet unable to get any large number of specimens.

The first spawning female was brought to me on the 3rd of July. It was a *Polypterus senegalus*, and, on holding it up by the head, it extruded successively twelve eggs. I had two males with which I tried to fertilize these eggs. One by one I tried them: first by placing them on the anal fin and on the vent of the male, then by mixing them with the fluid obtained from the seminal duct, and lastly by mixing them with the fluid obtained from the minced testis. These eggs were then transferred to muslin stretched on a frame and placed in a large quantity of river-water. They soon attached themselves to the muslin, but though I watched them until late into the night, no further change took place, and they one by one decomposed. On cutting open the female I found to my disappointment that there were no free eggs in the body-cavity, but that they were all attached to the ovary by their follicles.

The next spawning female was obtained on July 19th, when a fisherman brought me a female *Polypterus lapradei* which had two eggs in the oviduct and practically none in the ovary. There were no free eggs in the body-cavity.

On August the 9th a female *P. lapradei* was brought which had evidently spawned some time ago.

On August the 14th a female was brought which had nearly finished spawning. There were, however, no free eggs in the body-cavity. I tried to force out the eggs from the ovary, and in this manner several came away fairly easily, and I made every effort to fertilize them, but again without success.

On August the 15th another female *Polypterus* was brought, which had shed all its eggs.

In each case I took several natives to the spot where the female had been caught and made a very thorough search for the eggs in the neighbourhood, examining every blade of grass, but found not a single egg.

On August the 19th a small boy brought me a specimen of *Polypterus lapradei* only one inch and a quarter in length; it was a most beautiful object (Pl. IX. fig. 1). The upper surface is marked with black stripes on a golden ground, a conspicuous golden stripe runs on each side above the eye, across the spiracle, and along the dorsal surface of the external gill[1]. The external gills are at this stage of great size, reaching halfway to the tail, blood-red, and with a row of branches on either side. Each branch bears a row of pinnules on either side; the pinnules have the same structure as those of most Amphibian and Dipnoan external gill-filaments, being merely a long drawn-out blood-capillary loop. The afferent limb of the loop arises from the afferent artery of the gill-branch; the efferent limb of the loop joins the efferent artery of the gill-branch. Similarly the afferent artery of the gill-branch arises from the afferent artery of the external gill, while the efferent artery of the gill-branch joins the efferent artery of the external gill. Every alternate gill-branch is much smaller than the next gill-branch (Pl. IX. fig. 2). Each of these small gill-branches bends towards the surface of the body, while the large gill-branches extend parallel with the body. Thus space is economised, and the result is the same as four rows of branches on the external gill. Arising immediately behind the spiracle, the external gills may droop ventralwards posteriorly, and do not seem to be moved much by muscles, except just to straighten the shaft from the drooping position. The heart and blood-supply to the external gills can be seen with wonderful distinctness through the transparent ventral body-wall. The dorsal finlets are not differentiated from the tail, of which they seem to be only a forward prolongation. They are not distinct from one another, but form rather a continuous dorsal fin. The body is distinctly more truncate in the larva than in the adult, the head and tail-region being large. The eye is also very large in proportion.

The area of pigment ceases abruptly ventralwards in a line running from the tubular nasal opening under the eye dorsal to the shaft of the pectoral fin, thence to the base of the anal fin.

The larva was extraordinarily active, and, during the moments when it was at rest, supported the weight of its body on its pectoral fins, the blade of the fin being turned forwards and not backwards as is usually the case in the adult. The shape of the pectoral fin differs considerably from that of the adult. The ventral or postaxial border of the basal lobe is in this young larva much longer than the dorsal or preaxial border; while the fin-rays become successively longer in passing from the preaxial

---

[1] The young larva which I have described is about one-third of the length of any larval Crossopterygian which has, up to the present time, been obtained. The anatomy of this specimen I hope to describe in a future paper.

to the postaxial border. It follows that the shape of the fin is triangular, the apex being at the extremity of the postaxial border (Pl. IX. fig. 1).

Though the spot where this larva was caught was carefully searched, I did not succeed in capturing another.

Later on the same day, the 19th of August, I had another female *P. lapradei*, which must have finished laying its eggs some weeks before.

On the 21st of August, in my own fish-trap at the mouth of the small creek which led from the river to the swamp, I found a female *P. lapradei* which had finished laying its eggs, and it looked as though, in this case, it had spawned in the river or else at the mouth of the creek. I am inclined to believe, however, that it had temporarily returned to the river side of the trap after depositing its eggs in the swamp.

On this same day I had a *Polypterus senegalus*, still crammed with eggs, but not one free egg in the body-cavity.

During the last week in August and the first in September, I killed fifteen of my captive females; but in no case could I attempt artificial fertilization, as the ova would not come away from the ovaries, and in more than one case there were signs of degeneration.

On the 5th of September I left for England, leaving five pairs of *P. senegalus* in charge of a native, who was to preserve eggs for me if any should be laid.

Though I have little success in this direction to report, I have thought it well to put on record the difficulties which I encountered in the search for the eggs of *Polypterus* in order that any future investigator who may attempt to obtain developmental material of this fish may in being forewarned be also forearmed.

The main difficulties in obtaining the eggs seem to lie in the fact that *Polypterus* probably makes no nest, and certainly lays but few eggs at a time, these being scattered, probably broadcast, throughout the thick vegetation of the flooded grasslands. The eggs are minute, and therefore the chances of finding them in a state of nature are small in the extreme.

III. THE HABITS AND LIFE-HISTORY OF *PROTOPTERUS*.

a. *Nesting-habits.*

Although the development of *Polypterus* had been the chief aim and object of my second journey to the Gambia, I was also very anxious to obtain a series of the eggs and embryos of *Protopterus*. When I was on the Gambia the previous year, I had brought me a number of eggs of *Protopterus*, but I suspected that the way in which the native told me that they had been laid was quite abnormal or altogether untrue.

I had expected, in wading about the swamps, to come across deep holes in the ground similar to the nests of *Lepidosiren*, which I had become familiar with when in the Gran Chaco of Paraguay some three years ago with Mr Graham Kerr. However, I never found such holes, and was completely at a loss to know where to look for the nests of *Protopterus*, the natives being entirely ignorant of any but the most obvious facts of natural history, and having declared to me that the "Cambona," as they called *Protopterus*, was viviparous.

One day my head fisherman, Sory, came to me in a great state of excitement to say that he had found the children of the Cambona.   It was scorching mid-day in the height of the rainy season, the temperature 99° in the shade.   After crossing one deep swamp we came to the edge of another swamp, and there, about ten yards from the water's edge, on dry ground, was an oval-shaped hole filled with water, and in the water was a great commotion (text-fig. 8); the surface of the water was being continually lashed from side to side by the tail of a Cambona, the head of which was away down under the ground.   On being startled, the Cambona disappeared downwards, and the fisherman, putting his hand into the hole, drew forth a handful of larval *Protopteri*.

Fig. 8.   Nest of *Protopterus*.

Having now learned where to look for the nests of *Protopterus*, in a few days I found a number of similar nests, but never so far away from the water as the first one, which was found at the end of a period of drought, very unusual at this time of year.   I soon found a nest full of newly-laid eggs which must have numbered several thousands, for from the first day to the day the larvæ left the nest, twenty days later, I took fifty per day for preservation without perceptibly diminishing the numbers in the nest.

Throughout the period of the larvæ being in the nest, the male *Protopterus* stays with them and guards them jealously, severely biting the incautious intruder.

On one occasion the male was observed to leave the nest and to come out by a small opening which had hitherto been unnoticed, and wriggle off down to the water. This exit was always found about two feet from the main opening. Frequently there was a kind of pathway up to the entrance, where the grasses were bent aside. The main opening measured four to ten inches in diameter, while the exit rarely measured more than three inches. The depth of the nest was usually about a foot, and the shape of the nest was quite irregular. There was never any lining, and the eggs were laid on bare mud. All the males found in the nests measured about eighteen inches in length.

b. *Development of the Embryo.*

The eggs, which measure 3·5—4 mm. in diameter, begin to hatch about the eighth day, and by the tenth day the larvæ are all attached by their suckers to the side of the nest. The main features in development are remarkably like those of *Lepidosiren* lately described by Kerr, the larvæ being provided with a ventral sucker and four pairs of plumose external gills, one to each branchial arch. I have figured a few stages of the external features in development, most of which were drawn on the spot from life, in order that a comparison may be made with Kerr's excellent illustrations in the *Phil. Trans.* vol. 192, plates 8—12.

As all my specimens were procured from the same nest at twenty-four hours' interval, I am able to show the advancement made daily. As Kerr's material was obtained from a large number of nests, he was unable to say what was the age of each successive stage figured. Though in one nest were found a few specimens at least half a day in advancement of the rest, and a few also at least half a day behind the rest, yet the majority appeared to be at a uniform stage of development. When kept in shallow dishes I found that the development was much retarded.

Comparing Pl. VIII. fig. 1 with the corresponding stage in *Lepidosiren* (*op. cit.* plate 8, fig. *7 a*, *7 s*, & *7 b*), it is noteworthy that the egg is here divided into segments, which are more distinct from one another, the outer surfaces being rounder and not assuming the same curvature as the egg-capsule. In this the egg of *Protopterus* approaches the conditions of *Ceratodus*. This is the first stage of my series, so that I am not able to speak with regard to the appearance of the egg in the earliest stage of segmentation.

The subsequent down-growth of the epiblast over the invaginating yolk (as shown in Pl. VIII. figs. 2, 3, 4, & 5) is remarkably similar to the same process in *Lepidosiren* (*op. cit.* plate 8, figs. 10—14), the invaginating rim remaining a nearly straight line. This appears to me to be the more frequent method of invagination. Mr Kerr has himself, however, pointed out to me that the variations which frequently occur in *Protopterus* are very interesting. The invaginating rim is often curved, as in fig. *3 a*, rather than straight, as in fig. 3, while later the invaginating rim may become somewhat V-shaped, recalling a similar appearance in certain Amphibia. The invagination culminates in a crescentic blastopore. The yolk from the earliest stage onwards in *Protopterus* is light green in colour. During segmentation the epiblastic pole of the egg is pink, and this colour gradually replaces the green colour of the yolk, becoming, however, paler as invagination proceeds. In the later stages, where the tissues are becoming more transparent, the green-coloured yolk is again seen.

As in *Lepidosiren*, the medullary groove arises far forwards and grows back to the blastopore. In *Protopterus* (Pl. VIII. fig. 6) the medullary folds, though wider in proportion to the surface of the egg than in *Lepidosiren* (*op. cit.* figs. 17 *h* & 18 *h*), are not quite so definite, but undoubtedly do encircle the blastopore in the same way just before they close. From an external examination the blastopore seems to remain more widely open after closure of the medullary folds than in *Lepidosiren*.

Pl. VIII. figs 7 & 8, corresponding with *Lepidosiren* (*op. cit.* figs. 21, 22, & 23), show a very similar origin of the brain, optic outgrowths, branchial and pronephric eminences, but the pair of folds which will subsequently give rise to the mandibular and hyoidean visceral arches is much more marked in *Protopterus* (Pl. VIII. fig. 7, *M., H.*). The pronephric ducts have also an origin identical with *Lepidosiren* (*op. cit.* figs. 21 *m*, 22 *m*, 23 *m*). In Pl. VIII. fig. 8, which corresponds very nearly otherwise in development with *Lepidosiren* (*op. cit.* fig. 23 *l*), the whole embryo is not so flattened on the yolk, the head and tail-fold being much more conspicuous. At this stage is seen the first appearance of the crescent-shaped sucker (Pl. VIII. fig. 8, *c.o.*) first shown in *Lepidosiren* (*op. cit.* fig. 24).

In *Lepidosiren* the branchial arches arise on either side, first as one eminence (*op. cit.* fig. 22), later three eminences (fig. 23); the last of these then splits into two, and thus the four arches are formed. In *Protopterus* they arise first as one eminence (Pl. VIII. fig. 7, *br.*); later two eminences (Pl. VIII. fig. 8, *br.* I. & II., *br.* III. & IV.), these then each split into two[1] (Pl. VIII. fig. 9, *br.* I. II. III. IV.), thus giving rise to the four branchial arches.

In *Protopterus* (Pl. VIII. fig. 9, *M., H.*), anterior to the four branchial arches, there may be seen an indication of the mandibular and hyoidean arches, which in *Lepidosiren* (*op. cit.* fig. 24) are represented by a single eminence.

Protopterus hatches about the stage of Pl. VIII. fig. 10, often a little later, in some cases as late as Pl. VIII. fig. 11. Before hatching there appears to be a covering of cilia, for particles in the fluid within the egg-capsule stream down the sides of the embryo towards the tail end. At hatching the four pairs of external gills are a good deal in advance of the gills of *Lepidosiren*, the developing pinnæ being clearly seen. The rate and direction of growth of the first pair of external gills is very different to that of the succeeding pairs as

---

[1] The cleavage of the hindermost eminence to form the 3rd and 4th branchial arches occurs somewhat later than that of the foremost eminence.

shown in Pl. VIII. figs. 10 & 11, *Eg*. 1.   Here also may be seen, through the dorsal wall, the auditory cavities and the now large fourth ventricle.

Just before the stage of Pl. VIII. fig. 12 is reached, pigment begins to appear first in the retina, then on the surface of the head.   The fin-folds of the tail now begin to grow rapidly, and attain a much greater size than in *Lepidosiren* (*op. cit*. figs. 31, 32, & 33).   A copious network of blood-vessels spreads over the yolk.   The sucker is fully functional, and the larvæ hang vertically from the sides of the nest or vessel in which they may be confined. Although in *Lepidosiren* this organ is more conspicuous, yet the larvæ appear only to use it for clinging to the uppermost layer of débris in the nest, and so prevent their falling downwards and getting smothered[1].

A striking feature of *Protopterus* at this stage, compared with *Lepidosiren* (*op. cit*. figs. 31, 32), is the serial arrangement of the external gills, their roots being distinct from one another, and placed in a line along the dorsal surface of the deepest part of the yolk.   Anterior to the first branchial cleft there is a faint indication of a spiracular cleft between the mandibular and the hyoidean arches (Pl. VIII. fig. 12, *sp*.) of which in *Lepidosiren*, externally at least, there is no trace.

The roots of the external gills in *Protopterus* (Pl. VIII. fig. 13) remain longer separated from one another than in the *Lepidosiren* (*op. cit*. fig. 33).   The three posterior pairs also attain a greater proportionate size.   At this stage the tail and dorsal fin-fold are considerably more developed than in the corresponding stage of *Lepidosiren*.

As the external gills are reaching their maximum development, the origin of the gills becomes somewhat concentrated and rotates forwards, the hindermost gill becoming dorsal, the anterior becoming ventral.

For some days before leaving the nest, when the young larvæ are hanging suspended vertically from its walls by their suckers, the external gills are held stiffly out at right angles to the axis of the body, forming a radiating frill around the base of the head.   When the larva is lying in a small trough of water, the gills are not thus erected, and as the drawing (Pl. VIII. fig. 13) was taken from a living specimen, the gills are shown lying back along the sides of the body.

The pectoral and pelvic limbs develop synchronously as in *Lepidosiren*, and are just beginning to bud in Pl. VIII. fig. 12, *h.l*.   Correlated with the extension backwards of the roots of the external gills, the position of the bud of the pectoral fin is also far back, and lying immediately below the last external gill, is hidden by them.   In Pl. VIII. fig. 13 the pectoral limbs are of about the same size as the shafts of the external gills.   In one case a specimen had not developed the pinnæ of one external gill.   This bare shaft so much resembled the pectoral limb, that the larva appeared to have two pectoral limbs on one side.

[1] Kerr, *loc. cit*. p. 316.

In Pl. VIII. fig. 13 the operculum is growing back, the mouth is open, and the internal gills functional. The larvæ do not breathe air before leaving the nest. There is now a considerable development of pigment, especially in the anterior dorsal part of the body. The fin-rays are just making their appearance in the fin-folds of the tail. The sucker or cement-organ is at its maximum development. The tail is absolutely diphycercal from the first. Blood-vessels running in the track of the spiral valve shine through the body-wall (Pl. VIII. fig. 12, *s.v.g.*). The spiral valve is first indicated in fig. 12. The yolk remains chiefly massed in the original position close behind the sucker, and is not distributed along the gut to the same extent as in *Lepidosiren* (*op. cit.* figs. 33, 34, & 35). Wherever yolk is seen, it is of the original greenish colour.

The young *Protopterus* leaves the nest with practically the form of the adult (Pl. IX. fig. 3). The mass of food-yolk is not entirely absorbed as yet. The first pair of external gills has been lost, and the succeeding pairs have been much reduced in size. The tail ends in a very fine filament. The markings of the young *Protopterus* at this time are somewhat different from the adult. The general colour is dark brown, a conspicuous broad yellow band passing between the eyes. As with *Lepidosiren* so with *Protopterus*, the larvæ at this stage contract their black chromatophores at night and become blood-red, the eye shining out deep black in contrast.

It is here interesting to notice that the larval *Protopteri*, after leaving the nest, when kept in an aquarium the bottom of which was covered with seedling water-lilies, chara, &c., never show themselves by day, and if disturbed from their seclusion, hastily make their way back to their hiding-place. After dark, however, by the aid of a lantern, the larvæ may be seen swimming around in the most lively manner, but they do not come to the surface for air.

It seems, then, that the habit of expanding the chromatophores by day is of advantage to the larval *Protopterus*, making it almost invisible while lying passively on the dark soil. The chromatophores become contracted by night, not by reason of the darkness, but because this is the period of activity with the larvæ, and when swimming about they are certainly less conspicuous when transparent than when opaque, even at night.

Were it customary for the larval *Protopterus* to swim about in the daytime, they would probably then contract their chromatophores, becoming less visible with increased transparency. As a matter of fact, when in the daylight the larvæ were placed in a white porcelain dish, in a large number of cases they did contract their chromatophores. That this contraction on a light background did not always take place may possibly be accounted for by supposing that continued habit has produced a certain periodicity in the contraction and expansion of the chromatophores.

While on the Gambia, I kept a large number of young fry of about fifteen species of fishes, and I noticed that the nocturnal forms did become more transparent at night. The converse was naturally not noticed, since I know of no fishes which are only active in the daytime. With frogs, the case is quite different, for they are not aquatic, and would not therefore be made less conspicuous by being transparent. The chromatophores are often contracted by them in the daytime when exposed to strong sunlight, for the objects around them then become of brighter and lighter colour[1].

Soon after leaving the nest, the larvæ begin to feed on almost any animal matter they can get. For this reason, though I started homewards with a number of larvæ taken from the nest, only one reached England alive, having eaten all the others. On the voyage home, the young *Protopterus* began to move about in the daytime, ceased becoming transparent at night, lost the external gills, all but small vestiges, and began to come to the surface for air. This was about one month after leaving the nest, or about seven weeks after being laid. On reaching England it had quite the form of the adult.

In comparing the development of *Protopterus* with that of *Lepidosiren*, a very noticeable circumstance is the impossibility of comparing together a larva of each form as being exactly at the same stage of development. The various organs and features do not make their appearance in quite the same proportionate periods of time in the two forms; so that at any one stage, some set of organs in the one will not correspond in its state of development with the same set in the other.

Many of the differences noted in the external development of the two forms may, I think, be correlated with the presence in *Lepidosiren* of rather more food-yolk. The main differences are :—

1. A more complete separation in *Protopterus* of the cleavage-products.
2. The greater size of the medullary folds.
3. A more distinct remnant of the blastopore.
4. The earlier appearance of the cement-organ.
5. The earlier rising-up of the embryo off the yolk.
6. The appearance of two visceral folds in front of the four branchial folds.
7. A rudiment of a cleft between them.
8. The greater size of the gills at hatching.
9. The more complete separation of the external gills.
10. The rotation forwards of the external gills.
11. The concentration of the yolk forwards.

I have not thought it well to make any observations upon the bearings of the facts here described, since it is first necessary to know more of the development than can be learnt from a superficial examination. Mr Graham Kerr has undertaken to further study the development of *Protopterus* and to in-

[1] "Notes on the Batrachians of the Paraguayan Chaco," *Q.J.M.S.*, 1899, pp. 314, 327, 328.

corporate the results in his work on the development of *Lepidosiren*. I have here described the external features in development together with what I observed of the nesting-habits of *Protopterus*, as it would be difficult to treat either separately.

Comparing the nesting-habits of *Protopterus* with those of *Lepidosiren*, perhaps the most striking difference is the development by the male *Lepidosiren* in the breeding-season of the extraordinary vascular fringes of the pelvic fins, recently described by Kerr. Nothing of the kind is developed by *Protopterus*. Now, looking to the solution of the problem as to what is the function of these fringes in *Lepidosiren*, it is natural to look to see in what the habits of the latter differ from those of *Protopterus*.

The most striking difference is surely that, whereas *Lepidosiren* makes its nest several feet below the surface of the water, *Protopterus* makes its nest practically out of the water. I regard the habit of *Protopterus* of lashing the surface of the water at the entrance to its nest as a means of aerating the eggs in the nest. Now, it is tempting to regard the vascular fringes on the pelvic limbs of *Lepidosiren* as in some way connected with the aeration of the eggs, for it is obviously unable to make use of this method of aeration adopted by *Protopterus*. But the conditions under which this habit was observed in *Protopterus* were, as I have said, somewhat unusual, in that, owing to prolonged drought, the water in the nest was unconnected with the surrounding water. When this was not the case, the lashing of the tail on the surface of the water was not observed. Therefore I do not think this habit can be said to be quite characteristic of *Protopterus*. The entrances to the nests, however, were always only a few inches at most below the surface of the water, while with *Lepidosiren* the nests are made in deep water, and it seems more probable that the fringes on the pelvic limbs of *Lepidosiren* are, as Kerr holds, accessory organs of respiration, thus avoiding the necessity of frequent absence from the nest in order to visit the surface for air, and so perhaps risking loss of the entrance to the nest or the attacks of enemies. *Protopterus*, by reason of the shallowness of the water about the entrance to the nest, would not run these risks in seeking air, and therefore has no need of the accessory breathing-apparatus.

IV. The Nesting-habits of *Gymnarchus niloticus*.

While hunting for *Polypterus* eggs, I met with several large floating nests measuring in all two feet in length and one in breadth. The nests were made in the dense grasses of the swamp in three to four feet of water (text-fig. 9). The inside measurement was about a foot by six inches. Three sides of the nest projected from the water; the fourth side was several inches lower, being about two inches below the surface. The deepest part of the nest was opposite to that side where the wall was low, the bottom being about six inches below the surface of the water.

In this nest were deposited about a thousand large spherical amber-like eggs 10 mm. in diameter. The eggs hatched five days after being laid, and in eighteen days a thousand young fry of *Gymnarchus niloticus* left the nest when three inches long. This fish is called by the natives the "Suyo."

Though there are many interesting features in the development of these eggs, I do not intend to deal with them in detail here, but merely to mention that the development is exceedingly shark-like. The larvæ soon after hatching develop extremely long gill-filaments, which hang down in two blood-red branches from the gill-arches, of which there are four. The yolk-sac, at first spherical, later becomes drawn out into a long cylindrical bag, attached somewhat far behind for a Teleostean, and covered with a vascular network (Pl. IX. figs. 4 & 5).

FIG. 9. Floating nest of *Gymnarchus*.

The tail is from first to last perfectly diphycercal, and is at first provided with a dorsal and a ventral fin-fold reaching right to the tip of the tail.

Before leaving the nest, both outer gill-filaments and yolk-sac are absorbed and the mature form is reached.

Immediately after hatching, the larvæ commence their characteristic movements, throwing the head and fore part of the body from side to side incessantly. The larvæ are at first so small in proportion to the size of the yolk-sac, that they are quite unable to move it. By this constant movement the larvæ tend towards the surface, and the weight of the yolk tending downwards, the yolk-sac becomes gradually drawn out into the long appendage already mentioned. About three days after hatching, the larvæ are strong enough by their movements to raise the yolk-sac off the bottom of the nest for a moment, but it is quickly drawn back by its weight.

By the tenth day after hatching, the larvæ are able to drag their yolk-sac to the surface of the water, when they take a gulp of air into their lung-like swim-bladder and fall again to the bottom, on reaching which they again start for the surface with unceasing regularity, so that when looked at from above the nest of *Gymnarchus*, with its swarm of scarlet-bearded, yolk-hampered larvæ, presents a most amazing spectacle.

By the time the huge yolk-sac has been completely absorbed, the young larvæ are ready to leave the nest. They still, however, continue their ceaseless journeyings to the surface for air. It may now be noticed, however, that the passage back to the bottom of the nest is not merely a passive falling, but that the young larvæ actually dart backwards from the surface. When the young *Gymnarchus* leaves the nest it has fully developed the characteristic cylindrical tail of the adult, and in this connection its habits are very interesting.

The *Gymnarchus* propels itself through the water, not by the action of its paired fins, not by the motion of its tail, or the undulatory motion of the axis of its body, but entirely by the action of its dorsal fin. This fin extends nearly the whole length of the dorsal surface, ceasing abruptly at the commencement of the cylindrical tail. When *Gymnarchus* starts forwards, the motion is the result of a series of waves passing backwards along the dorsal fin. About five such waves are passing at a time. Suddenly the fish will proceed at the same rate in the opposite direction, and now the motion is the result of a series of waves passing forwards along the dorsal fin.

As the *Gymnarchus* swims rapidly backwards in this way, it may be seen to guide itself through the grasses by using this peculiar tail which it possesses as a feeler. Thus it appears to be quite immaterial to the fish which way it progresses, and it always appears to swim in comparatively straight lines.

How *Gymnarchus* constructs the wonderful floating nest in which it lays its eggs I have been unable to observe. The natives approach these nests with great caution, stating that the parent is at this time extremely fierce and has a very formidable bite. Both the adult fish and its eggs are greatly sought after as food.

A large number of the young fry of *Gymnarchus*, which I had caught immediately they left the nest, lived well on chopped-up worms. I tried to bring some of them to England alive, but every one died as we got into colder climes.

V. The Nesting-habits of *Heterotis niloticus* Cuv.

In the same swamps, during the month of July, a most striking feature is the presence of numbers of enormous nests, which proved to be those of *Heterotis niloticus* (text-fig. 10). These nests measured four feet in diameter, and were made in about two feet of water. In wading through the reed-choked swamp, when one came across one of these structures they appeared like miniature lagoons. The walls of the nest were about eight inches thick at the top and compact, being made of the stems of the grasses removed by the fish from the centre of the nest. The floor of the nest was the swamp-bottom, and was made perfectly smooth and bare.

Once I watched a "Fantang," as the natives call this fish, making its nest. It was circling round and round the wall of its nest, every now and then throwing its tail upwards and outwards, tossing on to the top of the wall the débris from the inside of the nest. Thus it toiled on until the wall reached the surface of the water and was complete. When the nest was finished, the water it contained was perfectly clean and clear, so that I could see with my water-telescope the eggs nearly covering the bottom of the nest. When all the eggs are laid, the fish leaves the nest by a hole at one side.

Fig. 10. Nest of *Heterotis*.

The eggs, which measure $2\frac{1}{2}$ mm., then appear to hatch in about two days, though, owing to the distance the nests were from my quarters, of this I am not certain. The nest appears to be used for at most four or five days. As soon as the larvæ are hatched, they begin to strike up from the bottom. The day after hatching they may be seen continually passing up and down, and are now provided with long external gill-filaments of a blood-red colour, but not so numerous or so long as in the case of *Gymnarchus* (Pl. IX. figs. 6 & 7).

The following day they cease to pass up and down, and converging to a swarm about one foot in diameter, form a deep continuous circle remarkable for its regularity and persistence. The swarm occupies the exact centre of the little lagoon. The young fry, which by now have lost the long external gill-filaments, are seen to be steadily careering round and round ever in the same direction for at least a day. About the fourth day the swarm becomes less persistent and regular, the larvæ swimming first to one side of the nest and then to the other, until about the fifth day they leave the nest by the exit for a few trial trips attended by the parent, and finally leave it altogether, swimming hither and thither in a dense swarm, from which the parent is never far distant. I kept a large number of the young fry for several weeks, but could not get them to feed, and eventually they all died.

The ova of *Heterotis* are shed into the cœlom as in the Salmon. *Heterotis* belongs to the group Osteoglossidæ, which has much the same distribution as the Dipnoi, though it seems doubtful whether this points to an antiquity of the group equal to that of the Dipnoi. Günther, however, regards the Osteoglossidæ as one of the earliest types of Teleostean fishes.

VI. The Nesting-habits of *Sarcodaces odoë* Bl.

In these same flooded grass-lands the eye is frequently caught by masses of white foam floating on the surface of the water. On close inspection it is seen to be filled with numerous transparent ova, about the same size as those of *Heterotis* ($2\frac{1}{2}$ mm.). Soon these eggs hatch, and on hatching make their way through the foam, in which they are laid, down to the surface of the water, and there the young larvæ hang holding to the surface of the water by a large adhesive organ situated on the front of the head (Pl. IX. figs. 8 & 9, *c.o.*).

The natives assured me that these were the eggs of the Sannko, more scientifically *Sarcodaces odoë*. On rearing some of these larvæ, I was able to confirm this statement.

Sarcodaces is one of the Characinidæ, of which family examples occur in Africa and South America.

VII. The Nesting-habits of *Hyperopisus bebe* Lacép.

It was a curious fact that of the six species of Mormyridæ which I obtained in the Gambia, only one besides *Gymnarchus* was found breeding in the swamps. This was *Hyperopisus bebe* Lacép.

Although I did not succeed in finding fertilized eggs of *Hyperopisus* this year, I obtained a number of females full of ripe eggs. I am practically certain that these ovarian eggs are identical with the eggs which I studied last year under the impression that they were the eggs of *Polypterus*.

These eggs were laid in shallow depressions of the swamp bottom, and attached to the rootlets of the grasses laid bare by the parent in scooping out the depression for the reception of the eggs (text-fig. 11, p. 135). The eggs are very small, about $1\frac{1}{4}$ mm. in diameter, and slightly oval, the long axis being rather over $1\frac{1}{4}$ mm. in length. They are yellowish in colour and semitransparent. The eggs hatch in four days, and the larvæ are provided with four large cement-glands situated at the top of

the head, and two smaller ones on the front of the head (Pl. IX. fig. 10, *c.o.*).
Immediately the larva is hatched it runs the upper part of its head against the
rootlets, and wriggling away again, draws out from the four cement-glands four fine

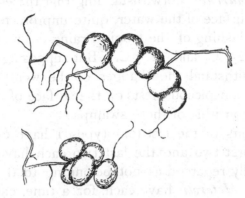

Fig. 11.   Eggs from nest, supposed to be that of *Hyperopisus bebe*.

threads of viscid mucus, which are hardened by contact with the water, and form
a minute rope about the length of the body of the larva.   By this the larva hangs
suspended for four or five days until the yolk is absorbed.   If the larva is detached

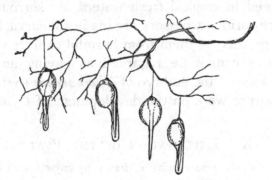

Fig. 12.   Larvæ, supposed to be those of *Hyperopisus bebe*, suspended from the rootlets in the nest.

meanwhile, a fresh rope is formed by a fresh secretion of mucus (text-fig. 12).
While hanging thus, each larva continually oscillates the whole length of its body
from side to side.   In one nest there are many thousands of these larvæ suspended
in this way, presenting the appearance of a shaking mass of jelly, for all the larvæ
oscillate themselves in unison.   I was unfortunately unable to rear any of these larvæ
to a stage old enough to be able to identify them.

VIII.   Conclusion.

I should here state that I had great difficulty in keeping alive any of the fish-
larvæ that I found for any length of time in any but the natural conditions,
Protopterus, however, excepted.   In the case of *Gymnarchus* a great number of

ways was tried, even floating perforated trays as an attempt to imitate the natural conditions. I do not so much wonder at my want of success in this as at the successful way in which the larvæ are hatched out in nature. I never found a dead larva in any nest of *Gymnarchus*, notwithstanding that the eggs and larvæ were lying within six inches of the surface of the water, quite unprotected from the burning rays of a tropical sun and the lashing of the tropical rains. The extremes of temperature taken in the nests were 25° C. and 32°·5 C. But supposing the larvæ to be so constituted that they can withstand the changes of the weather, how is it that large conspicuous eggs in very conspicuous nests on the surface of the water escape forming the food of the abundant bird-life of these swamps ?

In the breeding-habits of the last four types I have described, the interesting fact comes out that the first two and the last two each have, in common, organs in the larva which are usually regarded as not belonging to the Teleostean division of fishes. *Gymnarchus* and *Heterotis* have each, for a time, enormously elongated gill-filaments, structures which are so characteristic of Elasmobranch larvæ. Something of the kind was noticed in the development of the loach by Götte[1], but I think this is the only case of such organs in the Teleosteans. *Sarcodaces* and *Hyperopisus* have each well-developed cement-organs on the head. These structures are generally regarded as characteristic of the Ganoids. It seems, then, that the conditions by which fishes, which breed in tropical fresh waters, are surrounded, are conducive to the development of very various accessory organs in the larva, both for the purpose of respiration and also of preserving them from harmful contact with their surroundings, and that these structures cannot be regarded as having any great morphological meaning. The resemblance of the embryo of *Gymnarchus* to that of an Elasmobranch I hope to discuss in a future work on the development of *Gymnarchus*.

IX. EXPLANATION OF THE PLATES.

All the figures of Plate VIII. were originally drawn by myself with the aid of a camera lucida, and were then copied by Mr Edwin Wilson.

Figs. 2, 3, 4, 5, 12, & 13 were drawn on the Gambia from living specimens. Figs. 1, 6, 7, 8, 9, 10, & 11 were drawn from formalin specimens.

The magnification is 8 diameters.

The figures in Plate IX. were all drawn by myself except fig. 1, which was from a specimen preserved in formalin, drawn by Mr Edwin Wilson under my supervision and coloured from my notes by myself. Figs. 4, 5, 6, 7, & 8 were drawn from life on the Gambia. Fig. 3 was drawn from a specimen preserved with corrosive sublimate and acetic acid.

au., auditory sac; *bp.*, blastopore; *br.*, branchial eminences; *br.* I. II. &c., branchial arches; *c.o.*, cement-organ; *cl.*, position of cloaca; *Eg.* I. II. &c., external gills; *ep.e.*, growing edge of epiblast; *f.f.*, dorsal fin-fold; *H.*, hyoid arch; *H.br.*, hyobranchial cleft; *h.l.*, hind limb; *invag.*, line of invagination; *M.*, mandibular arch; *M.H.*, mandibulo-hyoid fold; *m.f.*, medullary folds; *o.c.*, optic outgrowth from brain; *op.*, operculum; *p.f.*, pectoral fin; *pn.*, pronephros; *s.v.g.*, groove marking rudiment of spiral valve; *sp.*, groove between mandibular and hyoid arches; *v.* IV., fourth ventricle; *y.k.*, yolk-cells.

[1] "Entwick. d. Teleostierkieme," *Zool. Anz.* No. 3, 1878.

PLATE VIII.

EXPLANATION OF PLATE VIII.

The figures of this Plate illustrate the external features in the development of
Protopterus annectens Ow.

FIG. 1.   Egg on the first day of observation, segmentation somewhat advanced: p. 125.

FIG. 2.   Egg on the morning of the second day, showing commencement of invagination: p. 125.

FIG. 3.   Egg on the evening of the second day, showing a further stage of invagination. Above is seen a transparent portion indicating the segmentation-cavity: p. 125.

FIG. 3 *a*.   A variation of the stage shown in Fig. 3. The area of the disappearing yolk-cells is viewed from a somewhat different aspect, in order to show that the line of invagination is not here straight but curved: p. 125.

FIG. 4.   Egg on the morning of the third day, showing the straight line of invagination and the gradual disappearance of the large-celled yolk: p. 125.

FIG. 4 *a*.   A variation of the stage shown in Fig. 4, showing that the line of invagination is not always straight: p. 125.

FIG. 5.   Egg on the evening of the third day, showing the last stage of invagination: p. 125.

FIG. 6.   Egg on the morning of the fourth day, showing medullary folds encircling the blastopore: p. 126.

FIG. 7.   Embryo on the evening of the fourth day, showing origin of optic outgrowths, mandibulo-hyoid fold, branchial eminence, and pronephros: p. 126.

FIG. 8.   Embryo on the fifth day, showing the segmentation of the mandibulo-hyoid fold and first segmentation of branchial eminence; also the remnant of the blastopore, and the first appearance of the cement-organ, and the development of the head and tail-folds: p. 126.

FIG. 9.   Embryo on the sixth day, showing the further segmentation of the branchial eminence into four: p. 126.

FIG. 10.   Embryo on the seventh day, showing the first pair of external gills strongly differentiated from the second, third, and fourth pairs. The trunk of the embryo has grown back from the main mass of yolk, taking much of the yolk with it. The true tail is not yet formed: p. 126.

FIG. 11.   Larva on the eighth day, just hatched, showing further growth of the external gills, and the auditory sacs showing through the tissues. The true tail-region has now begun to grow: p. 126.

FIG. 12.   Larva on the tenth day, showing the serial arrangement of the external gills, the indication of the spiracular cleft, the first trace of the hind limbs, being now anterior to the position of the cloaca, and the well-developed and functional cement-organ. The tail-fins are now well-developed. Pigment has appeared on the head: p. 127.

FIG. 13.   Larva on the seventeenth day, showing general advancement, the rotation of the external gill, the operculum, the vessel in the spiral valve groove, and the well-developed limbs: p. 127.

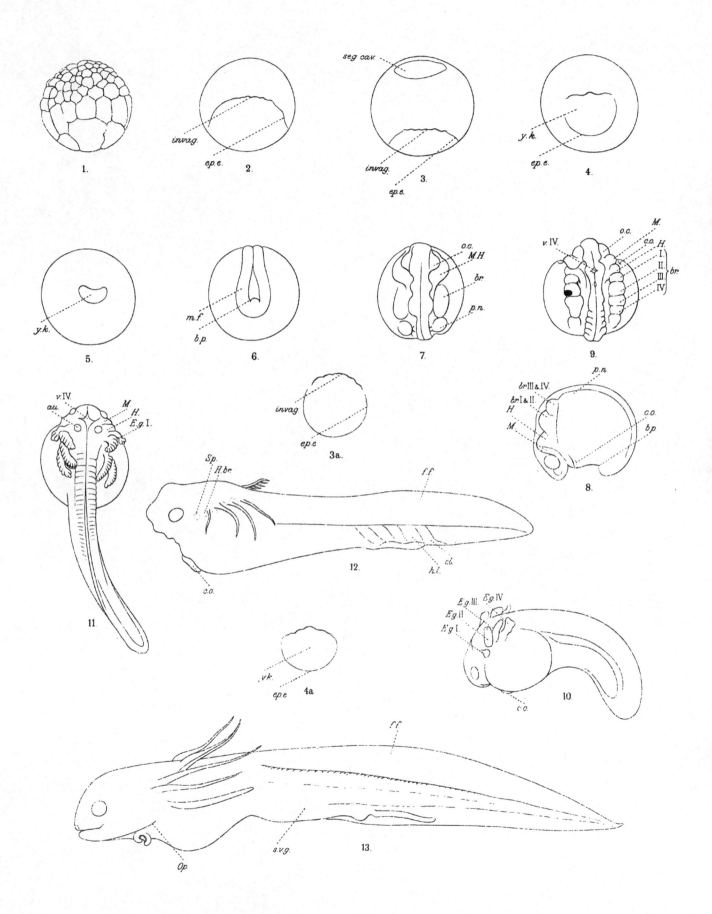

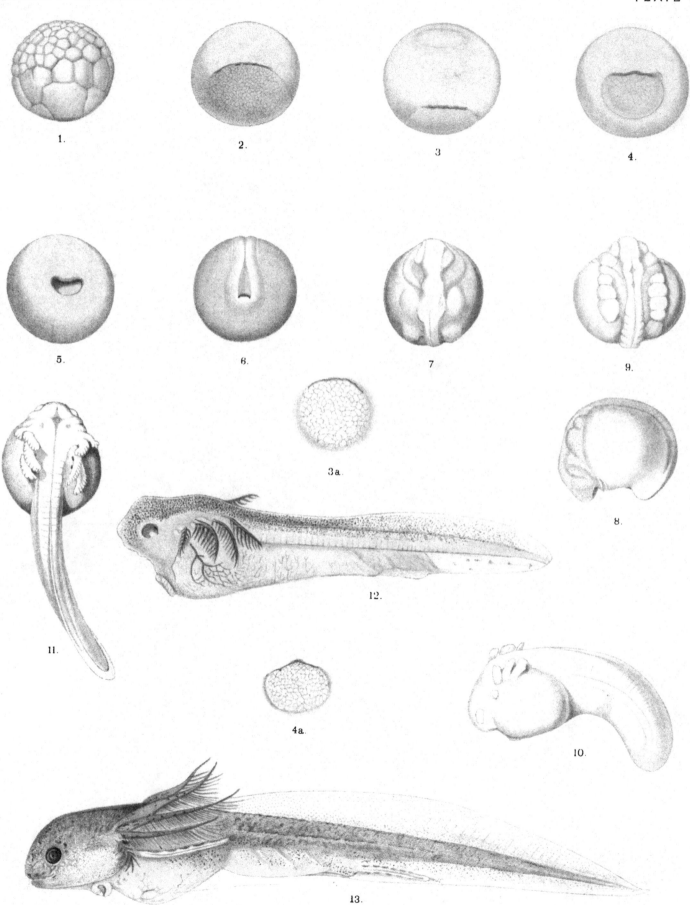

DEVELOPMENT OF WEST AFRICAN FISHES.

PLATE IX.

EXPLANATION OF PLATE IX.

FIG. 1. Larva of *Polypterus lapradei* Stein., $1\frac{1}{4}$ inch in length, magnification about four diameters, showing the very large external gill of the hyoid arch, situated immediately behind the spiracle, the blood-supply to the same from the ventral aorta, and the very large size of the head and tail-region of the body, also the dorsal fin not broken up into finlets. The larva is drawn in a very characteristic attitude as observed when in the aquarium. The smallness of the alternate branches of the gill and their different position are not shown.

FIG. 2. External gill of *Polypterus lapradei*, showing the difference in size of each gill-branch: p. 122.

FIG. 3. Larva of *Protopterus annectens* Ow., a few days after leaving the nest, about one month old. The larva was killed when its chromatophores were neither fully expanded nor fully contracted. The first pair of external gills has been lost and metamorphosis is nearly complete: p. 128.

FIG. 4. Larva of *Gymnarchus niloticus* Cuv., one week after it was laid, two days after hatching, showing external gill-filaments, blood-vessel to and from yolk-sac, and the completely diphycercal tail: p. 131.

FIG. 5. Larva of *Gymnarchus*, four days later, showing long drawn-out yolk-sac and abundant external gill-filaments. The fin-folds of the tail are shown, and the first trace of differentiation of the tail to form the whip-like feeler: p. 131.

FIG. 6. Larva of *Heterotis niloticus* Cuv., one day after hatching, showing external gill-filaments, backward extension of the yolk to the cloaca, the pectoral fins, four branchial arches, developing operculum, and heterocercal rays in the tail: p. 133.

FIG. 7. Larva of *Heterotis*, dorsal aspect, in order to show relations of operculum, gills, and pectoral fin: p. 133.

FIG. 8. Larva of *Sarcodaces odoë* Bl., just after hatching, showing the large adhesive organ borne on the front of the head: p. 134.

FIG. 9. The same, frontal aspect, to show the form of the adhesive organ: p. 134.

FIG. 10. Larva of *Hyperopisus bebe* Lacép., showing the four dorsal and two frontal cement-organs, the auditory sacs, and prominent gill-folds: p. 134.

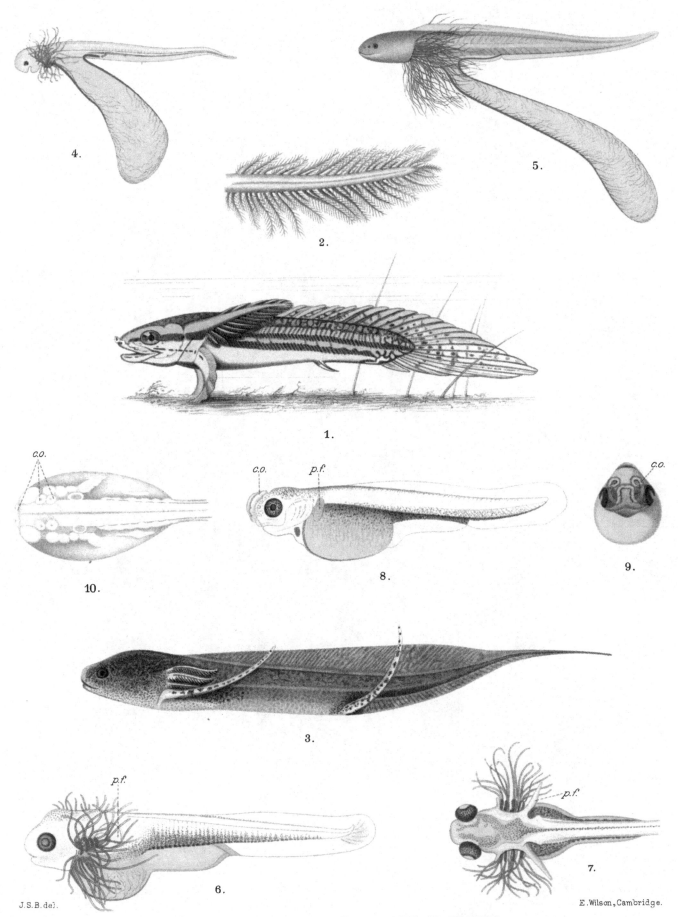

J.S.B.del.   E.Wilson,Cambridge.

DEVELOPMENT OF WEST AFRICAN FISHES.
1-2. POLYPTERUS. 3. PROTOPTERUS. 4-5. GYMNARCHUS. 6-7. HETEROTIS.

VII. THE HABITS AND DEVELOPMENT OF SOME WEST AFRICAN FISHES[1].

Read February 4, 1901.

Our knowledge of the fish fauna of the fresh waters of Africa has of late been greatly extended by the study of collections from the great African lakes, the Nile, and the rivers of the West Coast. Up to the present time, however, nothing has been known about the breeding habits and development of any of the most interesting forms, including *Polypterus, Protopterus, Gymnarchus, Mormyrus* and *Heterotis*. It was with a view to investigating the development of these fish and especially of *Polypterus* that I spent the summer of 1900 on McCarthy Island on the river Gambia. The flooded lands of this island I searched persistently from June to September, but failed to obtain the eggs of *Polypterus*; I did, however, obtain a very young larva of *Polypterus* measuring 1¼ inch in length.

In this larva the dermal bones are not yet developed over the general body surface, though some of the dermal bones of the head have already begun to ossify. The dorsal finlets at this stage are merely a continuation forwards of the finfold of the tail. The heterocercy of the caudal fin is scarcely more apparent (even in section) in this larva than in older specimens. The external gill is of very great size. The base of the shaft is situated immediately behind the spiracle, and is supported by a short segmented rod of cartilage borne upon the hyomandibular bar. Each pinna of the external gill bears a double row of pinnules. Alternate pinnae on each side are smaller and directed at a different angle from the intermediate pinnae, giving the appearance of two rows of pinnae on either side. The internal gills are very small and can as yet be of little functional importance, as the combined section of the arteries to the internal gills is certainly not a tenth part of the section of the artery supplying the external gill. The arteries to the two halves of the swim-bladder are likewise very small indeed. In this young larva the roof of the mouth is perforated by a duct from the pituitary body, as has been shown to be the case in *Calamoichthys*[2]. The oviduct appears to develop in a similar manner to that of *Lepidosteus*[3], being an included portion of the body cavity into which there still open a number of nephrostomes. These nephrostomes open upon a slight groove which will eventually become the dorsal wall of the oviduct.

While searching for the nests of *Polypterus* I discovered the underground nests of *Protopterus annectens* and obtained a complete series of eggs and larvae. The entrance to the burrow is in but a few inches of water, and when the water around the mouth of the nest dries up, the parent (who lives in the nest with the eggs and

---

[1] From the *Proceedings of the Cambridge Philosophical Society*, Vol. XI. p. 102.

[2] Bickford, Elizabeth E., "The Hypophysis of *Calamoichthys*," *Anat. Anz.* x. 1895.

[3] Balfour and Parker, "The Structure and Development of *Lepidosteus*," *Phil. Trans. Roy. Soc. London*, Part II. 1882.

larvae) is seen to lash the surface of the water with its tail. The larvae are provided with four pairs of plumose external gills and a ventral sucker as in Lepidosiren ; soon after hatching they attach themselves to the sides of the nest by the sucker and hang in a vertical position. The larvae hatch in eight days, and leave the nest in twenty days. The external features differ from those of the Lepidosiren only in unimportant details ; there is in the larva of *Protopterus*, however, an indication of a spiracular cleft.

I also found the nests of *Gymnarchus niloticus*. These are made in about three feet of water and float on the surface. The nests are two feet long and a foot wide, the wall of the nest standing several inches out of the water, except at one end where it is two or three inches below the surface, and leaves an entrance to the nest. The eggs measure 10 mm. in diameter ; the larvae hatch in five days, when they greatly resemble the embryos of Selachians. The gill arches are not covered by an operculum and bear rows of gill filaments which later become of great length and very numerous. The yolk-sac becomes drawn out into a long cylindrical bag which is completely absorbed by the time the larva leaves the nest. Each nest contains about 1000 eggs.

The nests of *Heterotis niloticus* were very abundant. They are built on the swamp bottom in two feet of water. They measure four feet across, the walls reaching the surface of the water. When completed this nest is perfectly round and the bottom is quite smooth. The eggs measure $2\frac{1}{2}$ mm. in diameter, are quite round and bright orange in colour. The larvae soon after hatching form a swarm in the centre of the nest and are provided with long protruding gill filaments.

In the same swamps *Sarcodaces odoë* Bl., lays its eggs in masses of froth on the surface of the water. The eggs measure 3 mm. and are transparent. The hatched larvae are provided with conspicuous adhesive organs on the front of the head with which they hang to the under side of the surface.

I also found nests containing eggs which apparently belong to *Hyperopisus bebe* Lacép., one of the Mormyridae. These nests are scooped out from the swamp bottom ; the eggs are attached to the rootlets thus laid bare. The hatched larvae are provided with six cement organs on the surface of the head. From them a delicate rope of mucus is spun often nearly the length of the body of the larva ; by this fine rope the larvae hang suspended from the rootlets until the yolk-sac is absorbed.

It is remarkable that the larvae of *Gymnarchus* and *Heterotis* are both provided with long protruding gill filaments which have hitherto, I believe, been only once recorded in the Teleostomi ; and that *Sarcodaces* and *Hyperopisus* are provided with conspicuous cement organs on the head ; these cement organs on the head of the larva have usually been regarded as characteristic of the Ganoidei.

It is thus seen that the conditions by which fishes, which breed in tropical fresh waters, are surrounded is conducive to the development of very various accessory organs in the larva, both for the purpose of respiration and also of preserving them from harmful contact with their surroundings.

VIII.   ON THE ORNITHOLOGY OF THE GAMBIA RIVER[1].

With text-figure 13.

In visiting the Gambia Colony between November, 1898, and July, 1899, my chief object was the study of its fishes, so that this paper by no means professes to give a complete list of the birds, though it is hoped that a record of such as were noticed may be found useful to my readers.   Apart from A. T. de Rochebrune's unsatisfactory "Faune de la Sénégambie" (1884) (which is generally supposed to be barely trustworthy), I have only been able to find two articles treating solely of the avifauna of the district, namely, "Notes on the Ornithology of the Gambia," by Dr Percy Rendall (*Ibis*, 1892, pp. 215–230), which gives an annotated list of birds collected within eight miles of Bathurst during a stay of 21 months, and the second Appendix (pp. 464–483) to Moloney's "Sketch of the Forestry of West Africa" '1887), which contains a list of the species known from the river, with references to the literature, by Capt. Shelley.

The majority of the birds that I observed are included in these lists, but a certain number are wanting; nor did I meet with all those noted by Dr Rendall; this, however, is easily explained by the situation of the districts which I explored.

In travelling up the river to my future headquarters, the island of McCarthy, we generally kept close to one bank or the other, and thus had fair opportunities of observing the avifauna.   In the lower reaches from Bathurst to Nianimaru impenetrable walls of mangroves line the shores and make it very difficult to obtain a view of the interior or to watch the birds.   Frequently, however, large parties of Pelicans and Marabou-birds crossed our course, while at low water there often appeared dense flocks of Crested Cranes, which, upon our approach, sailed away with their long necks and unwieldy heads far outstretched, uttering their loud hoarse cries.

As we neared Nianimaru, about 100 miles up the river, we left the mangroves behind and could see the actual banks teeming with bird-life amongst the dense tropical vegetation of every description, which extended as far as our destination and remained luxuriantly green throughout the year.   The country farther inland is, in the dry season, somewhat sparsely covered with almost leafless trees, and there being little to attract the ornithologist in the scorched-up plains or the stony plateaux away from the river, my hunting-grounds were somewhat restricted.   When we arrived at McCarthy Island on Nov. 10th the dry season was just beginning, but the swampy plains were still covered with dense jungle, and the foliage of the forest-covered plateaux formed a rich contrast

[1] From *The Ibis* for July, 1901.

to the red cliffs as they caught the last rays of the setting sun, while vast flocks of Egrets, Ducks, and Green Pigeons flew overhead from their feeding-grounds. Even before reaching McCarthy Island I was struck by the number of species of Rollers and King-fishers seen, though it was impossible to identify them from the colonial steamer. From Nov. 11 to Nov. 24 I was fully occupied with other work, and could shoot only a few birds daily for food. These were usually Pigeons, Spur-winged Plovers, the so-called Bush-fowl (*Francolinus bicalcaratus*), the Sand-Grouse (*Pterocles quadricinctus*), or occasionally a Whistling Duck. Other species of frequent occurrence were the Palm-bird (*Lophoceros nasutus*), usually seen in parties of from three to five, which continually flew in undulating fashion from tree to tree uttering their monotonous high-pitched

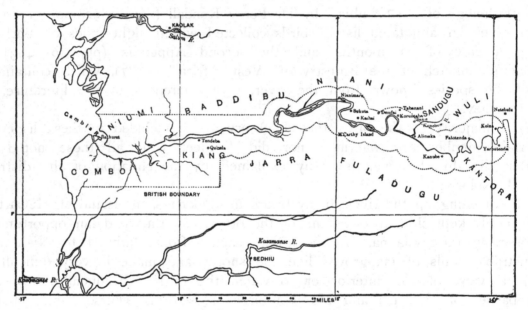

Fig. 13. Outline Map of the Gambia River.

cry of three descending notes; two species of Parrots; enormous flocks of *Lamprocolius purpureus*; as well as groups of *Lamprotornis caudatus* and Wood-Hoopoes (*Irrisor senegalensis*), ranging from ten to thirty, which vied one with another in their deafening chatter. The flocks of the last-named generally contained twice as many females as males, the latter being easily distinguished by their brilliant red beaks. The Foolish bird (*Centropus senegalensis*), various Drongos and Shrikes, and Weaver-birds in vast numbers, completed the tale. In the early days of my stay I noticed the call of our British Redstart, of which I saw both males and females throughout the dry season. The Grey Flycatcher (*Muscicapa grisola*) was also common along the river-banks at that time.

On Nov. 24 I went down the river to Nianimaru, and thence travelled in company with the Commissioner of the District up the north bank as far as the Kunchow Creek, returning to McCarthy Island on Dec. 20th. There was plenty of time for fishing and shooting, as we stayed at the important places for several days. The towns visited were Sukuta, Kaihai, Demfai, Tabanani, Same, and Koruntaba, most of them several miles from the river, as the natives almost always prefer to live in the higher country. On this trip I first became acquainted with the little flocks of the Long-tailed Shrike (*Corvinella corvina*), which were everywhere seen flying from bush to bush along the native roads, not in the least shy of our rather numerous string of porters and servants.

On reaching our destination we took up our quarters in native huts set aside for the Commissioner's special use, usually sharing them with the small Red-breasted Weaver-birds (*Lagonosticta senegala*), while in the trees overhead *Pycnonotus barbatus* reiterated its clear but plaintive little song. About the towns numbers of *Cryptorhina afra* acted as general scavengers. The males had brilliant red beaks, the females black. It was near Tabanani that I first saw perched on the top of some solitary tree-stump *Lanius auriculatus*, and also identified the Parrots previously noticed, viz. *Pœocephalus senegalus* in large flocks of over twenty, and *Palæornis docilis* in small parties. Here also I shot a number of the handsome Green Pigeon (*Vinago waalia*), which never alights on the ground and feeds chiefly on young figs, the trees being literally crowded with the birds. In the gullies the noisy chatter of parties of some half-a-dozen Babblers (*Crateropus platycercus*) attracted my attention as they ran up the tree-trunks in search of insects, in Woodpecker fashion.

At Same, on the Kunchow Creek, while fishing with trammel-net and line, I had an opportunity of identifying the King-fishers *Ceryle rudis*, *C. maxima*, and *Halcyon senegalensis*. The first-named afforded a most attractive spectacle as it poised itself in mid-air above its prey, with the neck and beak pointing downwards. Up and down the stream, skimming the surface as a Swallow does, flashed the gorgeous green, orange and crimson *Melittophaga bullocki*, while from the bushes along the banks came a noisy chatter that might have been made by monkeys, but which I soon learnt was uttered by parties of *Crateropus reinwardti*. On the neighbouring mud a solitary bird, marked like a Redstart but of the size of a Thrush, stealthily hopped along. A specimen was with difficulty bagged, and proved to be *Cossypha albicapilla*.

Much of the country hereabouts was of a varied nature, ranging from level plateaux covered with cane-brakes and small leguminous trees, to cultivated valleys nearer the towns, where the natives grow millet, maize, cotton, and ground-nuts ; here the soil was sandy, with only a solitary cotton-tree (*Bombax*), mahogany (*Khaya*), or fig-tree left standing. Nearer the river were level plains, flooded in the rainy season, where the natives plant a little rice.

As we came back through Koruntaba to the river-side, and thence to McCarthy Island again, I noticed a new bird, the Piebald Crow (*Corvus scapulatus*), and shot a Heron (*Ardea melanocephala*), a large Eagle (*Aquila wahlbergi*), and a Harrier (*Circus macrurus*).

I went down to Bathurst for the new year, and there engaged two native fishermen, arriving at McCarthy Island again on the 11th of January; thence, after a stay of eight days, I took my fishermen, cook, and nets down to Nianimaru for two weeks. A large number of the smaller birds were obtained at this time, including Shrikes, Woodpeckers, Bee-eaters, Honey-suckers, and Weaver-birds. All the Shrikes were found solitary or in pairs, except *Prionops plumatus*, which moved restlessly about the thickets of the stony plateaux in parties of a dozen. *Scoptelus aterrimus* was common here, though, unlike the large *Irrisor senegalensis*, it was always seen in pairs.

About this time the natives were burning the dead jungle on the lowlands, and I often watched with interest the enormous numbers of White Herons (*Ardea garzetta*) cautiously moving in front of the advancing flames and devouring the insects which were driven out of the grass by the heat of the fire. Above them twittered a vast number of *Merops nubicus*, literally showing a red cloud of wings as they greedily devoured the insects that attempted to escape by flight. This was the only Bee-eater observed in large flocks. I obtained many of the small Weaver-birds and Flycatchers during my fishing expeditions, while the canoe slid silently amongst the luxuriant vegetation of the little creeks; though sometimes we had to beat a hasty retreat, as, with a roar, a great cow hippopotamus splashed into the water ahead, warning us that it was dangerous to approach her young.

On Feb. 1st I returned to McCarthy Island, as there was some apprehension of an invasion by a neighbouring chief, Jimba ma Joula, but the arrival of H.M.S. "Alecto" caused him to change his mind.

During February and March I made several trips up and down the river, but got very little in the way of birds. The Harmattan winds, which were now at their worst, parched up all the vegetation, blowing sand and dust over everything and making life generally uncomfortable; the temperature at midday was always over 100° F., and often 106° F., though at night it became much cooler. At this time I procured specimens of *Buphaga africana*, no easy matter with a bird that clings so closely to the backs of cattle.

In the early part of March I shot a Cormorant which had all the markings of the very young *Phalacrocorax lucidus* described by Capt. Boyd Alexander (see Brit. Mus. Cat. B. xxvi. p. 351). I also obtained a fine specimen of *Musophaga violacea*, with the head and wing-patch of a particularly bright red. On the 24th of March I went down the river to spend a week at Quinela, on the south bank. Near that town there was a well, much used by the natives, under the moist shade of some mahogany-trees, and amongst the numberless birds that frequented it I first obtained *Oriolus galbula*, the uncommon *Pogonorhynchus vieilloti, Barbatula chrysocomis, Zosterops senegalensis*, and *Terpsiphone melanogaster*, remarkable in the males for a steel-blue crest and magnificent chestnut tail-streamers. One morning in the open plain I came across a large flock of *Glareola pratincola*, which had the peculiar habit of flying round for a minute or two and then settling quite close at hand, apparently trusting to its resemblance to the soil to escape observation. In the tree-tops overshadowing the town Pelicans and Marabou-birds build their nests, and at the end of March there were still some unfledged birds in them.

On April 1st I started in a cutter up the river from McCarthy Island to stay for a short time at the Kunchow Creek. The voyage occupied two days, and as we anchored when the tide was ebbing, I had several opportunities of going ashore to shoot. I found Guinea-hens (*Numida meleagris*) very plentiful, as were also the Bush-fowl and the Sand-Grouse. Arriving at the mouth of the creek on April 3rd, I rowed up it about a mile to Alimaka's wharf, where I had some huts built overlooking the river. At this time the air was everywhere heavily scented by the blossoms of a beautiful gardenia-like shrub, which teemed with various Sun-birds and Bee-eaters. The commonest Sun-bird met with here was *Nectarinia pulchella*, though I obtained three others. In the dense belt of forest by the river-side I saw several beautiful species of Flycatchers, and tried, without success, to obtain a specimen of *Elminia longicaudata*. *Parus leucomelas* was fairly common. About this time (April 10th) I first noticed the harsh cry of *Coccystes glandarius*, thenceforward very common. On leaving the Kunchow Creek, on a low mud-bank near its mouth I saw a large flock of *Œdicnemus senegalensis*, as a rule a solitary species.

On April 20th I joined company with two Englishmen who were on their way to inspect some gold-workings. We were taken up to Yarbutenda by a small steamboat, and thence we rowed up to Netebulu, in French territory, at the head of the navigable waters of the Gambia. After a stay of a few days, I parted from the Englishmen and made my way back overland towards McCarthy Island. I stayed about a week at Koina, where antelopes and game of all kinds were very abundant, and there I succeeded in obtaining two good specimens of the beautiful blue Flycatcher, *Elminia longicaudata*. This bird is extremely active and restless, flitting from tree to tree and continually spreading out its tail like a fan. The male and female seem quite similar. *Bucorax abyssinicus* was also seen several times.

On May 4th I reached Fatotenda, where I stayed a day or two near a small lake much frequented by birds. I saw there black and white Ibises and several Spoonbills. Near Darsilami I noticed several flocks of about twenty Leona Nightjars (*Macrodipteryx longipennis*). The males with the elongated wing-feathers were rather more numerous than the females. They flew continually round and then settled on the ground at my feet. I reached my headquarters on May the 11th, and then heard for the first time the call of a Cuckoo, the note of which resembled that of *Cuculus canorus*, but the bird I shot was *C. gularis*.

I now went down to Nianimaru for the last time (May 18th). The rainy season was just beginning, though unusually early. I obtained there *Halcyon chelicutensis* far away from the river, and my first Black-throated Weaver-bird (*Hyphantornis cucullatus*), which I afterwards found everywhere in thousands, building so closely in the large baobab-trees (*Adansonia digitata*) that the grass-nests, with their entrances underneath, were not more than a foot or two apart. The Bee-eater, *Dicrocercus furcatus*, now seen for the first time, was abundant. *Platystira cyanea* was fairly common in the dense forest-growth at the water's edge ; and *Pachyprora senegalensis* in the low bushes in the open. Honey-guides (*Indicator sparrmani*), perched on the tree-tops, were uttering a loud, melodious, but monotonous cry of two descending notes.

On June 1st I returned to McCarthy Island, where I stayed during the rains until the end of July. The Secretary-bird was not unfrequently observed, and birds had generally begun to build. *Laniarius barbarus* makes a shallow nest of twigs, not unlike that of the Bullfinch, in rather obvious positions. There were now abundant in the swamps two brilliant Weaver-birds, *Euplectes franciscanus* and *E. oryx*; they both make round grass-nests with side entrances, attached to two or three of the stems of the tall swamp-grasses. *Vidua principalis* was fairly common, and I often watched the male hovering over the female, rapidly opening and shutting the long tail-feathers, causing them to assume at times the shape of a lyre. In the swamps I came across the floating nursery of *Limnocorax niger*, made of flat blades of grass neatly woven together.

As I was unable to bring home a complete collection of the birds that I shot, I carefully measured and described those of which I did not preserve skins, and in the list given below such are marked with an asterisk. Birds which were merely seen I have not included ; most of them, however, belonged to well-known species[1].

I have referred in my list to one or two kinds of Eagles identified from skins obtained on the south bank. These were shot by Mr H. L. Pryce, the Travelling Commissioner of that district, who kindly allowed me to measure and take notes of them.

In identifying my descriptions I have received much assistance from Mr A. H. Evans, while the skins were named for me at the British Museum by Capt. Shelley.

*RUTICILLA PHŒNICURUS (Lath.). McCarthy Island, February 19, 1899. Common in the early months of the dry season.

SYLVIA CINEREA Bechst. Nianimaru, May 19, 1899. Common in the dry season.

COSSYPHA ALBICAPILLA (Swains). Nianimaru, February 28, 1899. Common, mostly in the thick bush.

---

[1] These included a Swallow, Wagtail, Hornbill, Snipe, Darter, several wading birds, and Ducks of various species. The last, though very frequently shot, were never identified.

CRATEROPUS PLATYCERCUS Swains. Koruntaba, December 19, 1898. Common.

CRATEROPUS REINWARDTI Swains. Kunchow Creek, December 17, 1898. Common near the river.

PARUS LEUCOMELAS Rüpp. ♂ ♀. Kunchow Creek, April 7, 1899. Common.

HEDYDIPNA PLATURA (Hartl.). ♂. Nianimaru, February 22, 1899. Common.

NECTARINIA PULCHELLA (L.). ♂ ad. Quinela, March 26, 1899; ♂ juv. Kunchow Creek, April 7, 1899. Common.

CHALCOMITRA SENEGALENSIS (Hahn). ♂. Kunchow Creek, April 4, 1899. Rare.

ANTHOTHREPTES LONGUEMARII (Less.). ♂. Nianimaru, January 22, 1899; ♀. Nianimaru, April 6, 1899. Fairly common.

ZOSTEROPS SENEGALENSIS Hartl. Quinela, March 2, 1899. Common at this place.

PYCNONOTUS BARBATUS Gray. Nianimaru, November 25, 1898. Very common.

ANTHUS GOULDI Fraser. Quinela, March 28, 1899.

SERINUS ICTERUS Bp. McCarthy Island, January 10, 1899. Common.

HYPOCHERA ÆNEA Hartl. ♂. Quinela, February 26, 1899. Common.

VIDUA PRINCIPALIS (L.). ♂. McCarthy Island, July 4, 1899. Fairly common at this time.

QUELEA QUELEA (L.). ♀. Nianimaru, January 21, 1899. Common.

ESTRILDA CÆRULESCENS Vieill. Nianimaru, February 21, 1899. Common.

*ESTRILDA PHŒNICOTIS Swains. McCarthy Island, February 20, 1899. Very common.

*LAGONOSTICTA SENEGALA (L.). McCarthy Island, February 20, 1899. Very common about the native huts.

SITAGRA LUTEOLA (Licht.). ♀ juv. Nianimaru, February 21, 1899. Common.

HYPHANTORNIS CUCULLATUS Hartl. ♂ ad. Nianimaru, May 18, 1899; ♂ juv. Nianimaru, March 25, 1899. Very common in the rainy season.

LAMPROCOLIUS PURPUREUS (P. L. S. Müll.). McCarthy Island, January 9, 1899. Very common.

*LAMPROTORNIS CAUDATUS (P. L. S. Müll.). McCarthy Island, December 17, 1898. Very common.

BUPHAGA AFRICANA L. McCarthy Island, March 12, 1899. Common.

ORIOLUS GALBULA L. ♂ ♀. Quinela, March 27, 1899. Fairly common.

DICRURUS ATRIPENNIS Swains. Nianimaru, May 21, 1899. Common.

DICRURUS AFER (Licht. sen.). Demfai, March 17, 1899. Very common.

*CRYPTORHINA AFRA Sharpe. McCarthy Island. Very common. The males have red beaks, the females black; they act as scavengers.

*CORVUS SCAPULATUS Daud. Tabanani, December 16, 1899. Common.

PRIONOPS PLUMATUS (Shaw). Nianimaru, January 21, 1899; Kunchow Creek, April 5, 1899. Fairly common in small parties.

CORVINELLA CORVINA (Shaw). Quinela, March 26, 1899. Very common in small parties.

LANIUS AURICULATUS (Müll.). Tabanani, December 15, 1898 (young); Quinela, March 25, 1899 (young). Fairly common in the open.

LANIARIUS BARBARUS (L.). McCarthy Island, January 9, 1899. Common in thick bush.

DRYOSCOPUS GAMBENSIS (Licht.). ♂. Nianimaru, January 22, 1899. Common in thick bush.

TELEPHONUS SENEGALUS (L.). Koina, May 1, 1899; Nianimaru, February 27, 1899. Common in the open.

MALACONOTUS POLIOCEPHALUS (Licht.). Nianimaru, January 20, 1899. Common.

BRADYORNIS PALLIDUS (v. Müll.). Demfai, March 17, 1899; Kunchow Creek, April 8, 1899. Common.

HYLIOTA FLAVIGASTRA Swains. Nianimaru, January 27, 1899. Common.

PACHYPRORA SENEGALENSIS (L.). Quinela, March 27, 1899; Nianimaru, May 28, 1899. Common.

ELMINIA LONGICAUDA (Swains.). Koina, May 2, 1899. Rare.

TERPSIPHONE CRISTATA (Hartl.). Nianimaru, May 30, 1899. Fairly common.

PLATYSTIRA CYANEA (P. L. S. Müll.). ♀. Nianimaru, May 21, 1899. Fairly common in the forest at the river-side.

*MUSCICAPA GRISOLA L. McCarthy Island, February 19, 1899. Common in the early months of the dry season.

HIRUNDO LUCIDA Verr. Nianimaru, February 26, 1899. Common at this time.

CYPSELUS AFFINIS (Frankl.). McCarthy Island, June 10, 1899. Common.

MACRODIPTERYX LONGIPENNIS Shaw. McCarthy Island, May 8, 1899. Common, often in small flocks.

SCOTORNIS CLIMACURUS (Hartl.). Fatotenda, April 10, 1899. Solitary.

SCOPTELUS ATERRIMUS (Steph.). Juv. Nianimaru, January 21, 1899. Fairly common in pairs.

*IRRISOR SENEGALENSIS Hartl. Nianimaru, December 26, 1898. Very common in parties of twenty to thirty.

MELITTOPHAGUS PUSILLUS (Sharpe). Nianimaru, January 10, 1899. Common, especially in the mangroves.

DICROCERCUS FURCATUS (Stanl.). Nianimaru, May 19, 1899. Common.

MELITTOPHAGUS BULLOCKI (Vieill.). Kunchow Creek, April 8, 1899; McCarthy Island, April 12, 1899. Fairly common.

MEROPS NUBICUS Gm. McCarthy Island, January 8, 1899. Very common.

*HALCYON SENEGALENSIS (L.). McCarthy Island, December 21, 1898. Common. The specimen obtained had a totally red beak, but otherwise answered to the description of this species.

HALCYON CHELICUTENSIS Finsch and Hartl. Nianimaru, May 18, 1899. Fairly common.

*CERYLE MAXIMA Gray. ♀. Kunchow River, December 16, 1899. Common.

*CERYLE RUDIS (L.). ♀. Kunchow River, December 17, 1899. Common.

LOPHOCEROS ERYTHRORHYNCHUS (Temm.). Quinela, March 27, 1899. Common.

*LOPHOCEROS NASUTUS (L.). McCarthy Island. Common.

*EURYSTOMUS AFER (Lath.). McCarthy Island, May 18, 1899. Common. Frequently seen on trees at the river-side. Very noisy and quarrelsome at beginning of the breeding-season.

*CORACIAS CYANOGASTER Sharpe. Tabanani, December 14, 1898. Fairly common.

*CORACIAS ABYSSINICA Gm. McCarthy Island, November, 25, 1898. Common.

*CORACIAS NÆVIUS Daud. McCarthy Island, November 23, 1898. Common.

MESOPICUS GOERTAN (P. L. S. Müll.). Quinela, March 29, 1899. Fairly common.

CAMPOTHERA PUNCTATA (Swains.). Nianimaru, January 22, 1899. Fairly common.

BARBATULA CHRYSOCOMA (Temm.). Quinela, March 27, 1899. Common.

MELANOBUCCO VIEILLOTI (Hartl.). Quinela, March 29, 1899. One specimen seen.

POGONORHYNCHUS DUBIUS (Hartl.). Nianimaru, February 25, 1899. Common.

INDICATOR SPARRMANI Steph. Nianimaru, February 2, 1899. Common.

CHRYSOCOCCYX SMARAGDINEUS (Swains.).   McCarthy Island, July 20, 1899. Rare; one specimen obtained.

*CUCULUS GULARIS Steph.   McCarthy Island, May 11, 1899.   Not heard or seen on the Gambia before this date after my arrival in November.

*COCCYSTES CAFER (Licht. sen.).   McCarthy Island, June 14, 1899.   Appeared first at this date.

*COCCYSTES GLANDARIUS (L.).   McCarthy Island, April 1, 1899.   Not noticed before this date.

*CENTROPUS ANSELLI Sharpe.   McCarthy Island, December 2, 1898.   Common all the year round.   Though the specimen shot undoubtedly belonged to this species, *C. senegalus* was probably the most common form.

*MUSOPHAGA VIOLACEA Isert.   McCarthy Island, March 10, 1899.   Not very common.

*SCHIZORHIS AFRICANA (Lath.).   McCarthy Island, March 5, 1899.

PALÆORNIS DOCILIS (Vieill.).   McCarthy Island, March 8, 1899.   Very common.

*PŒOCEPHALUS SENEGALUS (L.).   McCarthy Island, December 7, 1899.   Very common in flocks.

*STRIX FLAMMEA L.   McCarthy Island, June 13, 1899.

*SCOPS GIU (Scop.).   Nianimaru, February 25, 1899.   Common.

*HALIAËTUS VOCIFER (Daud.).   Common.   Identified from a skin obtained on the south bank.

*LOPHOAËTUS OCCIPITALIS (Daud.).   Not common.   Also identified by means of a skin from the south bank.

*CIRCAËTUS BEAUDOUINI J. Verr. et Des Murs.   McCarthy Island, December 20, 1898.

*AQUILA WAHLBERGI (Sundev.).   Koruntaba, December 19, 1898.

*HELOTARSUS ECAUDATUS (Daud.).   McCarthy Island.   Common.

*FALCO ARDESIACUS Bonn. et Vieill.   McCarthy Island, December 17, 1898.

*CIRCUS MACRURUS (Gm.).   McCarthy Island, December 16, 1898.

*ASTURINULA MONOGRAMMICA (Temm.).   Nianimaru, November 25, 1898.

*FRANCOLINUS BICALCARATUS (Linn.).   McCarthy Island, February 2, 1899. Very common.

*ŒNA CAPENSIS (L.).   ♂.   Quinela, March 28, 1899; ♀.   Tabanani, December 14, 1898.

*NUMIDA MELEAGRIS L.   Nianimaru, February 19, 1899.   Very common.

*COLUMBA GUINEA L.   Nianimaru, January 10, 1899.   Common.

*VINAGO WAALIA (Gm.). Demfai, December 13, 1898. *V. calva* was, I think, equally common.

*CHALCOPELIA AFRA (L.). McCarthy Island, November 24, 1898. Very common.

*TURTUR SEMITORQUATUS (Rüpp.). McCarthy Island, November 23, 1898. Common.

*TURTUR VINACEUS (Gm.). McCarthy Island, December 18, 1898. Common.

*TURTUR SENEGALENSIS (L.). McCarthy Island, December 18, 1898. Common.

*BUTORIDES ATRICAPILLUS (Afzel.). Nianimaru, January 23, 1899. Common.

*ARDEOLA RALLOIDES (Scop.). McCarthy Island, April 1, 1899. Common.

*GARZETTA GARZETTA (L.). Nianimaru, February 27, 1899. Very common.

*ARDEA MELANOCEPHALA Childr. Koruntaba, December 19, 1898. Common.

*SCOPUS UMBRETTA Gm. McCarthy Island, November 22, 1898. Very common.

*PTEROCLES QUADRICINCTUS Temm. ♂ ♀. McCarthy Island, December 16, 1898. Very common.

*LIMNOCORAX NIGER (Gm.). Nianimaru, February 26, 1899. Fairly common.

*ŒDICNEMUS SENEGALENSIS Swains. McCarthy Island, February 2, 1899. Common.

*SARCIOPHORUS TECTUS (Bodd.). McCarthy Island, March 12, 1899. Very common at this season.

*LOBIVANELLUS SENEGALUS (L.). McCarthy Island, November 13, 1898. Very common at this season.

GLAREOLA PRATINCOLA (L.). Quinela, March 29, 1899.

*PHALACROCORAX LUCIDUS (Licht.). Juv. McCarthy Island, March 5, 1899. Very common.

IX. ON THE STRUCTURE OF THE LARVAL POLYPTERUS[1].

Received and read December 17, 1901.

With Plates X—XII. and text-figures 14—18.

CONTENTS.

I. Introduction.

Although *Polypterus* is undoubtedly one of the most interesting forms of recent Vertebrata, in that it is one of the two survivors of a very ancient group of fishes which existed in an abundance of forms in the Devonian era, its anatomy was until recently only imperfectly known; while nothing was recorded of its life-history and development, the knowledge of which might be expected to throw much light on the relationships of the great groups of the lower Vertebrata. The study of the structure of a very young larval *Polypterus* which I had obtained seemed to me therefore of considerable interest.

The attempt to arrange recent groups of animals in a linear series or even in the form of a genealogical tree appears to me to be misleading. The affinities of recent groups should rather be represented by a chart of an archipelago of groups, the remains of a submerged continent, the configuration of which the study of palæontology alone can reveal.

The significance of the observations recorded in the following pages appears to me to be that *Polypterus* must be regarded as having affinities with several of the great groups of fishes on the one hand, and also with the amphibians on the other hand. From this I am led to conclude not that the modern Crossopterygians form a connecting-link in a phylogenetic sense, but rather that they form a central group amongst recent forms having affinities with, and so connecting together, a number of the great groups of Vertebrates; for, as will be seen, in parts of their structure they show undoubted affinities with the Teleostei and Ganoidei, while in other respects their structure reveals affinities

[1] From the *Transactions of the Zoological Society of London*, Vol. XVI. Part VII. October, 1902.

with the Elasmobranchii; in addition, however, they show affinities in a third direction, namely with the Stegocephali and through them to the Amphibia. That there is any very close connection between the living Crossopterygians and the living Dipnoi does not appear to be the case.

Some authors have attempted to show that the Amphibia may have been directly derived from some such forms as the recent Crossopterygii, others derive them from the Dipnoi: with both these views I venture to disagree. The Devonian Crossopterygii no doubt merged into the Devonian Dipnoi, and it was from forms probably intermediate between the Crossopterygii and the Dipnoi that the ancestral Amphibians originated. In these times likewise the Crossopterygii merged into the Elasmobranchii and thirdly into the Ganoidei, and it was from forms probably intermediate between the Crossopterygii and the Ganoidei that the Teleostei originated.

Thus the Amphibia appear to be related to the Elasmobranchii in two ways—first through the Dipnoi, second through the Crossopterygii. In like

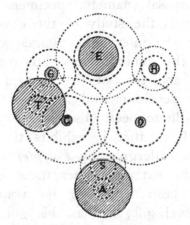

FIG. 14. Diagram to illustrate the affinities of Vertebrata and their relative importance in Primary, Secondary, and Tertiary times.

A. Amphibia and Amniota.
C. Crossopterygii.
D. Dipnoi.
E. Elasmobranchii.
G. Ganoidei.
H. Holocephali.
S. Stegocephali.
T. Teleostei.

Primary.

Secondary.

Tertiary.

manner, the Teleostei are related to the Elasmobranchii in two ways, the Crossopterygii and the Ganoidei.

To attempt to give a definite chart of the Ichthyopsida would be to assume a perfect knowledge of the structure, development, and palæontology of the various groups, which I do not pretend to possess. I will therefore give a mere diagram (text-fig. 14) in order to illustrate my meaning, and pass on to the facts of development as observed in this larval Crossopterygian, attempting such interpretation as they seem to warrant, in view of the absence of knowledge of the early stages in its development.

II. Material.

The external features of the larva of *Polypterus* which forms the subject of this paper, I have described in a former paper[1]. I have had at my disposal only a single specimen, measuring 30 mm., obtained during a collecting-trip to the Gambia in 1900[2]. In the present paper I have given an account of the structure of the cartilaginous skeleton in this larva, and somewhat of its subsequent development as studied in a number of specimens measuring from 9 cm. upwards. I have also given an account of the genital and excretory system of organs in this small larva, the subsequent development of which I have already described in a previous paper[3].

It will be seen that between the stage of this small larva (30 mm. in length) and the next stage at my disposal (namely, specimens 9 cm. in length) there is a considerable gap. As regards the study of the development of the skeleton, this is of no importance, as the intervening stages are easily interpreted. As regards the development of the genital ducts, however, an intermediate stage, between that here described and that of the 9 cm. larva already described, is much to be desired. Of the developmental stages of *Polypterus* preceding that here described nothing has hitherto been observed.

It may be said of this larva that it exhibits the condition of the last of the series of changes which is undergone by *Polypterus* in its development from the egg before it assumes the external appearances of the adult form. The term "larva" has frequently been employed for young *Polypteri* of all sizes which retain the external larval gill; but as this gill is retained frequently by adult specimens of some species of *Polypterus*, it is in no way justifiable to term such specimens as happen to possess this organ "larvæ," irrespective of their being in a larval condition otherwise.

I have laid some stress upon this, as an account of the excretory system of a *larval Calamoichthys* has been published, in which I am sure that the author has been influenced in his description by the supposition that the specimen was in a larval condition, though measuring 12 cm. in length. But to this I shall return later.

Methods.

The larva, which had been preserved in 5 per cent. formalin, was cut into transverse sections, of which the anterior half were $10\,\mu$ in thickness, while

[1] p. 122.

[2] In the account of the external features, this larva was said to be that of *Polypterus lapradii*. It has recently been shown by Mr Boulenger that the larva is that of *P. senegalus* and not that of *P. lapradii*.

[3] p. 100.

the posterior half were 7 μ thick. They were doubly stained with hæmatoxyline and eosine. All the figures of Plate X., and figs. 9 and 10 of Plate XI., were drawn from reconstructions by Kerr's method; that is, the sections were drawn on sheets of ground-glass by the aid of the camera lucida and a medium, of the same refractive index as the glass, run in between the sheets of glass[1].

I have found it extremely convenient to use this method in the following way:—Five dozen sheets of ground-glass 4 inches square and $\frac{1}{16}$ inch thick are ground and polished *en masse*, so that their cut edges when placed together form plane surfaces.

After having drawn the first section, the image of the second section is exactly superimposed on the drawing of the first, which is then removed, and the next glass plate being put exactly in its place and the drawing having been

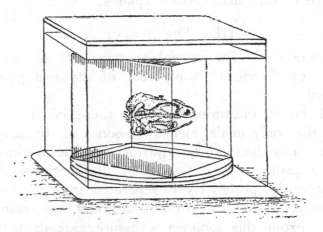

FIG. 15. Apparatus used in reconstructing.

made the image of the third section is exactly superimposed on the drawing of the second, and so on. In this way all the successive sections have the same register as the first, and when placed together in the form of a cube the magnified image of the object is formed.

The cube of glass plates is then transferred plate by plate to a cubical glass jar, the width of which is a little more than the diagonal of the plates, so that the cube of glass can just freely rotate within the square vessel. The vessel contains a mixture of fennel-oil and oil of cedar, which I have found to be sufficiently near the refractive index of the glass, and to have the advantage over clove-oil that it does not become dark with age.

The cube of plates is held in position by means of the glass turn-table shown in text-fig. 15. It consists of a circular disc of plate-glass, the diameter of which is very slightly less than the width of the vessel, so that it can rotate freely on the bottom of the vessel. A second and similar disc of glass

[1] *Quart. Journ. Micr. Science*, Vol. XLV. Part I. p. 6.

has four segments cut off it in such a way that when placed together and cemented to the first disc, so that their circumferences coincide, there is formed a trough which will exactly carry the cube of glass plates.

If they fit properly the plates must stand vertically, and the cube of plates can be rotated on the turn-table so that the reconstructed image can be seen from any aspect by an observer always looking normally at the surface of the fluid, *i.e.* the side of the square vessel.

No image, of course, can be seen when the cube is so turned that the observer looks between the surfaces of the plates, but with a very slight turning from this position the image at once comes into view.

If it is desired to reconstruct only a small series of sections, the remainder of the cube can conveniently be made up of squares of thick plate-glass (see text-fig. 15) in place of thin matt-surface squares.

III. The Skeleton.

The Chondrocranium.   The cranial anatomy of the adult *Polypterus* has been made known by Traquair[1], while that of the half-grown *Polypterus* was described by Pollard[2].

In this small larva, measuring 30 mm., there is as yet but slight traces of ossification, and that only in the membrane-bones of the head.   The primordial cartilaginous cranium may therefore be said to be at the height of its development.

The auditory capsule, though fused with the occipital and parietal region of the cranium anteriorly, posteriorly is separated from the exoccipital region by a very large foramen, giving exit to the xth and xith cranial nerves (Pl. X. fig. 4, X., XI.).   From this foramen a fissure extends between the auditory capsule and the supraoccipital region, and likewise a fissure between the auditory capsule and the basioccipital region.   The exoccipital cartilage, though fused dorsally with the supra- and ventrally with the basioccipital regions, is yet easy to recognize as a neural arch, which is becoming fused with the cranium (Pl. X. fig. 4, *Ex. occ.*).   The base of this neural arch has the same form and position as the base of the first true neural arch of the vertebral column, in front of which the first spinal nerve has exit (Pl. X. fig. 4, *Sp.i.*).

The basioccipital region does not completely envelop the anterior end of the notochord, but is composed of two halves (the parachordal cartilages) abutting on the two sides of the front end of this structure (Pl. X. fig. 2).   Posteriorly these two lateral masses of cartilage send wings ventrally, which meet and fuse below the dorsal aorta, enclosing it in a short canal which is roofed in by the notochord itself (Pl. X. figs. 2 & 4, *s.A.br.*).   Anteriorly to the bifurcation of the aorta, the basioccipital cartilages fuse below the notochord, sending forward a median narrow plate of cartilage which underlies the tip of the notochord

[1] *Journ. Anat. and Phys.* Vol. v. 1871.
[2] *Zool. Jahrb. Anat. u. Ont.* Bd. v. 1892.

(Pl. X. fig. 3, *s.Ch.br.*).   The fused auditory supra- and basioccipital regions form on either side an exceedingly massive cranial wall.   These walls are united dorsally by a fairly solid bridge of cartilage of shortest extent antero-posteriorly in the middle line (Pl. X. fig. 2, *s.oc.br.*).   Ventrally, however, they are only connected by the two little bridges of cartilage already mentioned—the one passing underneath the dorsal aorta behind its bifurcation, the other below the tip of the notochord in front of the bifurcation of the aorta.

In front of the tip of the notochord the floor of the cranium is formed merely of membrane, the bases of the lateral walls of the cranium being widely separated by a very large fontanelle in the posterior region of which lies the hypophysis[1].   The hypophysis is not at this stage enclosed in a special pocket of the cranial wall or "sella turcica," but curving backwards lies close under the hind end of the mid-brain (separated from it by thin membrane) and above the dermis of the roof of the mouth.   In a 9 cm. specimen the glandular part of the hypophysis is much larger in proportion and the membrane separating the hypophysis from the floor of the mid-brain has been converted into cartilage of considerable thickness, while the special cavity so formed remains in communication with the rest of the cranium by the opening for the passage of the infundibulum.

On the dorsal surface of the auditory region there extends forwards from the anterior end of the fissure separating the supraoccipital region from the posterior end of the auditory capsule a fairly pronounced ridge.   On the lateral surface, along the line of the horizontal semicircular canal, there is a very pronounced pterotic ridge bearing in its middle portion the articulatory surface for the head of the hyomandibular cartilage (Pl. X. figs. 1, 2, & 3, *Pt.r.*).   The front portion of this ridge gives off the sphenotic wing, which separating from the cranial wall passes forwards as a horizontally flattened bar of cartilage to fuse again with the cranial wall in the supraorbital region (Pl. X. figs. 1 & 3, *Sph.*).

On the ventral surface of the auditory region of the cranium, from the point in the middle line where the aorta bifurcates, on either side there extends forwards and outwards a deep groove in which runs the branch of the aorta, being the efferent artery from the hyoid external gill, and receiving a small branch from the first branchial arch.

Anterior to the bridge of cartilage uniting dorsally the two lateral masses of cartilage of the auditory region, the roof of the cranium is formed entirely of

---

[1] At this stage the glandular part of the hypophysis is very small and flattened against the dermis of the roof of the mouth; its tubules open anteriorly into a small cavity which is in communication with the mouth-cavity by a fairly wide duct, which persists until after the larva has grown to 9 cm. in length.   According to Bickford (*Anatomischer Anzeiger*, Bd. x. p. 469), this duct remains permanently open in *Calamoichthys*.   The nervous portion of the hypophysis has a very wide lumen and passing backwards ends in five finger-like processes.   In the 9 cm. specimen these have branched and become mingled with the tubules of the glandular portion posteriorly, but do not appear to be continued into them as stated by Waldschmidt (*Anatomischer Anzeiger*, Bd. 11. p. 318).

membrane to the front end of the cranial cavity, except for a delicate but continuous bridge of cartilage in the region of the pineal body (Pl. X. figs. 1 & 3, *Ep.br.*). This slender bridge expands towards the middle line into a horseshoe-shaped plate (Pl. XI. fig. 1, *Ep.br.*) lying over the vesicle of the third ventricle[1]. The pineal gland itself lies in the centre of the horseshoe, and is not covered by cartilage but lies in the dermis of the skin (Pl. XI. fig. 1, *Ep.*).

The lateral walls of the cranium extend forward from the auditory region on either side of the alisphenoid region as continuous vertical plates of cartilage, somewhat dumbbell-shaped in section and perforated by foramina for the iiird, vth, and viith nerves in the thinner middle portion (Pl. X. fig. 1, III., V., VII.). In the sphenethmoid region there is a large lateral fontanelle closed only by membrane through which passes the optic nerve, while above and below this membranous portion there pass the thickened upper and lower cartilaginous borders of the cranial wall, connecting on either side the ethmoidal region with the posterior region of the cranium (Pl. X. figs. 1 & 3). A section therefore passing through the middle of the orbital region cuts merely four bars of cartilage —two supraorbital bars separated from one another by a wide superior fontanelle, and two infraorbital bars (trabeculæ cranii) separated by a wide inferior fontanelle, the upper and lower of each side being separated by the large lateral fontanelle; the greater part of the "sphenethmoid" bone of *Polypterus* is therefore not preformed in cartilage.

In this region the section of the supraorbital bar of cartilage is V-shaped; the outer limb of the V is the anterior prolongation of the sphenotic wing, which is here fused with the supraorbital bar (Pl. X. fig. 3, *Sph.*). In the angle thus formed lies the supraorbital canal of the lateral line system (Pl. XI. fig. 1, *s.o.c.*). In front of the interorbital region the four bars of cartilage in the side walls of the cranial cavity expand and unite together, forming the anterior wall of the cranial cavity and being fused with the olfactory capsules (Pl. X. fig. 1). A canal for the passage of the olfactory branches of the vth and viith cranial nerves penetrates the outer portion of the supraorbital bar of cartilage at the point where it is fused with the olfactory capsule (Pl. X. fig. 1, *f.ol.*). Just anterior to this the capsule itself is penetrated by the same nerve. The olfactory capsules are very large and thin-walled, oval-shaped, and perforated anteriorly by a crescent-shaped nasal aperture (Pl. X. fig. 1). They are separated from one another to some extent anteriorly by a deep groove which ends above a spade-shaped rostral prolongation of the ventral portion of the nasal septum.

The ventral side of the olfactory capsule bears a facet for the articulation of the anterior end of the palato-quadrate bar of cartilage (Pl. X. fig. 1, *P.Qu.art.*).

---

[1] This vesicle is called by Waldschmidt the "epiphysis cerebri (pineal gland)"; in reality, however, it is the homologue of the median vesicle of *Lepidosteus* described by Balfour and Parker (*Phil. Trans. Roy. Soc. London*, Part II. 1882). The epiphysis itself lies over and behind this vesicle, and was apparently not found by Waldschmidt in *Polypterus* (*loc. cit.*).

The general form of the primordial cranium at this stage resembles very much that of the Selachii, especially the Scylliidæ, but in the absence of cartilage in the middle portion of the cranial walls, and the size of the great fontanelle, there is some similarity to the Urodela. The bridge of cartilage across the supracranial fontanelle seems to indicate a former roofing of the cranial cavity by cartilage, for, later, this bridge is represented by the small plate of cartilage noted by Pollard, and finally disappears. The absence of cartilage immediately over the pineal gland seems to indicate that the cranial roof was formerly perforated by a parietal foramen.

The Visceral Arches. The *palatoquadrate bar* is a pyramidal mass of cartilage, the anterior corner of which is drawn out into a vertically flattened bar, which anteriorly becomes almost circular in section, and finally flattened horizontally to form the articulation with the facet on the ventral side of the nasal capsule (Pl. X. fig. 1, *P.Qu.art.*). The quadrato-mandibular articulation is saddle-shaped, and it is in the region of this articulation that the palato-quadrate bar has its greatest transverse diameter (Pl. X. figs. 1, 2, 3, *P.Qu., Mk.*).

Behind this articulation the bar narrows to a point, which is concave vertically. This concavity rests upon the lower limb of the hyomandibular cartilage. The palatoquadrate bar receives no direct support from the cranial wall; the articulation is therefore hyostylic.

The *mandibular bar* has a similar pyramidal shape posteriorly, while its anterior portion forms a more regularly cylindrical rod than the palatoquadrate bar, and tapers to the symphysis with its fellow (Pl. X. figs. 1, 2, 3, *Mk.*).

The hyoidean arch, consisting of hyomandibular, stylohyal, ceratohyal, and hypohyal cartilages, articulates with the facet already mentioned on the underside of the lateral ridge of the auditory region (Pl. X. figs. 1, 2, 3, *Pt.r.*).

The *hyomandibular* consists of two rounded masses connected by a slender curved rod bearing upon its posterior convex surface the anterior end of a rod of cartilage segmented into two fairly equal portions, which constitute the skeletal axis of the base of the external gill (Pl. X. figs. 1, 2, 3, *Op.*). This rod persists as the small nodule of cartilage on the inner side of the opercular bone, which has been before noted by Van Wijhe and Traquair.

The *stylohyal* (Pl. X. figs. 1, 2, *St.Hy.*) is a rounded mass of cartilage which takes no part in the suspension of the jaws.

The *ceratohyal* (Pl. X. figs. 1, 2, 3, *C.Hy.*) is massive at either end, but constricted to a narrow bar in its middle portion.

The *hypohyal* (Pl. X. figs. 1, 2, 3, *H.Hy.*) is an almost spherical cartilage connecting the front end of the ceratohyal with that of the 1st branchial bar.

The *basibranchial* (Pl. X. figs. 2, 3, *B.Br.*) is a massive median cartilage on either side of which are articulated the enlarged lower ends of the four branchial bars. Each of these bars, in passing backwards and upwards, becomes very slender. Posteriorly these become slightly thickened again and curve inwards;

finally the upper end of the first branchial abuts upon the cranial wall just below the articulation of the hyomandibular, while the 2nd, 3rd, and 4th branchials tapering, curve downwards and end freely close under the auditory capsule (Pl. X. figs. 1, 2, 3, *Br.* 1, 2, 3, & 4).

Between the palatoquadrate and mandibular cartilage there lies a crescent-shaped labial cartilage with its convex surface forwards, the upper horn being more slender than the lower (Pl. X. figs. 1, 3, *Lb.*).

In the arrangement of the cartilages of the suspensorial apparatus the larval *Polypterus* exhibits a condition exactly intermediate between that of the hyostylic Selachians and the Teleostei, there being no very intimate union between the palatoquadrate and hyomandibular cartilage by means of a symplectic cartilage corresponding to that of Teleostei. The general correspondence of the arrangement with that of the larval Salmon on the one hand and the hyostylic Selachii on the other, and the relation of the spiracle to the hyomandibular in *Polypterus*, seem to me to go a long way towards dispelling doubts as to the homology of the suspensorium of the Teleostei with that of the hyostylic Selachii, such as have been entertained by some authors[1]. In this particular of course *Polypterus* differs widely from all living Amphibia, where the palatoquadrate bar has become fused with the cranial wall and the hyomandibular usually reduced.

On the other hand, the presence on the hyomandibular of *Polypterus* of a segmented rod of cartilage supporting the external gill is paralleled in the Aistopodous Stegocephali; for in *Dolichosoma longissimum*, according to Fritsch[2], there is found attached to the hyoid arch a skeletal structure which apparently supported an external gill. It is difficult, however, to make out whether this structure was attached to the periotic portion of the skull itself or to the upper end of the hyoid arch.

The Vertebral Column.

The *notochord* is at this stage of even calibre throughout the greater part of its length, tapering suddenly, however, anteriorly and gradually posteriorly towards its termination, where it shows a slight upward tendency (Pl. X. figs. 4, 5, 6, *Ch.*).

The structure of the notochord in *Polypterus* is very similar to that of *Lepidosteus* and the Teleosteans. The meshes between the large vacuoles enclose here and there a nucleus. The vacuoles towards the sheath become smaller and the nuclei more numerous, and pass rather suddenly into a definite nucleated epithelium lining the sheath of the notochord (Pl. XI. fig. 10, *Ch.ep.*). The sheath itself is very thick and almost structureless, being from $\frac{1}{10}$ to $\frac{1}{12}$ the thickness of the diameter of the vacuolated portion of notochord. A very delicate

[1] See H. B. Pollard, "The Suspension of the Jaws in Fish," *Anatomische Anzeiger*, x. p. 17. Also Parker and Haswell, *Text-book of Zoology*, Vol. ii. p. 202.

[2] *Fauna der Gaskohle und der Kalksteine der Permformation Böhmens.*

membrana elastica bounds the sheath externally (Pl. XI. fig. 10, *m.el.e.*) and separates it from the connective tissue surrounding the notochord with which the septa between the muscles are continuous.

At the junction of these septa with the notochordal sheath are found on either side three metameric series of cartilages—a dorsal row of stout pyramidal masses forming the bases of the neural arches with which they are continuous (Pl. X. figs. 4, 5, & Pl. XI. fig. 2, *n.pr.*); a lateral row of small pyramidal masses forming the foundations for the transverse processes and lateral ribs (Pl. X. figs. 4, 5, & Pl. XI. fig. 2, *l.pr.*); and the ventral row of still smaller nodules forming the foundations of the ventral series of ribs (Pl. X. figs. 4, 5, 6, & Pl. XI. figs. 2 & 10, *v.pr.*).

All these cartilages rest directly on the notochordal sheath, and in each metamere the three masses lie in nearly the same transverse plane, so that one section often passes through all three structures.

Neither the pyramidal bases nor their neural prolongations are fused across the mid-line, except in the tail-region. The bases are quite separate, while the upper ends of the two halves of the neural arch curve backwards, come into close contact with one another, and carry the lower end of the median dorsal spine (Pl. X. figs. 4, 5, 6, *d.sp.*).

In the anterior region of the vertebral column all three series are well-developed. In this region the lateral series are continued outwards along the septa, between the dorso-lateral and the ventro-lateral muscles, forming complete rib-like structures (Pl. X. fig. 4, *l.pr.*).

The ventral series are also well-developed and pass outwards between the ventro-lateral muscles and the kidneys (Pl. X. fig. 4, & Pl. XI. figs. 2 & 10, *v.pr.*). At this stage they do not in the anterior region reach the peritoneum.

In a section passing between the myomeres, and therefore through the bases of these cartilages, somewhat further back, it may be seen that the section cuts small nodules of cartilage at the extremity of the septa between the dorso-lateral and ventro-lateral muscles (Pl. XI. fig. 2, *o.l.r.c.*). These outer nodules lying near the nerve of the lateral line are not, however, connected with the cartilage of the central end of the septum. That is to say, the cartilage which forms the basis of the outer portion of the lateral ribs is developed independently of the cartilaginous basis of the transverse process.

In the middle region of the vertebral column both the lateral and ventral cartilages are small, while the dorsal cartilages of the two sides are seen to be placed in a slightly alternating position, so that a transverse section passing through the centre of a dorsal cartilage on the right side will pass through the posterior fringe of the dorsal cartilage of the left side (Pl. X. fig. 5, *n.pr.*).

In the caudal region the lateral series of cartilages are not found, while the ventral cartilages, though retaining their position, become the greatly enlarged hæmal arches. The first pair of enlarged ventral cartilages, however, do not unite ventrally, but each half bears at its extremity one of the last two interspinous cartilages of the anal fin[1] (Pl. X. fig. 6, *l.v.pr.*, *r.v.pr.*).   The two preceding interspinous cartilages of the anal fin are borne by similar rods of cartilage, which, however, do not unite with the ventral cartilage.

The most anterior cartilage of the anal fin is somewhat expanded antero-posteriorly, but is not supported by the hæmal cartilage.

Behind the anal fin the right and left ventral cartilages fuse together below the hæmal canal to form rod-shaped median cartilages, which expand distally and support directly the rays of the ventral fin of the diphycercal tail of *Polypterus* (Pl. X. fig. 6, *h.c.*).

In this region the right and left halves of the neural arches fuse together to form the dorsal spines (Pl. X. fig. 6, *d.sp.*), while wedged in between these are found the interspinous processes, which, expanding distally to trumpet-shaped masses of cartilage, support the rays of the dorsal fin of the tail.   There is no correspondence between the dorsal spines of the neural arches and these interspinous processes, for there may be one, two, or three of the latter intervening between two of the former.   More anteriorly these interspinous processes become much reduced and do not penetrate between the dorsal spines.   As the extremity of the tail is reached the neural and hæmal arches, together with the interspinous processes of the dorsal fin, become more and more oblique, lying almost parallel to the notochord (Pl. X. fig. 6, *d.sp.*, *l.sp.*).   Finally there lies upon the dorsal, and also upon the ventral, side of the notochord a flattened mass of cartilage, which appears to be composed of the rudiments of several neural and hæmal arches (Pl. X. fig. 6, *f.n.pr.*, *f.v.pr.*).   The notochord extends beyond this incomplete cartilaginous tube, a bare filament embedded in the fin-rays of the tail.

At this stage there is no trace of any intervertebral cartilages, neither is there any trace of a ring of cartilage enclosing the intervertebral regions of the notochord.

In older specimens of 9 cm. length considerable deposits of bone have appeared, producing important changes in the structure of the vertebral column and its processes.

In the vertebral regions the cartilaginous bases of the neural arches and lateral processes are still seen, though now surrounded by bone, deposited in the connective tissue adjoining the notochord (Pl. XI. figs. 3, 4, 8, *n.pr.*, *l.pr.*). The ventral cartilages have disappeared from the neighbourhood of the notochord. The bony deposits which have enclosed the lateral cartilages and formed the

---

[1] I have satisfied myself that my description of the way in which the anal fin is supported in this specimen is correct, but I am not quite sure that this will be found to be the normal arrangement.   It seems, at any rate to me, to be an additional proof that the hæmal arches are formed from the bending downwards and uniting of originally separate ventral ribs, which have secondarily assumed the function of supporting the anal fin.

transverse processes have caused the proximal ends of the ventral cartilages to recede outwards, and these are now found attached loosely to the underside of the transverse processes below their attachment to the lateral bony ribs, which have now become formed in the connective tissue of the septa between the dorso-lateral and ventro-lateral muscles (Pl. XI. figs. 3, 4, 7, *tr.pr.*). The ventral cartilages themselves form the heads of the ventral bony ribs which have now become formed at the junction of the septa between the successive ventro-lateral muscles with the peritoneum (Pl. XI. figs. 4, 7, *v.pr.*).

The shifting outwards of the heads of the ventral ribs or hæmal arches with the development of the large transverse bony processes of the vertebræ has resulted in allowing space for the greatly enlarged kidneys.

It will be seen that the heads of these ventral ribs do not now lie quite in the same transverse plane as the transverse processes, but crossing them diagonally project towards the muscles in front of the transverse processes (Pl. XI. fig. 7). In the full-grown *Polypterus* the heads of the ventral ribs articulate with the intervertebral regions of the vertebral column, that is, between two successive vertebræ, and it might be thought that this was their original place of attachment. It will, however, be understood, from the above description of their mode of growth, that they are originally attached to the ventral surface of the notochord in the same transverse plane as the bases of the transverse processes, that is, *vertebral* in position, and that they later lose their connection with the notochord, become twisted across the ends of the transverse processes, and with the subsequent development of the bony centra become attached again to the vertebral axis, and are finally *intervertebral* in position.

The centra themselves are at this stage very thin and hourglass-shaped, expanding widely anteriorly and posteriorly; they are not formed previously in cartilage (Pl. XI. fig. 8, *c.*) except for the small neural and lateral cartilages which they enclose; they directly surround the chordal sheath. There is, however, a single layer of cells separating them from the latter in the position where the membrana elastica externa should be, which somewhat resemble cartilage-cells; they do not, however, stain in the very characteristic way in which all other cartilage in this specimen has stained (Pl. XI. fig. 6, *m.c.*). The cartilages of the neural arches play but a very small part in the formation of the bony neural arches, which are also formed directly in connective tissue.

In specimens 1.3 cm. long the vacuolated portion of the notochord has become constricted vertebrally, while the sheath of the notochord, retaining a considerable thickness in the vertebral region, intervertebrally becomes greatly thickened (Pl. XI. fig. 6, *Ch.sh.*). It is this thickening of the notochordal sheath which is the chief cause of the great expansion of the bony vertebræ in the intervertebral regions. At this stage all trace of cartilage has disappeared from the composition of the vertebræ (Pl. XI. fig. 5).

The formation of the vertebral centra in *Polypterus* is very much on the same lines as that of *Lepidosteus* and the Teleostei. It differs from *Lepidosteus* in the absence of the intervertebral rings of cartilage which give rise to the procœlous articulations in that form, and from *Amia* in the absence of intercalated cartilages. In the possession of three pairs of vertebrally placed cartilages resting upon the notochordal sheath before the commencement of bone formation, *Polypterus* differs from all living Vertebrates.

That *Polypterus* is provided with a double set of ribs has long been known; and by some authors the upper series has been homologized with the ribs of Elasmobranchii, Amphibia, and Amniota, and the lower series with the ribs of the Dipnoi, Ganoidei, and Teleostei. Göppert especially[1] has developed this view regarding the reduction of the lower series in the former groups as due to the suppression of the ventro-lateral muscles. The starting-point which Göppert took was the condition in *Calamoichthys* at the stage where the lower ribs are attached to the under sides of the upper ribs shown in Pl. XI. fig. 4 of *Polypterus*. The discovery that all the lower ribs in *Polypterus* have their origin in a ventral series of cartilages resting directly on the notochordal sheath seems to me to very much strengthen his views. The absence of intercalated cartilages, such as are found in the majority of Vertebrates exclusive of the Teleostei or of the intervertebral cartilage-rings of *Lepidosteus*, may possibly be connected with the early and great development of membrane-bone which is so characteristic of the development of all the skeletal structures of *Polypterus*.

The Shoulder-girdle and Pectoral Fins. The cartilaginous shoulder-girdle consists of a compact mass of cartilage of considerable thickness anteriorly, and, though thinning out ventrally, of no great extent in a dorso-ventral direction (Pl. X. fig. 7). Posteriorly, it is prolonged as a thin plate of cartilage, towards the hinder border of which is the great coracoidal foramen (Pl. X. figs. 7 & 8, *Cor.f.*). Above this foramen is situated the protuberance bearing the convex facet which articulates with the proximal ends of the pro- and metapterygial cartilages of the pectoral fin. From behind, this protuberance appears as a distinct short rod of cartilage attached only at its proximal end to the flattened portion of the shoulder-girdle (Pl. X. fig. 7). In reality, however, it is continued forwards into the anterior thick portion of the girdle as a thickened ridge.

The pectoral fin of *Polypterus* is one of its most characteristic features, being unlike that of all other Vertebrates except its congener *Calamoichthys*. It is therefore a question of considerable interest as to what is the relation of this fin to other known types, and how it has been formed. Upon this question two views have been held:—(1) The fin of *Polypterus* has been derived from a biserial archipterygium of the type of *Ceratodus* by the shortening of the axis and the shifting of the proximal radials up the axis to form the meta- and propterygia. This view has been mainly held by Klaatsch[2] and Gegenbaur[3],

[1] *Morph. Jahrb.* Bd. XXIII.
[2] *Festschr. von C. Gegenbaur*, Part I. p. 261 (1896).   [3] *Jen. Zeitschr.* VII. 1873.

though Gegenbaur appears now[1] to derive it directly from forms with a very short axis. (2) The fin of *Polypterus* has been derived from a uniserial fin of the type of *Heptanchus* or, perhaps better, *Chlamydoselache* by the shortening of the metapterygial border and the fusion of the bases of the radials to form the expanded mesopterygium.

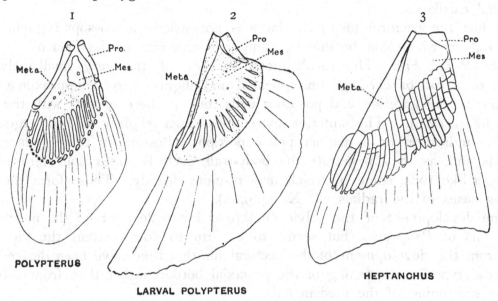

FIG. 16. Skeleton of pectoral fin of *Polypterus* and *Heptanchus*.

Pectoral fin of—1. Adult *Polypterus*.
2. Larval „
3. *Heptanchus*.

Meta. Metapterygium. *Mes.* Mesopterygium. *Pro.* Propterygium.

Now the development of the fin might be expected to throw some light on this question, and in the 30 mm. larva this is what we find:—The general shape of the fin is quite different from that of the adult. The shape of the fin-blade is triangular, the metapterygial border forming with the distal border an acute angle, while the propterygial border forms with the distal border an obtuse angle (text-fig. 16, 2).

The cartilaginous skeleton of the pectoral fin differs in many important points from the skeleton of the fin of the adult *Polypterus*. The propterygium extending along the preaxial border of the fin is stout but short, being scarcely more than one-third the length of the metapterygium extending along the postaxial border, and being situated almost at right angles to the latter (Pl. X. fig. 8, *Pro.*). The angle between these two rods is filled by a flattened blade of cartilage, which is, however, slightly thickened where it is in contact with the metapterygium (Pl. X. fig. 8, *Mes.*). Distally this flat blade or mesopterygium is split in a fan-like manner to form the cartilaginous bases for the radials. The distal ends of these re-unite to form a continuous border of cartilage. The mesopterygial blade is irregularly punctured for the passage of blood-vessels (Pl. X. fig. 8, *b.v.f.*). On the free edge of the metapterygium, at its distal end, is a slight flange of cartilage, seemingly forming a rudimentary continuation round the distal end of the radial cartilages.

In fact the pectoral fin of the larva is not strictly a crossopterygium, but is a uniserial fin which can be directly compared with that of the Selachii.

The Pelvic Fin. The cartilaginous skeleton of the very small pelvic fin consists of a rod of cartilage unsegmented and slightly curved, approximating to its fellow at the anterior and posterior ends, but nowhere uniting with the latter (Pl. X. fig. 9, *Bas.*). The anterior extremities taper slightly and diverge, there being no indication of a distinct pelvic portion. Posteriorly these rods expand outwards and, as they penetrate the body-wall, each becomes split into four or five finger-like rods, some of which may re-unite distally. These form the cartilaginous bases of the radials (Pl. X. fig. 9, *r.*).

The development of the pelvic fin throws but little light on the morphology of the fins of *Polypterus*, but seems to confirm to some extent the view here taken from the development of the pectoral fin, that they have been derived from a uniserial type by shortening of the postaxial border, rather than from a biserial type by shortening of the median axis.

IV. URINOGENITAL ORGANS.

The excretory system of this larva is of especial interest, for here we find that a regular metamorphosis is taking place. The larval pronephros, though still functional, shows signs of degenerating, while the mesonephros is still fairly simple in its arrangement, and in some respects is still in a larval condition.

The pronephros consists of a rather small glomus almost filling the pronephric chamber which lies close by the aorta (Pl. XI. fig. 9, *Gl.*); from this a narrow tube embedded in lymphoid tissue, the pronephric duct, passes outwards dorsal to the posterior cardinal vein and towards the body-wall (Pl. XI. fig. 9, *pn.d.*); on passing between the dorsal and ventral lateral muscles it suddenly dilates, turns headwards, loops round the brachial nerve, and, in a mass of lymphoid tissue lying in the angle between the anterior ends of the dorso-lateral and the ventro-lateral muscles and the skin, becomes greatly convoluted. In the successive convolutions the duct becomes narrower and narrower, occasionally becoming dilated and then suddenly narrower again; finally there leaves the convoluted mass a very fine duct (Pl. XI. fig. 9, *s.d.*), and this, running along the outer side of the posterior cardinal vein, passes backwards and inwards until it comes to lie just over the peritoneum. Two segments behind the glomus of the head-kidney it gives off the first kidney-tubule, and thence passes straight to the hind end of the body-cavity as the segmental duct, ever increasing in size until it meets its fellow to form the urinary sinus (text-fig. 17, *u.s.*).

The smallness of the pronephric chamber and the dilatation of the convoluted portion of the pronephric duct must be regarded as signs of degeneration. In fact the connection between the pronephros and mesonephros could not be traced on the left side, while on the right side it was done with the utmost difficulty. The glomus itself on the left side was very small, and the anterior end of the coiled pronephric duct could not be traced all the way to it.

On the right side there appeared to be traces of a peritoneal funnel leading to the pronephric chamber (Pl. XI. fig. 9, *P.f.*); but with the thick sections into which I had cut this region (10 μ) I was unable to satisfy myself that it actually opened into this chamber.

In specimens 9 cm. in length all trace of the pronephros had disappeared, there remaining in its place a small mass of lymphoid tissue.

When we compare the pronephros of *Polypterus* with that of the Ganoids and Teleosteans which have been described, there can be no hesitation in saying that it resembles most closely the pronephros of *Amia* described by Jungersen[1], where there is one pronephric chamber, one peritoneal funnel, and one opening of the pronephric duct into the pronephric chamber. To draw a close comparison between the pronephros of *Amia* and that of *Polypterus* is not possible, seeing that the stage I have described is undoubtedly only the last stage in the history of the pronephros.

Acipenser[2] has a very long pronephric chamber, with at first six, later four, funnels leading from it.

Lepidosteus, according to Beard[3], has also at first a long chamber with three funnels, later a shorter one with two, and finally, according to Balfour and Parker[4], only one funnel leads from the pronephric chamber.

It is of course conceivable that at an earlier stage *Polypterus* has more than one funnel leading from the pronephric chamber; but I think that this is not probable, seeing how widely the glomus is separated from the convoluted portion of the duct.

Lebedinsky[5] has described the pronephros of *Calamoichthys* as mingled with the anterior tubules of the mesonephros, somewhat as Semon describes the pronephros in *Ichthyophis*[6]. Seeing, however, that his "larvæ" measured 12 cm. and 15 cm. respectively, I think it would be surprising if in these young *Calamoichthys* there should remain traces of a pronephros which in *Polypterus* is already degenerating at 3 cm., and has completely disappeared at 9 cm. It is still more surprising that the larval head-kidneys of two creatures so closely allied as *Polypterus* and *Calamoichthys* bear no resemblance one to another. But it is not

---

[1] *Zool. Anz.* 1894, p. 246.
[2] Jungersen, *Zool. Anz.* 1893, pp. 464, 469.
[3] *Anat. Anz.* Bd. x. p. 94.
[4] *Loc. cit.*
[5] *Jen. Zeitschr.* 1891.
[6] *Arch. mikr. Anat.* XLIV. p. 216.

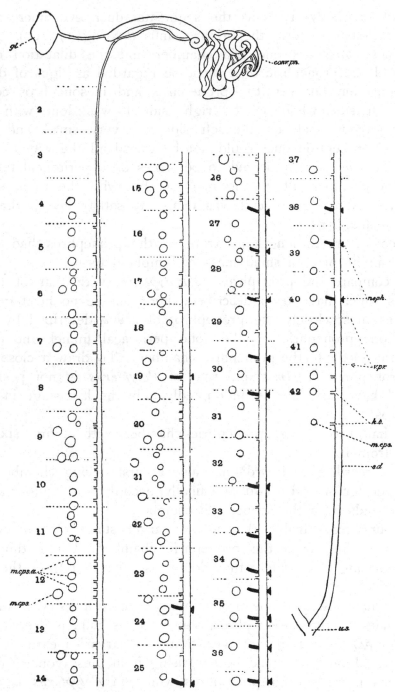

FIG. 17.   Diagram of kidney-system of larval *Polypterus*.

Projection from transverse sections of entire pronephros and mesonephros of the right side of
the larva of the female *Polypterus*, × 60.   The pronephros is a dorsal view of the same
reconstruction as that figured in Pl. XI. fig. 9.

gl. Glomus of pronephros.	*m.cps.* Glomerulus.
conv.pn. Convoluted pronephric duct.	*m.cps.a.* Accessory glomeruli formed between
neph. Nephrostome.	the original glomeruli and the seg-
v.pr. Position of ventral process marking	mental duct.
the metameres.	*s.d.* Segmental duct.
k.t. Kidney-tubule.	*u.s.* Urinary sinus.

clear to me by what means the author of this account has distinguished the tubules, "outer and inner funnels," &c. of the pronephros from those of the mesonephros. I am bound to conclude therefore that what Lebedinsky has described as "Vorniere" in *Calamoichthys* are some of the anterior tubules of the mesonephros, possibly accessory undeveloped ones, and that the pronephros ("Vorniere") has already disappeared.

At this stage the mesonephric tubules are already becoming complicated in the fore part of the body. There are from two to five glomeruli to each body-segment, and about the same number of tubules opening into the segmental duct (text-fig. 17). The accessory glomeruli and tubules arise close to the segmental duct, while the older formed ones become displaced and are found nearer the dorsal aorta towards the middle line (text-fig. 17, *m.cps.*). In the hinder half of the mesonephros the glomeruli become fewer and fewer, and the arrangement of the tubules simpler, while to each body-segment there is now found one glomerulus and one tubule opening into the segmental duct (text-fig. 17, *k.t.*). Throughout this portion of the body there is found lying immediately to the outer side of the genital ridge (ovary) and ventral to the segmental duct a strip of thickened peritoneal epithelium (Pl. XII. figs. 1, 2, 3, 4, 5, 6, 7, 8, *ov.d.*); this will eventually form the dorsal wall of the oviduct. *From the twenty-fourth to the fortieth segment there is found opening upon this thickened ridge in each segment a nephrostome communicating with the glomerulus* (Pl. XII. figs. 3, 4, *neph.*, & text-fig. 18, 1, p. 173).

Section I. (Pl. XII. fig. 1) is taken through the twenty-fourth segment, and shows the thickened strip of peritoneum in section lying to the outer side of the mesentery of the genital ridge (ovary) and the more complicated arrangement of tubules.

Section II. (Pl. XII. fig. 2) is taken through the twenty-ninth segment, and shows the thickened ridge (*ov.d.*) in section lying to the outer side of a small projection which here represents the genital ridge (*g.r.*). It also shows the accessory tubules on either side and the accessory glomerulus of the right side (*k.t.a., m.cps.a.*). The arrangement of the older tubules is here less complicated.

Section III. (Pl. XII. fig. 3) is through the thirtieth segment, and shows on the left side a nephrostome opening upon the thickened ridge and communicating with the glomerulus (*neph., m.cps.*). On the right side accessory tubules are shown (*k.t.a.*).

Section IV. (Pl. XII. fig. 4) is through the thirty-fourth segment; on the left side the edge of a nephrostome and also of a glomerulus is cut through (*neph., m.cps.*), while on the right side is shown a single large glomerulus from which a kidney-tubule leads (*m.cps.*).

Section V. (Pl. XII. fig. 5) passes through the thirty-seventh segment, and shows the openings of two kidney-tubules into the right and left segmental duct (*k.t.*). The thickened ridge still persists, though the genital ridge does not extend so far back.

Section VI. (Pl. XII. fig. 6) is very near the hind end of the body-cavity behind the last kidney-tubules. The thickened ridge of peritoneum is still seen lying under the segmental duct (*ov.d., s.d.*).

Section VII. (Pl. XII. fig. 7) is right at the end of the body-cavity; the segmental ducts are now turning ventralwards and approaching each other (*s.d.*), while the thickened ridge still lies just ventral to them (*ov.d.*).

Section VIII. (Pl. XII. fig. 8) passes just in front of the vent, behind the point of union of the segmental ducts. The body-cavity is seen here to be prolonged backwards in two portions on either side divided by a septum (*b.c.*); in the upper portions is still seen the thickened ridge of peritoneum (*ov.d.*). This upper portion is the commencement of the folding-off of the oviduct from the body-cavity by the forward growth of the septum, and it is at the point *x* that the oviducts eventually open into the urinogenital sinus. The lower portion of the body-cavity ends posteriorly at the point where the abdominal pores later break through (*a.p.*).

The further stages in the folding-off of the oviducts from the body-cavity I have been unable to observe, as in my next stage (9 cm.) the formation of the oviducts is already complete, though their posterior ends still end blindly in the wall of the urinary sinus. It is hardly possible to doubt, however, that the thickened ridge which we have traced through the posterior portion of the body-cavity forms eventually the dorsal wall of the oviduct. That is to say, we have, in the *Polypterus* larva, a row of nephrostomes opening into that portion of the body-cavity which will eventually become the genital duct.

In *Lepidosteus*, Balfour and Parker[1] found that the hollow ovary was formed by the ventral edge of the ovary growing round and fusing with the peritoneum, thus including a portion of the body-cavity. Into this cavity in the larval *Lepidosteus* they found nephrostomes opening; but Balfour says that he could not observe how the hind ends of the oviducts were formed, but admitted that if there should be found nephrostomes opening into this portion of the duct, it would prove that the oviducts of *Lepidosteus* are not true Müllerian ducts. Now this is precisely what we have in *Polypterus*. Jungersen[2] has shown that the ovary of many Teleosteans is formed by the inclusion of a portion of the body-cavity much as Balfour and Parker described the process in *Lepidosteus*, but, so far as I am aware, no one has found nephrostomes opening into the ducts of the ovary. It seems fairly certain, then, that in the female *Polypterus* the genital duct is not a Müllerian duct, that is, is not developed from the segmental duct.

In a previous paper[3] I described the genital duct in the male *Polypterus*, and gave reasons for regarding the genital ducts in the male and female as

[1] *Loc. cit.*   [2] *Arb. Inst. Würzburg*, Bd. IX. 1889.   [3] p. 101.

homologous structures, showing that in young specimens 9 cm. in length the hinder ends of the ducts in the two sexes had a very similar appearance, but I was unable to show their actual origin. I have now indicated the origin of the genital duct in the female.

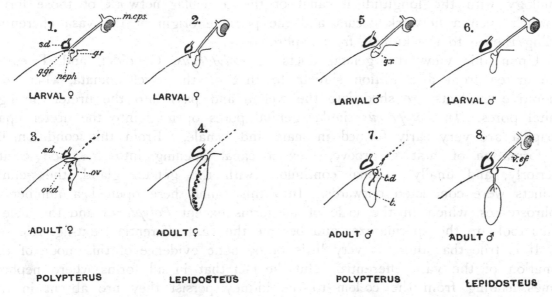

FIG. 18. Series of diagrams illustrating the suggested homologies of the genital ducts and vasa efferentia in *Polypterus* and *Lepidosteus*.

neph., nephrostrome of tubule passing to glomerulus; *m.cps.*, glomerulus; *g.gr.*, thickened strip of peritoneum or genital groove; *g.r.*, genital ridge; *s.d.*, segmental duct; *t.d.*, testis duct; *t.*, testis; *ov.*, ovary; *v.ef.*, vas efferens; *ov.d.*, oviduct.

The dotted line in 3, 4 and 7 indicates that the nephrostomes no longer persist. 5 and 6 indicate a condition believed to exist, but not yet proved.

On re-examining my sections of the young male 9 cm. in length, I find an indication that the posterior part of the genital duct is formed in the same way, namely, by a folding-off of a groove in the cœlomic wall on the outer side of the genital ridge (Pl. XII. figs. 9, 10, *g.d.*).

Now this is precisely the region where in the female I have found the nephrostomes opening, and I can but suppose that the larval male possesses them too in early life. That is to say, in all probability, in the larva of the male *Polypterus* there will be found nephrostomes opening into that portion of the cœlom which becomes folded off to form the male genital duct. If, then, nephrostomes were retained in the adult, we should have formed a series of tubules connecting the genital duct with the glomeruli of the kidney; and if the genital duct ceased to perforate the wall of the urinary sinus the genital products could find their way out through the kidney, and the nephrostomes would be converted into vasa efferentia, similar to those of *Lepidosteus* and other Ganoidei, of Elasmobranchii, Dipnoi, Amphibia, and Amniota.

I suggest, then, that since I have previously given reasons for supposing that the genital ducts in the male and female *Polypterus* are homologous, the discovery of these nephrostomes opening into the female duct precludes the possibility of their being homologous with Müllerian ducts, but indicates a possible homology with the longitudinal canal of the testicular network of those forms in which such a network occurs, and the possible origin of the vasa efferentia, leading thence to the kidney, from nephrostomes.

Upon this view the genital ducts of *Polypterus*, Ganoidei, and Teleostei have arisen from a condition similar to that of the Cyclostomata, where the generative products are shed into the cœlom and pass into the ureter through genital pores. In *Polypterus* similar genital pores opening into the ureter upon a papilla are very early formed in male and female. From this condition, by the formation of first a groove, then a canal opening into the body-cavity anteriorly, and finally a canal continuous with the genital gland, the genital products were conducted outwards. Into this canal there opened a number of nephrostomes, which in the male of all forms except *Polypterus* and the Teleosteans took up the spermatozoa and became the vasa efferentia (text-fig. 18).

It is true that there is very little ontogenetic evidence of this mode of the formation of the vasa efferentia. But the fact that in all forms where nephrostomes leading from the cœlom to the kidney persist they are absent in the region of the testicular network, would fit in with the view that the nephrostomes in this region have been used to form the vasa efferentia.

Jungersen in his work on the genital ducts of Teleosteans[1] cites evidence brought forward by Siebold and by Brock, that in certan Teleosteans in which the testes are hollow, the cavity of the testis, and therefore probably also of the testis-duct, is formed by a folding-off of a portion of the cœlom.

Jungersen came to the conclusion that in all Vertebrata except Elasmobranchs the oviducts were homologous and formed by a folding-off of a groove in the cœlomic wall. If this conclusion is correct, then there is a difficulty in the way of the view I have put forward which it is not easy to overcome. In many forms which possess a testicular network there is also found, in the male, a more or less rudimentary female genital duct usually called a Müllerian duct. It seems unlikely, then, to say the least of it, that, in these forms, the longitudinal canal of the network in the male and the genital duct of the female can be homologous, for these structures occur side by side in the males.

If the ducts of Ganoidei and Teleostei really are homologous in the male and female, while in the other Gnathostomata they are not homologous, then either the female duct or the male duct cannot be homologous throughout the Gnathostomata.

It can, I think, hardly be doubted that the male and female ducts in Teleostei are homologous. I have before given reasons for thinking that they are

[1] *Loc. cit.*

so in *Polypterus*. It remains, then, to decide whether it is more likely that the male duct in *Polypterus* is homologous with the longitudinal canal of *Lepidosteus* and so with that of other forms in which a testicular network occurs, or that the female duct in Ganoidei, Crossopterygii, and Teleostei is homologous with the female duct of other Gnathostomata, as held by Jungersen.

The male duct of *Polypterus* lies in exactly the same position as the blind longitudinal canal of *Lepidosteus*, and it is difficult to avoid the conclusion that these ducts are homologous with one another, and again that this canal in *Lepidosteus* is the longitudinal canal of the testicular network of other forms.

Jungersen[1] admits this comparison, but derives the male duct of *Polypterus* and Teleostei from the fusion of a testicular network with the loss of its connection with the kidney, which he, with Semon[2] and Kerr[3] and many other authors, regards as the primitive condition. But Jungersen[4] also holds that the female ducts are homologous throughout the Gnathostomata. The presence of the rudimentary Müllerian duct in many forms with a testicular network makes it impossible on this view to regard the ducts of male and female in any vertebrates as homologous structures, as they appear to be in Teleostei.

Supposing, however, that the female ducts in *Polypterus*, Ganoidei, and Teleostei are not homologous with those of Elasmobranchii and Amniota, as their position in the hind end of the body-cavity, and the presence in *Polypterus* and *Lepidosteus* of nephrostomes opening into them, makes very probable, then we may regard the male and female ducts of Cyclostomata, Crossopterygii, Ganoidei, and Teleostei as homologous, and the female ducts of other Gnathostomata as independent structures, *i.e.* Müllerian ducts. It is now known that these Müllerian ducts arise as a splitting off from the archinephric duct only in Elasmobranchs, but this difference in the mode of origin of the oviducts is not considered by many authors to constitute a proof that they are not homologous. The peculiarity of the oviducts of these groups (Elasmobranchii, Dipnoi, Amphibia, and Amniota) in tending to appear in the males as more or less rudimentary structures, seems to me to be a strong argument in favour of their being homologous in these groups; and it is very noticeable that these rudiments are usually, if not invariably, absent in the Crossopterygii, Ganoidei, and Teleostei, those groups, namely, in which the oviduct in the female appears not to be homologous with the Müllerian ducts but rather with the genital ducts of the male.

Summary of the Structure and Development of the Urinogenital Organs.

The pronephros of *Polypterus* is beginning to degenerate in the larva 30 mm. in length. At this stage it consists of a small pronephric chamber containing the glomus and of a duct leading from the chamber to pass to and become the

---

[1] *Zool. Anz.* Bd. XXIII. No. 617.   [2] *Jen. Zeitschr.* 1891.
[3] *Proc. Zool. Soc.* 1901, Vol. II. p. 493.   [4] *Arb. Inst. Würzburg*, Bd. IX. 1889.

duct of the mesonephros, and consisting of three portions—a very narrow portion leading outwards, a dilated and much convoluted portion lying close under the skin, and a very narrow portion leading inwards and backwards to become the segmental duct of the mesonephros.

The pronephros of *Polypterus* resembles very much that of *Amia*.

The mesonephros anteriorly is at this stage becoming complicated. Posteriorly it has still a segmental arrangement, each segment being provided with one Malpighian corpuscle with its tubule leading to the segmental duct and its peritoneal funnel. The peritoneal funnels open upon the rudiment of the oviduct, which is not yet folded off from the cœlom. The genital pores do not yet penetrate the walls of the urinary sinus.

From examination of somewhat older specimens, the male and female ducts appear to arise in a somewhat similar manner, and it is regarded as probable that in the young male larva the peritoneal funnels open upon the rudiment of the male duct. It is suggested that these peritoneal funnels, if they persisted, would then answer to the vasa efferentia passing from the longitudinal canal to the Malpighian corpuscles of *Lepidosteus*.

The male and female ducts of Crossopterygii and Teleostei are regarded as homologous structures, which in the males of other Gnathostomata have been gradually absorbed in the testicular network, and in the females of other Gnathostomata except the Ganoidei are replaced by Müllerian ducts.

Whatever be the true interpretation of the genital ducts in the Ichthyopsida, the Crossopterygii, Teleostei, and Ganoidei are closely united in this respect, the latter group appearing to stand in closer relationship to the rest of the Gnathostomata than the former groups.

V. Conclusion.

We have seen, then, that in its development *Polypterus* shows affinities in various directions. The development of the pectoral fin and of the cranium and visceral arches has some resemblance to that of the Elasmobranchii, while of course the spiral valve, the conus and optic chiasma in the adult are other structures common to the two groups. The development of the vertebral column is not very dissimilar to that of the Teleostei, though in the possession of lateral cartilaginous processes in addition to the neural and ventral processes of the Dipnoi, Ganoidei, and Teleostei, *Polypterus* presents a very generalized condition; while there can be no doubt that from the standpoint of urinogenital organs both the larval and adult *Polypterus* are as closely connected with the Teleostei as even *Amia* itself. It has, however, been admitted by the most competent palæontologists that the structure of the dermal bones of the head and of the shoulder-girdle of *Polypterus* is so like that of certain Stegocephali, that it must be regarded as more than a mere resemblance, while there are many points in the development of the skeleton that distinctly approach the condition of the Amphibia. The only possible interpretation of these facts appears to me to be that the living Crossopterygians form a central group among recent forms having some characters in common with most of the great groups. Seeing that *Polypterus* is thus a generalized form and is not very different from the ancient Devonian Crossopterygii, it seems to me not improbable that the particulars of structure in which it resembles the admittedly recent Teleostei are in both groups of a primitive nature. I do not pretend that upon this point I feel very certain, but I think I have put forward some evidence in favour of the view I have taken.

It is with reluctance that I publish a piece of work the material for which consisted of a single specimen; but as there seemed to be no prospect in the immediate future of an abundant supply of the early stages in the development of *Polypterus* being obtained, and as upon examination I found that there were points of considerable interest in this small larva, it seemed only right that they should be known. I trust, however, that when more material comes to hand it will be found that what has been seen in this larva has been accurately described.

VI. Lettering of the Figures.

ad.	Adipose tissue.
al.	Alimentary canal.
An.f.	Anal fin.
Ao.	Dorsal aorta.
Ao.s.	Aortic supports.
a.p.	Abdominal pore.
Au.	Auditory capsule.
Bas.	Basipterygium.
b.c.	Body-cavity.
B.Br.	Basibranchial cartilage.
Br. 1, 2, 3, 4.	Branchial cartilages 1, 2, 3, 4.
Br.n.	Brachial nerve.
b.v.f.	Blood-vessel foramen.
c.	Centrum.
Ch.	Notochord.
Ch.ep.	Chorda epithelium.
Ch.sh.	Sheath of notochord.
C.Hy.	Ceratohyal.
conv.pn.d.	Convoluted pronephric duct.
Cor.	Coracoid.
Cor.f.	Coracoidal foramen.
d.Ao.	Dorsal aorta.
d.l.m.	Dorso-lateral muscles.
d.sp.	Dorsal spines.
eg.af.	Afferent artery to external gill.
eg.ef.	Efferent artery to external gill.
Ep.	Epiphysis.
Ep.br.	Bridge of cartilage in region of epiphysis.
Ex.occ.	Exoccipital cartilage.
f.n.pr.	Fused neural processes.
f.ol.	Foramen for olfactory branches of vth and viith nerves.
f.v.pr.	Fused ventral processes.
G.X.	Vagus ganglion.
g.d.	Genital duct.
Gl.	Glomus.
g.r.	Genital ridge.
gr.s.o.c.	Groove for the supraorbital canal.
h.c.	Hæmal cartilage.
H.Hy.	Hypohyal.
Hy.M.	Hyomandibular.
i.s.	Intermuscular septum.
k.	Kidney.
k.t.	Kidney-tubule.
k.t.a.	Accessory kidney-tubule.
Lb.	Labial cartilage.
l.bl.	Left swim-bladder.
l.pr.	Lateral process.
l.r.	Lateral rib.
l.sp.	Interspinous cartilages.
l.v.pr.	Left ventral process.
m.cps.	Glomerulus.
m.cps.a.	Accessory glomerulus.
m.c.	Cartilage-like cells of membrana elastica externa.
m.el.e.	Membrana elastica externa.
Mes.	Mesopterygium.
Meta.	Metapterygium.
Mk.	Mandibular cartilage.
m.v.	Median vesicle of brain.
n.a.	Neural arch.
neph.	Nephrostome.
n.pr.	Neural process.
n.sp.	Neural spine.
o.l.r.c.	Outer cartilages of the lateral ribs.
Op.	Operculum.
ov.	Ovary.
ov.d.	Oviduct.
Pa.	Pallium.
p.c.v.	Posterior cardinal vein.
P.f.	Peritoneal funnel (traces of).
pn.c.	Pronephric chamber.
pn.d.	Pronephric duct.
Pr.	Rudimentary continuation of radials.
Pro.	Propterygium.
P.Qu.	Palato-quadrate cartilage.
P.Qu.art.	Palato-quadrate articulation.
Pt.r.	Pterotic ridge.
r.	Cartilaginous bases of radials.
r.bl.	Right swim-bladder.
rect.	Rectum.
r.v.pr.	Right ventral process.
s.A.br.	Subaortic bridge.
Scp.	Scapula.
s.Ch.br.	Subnotochordal bridge.
s.d.	Segmental duct.
s.o.c.	Supraorbital canal.
S.occ.	Supraoccipital region.
s.oc.br.	Supraoccipital bridge.
Sph.	Sphenotic.
Sph.Eth.	Cartilage of sphenethmoid region.
Sp.i.	First spinal nerve.
st.	Stomach.
St.Hy.	Stylohyal.
T.	Testis.
Tr.	Trabeculæ cranii.
tr.pr.	Transverse bony process.
u.s.	Urinary sinus.
V.Ao.	Ventral aorta.
v.l.m.	Ventro-lateral muscles.
v.pr.	Ventral process.
v.r.	Ventral rib.
x.	Point where oviduct will open into urinary sinus.
II., III., V. &c.	Positions of exit of cranial nerves.

PLATE X.

EXPLANATION OF PLATE X.

FIG. 1. A side view of a reconstruction from transverse sections, 10 μ in thickness, of the right half of the primordial cranium and visceral arches of the larval *Polypterus*. The great afferent and efferent blood-vessels to the external gill are also shown. The last section of the reconstruction cuts off the exoccipital region, the third and fourth branchial arches, and the afferent and efferent vessels to the external gill.

FIG. 2. A posterior view of the same, reconstructed. The cut surfaces of the cartilage where the reconstruction ends posteriorly are dotted. The anterior region of this reconstruction has also been cut off just behind the nasal capsules.

FIG. 3. Anterior aspect of the same. The cut surfaces are dotted. The portion reconstructed is the same as for fig. 2.

FIG. 4. Reconstruction of the left side of the occipital region and anterior end of the vertebral column, showing that the exoccipital cartilage has still the form of a neural arch, though already fused to some extent above and below with the cranium. The lateral and ventral processes are seen to be very long in this region. The parachordal cartilages are shown to send a bridge of cartilage under the aorta.

FIG. 5. A dorsal view of a reconstruction of a portion of the mid-region of the vertebral column[1]. The dorsal spines have been cut off.

FIG. 6. Reconstruction of the caudal portion of the axial skeleton showing the form of the neural and ventral arches at this stage and the manner in which the anal fin is supported; also the very slight upward tendency of the termination of the notochord.

FIG. 7. Reconstruction of the cartilage of the left half of the shoulder-girdle, seen from behind.

FIG. 8. Reconstruction of the shoulder-girdle and left pectoral fin, seen from the front, showing the great length of the metapterygium as compared with that of the propterygium and the formation of the radials from one continuous plate of cartilage.

(Fig. 8 of this Plate is × 20; all the other figures are × 30.)

---

[1] The *anterior* end of this portion is to the *right*.

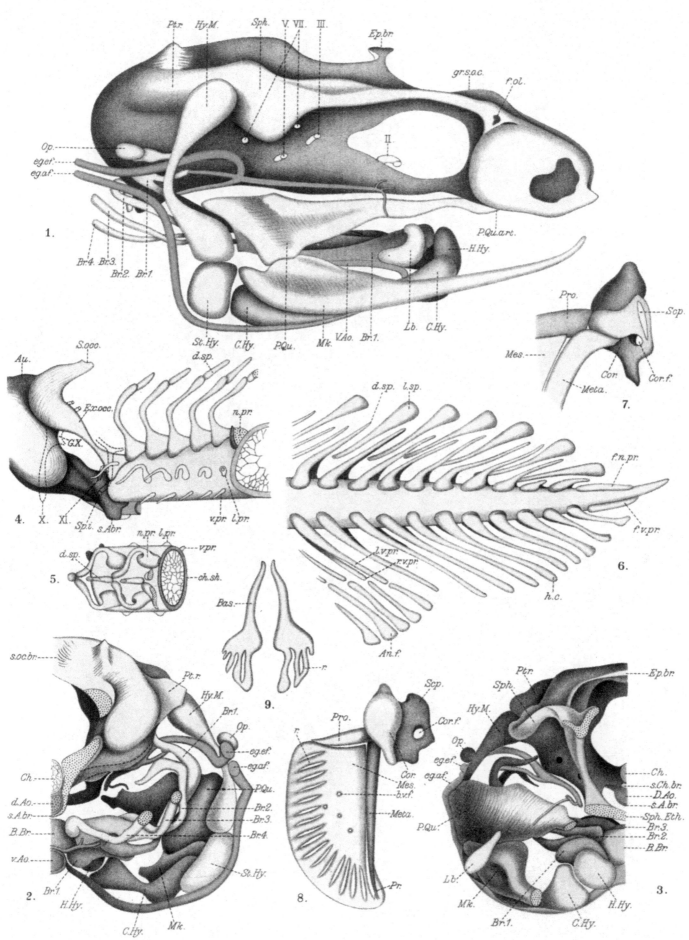

PLATE XI.

EXPLANATION OF PLATE XI.

Fig. 1. A section through the alisphenoid region of the head cutting the median vesicle, epiphysis, and cartilaginous bridge in the region of the epiphysis, also cutting the trabeculæ cranii, the sphenotic wings, and palato-quadrate bars. × 30.

Fig. 2. A section through the vertebral region of the axial skeleton in the anterior part of the body of the 30 mm. larva, showing the neural, lateral, and ventral processes abutting on the notochord, also the outer portion of the cartilages of the lateral ribs. × 30.

Fig. 3. A section through the same region in a 9 cm. specimen, showing the encrusting bony deposits forming the vertebræ with their large transverse process. The ventral processes no longer abut on the notochord. × 22.

Fig. 4. A similar section in the hind part of the body of a 9 cm. specimen, showing the cartilaginous ventral processes, now forming the heads of the ventral ribs, attached to the undersides of the extremities of the bony transverse processes. The ventro-lateral muscles likewise have retreated from contact with the notochord. × 22.

Fig. 5. A section through the vertebral region of the axial skeleton in the anterior part of the body of a 13 cm. specimen, showing the increased development of the bony vertebræ, the complete suppression of the cartilages of the axial skeleton, and the vertebrally constricted notochord. × 15.

Fig. 6. A section through the intervertebral region of the same series as fig. 5, showing the intervertebral dilatation of the notochord, especially of its sheath, which is enclosed in a single layer of cartilage-like cells. × 15.

Fig. 7. A slightly oblique longitudinal section through the ventro-lateral muscles in a 9 cm. specimen, showing the head of the ventral rib to the left containing the cartilage of the ventral process and lying in front of the transverse process, and the continuation of the ventral ribs outwards and backwards towards the right. The figure has been put on the Plate in two portions. × 20.

Fig. 8. A section of the same series as fig. 7, but in the middle of the vertebral centra on the left, sloping outwards through the sides of the vertebræ towards the right. Two cartilaginous neural processes are seen towards the middle of the figure embedded in the bony centra, further to the right the section passes through the bony neural arch, while on the right of the figure are seen two lateral cartilaginous processes. × 20.

Fig. 9. Is a reconstruction of the pronephros of the right side of the 30 mm. larva, and viewed from the headward side, with the glomus lying under the notochord and the coiled portion of the pronephric duct lying between the brachial nerve and the skin. The notochord, dorso-lateral muscles, and ventro-lateral muscles are seen only in section. × 70.

Fig. 10. Is a similar reconstruction of the most anterior mesonephric tubules of the right side, showing that they do not lie in the same region of the body as the coiled duct of the pronephros, but much nearer the middle line close under the notochord, and that they end anteriorly quite suddenly. × 70.

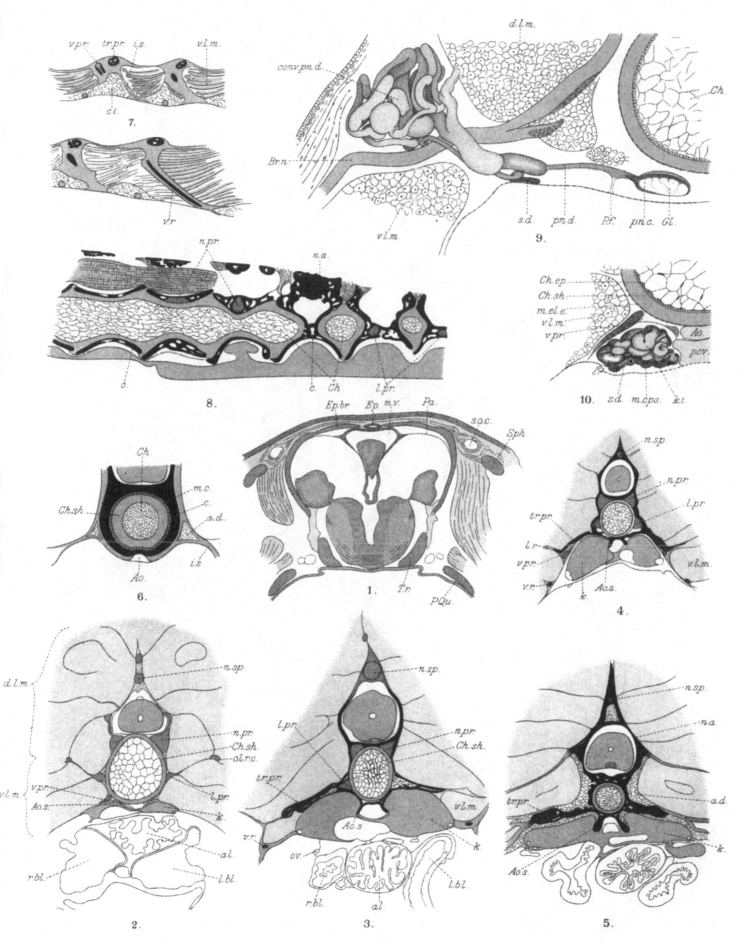

PLATE XII.

EXPLANATION OF PLATE XII.

FIG. 1. A section through the mesonephros of the 24th body-segment, showing the thickened strip of peritoneum in section lying to the outer side of the mesentery of the genital ridge (ovary) and the more complicated arrangement of the tubules in the anterior region.

FIG. 2. A section through the 29th body-segment, showing the thickened ridge lying to the outer side of a small projection, which here represents the genital ridge. It also shows the accessory tubules on either side and the accessory glomerulus on the right side. The arrangement of the older tubules is here less complicated.

FIG. 3. A section through the 30th body-segment, showing on the left side a nephrostomic opening upon the thickened ridge and communicating with the glomerulus. On the right side accessory tubules are shown.

FIG. 4. A segment through the 34th body-segment; on the left side the edge of a nephrostome and also of a glomerulus is cut through, while on the right side is shown a single large glomerulus from which a kidney-tubule leads.

FIG. 5. A section through the 37th body-segment, showing the opening of two kidney-tubules into the right and left segmental duct. The thickened strip of peritoneum still persists, though the genital ridge does not extend so far back.

FIG. 6. A section passing near the hind end of the body-cavity behind the last kidney-tubules. The thickened strip of peritoneum is still seen lying under the segmental duct.

FIG. 7. A section passing through the very end of the body-cavity; the segmental ducts are now turning ventralwards and approaching one another, while the thickened strip lies just ventral to them.

FIG. 8. A section passing just in front of the vent, behind the point of union of the segmental ducts. The body-cavity is seen here to be prolonged backwards in two portions on either side divided by a septum. The thickened strip of peritoneum is here seen on the inner side of the dorsal portion.

FIG. 9. A section through the posterior end of the kidneys of a 9 cm. male specimen, showing the genital duct still in connection with the peritoneal epithelium.

FIG. 10. A section of the same specimen further forwards, showing the large size of the genital duct at the base of the testis.

(Magnification 80.)

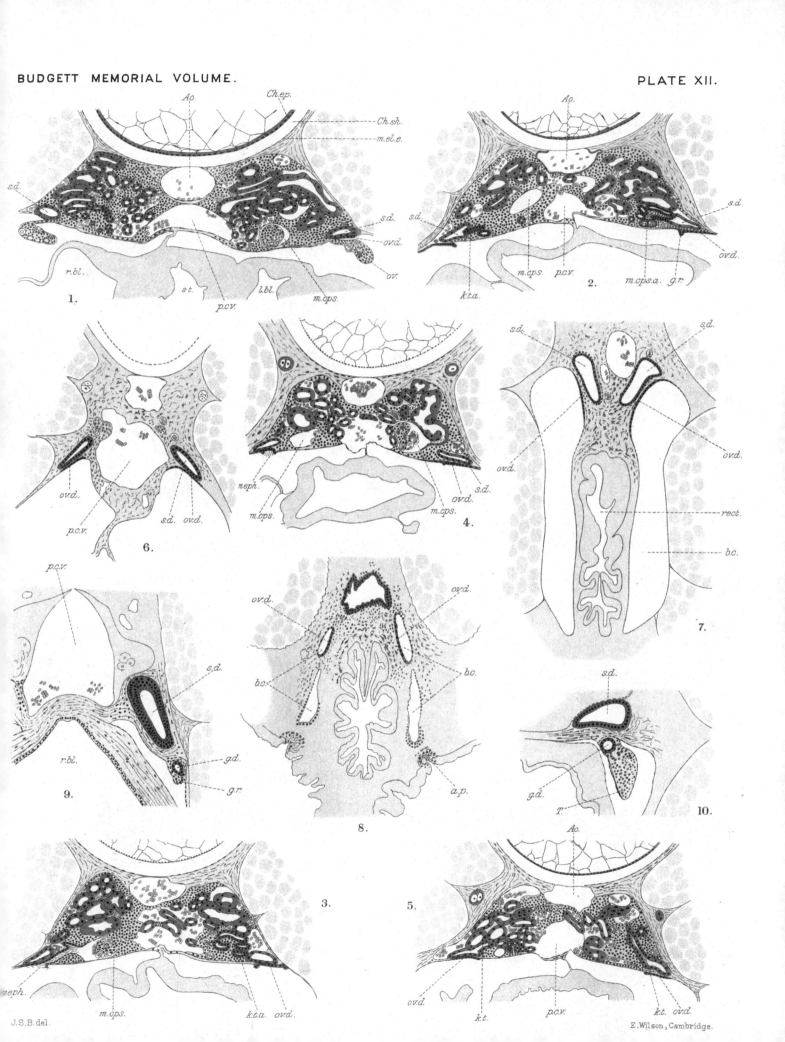

X. ACCOUNT OF JOURNEY TO UGANDA[1].

Mr J. S. Budgett, M.A., F.Z.S., gave an account of his recent journey to Uganda and return by the Nile, which was illustrated by a large series of photographs taken by him as he went along.

Mr Budgett made the following remarks :—

The special object of my journey to Uganda was to continue investigations on the life-history of African fishes and especially of *Polypterus*. Having made previous attempts to solve this problem under certain conditions in a confined area, it was thought that by observing it under varied conditions of latitude and altitude, new light might be brought to bear upon it.

From Uganda it was possible either to work down the Congo from its source or to return northwards down the Nile.

On my way to Uganda, and in Uganda, I gathered what information I could about the two routes. The southern end of the Nile Valley really lies at the foot of Ruwenzori, while a short journey westwards from this point would bring one into the head waters of the Congo.

A special inducement to take the Congo route was that information might in this journey be obtained about the new animal *Okapia johnstoni*, and other interesting forms, believed to exist in the Semliki Forest.

From information gathered in Uganda, it was clear that it was useless to hope to meet with Okapi in British territory, and, moreover, I here learned that the Belgians had found the *Okapi* in large numbers in the Welle country.

I found also, what had been very difficult to learn before leaving home, that the season of the rains and the breeding of *Polypterus* were considerably earlier at the source of the Nile than they were further northwards ; that the Semliki Valley was a most inconvenient place at which to make a permanent camp by the river-banks, owing chiefly to the scarcity of food ; and that only one species of *Polypterus* was to be found there, while at least three species were to be found in the Nile farther to the north.

The difficulty of taking delicate apparatus through the Congo Forest to the upper waters of the Congo was incomparably greater than to the upper Nile.

And, lastly, it was to be borne in mind that the time of year at which one might hope to be successful in the main object was that at which it was well nigh impossible to do much in the way of collecting the higher Vertebrata which might be supposed to be of interest in the Semliki Valley ; for at this time the grass is so high that moving away from beaten tracks is almost impossible, while anything smaller than giraffe or elephant is seldom seen. Bearing these facts in mind, I

[1] From the *Proceedings of the Zoological Society of London*, January 20, 1903.

had little hesitation in deciding to work down the Nile, striking it at a point farther northwards than the Semliki River, in order to take advantage of the lateness of the season in that region. Accordingly, having fitted out my safari or caravan at Entebbe, I started for Butyaba, on the west coast of Lake Albert, on July 10th. I had the advantage of starting thus along a good road which had just been cleared for the greater part of the way, and along which the rest-houses had been repaired for the convenience of the Commissioner and Consul-General of British East Africa, Sir Charles Eliot, who has recently made such a remarkably rapid journey from Entebbe to Gondokoro, the frontier station of Uganda on the upper Nile. At Kampala I diverged, however, for a few days along the old road to Massindi to the eastward.

I had with me at the start 50 men and boys and my bicycle.

So many books have been written on Uganda, that there is little need to describe the scenery of these tropical highlands, especially as Sir Harry Johnston's wonderfully complete book is now in everybody's hands.

Shortly, one may say that, on going northwards from Lake Victoria, forest is hardly seen after leaving Kampala. We passed day after day through almost endless elephant-grass, with palm-groves and papyrus-swamps in the lower parts. The hills are clothed with clumps and patches of acacia and euphorbias, while their summits are very frequently covered with huge granite boulders. There were thunderstorms and rain every afternoon, and for the first few days I saw little in the way of animal life: occasionally a Civet cat would cross the path, while overhead Hornbills and Plantain-eaters of various species were common. In the valley of the Maangia for a time we were rid of the everlasting elephant-grass, and here *Cobus thomasi* and Zebra were plentiful. The *Cobus thomasi* of this region is somewhat different to that met with in the valley of the Nile, the horns having a wider curve and being stouter and of a lighter colour, while the animal itself is of a larger build and has more brilliant markings.

The Maangia River flows northward through wide undulating plains, covered at this time with hay-like grasses upwards of five feet in height, and dotted over with very fine acacias, of a cedar-of-Lebanon appearance and of richest deep-green colour. The grasses and bushes of the roadside teem with bird-life: as we rode along, the little *Vidua principalis*, with his dozen sombre wives, was a constant companion, flitting just ahead of us for a mile or more along the road. Likewise the Common Shrike of these parts (*Lanius excubitorius*) has exactly the same habit of driving along in flocks with a caravan as the *Corvinella corvina* of the West Coast. In the marshy parts, *Scopus umbretta* was often seen.

Then, leaving the plain, we struck up over the Bukamva hills, and at the crest dived into the dense grass at the side of the road, to travel for four weary hours over a wretched and little-used track, often obliterated by the tramp of elephants, and where it was quite impossible to make any progress with the bicycle. Then, descending rapidly by swampy valleys and thick jungle, we came suddenly into the new road to Hoima and Butyaba. From here to Hoima the road crossed the steep hills which form the boundary between Uganda and Unyoro, passing by abundant plantations of bananas and through many a lovely valley, at the bottom of which a stream ran through the richest vegetation, the banks carpeted with Cannas, winter cherries, and hemlock, while overhead were many *Pterocarpus*-trees with blossoms like the Alamanda flower. As I did not care to get too far from my safari, I would often bicycle on for an hour, and then sit down in one of these shady spots, and watch the Mouse-birds hanging like acrobats in attitudes most quaint, and Sun-birds darting in and out of the great red blossoms of the *Spathodea*, while often a noisy flock of *Prionops plumatus* passed hurriedly along. In the more open parts I often saw several pairs of Ground Hornbills, and each time one had a blue throat, the other a red one. I can but think that this is a sexual character, though which is the female and which is the male I was unable to determine. Other birds seen were *Pæocephalus meyeri*, a species of *Macronyx*, *Irrisor erythrorhynchus*, *Hirundo rustica*, and the handsome Snipe *Rhynchæa capensis*.

A very amusing bird that I watched was *Erythropygia ruficauda*, which is most assiduous in its courting of the female, spreading its tail before her like a fan.

In the more shady parts one might often see the butterflies, as I remember seeing them in the forests of Paraguay, covering the ground with large patches of colour, in flocks according to their species.

Shortly before reaching Hoima the river Kafu is crossed ; here a network of papyrus-swamps with good causeways over them abound in duck, geese, and king-fishers of various kinds. Then, winding upwards, a high point is reached from which the village of Hoima is seen, and in the very far distance one can make out the Blue Mountains on the other side of Lake Albert. At this high point I saw a charming little Widow-bird (*Vidua hypocherina*). Lions round this part are plentiful and somewhat dangerous, as they usually are in countries where game is not abundant. At Hoima I heard that several natives had recently been carried off by the Lions.

From here, two days' march through the so-called Budonga Forest brings one to the shores of Lake Albert. This Budonga Forest is nothing more than rather heavily-wooded scrub. It is true that in the ravines and gorges there are strips of real forest, but it is not in any way comparable with the real forest of the tropics, where the sky can scarcely be seen.

This Budonga woodland teems with herds of Elephant—I myself calculated that there were over 200 in one herd which we came across. Some of the males had enormous tusks, and these big fellows seem to keep slightly aloof from the rest of the herd. I knew that we were quite close to this herd, as there were great roadways through the jungle with quite fresh, smoking dung ; and here I first noticed what struck me many times subsequently, that when elephant-dung falls on a pathway or clearing, there within half an hour you will constantly find, heaped up all round the dung, the earth-workings of a shrew or mole. What is it the shrew seeks in the dung? Is it the fly-larvæ that have been blown upon the dung, or is it the dung itself? Frequently elephants in these parts appear of a bright red colour, having covered their bodies with the dust of crushed-up termite hills.

On July 29th I looked from a high point on the road on to Lake Albert, a vast sheet of glistening water, 1000 feet below, bordered on this side with level plains of park-land, broken here and there by lagoons and swamps, where I was to try first for the *Polypterus*.

Of the results of my *Polypterus* work during this journey I shall say nothing here ; suffice it that I stayed down by the lake-side from July 30th to August 15th, trapping, netting, and shooting. During this time the fishes most abundantly met with were *Hydrocyon forskalii, Alestes baremose, Distichodus niloticus, Labeo hosei, Bagrus bayad, Eutropius niloticus, Synodontis nigritus, Tilapia nilotica*, also very large specimens of a *Citharinus*.

Lates niloticus is frequently caught by the natives here five and six feet long, usually with the spear. *Protopterus* and *Polypterus* were both obtained here. The River-Tortoise (*Trionyx triunguis*), 28 inches in length, and very large specimens of *Rana occipitalis* were also common here.

The common Antelopes were *Cobus thomasi, Cobus defassa, Tragelaphus scriptus*, Oribi, and *Cephalophus æquatorialis*. Down on these lake-side flats the avifauna differs in a marked manner from that in the highlands. *Hyphantornis cucullatus* was now building in hundreds in the water-side bushes ; *Laniarius barbarus* and *Telephonus senegalus* in the low bushes, with *Merops albicollis, Lamprocolius purpureus, Pyromelana flammiceps, Terpsiphone perspicillata*, and *Dicrurus assimilis*, were the birds most frequently met with. These birds were seldom seen in the highlands.

Lanius excubitorius seems to have a curious habit of giving a peculiar chattering call whenever a wounded animal is near. We often made use of this indication when tracking wounded beasts, and I have no doubt of the truth of this fact.

From here I struck due east through the Budonga Forest again to the Victoria Nile. The journey through this woodland country was at this time of year most arduous, all the paths being densely overgrown with rank grass, while in the ravines the creepers and hanging lianas were a great hindrance to the porters. During the four days I was in this rank jungle I saw very little in the way of animal-life except Elephants, a few Baboons, and an occasional Puff-Adder, one of them 4 ft 5 in. in length, and a few interesting insects—Phasmidæ and Mantidæ.

Plant-life was much more interesting, and almost overwhelming with its abundance of variety and its beauty.

The handsome Nightjar (*Cosmetornis vexillarius*) was often seen at sunset in these forest-camps.

At length we struck the old road from Masindi to Wadelai, and the bicycle came into use again. At my first camp along this road there were large numbers of a golden-eyed black Weaver-bird (*Ploceus nigerrimus*), which I saw nowhere else. In its size and shape, courting- and nesting-habits, it resembles very closely the gregarious *Hyphantornis cucullatus*.

The Masindi road now made straight for a high conical hill, from the shoulder of which we had this part of Africa laid out as a map before us. To the south, the Budonga Forest; to the west, the north end of Lake Albert, with the valley of the White Nile extending northwards; and immediately below, from east to west, the valley of the Victoria Nile. Descending from terrace to terrace, we at length arrived at the village of Fajao, just below the Murchison Falls, on the 22nd of August.

This wonderful gorge has been described by Baker, Vandeleur, and others, and their descriptions are no exaggeration. One looks down on the swirling, surging water, that, leaving the base of the falls, sweeps round the hill on which the old fort used to stand, with a feeling of utter amazement at the vast numbers of leaping fishes, crocodiles, and hippopotami that have found their way into this *cul-de-sac* of the Nile system.

Here I continued my work with more success than on Lake Albert, the commonest fishes being *Alestes baremose*, *A. macrolepidotus*, *Lates niloticus*, *Clarias lazera*, *Tilapia ziblii*.

The natives here use enormous wattle-traps, which they set in certain fixed spots, usually out of the main force of the current, and often catch very large fish in them.

On August 29th I started again, as this is a most unhealthy place; many of my men were on the sick-list, and food was getting scarce.

Once out of the gorge of the Victoria Nile, we came into open rolling savannah country of grass and Borassus-palms, baobabs, and scrubby acacias. Then crossing several rivers with difficulty, we arrived at Wadelai on September 1st.

During this stage of the journey I noticed several birds not seen in this part of Africa before: there was *Corvinella affinis*, *Parus leucopterus*, and several species of Capitonidæ, all reminding me, as did the landscape, of the Gambia on the West Coast. Here also were *Melittophagus bullockoides*, *Macronyx croceus*, *Urolestes æquatorialis*, *Telephonus* and *Crateropus*.

From Wadelai I sent my porters on to Nimule, about 100 miles distant, taking my loads and servants down the river by boat.

After a few days' work at Wadelai, we started down the river on the 8th of September. The scenery on this part of the Nile is very charming, the hills in many places coming right down to the water's edge. Here one sees the process of the growth of the *sud* in every stage. Beginning with the floating separate plants of *Pistia stratiotes*, the seeds of, first a small floating rush, then of the "*oom soof*" grass, settle on and gradually bind together this carpet of separate plants into a floating island of grass. So abundant are these floating islands that often we appeared to be stationary, even when moving at five or six miles an hour, for all the visible banks were moving too. Once, however, the mass lodges against the stationary papyrus, it quickly becomes overgrown by this, and is converted into permanent *sud*.

Fishing villages are numerous on this part of the Nile. The natives make very good traps of papyrus-grass, and also hunt the hippopotamus with long spears with a rope and float of ambatch-wood attached. Amongst other fishes caught here were *Mormyrops*, *Mormyrus*, *Hyperopisus*, and *Malapterurus*.

Much of this way the hills retreat, and there is nothing seen but grass floating and grass stationary, not even bird-life to relieve the monotony. The last 20 miles, however, before reaching the garrison town of Nimule it is very different. The Nile flows straight towards the mountains above Nimule, and here widens into beautiful lagoons covered with water-lilies, in the foreground sheets of *Pistia* of the most vivid green, in the background bold wooded hills. Here and there are rocky islands with schools of hippopotami basking in the sun; Bee-eaters (*Melittophagus pusillus*), the Jacana (*Parra africana*), and the gorgeous little kingfishers (*Corythornis cyanostygma*) abound. And then the Nile plunges into the great Nimule gorge, to tumble down cataract after cataract, breaking up and pulverizing the floating vegetation, and issuing again at Fort Berkley free from *sud*.

I was now getting anxious about catching the Sudan Government steamer, which comes up once a month to Gondokoro, and determined to leave the Nile and go straight overland to Gondokoro. The actual distance was little over 100 miles, but at this time of year the difficulties of travelling and crossing over rivers in flood were such that one could not tell at all how long the journey would take. After passing through the Nimule gorge, we came to the affluence of the Assua with the Nile. The Assua was now in flood, the only way of crossing being by means of small rafts of ambatch-wood equal to taking one load at a time. None of my porters were able to swim, and all had likewise to be taken across on the rafts. After very nearly losing two men down the rapids, the crossing was completed after eleven hours' hard work. Here, again, the bird-life was different. I saw many birds while on the march that I was unable to identify. There were great numbers of a Weaver-bird of brownish colour with a white crown, building innumerable star-like nests made of straight wiry grasses woven in at a tangent to the nest. There were also seen in these parts, for the first time, *Scoptelus notatus* and *Cryptorhina afra*, though amongst these were not seen specimens with red beaks as was the case on the Gambia.

Just below its confluence with the Assua river the Nile flows on two sides of a high hill; a fact which strikes one as remarkable, for the two branches were mountain-torrents of very little depth of water.

There we left the river, and passed through country with many villages and a good deal of cultivation, especially ground-nuts and millet. The aspect of this country of the Madis struck me as remarkably similar to that of the Gambia: the soil was rich and sandy, and the nuts produced were of great size. In some of the valleys we saw quantities of very fine bamboo, while many of the trees were almost smothered by the beautiful creeping lily *Gloriosa superba*.

In one of these villages quite 10 per cent. of the natives had marked elephantiasis. They were very friendly, and provided me with whatever I wanted. After three days' wandering by winding paths from village to village, we came back to the main path by the side of the Nile, which for over sixty miles runs along close under a range of mountains on its western bank. From this point northwards for some time the beautiful little Parrot *Palæornis docilis* was common. The only Antelopes seen in this part of the journey were *Cobus leucotis* and a species of *Damaliscus*. The grass seemed to get longer and longer, and marching in the early morning, when the heavy dew was hanging from every blade of grass in great drops, was most disagreeable.

On September 19th we reached the flourishing village of a well-known chief named Adimadi. This village was situated in a hollow on the top of a high hill, with natural rocky fortifications surrounding it, and overlooking a fertile valley to the east. On the heights above the village I saw considerable numbers of what appeared to be a large red *Colobus*-monkey, a specimen of which I failed to secure.

Long-horned cattle were plentiful here, and are probably the same race as the long-horned cattle of Ankoli.

On the hill-sides were numbers of very fine African mahogany-trees (*Khaya*) and springs of good water. In these trees were many kinds of Plantain-eaters and Rollers (*Coracias caudatus*). This was the first place during the whole journey that I met with any Rollers. The hitherto daily rainstorms were getting less frequent, and the dry season was setting in.

From here we marched through undulating park-like country with small trees, to a similar isolated group of hills, with the village of Leju nestling beneath. Here, again, were fine spreading trees, in which were numbers of beautiful glossy Starlings (*Spreo superbus*), and also the King of the Sparrows (*Dinemelia dinemelia*). Passing down from the Leju hills again, we marched through country of a rather barren nature, of rank grass and small acacias. The whole way the elephant-tracts were very numerous, and we came suddenly on a herd of twelve with two old tuskers among them.

The country now became more and more barren, and on September 22nd we reached the hills again, opposite the Belgian station of Redjaf. Here I saw several birds I had not seen before, including *Merops nubicus*, *Vinago waalia*, *Laniarius erythrogaster*, and, I think, *Lanius collurio*, though it may have been a different species.

At Gondokoro I sent back all my porters and Uganda servants, and after a few days' work fishing &c., I started northwards, on the Sudan Government steamer "Abuklea," for Khartoum, on September 27th.

The first few days the steamer passes through firm banks, on which, notwithstanding the grass, we saw several water-buck and some buffalo. Many small villages line the banks, while several old Dervish forts are passed.

At Canissa, about 100 miles north of Gondokoro, I changed into the " Kaibar," the post-boat to Khartoum ; then, passing through the *sud* region in three days, we came to the mouth of the Sobat and the land of the Shelluks.

On my arrival at Khartoum, I set to work to get Arab fishermen and servants, fishing-tackle, provisions, &c., and returned in a few days to Fashoda.

Here I made my final attack on the *Polypterus* problem. I had three species of *Polypterus* to work with, while material was fairly abundant. However, after several weeks' work, I finally packed up my things, and disconsolately returned to England ; having got a good deal of side-light on the life and habits of *Polypterus*, having seen something of the Fauna and Flora of the most wonderful river in the world, but having again failed in my principal object— namely, to obtain the early stages in the development of *Polypterus*.

In conclusion, I should like to say that, throughout the journey, I received at the hands of the Uganda and Sudan officials the most courteous and liberal assistance on all occasions.

XI. NOTE ON THE SPIRACLES OF POLYPTERUS[1].

On seeing a letter in *The Field* for November 8th, 1902, by Mr Boulenger, in which he says that, after observing *Polypterus* in captivity for more than a year, he had not been able to learn anything concerning the use of the spiracles to *Polypterus*, I determined to go over my former observations concerning these structures and see whether I had by chance been mistaken as to their use.

I have in captivity a pair of *Polypterus senegalus* kept in an aquarium at a temperature of 75° to 80° F. They are quite tame and regularly take food from a fork.

On December 2nd I watched them for one hour after feeding. While eating, the spiracles were repeatedly rapidly opened and closed, though not widely. The movement was apparently independent of other masticatory movements. Within the hour each of the pair came to the surface three times at irregular intervals.

1. Specimen A came to the surface and gulped air with the mouth; immediately after leaving the surface two large bubbles of air were discharged from under the opercula. During the descent to the bottom the two spiracles slightly opened and from each a minute bubble of air issued.

2. Specimen B performed the same movement, but no air was seen to issue from the spiracles.

3. Specimen B repeated the movement, and small bubbles of air did issue from the spiracles.

4. Specimen A repeated the movement, but no air was seen to issue from the spiracles.

5. Specimen B repeated the movement violently after some excitement, and as it met the surface *widely opened* the spiracles, forming a triangular aperture, one side being the side of the head and the two other sides being the two plates of bone which form the spiracular flap; whether air passed in or out of the spiracles was impossible to see as the top of the head was out of the water. No bubbles passed from the spiracle during descent.

6. Specimen A repeated the movement, the spiracles did not open, and no air was seen to issue from them during descent.

[1] From the *Proceedings of the Zoological Society of London*, January 20, 1903.

On December 8th I watched them again for an hour. *Polypterus* A and B came to the surface for air eight times, and four times the spiracles were *widely opened* above the surface of the water, and a sound produced as of the sucking in of air.

I have often found it convenient to kill *Polypterus* by piercing the cranial roof and destroying the brain. During the operation it is quite easy to stimulate the brain-centres in such a way that the spiracles are widely opened as described above. It is possible to stimulate continuously so that the spiracles are retained in the widely opened condition. I believe, then, that the spiracles are used to take in and to give out air from the swim-bladder. At certain times the fish rises quite slowly to the surface in the horizontal position, when it would be easier for it to exchange the air in the swim-bladder from the surface of the head than to turn its head upwards in order to take air by the mouth. By closing the mouth and opercula, distending the body-wall and opening the spiracles, I believe the fish is able to inhale air, and I should suppose that it expires previously during the same movement, as does *Protopterus*. I think it is possible also that in the very shallow water which this fish frequents at certain times of year, it may be of use to the fish to change the air in its swim-bladder in this way. I have often noticed, in changing the water in a tank in which numbers of these fish are confined, that when the water is exhausted the spiracles are frequently opened.

The position of the spiracles almost immediately over the long slit-like glottis is in favour of the view that they are connected in their functions with the latter. They seem also to be used, as I at first believed, to let out the excess of air from the pharynx after the fish has taken air into the swim-bladder, either with the mouth or with the spiracles.

Observation upon these points is very difficult owing to the rapidity with which the movement takes place; but the fish I have been watching have become very tame, after three years of captivity, and these movements are now more slow and much more easily watched.

XII. THE DEVELOPMENT OF POLYPTERUS SENEGALUS CUV. BY J. GRAHAM KERR, UNIVERSITY OF GLASGOW.

With Plates XIII—XV. and text-figures 19—85.

CONTENTS.

I. INTRODUCTION.

The tragic story of Budgett's expeditions in search of embryological material of *Polypterus* has already been told. It is my purpose in the following pages to give a sketch of the general features of the development of *Polypterus* as determined by the investigation of the material which Budgett collected. As a preface to my description, I reproduce in the following few pages such passages in Budgett's diaries as have to do with *Polypterus*.

PASSAGES FROM JOURNAL.

First Gambia Expedition, 1898—99.

Nov. 15. When I got back I found that a very fine *Polypterus* had been brought for me and was now in my aquarium. He was about 18 inches in length, and spread his great pectoral fins on the bottom, using them like a seal's paddle. He snapped up greedily two beetles floating on the surface; but shortly after slowly died. The natural colour is a dirty straw colour with shadings of black above. While alive in my aquarium he was extremely lethargic.

Nov. 17. There seems to be no doubt that *Polypterus* as well as *Protopterus* begins to breed in May and June, when the rain begins, and both of them in the rice swamps.

Nov. 18.   The largest *Polypterus* I have yet had was brought in this afternoon, measuring 19 inches.   I examined the ovaries and found that the ova were scarcely visible to the naked eye, but globular and transparent under the microscope.

Nov. 23.   A third *Polypterus* (*lapradii*) with an opercular gill was brought to me, it is peculiarly marked and about ten inches long.

Jan. 16, 1899.   My aquarium is now in a first rate state, and the *Polypterus* look quite happy, and I catch some small fry for them most days, and they greedily devour them.   This is always at night, and then they appear to come very slowly and stealthily nearer their prey, actually touching the young fry with their tentacular nostrils from underneath, and then making the final snap and a gulp at the fish.

Now that there are plants and food in the aquarium they come regularly to the surface at night time and lie thus in wait for prey.

I set the trammel at night.   I have had a most circumstantial account of the habits of breeding of the Sayo.   He is said to make a floating nest in which he places strings of black eggs, the size of large Kuss seeds.   When the young leave the nest they are said to follow the parent about at the surface while he lies down at the bottom.

Jan. 28.   The Sayos seem almost periodically to come to the surface to breathe, and then swim round in a curve with the dorsal fins projecting from the surface. Otherwise they lie half buried in the mud at the bottom, and quite motionless, then paddling themselves along with their pectorals, they slowly move to another spot and lie there motionless.

March 2.   Set things straight.   Found fish all well.   Observed that *Polypterus* used the spiracle to let out the air accumulated in the gill cavity.

March 3.   Made cage for Chameleon.   Tried experiment with Sayos, putting zinc perforated below the surface with same water as large tank.   Sayo dead in two hours.

March 4.   Noticed a most brilliant reflection from the eyes of my Sayos.   I noticed to-day that the eyes of Sayo appear to be luminous and glow like coals.

March 5.   SUMMARY OF INFORMATION REGARDING *POLYPTERUS* SENT BY LETTER TO GRAHAM KERR.

1.   At the beginning of the dry season I have observed thousands of young fry trying to leave the creeks and swamps and to get to the river.   All these are approximately the same length, namely, 10—14 cm., and without exception devoid of external gills.

I have seen them during the end of November and December, when the natives have made dams across a swamp stream, and bailed the water out for fish.

2.   During November, December, January and February, larger specimens have been caught with a hook (in the river) and also with the native cast net, which are mostly of two sizes, either 25—30 cm. or 40—55 cm.

About 70 °/₀ of the smaller sized ones have external gills of different degrees of development.   Those that have none, differ slightly in other ways from those that have, both in colour and markings and in form, but I hardly think these differences are specific, but rather constitutional, depending upon the exact habitat.

The larger sized *Polypteri* have no external gills, and are generally of a more uniform colour.

3.   As regards the spiracle, I had several large *Polypteri* in a pond in connection with the river, and emptying this one day until only a few inches remained, the *Polypteri* wriggled about in the shallow water, and on reaching the deeper water, opened the spiracles and discharged two large bubbles of air.

I should therefore look upon the peculiar arrangement of the spiracle in *Polypterus* as favourable to life in shallow water, enabling the fish to discharge the air which has collected in the gill chambers.

4.   *Polypterus* comes to the surface to breathe every few minutes, and then discharges a bubble of air. If a *Polypterus* is confined in water where it is prevented from reaching the surface, it will be dead in a few hours.

5.   In the stomachs of *Polypterus* I have found insects, crustaceans, etc., but no vegetable matter. Captive *Polypteri* will feed on any kind of raw meat, also on chopped up boiled white of egg.

6.   The eyes, which are reddish, glow in the shade like sparks of fire, but are *not* luminous.

7.   The largest specimen obtained is 70 cms.

8.   *Polypterus*, though sluggish and tame if undisturbed, if irritated becomes most active and fierce. The wound from the dorsal spines is very painful.

9.   The ovaries, in December, contained ova which were just visible to the naked eye. In March, these have reached the size of small shot, and are surrounded with masses of pigment.

10.   According to native information, not yet confirmed by me, *Polypterus* leaves the river and works up the creeks and into the rice swamps as the water of the river rises at the first rains. Two weeks after the rains have flooded the rice farms, *Polypterus* begins to make its nest, seeking out a favourable spot amongst the thick grass. Here it cuts off the grass blades below the level of the water, and forms a shallow nest which floats on the surface amongst the thick grass.

The eggs are laid in a row in the nest, and are at first white, later red, and afterwards burst, setting free the contained larvae.

After leaving the nest they swim about in a dense shoal, the mother swimming below and guarding them.

11.   According to native information, all sizes of *Polypterus* breed each year. Some of the specimens which work their way up the channels are five feet long.

April 14.   The Sayo, I learn, comes into the swamps with this overflow of the river, and the female at this time is heavy with eggs. She chooses a spot where the grass is very thick, and makes a kind of raft with the heads of the grasses, upon this she lays her eggs in rows. When hatched, the larvae soon leave the nest and swim about in a shoal, guarded by the mother, who swims below. I find that there are two distinct varieties of *Polypterus*. Both are equally common, they may be described as the long[1] and the short[2] headed form.

P. senegalus, the *long headed* form, has the operculum extending more than half-way to the first dorsal finlet. The head is not so depressed, and the opercular bone is almost of the same diameter each way. The first dorsal finlet is on the 18th to the 20th row of scales, and there are from eight to ten free dorsal finlets. There is little dark pigment. The young do not retain the external gill.

April 28. I was down at the wharf this afternoon and saw numbers of Sayos swimming along the edges of the river. One seized a largish crustacean and the rest gave chase. I have not before seen them swimming about in their native state. All these were the short headed form (*P. lapradii*).

May 1. This afternoon I examined a dozen Sayos caught here, there were nine females and three males ; all the ovaries were immature, but the testes seem to be coming on ; the natives say, "eggs next month." All these were of the long headed type (*P. senegalus*).

June 2. Had a very large Sayo brought me, which turned out to be a male. Dissected kidney and genital ducts. The testes are slightly lobulated and hang freely in the body cavity : a large genital duct leads to the urino-genital sinus. The kidneys open by numerous pores into urino-genital sinus. The sinuses open by common apertures just behind the vent.

June 5. Killed and dissected one of the smallest Sayos. To my surprise, although barely six inches in length, it contained a small number of ripening ova. The upper hemisphere is heavily pigmented. They seem to be about the size of the frog's eggs of S. America.

June 9. Dissected pregnant *Polypterus*, the ova as large as number 8 shot. Apparently the ducts are a continuation of the ovarian sacs and lead straight to the vent. The ova are pigmented above and white below. The males have a thickened base to the anal fin.

June 19. In *Protopterus* the spermatozoa are about twice the length of the red corpuscles : in *Polypterus* they are minute and about one-sixth the length of their red corpuscles.

June 30. I opened another of my female Sayos. The ova appear much as they were a week ago, but for the first time I discovered the oviduct funnel and made a sketch of it.

July 2. The ova appear to be assuming an ovoid shape, the pigment, though mainly restricted to the upper pole, extends in a diffusing network to the lower pole. There is a distinct micropyle.

July 3. Last night dissected female Sayo, there seems every probability that

----

[1] *P. senegalus.*   [2] *P. lapradii.*

artificial fertilization will be possible. This one had evidently shed a portion of the ova, though none were to be found in the oviduct, and only a very few in the body cavity.

July 4. Twenty Sayos (*P. lapradii*) brought. All contain ripe ova. I am attempting artificial fertilization.

July 15. In the evening I dissected the weekly female *Polypterus*, and found the ova distinctly more advanced. They could be separated from their follicles, and were then seen to be enveloped with a viscous coat which swelled up in contact with water.

July 17. Dissected female *P. lapradii*, the ova distinctly more ripe and come away easily from the follicle, and also from the next viscous layer. Then comes a clear viscous layer and then the egg membrane.

Second Gambia Expedition, 1900.

1900. *June* 24. Dissected male Sayo; found spermatozoa on the anal fin and in teased testis, but there are also minute vibrating bodies in the testicular fluid which I mistook for the spermatozoa last year.

July 2. Hot fine day. Man brought me six *P. senegalus*. On holding up a female she extruded about a dozen jelly coated eggs. I tried to fertilize them, and put them on a muslin frame. They would not develope. The other eggs in the ovary were not ripe, and would not free from the ovary. The male did not seem to have ripe spermatozoa in any number.

July 17. Examined a male Sayo this evening, spermatozoa are ripe in the testis, but hardly a trace on the anal fin or vent,—though the anal fin in a pithed fish kept being erected into a spoon-shaped arrangement.

July 19. Went out to swamp inlet, and met there a man who had been fishing and had caught two Fantang (*Heterotis*), a Cambona (*Protopterus*), and a Sayo. I examined the latter and found an egg or two in the cloaca. On cutting open, found all eggs shed. To-morrow I go to the spot from whence it came to look for eggs.

Aug. 13. In the afternoon I had a pair of Sayos brought, the result of three days' fishing, which has cost me nearly a pound. The female had shed almost all her eggs. Many remained; however these came away fairly easily from their follicle, and I made every effort to fertilize them. Four attempts were made, two with chopped testes, two with the contents of the testis duct. Whether any are fertilized or no I cannot say. About half have already gone bad.

Aug. 14. Had a fine young *Polypterus* brought with very large external gills, of which one branched. The Sayos undoubtedly become a lighter shade in the dark, as do the Cambonas.

Aug. 15. In the afternoon had a female Sayo brought which had shed *all* its eggs. This caught near wire enclosure. Preserved the Sayo.

Aug. 18. The Sayos are now grubbing about the roots of the water-plants. The *lapradii* in enclosure are now getting quite tame and come to feed almost as quickly as the *senegali*.

Aug. 19. Jambanding brought four *Protopteri* and a larva of *Polypterus lapradii*[1], $1\frac{1}{4}$ inches in length. A most beautiful object. The external gills reached half-way down to the tail, blood red, and with secondary and tertiary branches. Arising more dorsally in this larva than in the adult, they droop ventralwards posteriorly, and do not seem to be moved much by the muscles.

The blood supply can be seen with wonderful distinctness through the transparent body wall. The pectoral fins are perfectly formed, and large in proportion to length of body. Pelvic fins somewhat small. Tail not very large and not differentiated from the dorsal finlets, of which they seem to be only a forward prolongation. The dorsal finlets are not distinct from one another, but form rather a continuous dorsal fin. Of these finlets, but *six to eight* can be said to belong to the back, the remainder being merged in the tail. (The adult has 13 free dorsal finlets.)

The body is distinctly more truncate in the larva than in the adult, the head and tail region being large. The eye is large in proportion. Colouring—transparent flesh below; above, golden where not darkened by black pigment cells, i.e. a pair of broad bands running through the tubular nasal opening over the eye, and along the dorsal surface of the external gill. Two pairs of gold bands run down the sides of the body, parallel as in the adult.

The area of the pigment ceases suddenly ventralwards in a line running from tubular nares, under eye, dorsal to the insertion of shaft of pectoral fin.

Aug. 21. " Kaya " caught a fine female Sayo. This had already laid eggs, and must, I think, have temporarily returned to the riverside of the " Fa." A *senegalus* brought in afternoon; this was full of eggs which were not quite ripe. I tried fertilizing, but no use. I have never seen a female with loose eggs in body cavity. The water in the swamp is terribly low; several small Sayos in the cage have died from smallness of water and heat of sun.

Aug. 22. Tried artificial fertilization on Sayos from cage. Eggs nearly ready to leave ovary, some free ; these would not be fertilized, owing, I think, to smallness of the only available male.

Aug. 30. Went down to enclosure and caught a number of *senegalus*. Tried artificial fertilization on two pairs.

Aug. 31. Attempted artificial fertilization on seven females. Several seemed to have almost quite ripe ova; all were crammed with eggs, but the ova would not come away easily from their follicle, and fertilization failed again. There were in one specimen signs of degeneration of the ova. It seems, however, possible that they may even yet ripen and be laid in those I have left in enclosure No. 3.

---

[1] Really *P. senegalus*, vide p. 156.

Uganda—Nile Expedition, 1902.

July 31. Was just starting to new camp by steel boat, my tent was packed, loads all tied up, and we had started towards the boat, when head-man brought several fish, and among them *Polypterus senegalus*, 17 inches in length, they said more common here than new camp, so am stopping a day or two. Female had eggs but mostly shed. When I got back had five more "Intonto" brought, one female full of eggs, but evidently they had just laid. *The males do not seem to have milt.*

Aug. 1. Two more "Intonto" brought. Female most eggs laid. Both about 13 inches.

Niger Expedition, 1903.

Aug. 12*th.* Several showers in the afternoon, landed at Assay at 5.30.

Aug. 13*th.* Early morning setting house in order. I have a fine veranda to work in, overlooking the Forçados river, an island in front where shallows and creeks of good prospect. A few light showers and sunshine in the afternoon.

Took canoe up to small village, where I was much interested in the fish traps and nets.

There is a large lagoon and village somewhere at the back, but it seems difficult to get to at this time. I had some small specimens of *Ophiocephalus* brought, and also a small fish with a head like a pipe-fish, and long eel-like body which I do not know.

The natives are of a very low type, but I think in a short time I shall get on with them all right.

The Niger Company have quite a good garden with oranges, mangoes and plums. The river is nothing like up yet. Though I think the natives will probably bring me small specimens of *Polypterus*, I fancy I shall have to catch the adults myself, unless I can get to an Igabo village. *Hyperopisus* seems exceedingly common here, and I saw one *Gnathognemus* with brilliant yellow colouration. I hope to get small specimens alive.

It seems to be impenetrable forest all around, though there is open ground on the island.

On the whole, I feel much more hopeful here than up river, chiefly, I think, because of the innumerable rain-water swamps here.

Aug. 14*th.* Prepared little traps in the morning. Had large number of young fry brought alive. Among these, two pipe-fish like a brilliant Tilapia ; altogether about eight kinds. Also one large specimen of *Xenopus* and small ones.

I set the bolter, trap net, and ten bottle traps in the island water.

Aug. 15*th.* Got on the line one *Gymnarchus*, none in the net. A few small fry and some prawns in the bottle trap. Pouring rain all day. It more or less stopped raining at 3 p.m., so I went out and set the night-line. Rain poured all night.

Aug 16*th.* Government launch started 7.30. Boys took in night-line with two *Gymnarchus* and a *Synodontis*. Preserved their heads. Then went to try

26

to find entrance to lagoon; pretty creek; talked to the bush Agabo, but no entrance. Took some fish from natives and talked about *Polypterus* (acāta) and *Calamichthys* (cansisse).

In the afternoon walked out to lagoon at the back. Road very bad, partly through virgin forest undergrowth. Saw fine green Hyla. The lagoon was muddy with leafy growth; not, I should say, suitable to *Polypterus* for spawning. Medicine at night.

Aug. 17*th*. Cheap. Repaired bottle traps. Had *P. senegalus* brought. The bell-jars are going very well. Set night-line in creek at the mouth of the Assay river.

Aug. 18*th*. Still pouring. Two large *Gymnarchus* on night-lines, and a large *Synodontis*, five hooks broken. Had more *senegalus* brought. Set eleven traps in bush-stream.

Aug. 19*th*. Nothing in traps; had collection of fry brought, none of the pipe-like fish, another transparent small fish, many brilliant *Cichlidae*, *Ophiocephalus*, and a small *senegalus*, three inches, with external gill.

P. lapradii brought dead. Tried fertilization. The males were all absolutely ripe. One female had completely lost eggs. One yet crammed. One *P. lapradii* partly shed.

In the males the tubules of the testis ridge greatly dilated with sperm, which was seen to travel down the testis duct. This becomes more clearly defined the more dilated.

Endean came back with a bottled *Calamichthys* and three live ones. They are very dark above, and a dark orange below. E. said he could have brought a hundred. Set the night-lines and eleven bottle traps.

Aug. 20*th*. Had six *P. senegalus* between $2\frac{1}{2}$ and 3 inches, with large external gills! Preserved in Bles' Fluid. Also two small *Protopterus*, and some remarkable tadpoles, probably Hyla. They are very large, wonderfully transparent, the flesh a delicate orange colour, eye metallic, six large black spots, and several small ones symmetrically arranged. A large number of small *Cyprinoids* of a wonderful colouring of two kinds. A large female *P. lapradii*, two feet, and a fairly large *P. delhezi*, both put into swamp water, for males. Many kinds of frogs.

The day has been almost fine. Why the natives should be able to get *Polypterus* $2\frac{1}{2}$ inches and no smaller, I cannot make out, but I am hopeful. Several *Polypterus senegalus* full grown.

Aug. 21*st*. Tried to fertilize four females, no success. Had two fine *P. delhezi*.

Aug. 22*nd*. Went down to see nest of *Gymnarchus*. Opposite Bari entered creek, then a small stream, very winding, full of natives fishing and fish traps. Took many *P.'s* from canoes. After half-a-mile came to swamp choked with *Pistia* and floating grass. Here found old disused nest of *Gymnarchus*. Then crossing entered another small stream, all the way through dense forest carpeted with ferns; after quarter-mile came to large lagoon free from sand, in small creek from this on far side, found nest of *Gymnarchus* full of young ready to leave nest. Meanwhile talked to guide, he knows the eggs of *Polypterus*; has seen them spawning at the beginning and end of rise of Niger, always at west end of lagoon, attaching eggs to sticks under water in great quantity. He says they are hatched in two or three days. I must wait until water begins to go down. Have offered him £2 for eggs. In the evening had small *P. senegalus* brought with huge gills.

Aug. 24th. Repacked N. Nigerian fish, attended to S. Nigerian collection. Attempted to fertilize more *P.'s*, none with eggs in oviduct.

Aug. 25th. Revisited the lagoon. Tow netted all morning, many kinds of small fish and some jelly fishes[1]. No *Polypteri* caught. Fine hot day. Price of *P. senegalus* has come down to 3*d*.

In afternoon I opened three female *P. senegalus*, each of which had the oviducts crammed. Two of them appeared to have been captured some days ago, and the ova to have begun to decompose. The third one was quite healthy. I fertilized with teazed testes, and spread on the bottom of small glass dishes, to which they firmly adhered. Fine day, with a few showers.

Aug. 26th. The ova appear to be segmenting, about 60 °/₀ have decomposed. The pigment seems to be redistributed irregularly. One egg I examined, as it looked more normal than the rest.

The upper pole was covered with fine cells of a light brown colour, the lower pole cells were white and about twice the size. The upper cells easily visible with a Leitz (× 8). Round one half of the egg a deep constriction between the brown cells and the white cells. In attempting to free from envelope, egg broken, but preserved in formalin.

Later (*mid-day*) another egg was examined, the constriction surrounded the whole egg, about quarter white, three-quarters brown. Preserved in formalin.

4 *p.m.* Another examined; the white area is now a mere plug, evidently the "yolk plug of the blastopore." The shape of the egg is oval, the gelatinous envelope does not fit the egg closely, and is pretty firm. Preserved in formalin.

Most of them appear to be backward or not properly fertilized. They are clearly segmenting, but irregularly; the pigment is scattered in patches and streaks, chiefly following the segmental furrows.

At five to-day a Bari fisherman brought me two female *senegalus*: one had the oviducts full, the other had also a large portion of the ova free in the body cavity. I had two good males, and tried to thoroughly fertilize them, then spreading them on hatching tray, set them in hatch-box, in the river. I noticed that many of the ova from the oviducts appeared to have begun segmenting, the body cavity was full of fluid, and I am inclined to think that there is internal copulation. The vent of the female was swollen and protruding, suggesting that the female receives the milt from the anal fin of the male, together with a certain amount of water, though there must also be sufficient serum added to prevent coagulation of the

---

[1] *Limnocnida tanganyicae*, vide XV.

gelatinous envelope.   The tubules of the testis of the male used were greatly distended, and the sperm was clear and not opaque as in the other males.

Hot sunshine in morning, downpour of rain in the afternoon.   Temperature mid-day, 28° C.   Temperature of water in river, 28° C.   8 a.m.

Am not making drawings until material more abundant and normal.

Aug. 27th.   About 70°/₀ of the ova fertilized yesterday are developing.   At 8 a.m. the blastopore was just closed, while the embryonic plate extended in a pear-shaped manner from the latter.

A series during the day were put in formalin.   From this time on the ova are easily extracted after treating for a short time in formalin.   They were then put into Glacial Corrosive: about ten of these stages.

First, 10 a.m.   [Stage 14¹.]   First appearance of the embryo.

Second, 2.30 p.m.   [Stage 15.]   Uprising of neural folds around the plate.

Third, 8 p.m.   [Stages 16—19.]   Closing over of the folds over the plate. Brown pigment is irregularly scattered, chiefly on the upper pole of the eggs.   Much pigment sinks into the neural groove.   The head portion of the neural plate is last closed in, and then is little broader than the body portion.   The yolk plug does not seem to be included in the neural groove.

Had several female *senegalus* brought, but none had ripe ova.   Young *Xenopus* tadpoles swarming round the hatching-box.

Aug. 28th.   Fourth stage.   [Stages 20 and 21.]   2 a.m., beginning of tail lengthening.

Fifth stage, 8 a.m.   [Stage 23.]   Slight protuberance in gill region.

Sixth stage, 2 p.m.   [Stage 24.]   Tail and head well lengthened.

Seventh stage.   [Stage 25.]   Beginning of suckers and external gill ; 8 p.m.

A very beautiful tadpole in rain swamp.   Back, velvety black with gold stars ; before and in front, a bar of metallic blue.   The mouth a protrusible funnel, no teeth.

Aug. 29th.   Eighth stage.   [Stage 26.]   External gill just before beginning to branch, suckers well developed.   Indication of anus near the end of the tail.   No sign of gill slits or eyes or pronephros ; 8 a.m.   Am now taking ten specimens twice daily.

Had six *Calamichthys* alive from Ackow.   Several female *P.'s,* but none ripe.

Ninth stage, 8 p.m.   Gills beginning to branch.   Dorsal and ventral fin fold beginning to be formed.

Finger very bad.

Aug. 30th.   Tenth stage, 10 a.m.   [Stage 27.]   Not much change, gills branching, brain cavities forming, and also ear sac.   Circulation begins.

From now one stage daily.

Made framework covered with calico, and put hatching-box inside as safeguard.

One specimen in watch glass for two days developing well and rapidly, shall bring up a large number.

Opened dead female *Calamichthys* ; full of large eggs of oval shape, though

¹ I have inserted in brackets the number by which the stage is designated in my description.   J. G. K.

probably more heavily yolked than *Polypterus*. Suckers point almost directly forwards. If larvae taken from capsules, they will live well, and can hang by suckers from surface of water. No sign of gill clefts.

Aug. 31*st*. On examining hatching-box found that practically all, except those that had hatched, were dead, probably from overcrowding. Collected about 70 from bottom of hatching-box. Put them into watch glasses.

Eleventh stage. Not much change, increase in length of gills and curvature of head. The specimen in watch glass for three days has gone ahead of the rest. Soon after the blood becomes red the gills though more branched are shorter. The gut is clearly lined with yolk cells in the tail region. The notochord is slightly turned up posteriorly. The body is semi-transparent. Gold pigment appears in head region. Black pigment in the iris. No sign of gill clefts. The suckers now point ventralwards. The tail is sharply pointed, the larva is 5 mm. in length, has grown $\frac{1}{2}$ a mm. a day.

Had about 20 *Polypteri* brought, eight females but none ripe. Incessant rain.

Sept. 1*st*. The larva is dead. All the others are attacked by fungus.

No real advance in them. Preserved five in morning about 30 in afternoon. Fixed all the dead ones for experiments in section cutting. The fungus caused a gelatinous encasement over the whole larva which could be slipped off but then the larva quickly died. The one in advance of the rest was quite decomposed in the morning, and I have no drawing of it. I hope to be able to try again.

Fine day. Shall now hunt the bush for young fry.

Three females brought, none ripe. Went out early to small creek, hunted for eggs, etc., without success. Collected large numbers of *Polypteri* mostly males. Examined 6 females, all eggs still attached.

The fish are caught by accident in traps which the natives set for shrimps. They say that in a few weeks' time they catch numbers of young *Polypteri*.

I find that of the six larvae placed in bell-jar one is still alive; it is not yet so far advanced as the single one which died. But is hanging by suckers from *Myrio*(*phyllum*) and looks healthy.

Tornado in the evening. Very fine day.

Sept. 2*nd*, 3*rd*, and 4*th*. Fingers very bad. Examined over 50 *Polypteri* of three species; tried to fertilize ova from oviducts of *P. lapradii* which were crammed in two specimens. Males brought with them. Of all only one egg found to develop, have put this in tube in the river. Why no more fertilize I cannot say. Perhaps damage to ova during transport. I have found none free in body cavity as in last successful trial.

Sept. 5*th*. Boy Apia has found new nest of *Gymnarchus*, have told him to get eggs in formalin. Natives from another village brought a huge *Protopterus* about four ft. long. They say they got many small *Polypteri* to-day but did not know I wanted them.

Grant came down to attend to my fingers to-day.

Sept. 6th. Apia brought a batch of *Gymnarchus* eggs at a stage just before hatching, in formalin. I wanted them earlier. I find this an excellent method to give the boy a bottle of formalin and when he brings the eggs to put them in Corrosive Sublimate Solution.

I examined six female *senegalus* still crowded with eggs, none free. Rain all day, still unable to use fingers.

Sept. 7th. Tried to fertilize eggs from oviduct of *P. endlicheri*. Also of two *Calamichthys*, the latter look promising. Hunted for *Polypterus* eggs in small creek opposite. My system now is to examine only promising females and what males I want, giving back all and paying the natives 6*d.* a piece for opening them.

Had tin of fish soldered up by engineer who made a horrid mess of it.

Sept. 8th. Examined four *Calamichthys* ♀, 32 *P. senegalus* ♀. Two *Calamichthys* had ova in ducts. Tried these but not hopeful.

Most of the *P. senegalus* had not yet ripened, some had apparently already spawned.

Sept. 9th. Examined six *senegalus* and two *Calamichthys*: set oviduct eggs in box. I believe that when oviducts found full of eggs without free eggs in body cavity, these are the remnants of a spawning set free after copulation and not fertilized. Only in one or two cases have I succeeded in fertilizing oviduct eggs, while practically all the body cavity eggs were either already fertilized or were easily artificially fertilized. The "run" of *P. endlicheri* seems to be quite over, that of *senegalus* nearly over, while that of *lapradii* is hardly begun. In fact the boy Apia still sticks to it that *lapradii* does not spawn in numbers until the water begins to go down. This is not until the middle of October.

The young of *Calamichthys* and *endlicheri* possibly also *senegalus* are said to be caught in great numbers at high water, i.e., in about two weeks' time.

Sept. 10th. Examined two *P. lapradii*, 26 *P. senegalus*, four *Calamichthys*, set ova from oviducts of two *Calamichthys* and one *senegalus* in box. Took photos of gallery and house.

Sept. 11th. Examined four *Calamichthys*, 12 *P. senegalus*: set ova from three *Calamichthys* and one *senegalus* from oviducts.

Sept. 12th. Examined 11 *P. senegalus*. One had oviducts full. These were put after fertilizing into bell-jar on hatching frame at 1 p.m.; by 4 p.m. most of the eggs were segmenting. One egg which I unfortunately burst was very perfectly segmenting in four equal portions, the constrictions being deeper than in the frog.

In the segmentation generally the segments appear to be almost separated from the rest. The pigment though mainly restricted to the upper pole drifts about with segmentation creating the semblance of marbling. The egg capsule is not closely attached to the egg and in some small pigmented bodies (?polar) were floating freely.   By midnight segmentation had so far advanced that the upper pole cells were not easily visible with × 8 Leitz.

Sept. 13*th.*   8 a.m., the upper pole had overgrown the lower pole to $\frac{3}{4}$, by 1 p.m. a very small knob of yolk cells protruded from blastopore.

I recovered the hatching-box and set the eggs in river water.

Sept. 14*th.*   Some of the eggs set in hatching-box seem to be going on all right.

Examined 20 *Calamichthys*, 12 *P. senegalus*; about two-thirds of the *Calamichthys* had ova in the ducts.   I tried to fertilize them.   A few have certainly started to segment.   Not one of the *P. senegalus* had ova.   Most, however, were crammed with eggs.

Sept. 15*th.* Had no fish brought. A little cheap. Mr Maidman has disappeared for two days.

Sept. 16*th.*   All eggs bad.   Went off in canoe up creek, found water all over the bush.   Had set bottle-traps but caught nothing.   Trawled lagoon but caught only jelly fish.   Examined eight *Calamichthys* and 20 *P. senegalus*.   Three *Calamichthys* had ova in ducts but could not fertilize, one *senegalus* but bad.

Had long talk with Apia.   He saw many times *Polypterus* ova attached to grasses at West end of lagoon; all of one fish near together.   Also he saw many times hatched young about $\frac{3}{4}$ inch long in a swarm, above which the parent fish remained, these he used to kill and then the young fry dispersed.   This was apparently at the beginning of November every year.   The eggs covered the sticks to which they were attached.

Sept. 17*th.*   Had nine *P. senegalus* brought, none with eggs.   Heavy rain. Soldered, etc.

Sept. 18*th.*   Made sketches in the morning of early development of *Polypterus*. In the afternoon examined ten *P. senegalus*.   Two of these had oviducts full and also some portion of the ovaries free.   Two had oviducts full.   I tried to fertilize them but had only captive males and am very doubtful if successful.

On examining hatching-box found three live *Polypterus* $7\frac{1}{2}$ mm. long, they were at an intensely interesting stage, six days old.   One I preserved in formalin and later Corrosive Sublimate.   The other two I put in bell-jar with river water and water weeds.   There was also a single unhatched egg of, I believe, *Calamichthys*.

In larval *Polypteri* no sign of mouth or spiracles or nasal tubes.   The pectoral fins are developed.   The operculum is growing back, but internal gills not yet developed.

Sept. 19*th.*   Half-a-dozen *P. senegalus*, 3—4 inches with large external gills. The eggs put down yesterday seem all to be bad.   The two larval *Polypteri* and the *Calamichthys* going on all right.   One of each put in river in rearing cage.

The larval *Polypterus* under observation was, at 9 p.m., 9 mm. in length.

The mouth was forming and there was a pit, apparently the beginning of the spiracle at the base of the gill. Pronephros conspicuous, red mass just under the gill. Suckers in front of the mouth. Colour black above with gold spots. Yolk mass not yet exhausted. External gill very short base, radiating branches five on either side. Tail absolutely diphycercal.

Had four *P. senegalus*. Soldered and put 100 *P. senegalus* in tin box.

Sept. 20th. Sunday. Made coloured figure (Pl. XV. fig. 34 *l*) of larval *P. senegalus* 9½ mm. and then preserved it.

Had young larva 5 cm. brought with splendid gills which it carried turned vertically over the head, caught in trap by my boy. Also 12 3—4 in. larvae with external gills.

Had 20 *P. senegalus* brought, one had large numbers of free eggs in the body cavity and also ducts full. Most of these were almost certainly fertilized but had been kept too long in body cavity. They were clearly segmenting but were all looking unhealthy. Most of them burst up as soon as put into water.

Sept. 21st. Of all the eggs put in hatch-box on the 18th only three have hatched. The *Calamichthys* larva was attacked by fungus and died. The *P. senegalus* larva nine days old has the yolk extending half-way to the vent from end of original yolk sac. The pectoral fins are greatly enlarged. There is a slight indication of the formation of the dorsal finlets along the extreme base of the dorsal fin-fold.

Sept. 22nd. Made dorsal sketch of *P. senegalus* larva tenth day. (Pl. XV. fig. 35 *m*.)

The body is growing greatly, the tail diminishing, hence total length not greatly increased. The pectoral fins are of great size and are much used as shown in my figure of the 1¼ inch larva. (Pl. IX. fig. 1.) The yolk has practically disappeared. Suckers gone. Notochord extends to extreme tip of tail which is now for the first time sharply pointed.

Had about 30 *P. senegalus* females brought, only two had eggs free, these not in good state. Examined 10 female *Calamichthys* one of which had large number of free eggs. I had a good male and these look hopeful. The three young larvae of *P. senegalus* are doing well. In the evening had about 50 gilled *senegalus*, two were under two inches, one was 1¾.

Sept. 23rd. Slight changes in *P. senegalus* larva to-day. Fin rays forming. Pectoral fins becoming lengthened along posterior border. Body more developed. Tail fin-fold reduced. Notochord protruding.

Of the three larvae left in hatching frame yesterday only one found to-day. This was five days old and in exactly the stage to which I reared the eldest of the original batch and which died. So that I now have a continuous series. This one killed in formalin preserved in corrosive. If put straight into corrosive too much contraction. Four *Calamichthys* brought, one had free eggs. Eight more gilled larvae 3—6 inches.

Sept. 24th.   Killed the *P. senegalus* larva reared to stage preceding that described in Zoo. paper.   Made coloured sketch (Pl. XV. fig. 36 *l*).   The stomach, liver, etc., well seen through transparent body wall, intestine full of green matter, apparently at this stage a vegetable feeder.   The vent has shifted back a good deal.

The gilled young never come to surface of aquarium to breathe.

Sept. 25th.   "Kampala" arrived with Captain and Mrs Larymore on board. He undertook to send the mail boat to call for me on the way down to meet the "Nigeria" on the 4th, sailing on the 6th.

Sept. 26th.   Soldering up and packing.

Sept. 28th.   Range finding.   River 1900 yards wide here.

Cat-fish make the cry-like noise by means of the joint of the pectoral fin.

Have had a good many young *P. senegalus* about three inches, but cannot get young fry.

Apia says that the eggs are attached in great numbers to stems in the water.   The eggs are found at the surface and down to six inches from it.

Sept. 29th.   Packing up.

Oct. 1st.   Have put a stop to supply of *Polypterus*.   I have examined over 300 female *P. senegalus*.   Nearly 100 female *Calamichthys* and about 50 female *P. lapradii* besides about 50 female *P. endlicheri* all of which had spawned.

Of the *P. senegalus* 40 had oviducts full, three had a portion of the body cavity also full.   Of *P. lapradii* two had the oviducts full.   Of *P. endlicheri* three had the oviducts full.   Of *Calamichthys* 20 had oviducts full.   Of 40 *P. senegalus* one developed 1000 eggs, one developed 30.   Of two *P. lapradii* one developed one egg.   Of three *P. endlicheri* no eggs developed.   Of *Calamichthys* one egg developed.

II.   MATERIAL AND METHODS.

Budgett's material of developmental stages of *Polypterus* consisted when it came into my hands of about 200 eggs and embryos, a number of which were more or less damaged.   Of these 33 were preserved in formalin, amongst them five eggs apparently at different stages of segmentation.   As these eggs are contained in the egg capsule, and the removal of this is practically impossible except in the fresh state without risking the destruction of the egg, I have not made use of them in this investigation.   The main part of the material consists of eggs and larvae from stage 14[1] onwards which had been extracted from the egg capsule and preserved in alcohol by Budgett at Assé.   There were no labels except a simple number in each tube so that we are without information as to the details of fixation.   Budgett mentions in his diary (cf. p. 204)

---

[1] By stage *n* is meant the stage of which the external features are shown in fig. *n* of the lithographed plates.

that he fixed certain eggs in Corrosive Sublimate and Acetic Solution and it seems pretty certain from the histological condition of the eggs and larvae that this fixative was used all through for the alcohol material.

The series is a very complete one from stage 14 up to stage 27, and the eggs and larvae are in good condition. At about the last mentioned stage, however, the larvae appear to have become sickly and begun to die off rapidly, and the large number of specimens ranging from stage 27 to 29 appear to have been in great part either dying or dead when placed in the fixing fluid. Stages subsequent to 29 are represented by five larvae of stages 30, 31, 32, 33 and 36. Of these 32 is a fragmentary specimen and 36 is unfortunately badly macerated. It will be seen that a large gap separates stage 36 from the 30 mm. larva obtained by Budgett on the Gambia.

The material as a whole has been found to be sufficient to give a pretty complete picture of the general course of development of a Crossopterygian fish, and this has of course been a great desideratum to embryologists, though it is lamentable to think of the price that has been paid for it.

Various important details must be left until more material is available. This applies particularly to the processes of segmentation and gastrulation and to the development of the skeletal system and of the genital ducts—which are hardly touched upon in the present account.

The material has been investigated by means of paraffin sections. I have not felt justified in sacrificing any considerable part of this valuable material in developing a method of celloidin imbedding specially suited to it. This is greatly to be regretted as with greater experience of embryological technique I feel more and more convinced of the absolute necessity of employing the two methods of section-cutting side by side if really reliable results are to be obtained. As pointed out in a previous paper the paraffin method when used alone is liable to be absolutely misleading where yolk laden tissues are being investigated. The individual yolk grains are not held sufficiently firmly in position by paraffin: they stick on the edge of the knife and go ploughing through the tissues obliterating some of the fine details and bringing others into view which had no real existence before the section was cut[1].

What reconstructing has been necessary has been carried out by means of the ground glass plate method[2] which, in my opinion, possesses important

[1] To illustrate the kind of error brought about by the study of paraffin sections uncontrolled by the examination of sections cut in celloidin I may instance the development of the wall of the buccal cavity in Dipneumonic Dipnoans and Urodele Amphibians. Here the appearances seen in paraffin sections can be interpreted quite well on the generally accepted view that the buccal cavity is lined by an actual ingrowth of ectoderm. That no such ingrowth takes place is in my opinion *demonstrated* by the examination of well prepared celloidin sections in which the yolk granules have been held firmly in position with their surrounding matrix of protoplasm. [Cf. Graham Kerr in *Quart. Journ. Micr. Science*, Vol. XLVI. 1903, p. 423, and on the other side, Greil and others.]

[2] *Quart. Journ. Micr. Science*, Vol. XLV. 1902, p. 5.

advantages over the ordinary methods in which use is made of plates of wax or clear glass or celluloid.

Budgett had, up till the onset of his last illness, studied only the external features of development. Of these studies we have most unfortunately no manuscript record whatever—not even a single rough note. The only record we have is contained in the series of beautiful drawings which he had made and which are reproduced in the plates accompanying this Memoir. It will be seen that the majority of the drawings of external features were made by Budgett himself: for the remainder—which have been drawn under my own supervision and for the most part from camera outlines—we are indebted to the facile brush of Mr A. K. Maxwell. I wish to make special acknowledgment of the great care and accuracy of Mr Maxwell's work: its artistic qualities need no word from me.

In conclusion, I desire to acknowledge the debt I owe to the admirable technical skill of my laboratory assistant, Mr P. Jamieson, in preparing the numerous series of sections.

III. BREEDING HABITS OF *Polypterus senegalus*.

Unfortunately we are still without definite observations regarding oviposition and fertilization in *Polypterus*. From Budgett's observations and from the information given to him by natives we may conclude that the eggs are deposited in the shallow lagoons connected with the main river early in the rainy season. They apparently adhere strongly to the surface of submerged twigs or water-plants. (See pp. 203, 207.)

As regards the mode of fertilization there are certain points—particularly the modified and erectile character of the anal fin of the male—which point in the direction of internal fertilization. (See p. 199.)

After hatching the young fish apparently accompany the parent (probably the male) in a dense swarm very much as we know is the case in Actinopterygian bony Ganoids. (See p. 207.)

IV. EXTERNAL FEATURES.

A single specimen of unsegmented egg was found amongst Budgett's material and is figured in Pl. XIII. fig. 1 *m* and Pl. XIV. fig. 1 *l*. The egg measures 1·3 mm. in horizontal diameter by ·9 mm. in vertical diameter. The upper pole is much flattened. The egg is richly pigmented. The deep brownish black pigment is absent from an area in the region of the upper pole where the nucleus is present. Round this it forms a broad continuous band, while further out towards the equator of the egg it is distributed in blotches of various shapes.

Segmentation.

From Budgett's pen and ink sketch (text-fig. 19) we see that the segmentation is at first characterised by its almost absolutely equal character. We may infer with considerable probability that two meridional furrows are succeeded by a latitudinal one which is practically equatorial.

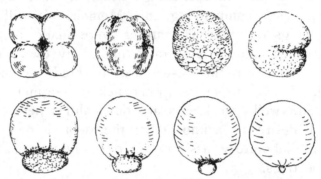

FIG. 19.   Stages 3 (view from above and from the side), 4, 5, 7, 8, 9 and 10.   Segmentation and gastrulation in *Polypterus*.   (From pen and ink sketches by Budgett.)

Gastrulation.

An early stage in gastrulation is shown in text-fig. 19 (top right hand figure). A deep groove has appeared cutting into the egg about the level of the equator and extending about half way round the circumference.   At this stage the line of invagination is practically part of a great circle of the egg.   As the line of invagination increases in length its curvature becomes greater and eventually the ends meet so that the line of invagination is now a complete circle slightly below the equator of the egg (Pl. XIV. figs. 6 *l* and 6 *h*).   The deep sharp character of the invaginated furrow causes the unpigmented mass within it to stand out conspicuously above the general contour of the egg forming an enormous "yolk plug."   The circle surrounding this becomes smaller and smaller as the lip forming it advances over the yolk (text-fig. 19, and Pl. XIV. figs. 6 *l*, 7 *l*, and 8 *l*).   There is no material for determining the relative extent of this advance at different points beyond the fact that the curve remains circular which makes it probable that the advance goes on with approximately equal rapidity all round.   The yolk plug as it diminishes in size stands out with great prominence above the general egg surface, and this great prominence of the yolk plug is in fact one of the most characteristic features of the *Polypterus* egg, as is well seen in Budgett's pen and ink sketches of the living egg (text-fig. 19).

When the yolk plug is reduced to a small vestige the first appearance of the embryonic organs takes place.   The medullary plate appears in the form of a broad flattened area in front of the blastopore (text-fig. 20, stage 12, and Pl. XIII. figs. 12 *m*, 13 *m*).   In some specimens (fig. 12 *m*) a shallow groove runs forwards along the middle of the medullary plate with a rather less distinct groove on either side, but in none of the specimens was any definite protostomal seam visible. The edges of the medullary plate are sharply marked, the whole plate being well raised up above the general surface (fig. 12 *l*). Posteriorly they converge to the region of the blastopore, but the material is not sufficient to show whether at any period they are traceable round behind the blastopore.

FIG. 20.   FIG. 21.   FIG. 22.

FIG. 20. Stages 12 and 13. Formation of medullary plate as seen from dorsal side.
(Pen and ink sketch by Budgett.)
FIGS. 21 and 22. Stages 15 and 16. Dorsal view of medullary plate. (Pen and ink sketch by Budgett.)

The medullary plate becomes considerably narrowed, assumes a pear shape, and its margins begin to project more distinctly, foreshadowing the uprising of the medullary folds (text-fig. 21, and Pl. XIII. fig. 15 *m*).

This latter calls for no special mention except that at a comparatively early stage, while the medullary groove is still widely open a conspicuous depression of the floor of the medullary groove arises anteriorly forming the cavity of the Thalamencephalon (Pl. XIII. figs. 16 *m*, 17 *m*, 18 *m*). As shown in figs. 16 *m* and 17 *m*, faint transverse grooves may appear on the floor of the medullary groove. In the eggs shown in figs. 16 *m* and 18 *m* two such grooves are distinct and in corresponding positions in the two eggs.

The medullary folds arch over in typical fashion (text-fig. 22, and Pl. XIII. figs. 17 *m*, 18 *m*, and 19 *m*) and fuse: the fusion commencing posteriorly and being delayed somewhat in the brain region. As soon as the fusion has taken place posteriorly it causes a rounded bulging to project just in front of the anus, the first rudiment of the tail fold (fig. 17 *h*). When the process of fusion is finished the neural rudiment forms a conspicuous projection above the egg's surface (Pl. XIII. figs. 20 *m*, 21 *m* : text-fig. 23). The head and the tail "fold" become equally conspicuous for a time but soon the tail projection begins to grow rapidly, and to project to a much greater extent than the head region (Pl. XIV. fig. 23 *l*).

FIG. 23. *Polypterus* embryo (stage 21) after closure of medullary folds, as seen from the right side and from above. (Pen and ink sketch by Budgett.)

At the stage shown in Pl. XIV. figs. 22 *l* and 23 *l*, a faint elevation (*pn.*) is seen on each side in certain embryos—caused by the underlying pronephros, while a smaller bulging of the surface further forwards indicates the first rudiment of the external gill.

Stage 24. When the tail projection has attained a length of about half the diameter of the egg (Pl. XIII. fig. 24 *m*) the head region of the embryo begins to broaden out considerably from side to side.

The external gill rudiment becomes more conspicuous and anteriorly the front end of the embryo forms a conspicuous blunt projection on each side —the rudiment of the cement organ.

Stage 25. The tail outgrowth is seen (Pl. XIV. fig. 25 *l*, and text-fig. 24)

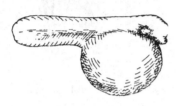

to have increased in length : it is more flattened and a thinning dorsally and ventrally indicates the development of the median fin. The cement organ (*c.o.*) now forms a conspicuous short cylindrical projection on each side of the head region ventrally. The external gill rudiment (*e.g.*) has become slightly flattened : is in an oblique position and is beginning to show the first trace of projections which will later form the pinnae.

Stage 26. The chief change is to be seen (Pl. XIV. fig. 26 *l*) in the greater length of the "tail" outgrowth. It will be borne in mind that the expression "tail" is not strictly accurate, for at this stage the position of the cloaca has become clearly visible in external view, and the cloacal aperture is seen to be situated far back near the apex of the so-called tail outgrowth. This posterior region of the body is by now becoming greatly flattened, there being present

FIG. 24. Lateral and dorsal view of young *Polypterus* (stage 25) showing rudiments of cement organs and external gills. (Pen and ink sketch by Budgett.)

a broad median fin which increases the dorsoventral diameter. In the head region the hind brain is distinctly marked off by the thinning of its roof (*h.b.*).

Stage 27. Up till now the lower half of the egg has retained its original form as part of a sphere, but about this time (Pl. XIV. fig. 27 *l*) it becomes drawn out in an anteroposterior direction so as to project less prominently than

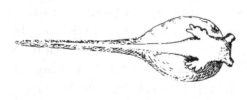

it has done hitherto. The median fin fold is still increasing in breadth and conspicuousness. The external gill rudiment now shows a somewhat palmate arrangement, short lobes beginning to project conspicuously from it.

In stages 28 (text-fig. 25) and 29 (Pl. XIII. fig. 29 *m*) these are seen to be growing out into long finger-like projections. Scattered pigment cells are now visible in the dermis.

Stage 30. Length slightly over 4 mm. The drawing out of the yolk mass in an anteroposterior direction which was noticed in stage 27 has now (Pl. XIV. fig. 30 *l*) gone on to a considerably greater extent, so that

FIG. 25. Dorsal and lateral view of young *Polypterus* of stage 28, showing the commencing pinnate development of the external gills. (Pen and ink sketch by Budgett.)

the larva is losing its tadpole shape. The external gill is increasing rapidly in size and its pinnae are becoming much more long and slender. The cement organs have altered somewhat both in shape and position. They are now cup-like in form with a conspicuous cavity and they have taken up a more ventral position (text-fig. 26, *c.o.*). The median fin fold still increases in prominence and we may now speak of an increase in the size of the true tail region—i.e. the region behind the cloacal opening.

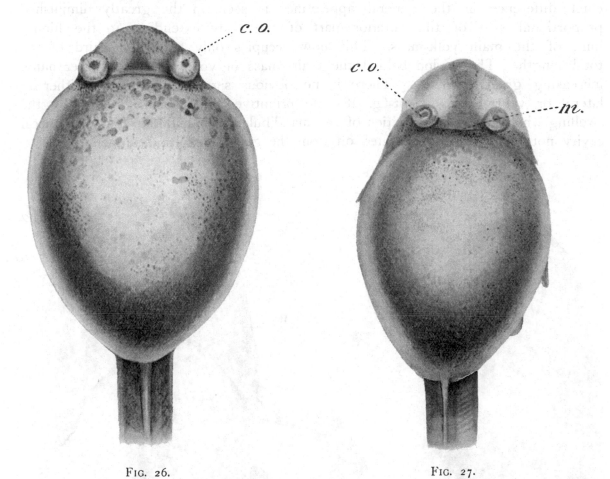

FIG. 26.   FIG. 27.

FIGS. 26 and 27. Ventral view of stages 30 and 31, showing cement organs.
c.o. Cement organ; *m.* Position of mouth.

In the single specimen of this stage there is no trace of paired limb rudiments visible in external view.

Stage 31. Length 5·5 mm. With the further elongation and shrinkage of the yolk mass the former disproportion in size between anterior and posterior body regions has now almost completely gone (Pl. XV. fig. 31 *l*). The true tail region is growing rapidly so as now to be about a third of the total length. The cement organs (*c.o.*) have become still more ventral in position and now look almost directly downwards. At their apex is a deep pit (text-fig. 27, *c.o.*).

The pinnae of the external gill have increased much in length and many of them branch. The head region has grown forwards considerably, and is becoming moulded into a distinct character. The operculum (*o.p.*) has appeared as a nearly vertical fold ventral to the attachment of the external gill. The pectoral limb (*p.f.*) is now visible in the form of a vertical flattened plate with round posterior edge.

Stage 32. (Pl. XV. fig. 32 *l.*) Length 6·75 mm. Apart from the mottled pattern due to the chromatophores segregating together in special areas, the chief difference in the general appearance is seen in the greatly diminished proportional size of the anterior part of the body extending to the hinder limit of the main yolk-mass. This now occupies only about one-third of the total length. The region behind the main mass of yolk is, on the other hand, increasing greatly in size. There is no obvious sign of the mouth either in lateral or ventral view (text-fig. 28) the primitive mouth being hidden by the swelling which marks the position of the mandibular arch, and the definitive buccal cavity not being as yet separated off from the general external surface.

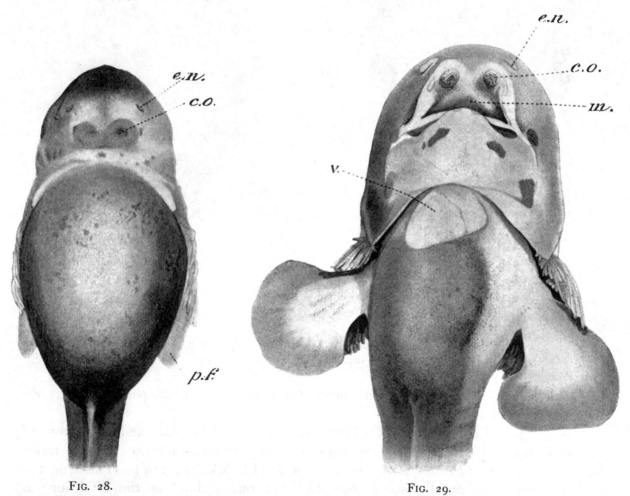

FIG. 28.   FIG. 29.

FIGS. 28 and 29. Ventral view of stages 32 and 33, showing cement organs.
c.o. Cement organ; *e.n.* Nasal involution; *m.* Mouth; *p.f.* Pectoral fin; *v.* Ventricle.

The cement organs (*c.o.*) are now much less conspicuous in side view. In ventral view (text-fig. 28) they are seen to be nearly in contact.

In front of and external to them the nasal involutions (*e.n.*) have made their appearance, no groove connecting them with the mouth. The opercular fold is more conspicuous and is growing backwards (Pl. XV. fig. 32 *l*). The pinnae of the external gill have become secondarily pinnate. The pectoral limb rudiment is growing rapidly backwards and now stretches considerably behind its base of attachment. On its outer surface it shows a curious radiating arrangement of chromatophores.

Stage 33. (Pl. XV. fig. 33 *l*.) Length 8 mm. The shape of the head as seen in profile shows a great change from stage 32, due mainly to the buccal cavity being now sharply defined. A prominent lip fold bounds the mouth opening dorsally and ventrally. Just behind the outer end of the ventral fold is a deep slit-like pit. The narial openings (*e.n.*) are very distinct and elongated anteroposteriorly (cf. text-fig. 29). The cement organs are seen situated upon the upper lip, one on each side (Pl. XV. fig. 33 *l* and text-fig. 29, *c.o.*). The opercular folds may be said to be completely formed, extending back freely for ·33 mm. and nearly meeting one another ventrally. The pectoral fin has changed greatly in shape, its hinder end having become prolonged into a sharp point. (Owing to the position of the fin this does not appear in the text-figure.) The line of attachment of the limb has altered in position. While it originally ran in an anteroposterior direction it now runs from in front outwards as well as backwards so that its surface, which originally faced directly towards the mesial plane, now faces somewhat backwards as well.

Stage 34. Length 9·5 mm. Age eight days from fertilization. Of this stage we possess only Budgett's coloured sketch (Pl. XV. fig. 34 *l*), which does not give minute detail but which shows the general coloration of the larva—reddish brown with small scattered spots and a large ⊂-shaped blotch of a bright yellow colour on each side of the head.

Stage 35. Age ten days from fertilization. This stage also is represented only by a coloured sketch which is reproduced as fig. 35 *m* on Pl. XV. The figure is of special interest as showing the position in which the pectoral fins are held at this stage—with the straight originally ventral border posterior, and the originally mesiad surface dorsal.

Stage 36. Length 9·3 mm. Age twelve days from fertilization. Of this stage we have a coloured sketch by Budgett (Pl. XV. fig. 36 *l*), and amongst the material was found apparently the actual larva figured—unfortunately much shrivelled and in bad histological condition.

The general colour effect remains the same. The gut however shows a bright green colour from ingested food. The mouth has assumed its definitive character and the lips are closely apposed. The cement organs have disappeared. The olfactory opening is now double, composed of two widely separated pores on each side. There is as yet no trace of the tubular narial outgrowths found in the adult.

With this larva the series comes to an end, and we must remain for the present in ignorance regarding the developmental phenomena shown in larvae between this stage and that of the 30 mm. larva.

INTERNAL FEATURES OF DEVELOPMENT.

V. Structure of the Unsegmented Egg.

A vertical section through the egg of stage 1 is represented in text-fig. 30.

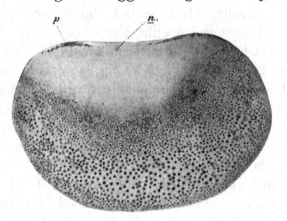

FIG. 30.   Vertical section through unsegmented egg.

n., nucleus; *p.*, pigment.

It will be seen that about a third part of the egg towards the upper pole possesses finely granular yolk which passes by comparatively sudden transition into the coarse yolk of the lower part. Lying superficially in the fine grained yolk towards the upper pole is the nucleus (*n.*), a faintly staining structure measuring ·03 mm. in diameter. The fine grained yolk shows near its surface an irregular layer (*p.*) of densely aggregated fine particles of dark brown, almost black, melanin. Immediately over the egg nucleus and its immediate neighbourhood this distinct pigment layer is absent, but here, as indeed throughout the whole of the finely yolked part of the egg, there is to be made out with the high power an evenly distributed kind of melanin dust composed of extremely fine particles just visible under the *D*-objective of Zeiss. In the presence of these fine particles of the excretory melanin we probably see, as in the finely broken up yolk of the upper part of the egg, an evidence of the relatively greater metabolic activity of that part of the egg, the pigment being formed originally through the disturbance of metabolism due to the harmful effect of light rays. The melanin deposit in the surface layers of this and other eggs has been probably made use of for protective purposes—as in the case of so many other primarily excretory products—to keep out these harmful rays[1].

There is no material to illustrate the process of segmentation.

VI. Features of development as shown by Sagittal Sections.

Before proceeding to describe the development of the various systems of organs it will be instructive to examine the appearance of sagittal sections at various stages.

Gastrulation.

Unfortunately the material during the stages of gastrulation is extremely scanty.

The egg figured in Plate XIV. fig. 7 *l* was unfortunately slightly injured along the projecting lip of the invagination groove. It was converted into sections as nearly as possible parallel to the sagittal plane, and one of these sections nearly median is shown in text-fig. 31. The groove is seen to be deepest at one side. The segmentation cavity is still visible, but has become narrowed by the rising up of the yolk cells of its floor. The roof of the segmentation cavity is as in *Lepidosiren* composed of two layers of cells, somewhat cubical in form, with finely broken up yolk and containing minute melanin granules.

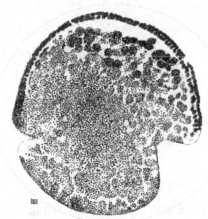

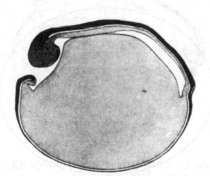

Fig. 31. Sagittal section through stage 7. The portions shown by the dotted outline close to the invagination groove were destroyed in the specimen.

Fig. 32. Sagittal section through stage 14.

Stage 14. This is the earliest stage represented in Budgett's continuous series of embryos. There were eight eggs and fragments of three others. The more perfect ones I have labelled 1 A—1 H. They all correspond closely with Plate XIII. fig. 14 *m* in external appearance. As regards internal structure

[1] The absence of pigment is well seen in those allied forms where the eggs are shielded by other means from the light, as in the case of those Dipnoans whose eggs develop in burrows, and those numerous anurous amphibians where the eggs are embedded in light-proof foam.

the chief points are brought out by text-fig. 32. The sagittal section shows a widely open blastopore. The thick yolk laden floor of the archenteron no longer projects outwards to form a yolk plug. The medullary plate is marked off from the surrounding epiblast by a distinct ridgelike thickening. The plate itself consists of a single layer of columnar cells, the superficial layer of the epiblast stopping short about the summit of the bounding ridge.

Stage 15 + (*Egg* 3 B). The appearance of a sagittal section of an egg slightly older than that shown in Plate XIII. fig. 15 *m* is shown by text-fig. 33. The blastopore is reduced to a very narrow opening, patent only in one section, and even here the appearance is such as to suggest the possibility of there being *no* opening in the living egg at this stage. The ventral aspect of the embryo is covered with a thin layer of ectoderm, showing only here and there two layers of cells. Passing dorsalwards along the anterior surface the two layered condition becomes marked. Approaching the medullary fold the deep layer thickens and shows mitotic figures. The superficial layer can be traced only as far as the top of the medullary fold, where it rapidly tapers off—the ectoderm within the medullary folds, i.e. the ectoderm of the medullary plate, consisting only of the deep layer—a single stratum of tall columnar cells. Posteriorly this passes into the thick undifferentiated mass of cells forming the dorsal lip of the blastopore.

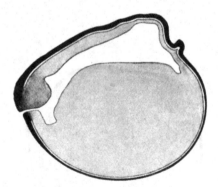

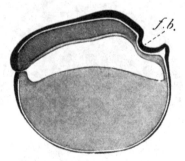

FIG. 33. Sagittal section through egg slightly older than stage 15.

FIG. 34. Sagittal section through egg of stage 17. *f.b.*, cavity of primitive forebrain.

Stage 17 (*Egg* 3 E). In sagittal section (text-fig. 34) the chief advance beyond the condition of stage 15 + is found in the anterior part of the medullary plate. The depression of the floor of the medullary groove is seen in sagittal sections to be quite definitely the cavity of the forebrain. The anterior edge of the depression is formed by the anterior part of the medullary fold, the floor being formed by the region of the medullary plate behind this: the tip of the infundibulum would then correspond not to the front end of the medullary plate but to a point considerably farther back. The floor of the medullary groove appears to be still one cell thick. Posteriorly the medullary plate passes into the mass of undifferentiated cells lying dorsal to the blastopore, which is already bulging out somewhat—foreshadowing the development of the tail. The notochord can just be made out in process of being split off from the dorsal endoderm. The blastopore is much narrowed: it is again questionable whether there was actually a patent opening at this stage in the fresh condition. The ventral ectoderm is now seen to be two layered for some distance in front of the blastopore.

Stage 20 (*Egg* 4 B). Text-fig. 35. The general appearance is much altered by the closure of the medullary groove, which in sagittal sections appears to take place quite suddenly. The neural rudiment is seen as a wide tube, both roof and floor composed of a single layer of columnar cells, dilated anteriorly to form the primitive forebrain. In this region the wall becomes thicker and more than one cell layer thick. There is a considerable deposit of melanin in the protoplasm of the cells forming the wall of the neural tube as well as in the external ectoderm. In the former the pigment is concentrated at the inner ends of the cell bodies, i.e. in what were originally their outer or superficial ends before the neural tube became closed.

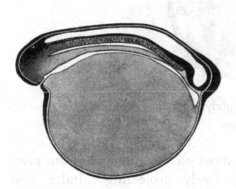

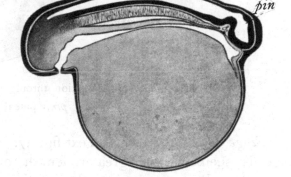

FIG. 35.   Sagittal section through egg of stage 20.   FIG. 36.   Sagittal section through egg of stage 23.

pin., pineal rudiment.

The notochord is now sharply marked off, its cells having become flattened. The primary sheath is distinctly visible as a fine line bounding the section of the notochord. The blastopore or anus is distinctly open: the tail knob projects decidedly.

Stage 23 (*Egg* 5 B). Text-fig. 36. The general ectoderm is distinctly two layered.

It is thin over the ventral surface, thicker elsewhere—especially over the front of the head.

In the brain the upgrowth of the floor which marks off the forebrain region has become much more prominent and in the roof it is possible to make out the thinning process which indicates the hindbrain, the slight depression which marks off the midbrain from the thalamencephalon, and just in front of this a very slight bulging outwards of the roof—the rudiment of the pineal body.

The notochord has its cells more flattened than in the previous stage: it does not extend quite to the infundibulum, its tip being separated from it by a few large heavily yolked cells.

Beneath the notochord the subnotochordal rod is visible in the actual sections, not however extending forwards to the tip of the notochord.

The gut cavity is widened out in front and then passes down ventrally as a narrow chink. The ventral wall of the gut is of course enormously thick, formed of the main mass of yolk. Dorsally its wall is in front quite thin, composed of a single layer of flattened cells: posteriorly these become large and columnar, and the wall becomes thickened accordingly.

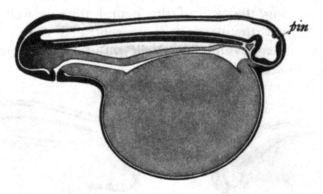

FIG. 37. Sagittal section through egg slightly older than stage 24.
pin., pineal rudiment.

Stage 24 + (*Egg* 6 G). Text-fig. 37. The most obvious advance from previous stage is seen in the greater length of the freely projecting hinder end of the body. By superposing the outlines of the sagittal sections of preceding stages it is seen that while the postanal or true "tail" region shows a slight increase in length, by far the greatest increase is praeanal, so that the anus has been carried back and is still close to the tip of the hind end. The ectoderm is still conspicuously thickened over the front of the head and also over the tip of the tail. In the brain the pineal rudiment has become more distinct. In front of the notochord there is a compact mass of heavily yolked protoplasm between it and the infundibulum.

In the alimentary canal rudiment the chief point to notice is that the anterior end of the alimentary canal is now becoming cut off from the rest of the hypoblast, the split between the two showing a few yolky mesenchyme cells. It is still solid, the lumen of the gut extending at this stage from its hinder limit back to the cloacal opening. Behind the level of the cloacal opening the lumen projects slightly into the undifferentiated mass which forms the tip of the tail.

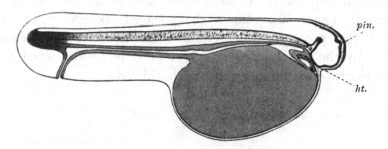

FIG. 38. Sagittal section.
pin., pineal rudiment; *ht.*, heart rudiment.

Stage 26 (*Egg* 8 B). Text-fig. 38. The leading feature here is again the increase in length of the praecloacal portion of the "tail." In the brain the pineal rudiment projects more freely and in its lumen are seen granules of dark pigment. The thinness of the hindbrain roof has become much more marked in comparison with that of the mesencephalon in front of it.

The notochord cells have become greatly vacuolated, their yolk has been greatly diminished and the tip of the notochord now extends further forwards. The mass of yolk-laden protoplasm between it and the infundibulum appears now to be assuming the characters of loose mesenchyme.

The folding off of the gut has gone a step further, the folded off anterior portion now extending so far back as to include part of the lumen. In the space beneath the folded off part of the gut irregular masses of mesenchyme (*ht.*) indicate the material from which the heart will develop later.

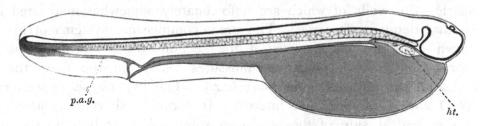

FIG. 39. Sagittal section; stage 29.
ht., heart rudiment; *p.a.g.*, postanal gut.

Stage 29 (*Egg* 13 G). Text-fig. 39. The posterior region of the body is seen to have elongated considerably, and the lengthening process is now affecting the postanal portion so that we may now speak of a true tail region being formed. With this elongation of the tail region the mass of undifferentiated material of the tail tip, which in earlier stages lay close to the cloaca, has become removed to a considerable distance from it. It remains connected with the cloacal wall however by a drawn out strand of hypoblast forming a well marked postanal gut (*p.a.g.*). Above it lies the aorta and below it a well marked "subintestinal" vein. The postanal gut is quite solid, and indeed at this stage the lumen of the true alimentary canal has almost disappeared except in the cloacal region, where it receives the pronephric secretion. The heart and pericardium are now well developed. In the central nervous system it is to be noted that the central canal ends posteriorly in a dilated bulb with thin roof.

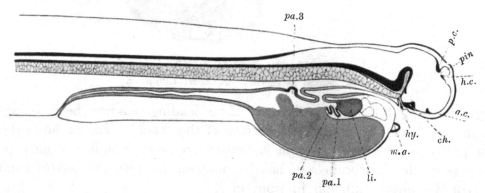

FIG. 40. Sagittal section: stage 32.

a.c., anterior commissure; *ch.*, optic chiasma; *h.c.*, habenular commissure; *hy.*, pituitary involution; *li.*, liver; *m.a.*, ridge forming ventral boundary of mouth; *pa.* 1, 2, 3, pancreatic rudiments; *p.c.*, posterior commissure; *pin.*, pineal rudiment.

Stage 32. *Larva* 18. Text-fig. 40. There is unfortunately a big gap between this and the last mentioned sagittal series. The larva has advanced far in its development and there is a correspondingly great change in the appearance of the sagittal section. The postanal part of the body has much increased. The brain has increased in size. The pineal body has also increased to form a large rounded vesicle—the walls of which are unfortunately somewhat macerated in the specimen so that detail is obscured. The great commissural systems of fibres are now laid down—and are conspicuous in the section—on the roof of the brain the posterior (*p.c.*) and habenular commissures (*h.c.*), on the floor the optic chiasma (*ch.*) and the anterior commissure (*a.c.*). The hypophysis is seen for the first time (*hy.*) and is of particular interest. It forms a short saccular structure opening on the ventral side of the head—on what will later become the roof of the buccal cavity or stomodaeum. This is bounded below by a ridge into which pass laterally the mandibular arches (*m.a.*).

The notochord has increased much in size, its cells except for the superficial chorda epithelium having now reached their maximum of vacuolation. The main mass of yolk is much reduced in size while the pericardium and heart and liver rudiment (*li.*) in front of it show a corresponding increase.

VII. GENERAL DEVELOPMENT OF MESODERM, NOTOCHORD AND HYPOCHORDA.

MESODERM. The mesoderm in *Polypterus* arises in the same way as it does in *Lepidosiren* and *Protopterus* and *Petromyzon*, i.e. it becomes separated off from the yolk or main mass of hypoblast by the formation of a split which first separates mesoderm from notochord and then spreading outwards splits off the mesoderm from the definitive endoderm. Already before there is any actual separation the superficial portion of the primitive entoderm which will become mesoderm becomes distinguishable from the underlying definitive endoderm by its yolk assuming a more finely grained character. In *Polypterus* as elsewhere the breaking up of the yolk into finer grains is constantly seen where active development is about to take place, it being no doubt the first stage of the digestive process by which the yolk is rendered available for use as food in the active metabolic processes associated with development. There takes place a considerable growth in bulk of the part of the mesoderm rudiment lying adjacent to the notochord with the result that the ectoderm of the medullary plate becomes raised up into a low rounded swelling on either side of the middle line (cf. Pl. XIII. figs. 12 *m* and 13 *m*). As regards segmentation of the mesoderm rudiment: in stage 15 there are signs of segmentation at the anterior end but they are not distinct enough to make sure how many segments are marked off. At stage 16 an embryo cut into longitudinal section showed seven protovertebrae on one side.

On account of the unfavourable nature of the material I do not propose to attempt to trace out the details of development of the myotoms.

NOTOCHORD. The notochord arises as usual from a thickening of the archenteric wall in the mid-dorsal line. Along its ventral side there runs a distinct groove in early stages, pointing to the ancestral mode of formation from a longitudinal evagination of the gut wall, as is still seen in *Amphioxus*. By stage 16 the notochordal rudiment is separated off from the gut throughout the greater part of its length. A distinct line bounding it shows that the primary sheath is already present as a thin cuticle. In sagittal sections it is seen that the cells of the chorda rudiment are becoming flattened. By about stage 20 the flattening of the cells is more marked and they have become platelike in character. This does not apply to the extreme front end of the notochord. Here the cells lag behind in development and in front they pass by insensible gradations into the ordinary endoderm cells. In stage 23 the continuity with hypoblast is still seen anteriorly, but further back the notochord is now widely separated from the hypoblast. The primary sheath is now conspicuous. Vacuolation of the chorda rudiment begins about stage 24 or 25, large isolated vacuoles appearing here and there. What in previous stages looked like the front end of the chorda rudiment is now seen to have separated off, to form a mass of yolky protoplasm lying between the tip of the notochord and the infundibulum. In stage 25 the notochordal substance is seen to be more highly vacuolated except at its front end where the cells are dense and much flattened as if by pressure from behind. The mass of protoplasm in front of the chorda is obviously being somewhat squashed between chorda and infundibular wall. As development goes on the vacuolation of the notochord becomes more and more pronounced. The mass of protoplasm lying in front of the tip of the notochord loosens out and becomes converted (stages 26 and 27) into ordinary loose spongy mesenchyme. The notochord increases much in bulk—mainly by increase of the vacuoles. In stage 30 there is seen a little mass of dense, not vacuolated, protoplasm right at its tip. This also, however, has become vacuolated in stage 32, bringing about a sudden increase in length which brings the tip of the notochord into close contact with the infundibular wall.

In the later stages the notochordal sheath shows marked increase in thickness due to the commencing deposition of the secondary sheath internal to the primary. In stage 36 this is about $3\,\mu$ thick. There is a thin nucleated layer of chordal protoplasm lining the sheath but no obvious tendency to form a columnar chordal epithelium. In the 30 mm. larva the secondary sheath is about $25\,\mu$ in thickness. It is clear and jellylike, and there are no signs of attempts to colonize it on the part of the cartilage cells lying outside.

HYPOCHORDA. That enigmatic structure the subnotochordal rod duly makes its appearance in *Polypterus*. It is seen at the height of its development about stage 24, forming a cylindrical rod of heavily yolked cells lying immediately beneath the notochord. Anteriorly—about the level of nephrostome B—and posteriorly—a little anterior to the cloaca—it merges into the mass of hypoblast. Two points of interest are to be noticed, (1) that its surface is covered by a fine cuticular sheath resembling the primary sheath of the notochord, and (2) that its cells show a slight though quite distinct tendency to vacuolation like the chorda cells.

As regards the development of the Hypochorda there is no material to show its precise mode of separation from the main mass of hypoblast. In embryo 4 B (stage 20) it is already separated off throughout the greater part of its extent. As regards its fate : in 8 A (about stage 26) the hypochorda is becoming flattened out between notochord and the developing dorsal aorta in its anterior portions. Quite in front it has broken up into mesenchyme and no longer exists as a definite organ. Posteriorly it remains unchanged. In succeeding stages it gradually breaks up and disappears from before backwards. In 11 A (stage 28) it has almost completely gone in the trunk, though it still persists in the postcloacal region.

It appears somewhat vain to speculate as to the phylogenetic meaning of the Hypochorda. That it has played an important *rôle* in phylogeny is indicated by its wide occurrence as a vestigial structure particularly amongst the anamnia. Its general mode of development, its possession of a cuticular sheath in *Polypterus* (as in Amphibia—Field[1], Bergfeldt[2]), and the development of a vacuolar condition in *Polypterus* appear to point in the direction of its representing a second notochord which has now been reduced to a functionless vestige. There is no mechanical objection to the idea that there may formerly have existed more than one notochord, provided that they were situated in the mesial plane, as their presence there would be quite compatible with the movements of lateral flexure for which the mechanical arrangements of the primitive vertebrate with its laterally situated muscle segments are specially adapted.

The extension backwards to the level of the cloaca or even into the tail appears to be a grave difficulty in the way of the homologizing of Hypochorda with the epibranchial groove of *Amphioxus*.

VIII. DEVELOPMENT OF ALIMENTARY CANAL.

1. General features.   2. Cement organs.   3. Buccal cavity—Hypophysis.   4. Pharynx and gill clefts, Lung.   5. Intestine and spiral valve.   6. Cloaca.   7. Liver and Pancreas.

1. The general features in the evolution of the alimentary canal are seen by examining the series of sagittal sections.

In stage 3 B (text-fig. 33) a large archenteric cavity is seen to be present, its lower wall greatly thickened. A little way in from the external opening or blastopore the floor dips down into a well-marked recess. Anteriorly the cavity is seen to extend downwards to some distance below the level at which the mouth will appear later.

In stage 3 E (text-fig. 34) the dome shaped floor of the enteric cavity is seen to be flattening down considerably as the yolk is consumed.

In stage 5 B (text-fig. 36) the tail outgrowth is beginning to form, and with it the blastopore or cloacal opening is carried backwards and now looks almost directly downwards. Up till now the enteric cavity has extended forwards close to the skin at the point where the mouth will appear. At this stage however the precociously developing pericardiac coelom begins to develop, and as it does so there is formed dorsal to it, squeezed as it were in between the expanding pericardium and the floor of the Thalamencephalon, a solid plate-like structure of hypoblast which will later give rise to the pharynx. With the development of this pharyngeal mass the anterior end of the enteric cavity comes to lie further and further back until it is far behind the tip of the notochord (cf. text-figs. 37, 38, 39).

At the other end of the alimentary canal the cloacal opening is seen to be carried further and further back as the hinder trunk region extends in length. The cavity of the cloaca remains widely open, correlated with the early functional activity of the pronephros whose ducts open into it. In front of the cloaca the

---

[1] *Morphol. Jahrb.* XXII. 1895, S. 349.   [2] *Anat. Hefte* (1), VII. 1896, S. 95.

lumen becomes obliterated in parts for a time by the walls coming together (text-fig. 39).

At about this period active growth begins in the tail or postcloacal region. As this goes on the mass of indifferent yolky tissue constituting its tip becomes pushed further and further back, its anterior portion undergoing differentiation from before backwards into continuations of the various tissue systems, neural, notochordal and others lying anterior to it. In this way there is differentiated a well-marked backward prolongation of the tissue forming the gut wall, in other words a "postanal" gut. As seen in text-fig. 39 this structure is particularly conspicuous in *Polypterus*. It forms a solid cord of yolk-laden cells which at no time develops a lumen and which as the tail increases in length gradually thins out and eventually disappears completely (about stage 31—32). In the last of the sagittal series important advances are seen in the structure of the alimentary canal. Now for the first time the lumen is patent throughout. The buccal cavity is only on its ventral side marked off sharply from the extrabuccal skin surface—by a prominent fold containing the rudiment of the mandibular arch (text-fig. 40, *m.a.*). Dorsally is seen the pituitary involution with thick yolky walls and having a distinct narrow lumen opening freely to the exterior (*hy.*).

Passing backwards along the alimentary canal the most prominent landmarks in its topography are seen to have been already laid down. The stomach is seen projecting back considerably behind the pyloric region. The anterior region of the intestine has its ventral wall still formed by a great mass of yolk except at its extreme front end. The dorsal wall on the other hand has become thin and its hypoblast forms a distinct epithelium. The liver rudiment (*li.*) now forms a large rounded projection of the gut wall, conspicuous in sections from its yolk being practically entirely used up. Just posterior to the liver two tubular outgrowths of the ventral enteric wall are seen—the ventral pancreatic rudiments (*pa.* 1, *pa.* 2). Far removed from these, behind the tip of the stomach, the dorsal wall of the intestine gives off a rounded caecum which must be interpreted as a dorsal pancreatic rudiment (*pa.* 3). A little way behind this the yolk mass terminates, and from this point onwards the gut forms a tapering tube lined by definite epithelium back to the anal or rather cloacal opening.

2. *Cement organs.* One of the most interesting derivatives of the primitive gut wall, in fact one of the most interesting of the organs whose development has been investigated in *Polypterus*, is the cement organ which forms so conspicuous a feature on each side of the head region of the larva.

The rudiment of the cement organ is first distinct about stage 20 or 21. In horizontal sections through such a stage (text-fig. 41) a rounded pocket-like diverticulum (*c.o.*) of the gut wall is seen on either side projecting towards the epiblast and pushing this up into a slight rounded elevation. Text-fig. 42 shows a transverse section through about the same stage. In sections through succeeding stages (text-figs. 43 and 44) the diverticulum is found to project more decidedly from the general hypoblast: it projects upwards as well as outwards. It is still in

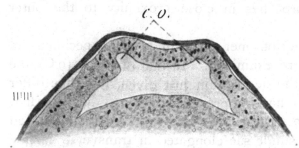

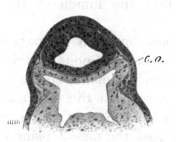

FIG. 41.   Horizontal section through rudiments of cement organs (*c.o.*) in stage 20.

FIG. 42.   Part of transverse section of embryo of stage 20 showing rudiments of cement organs.

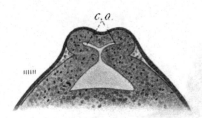

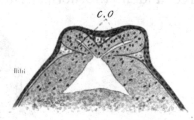

FIG. 43.   Horizontal section showing cement organ (stage 23).

FIG. 44.   Horizontal section showing cement organ rudiments (stage 23).

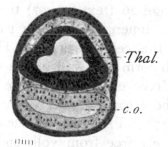

FIG. 45.   Horizontal section through cement organ rudiment of stage 24.

FIG. 46.   Transverse section through anterior end of head (stage 25) showing fusion of the two cement organ rudiments (*c.o.*).   *Thal.*, thalamencephalon.

stage 24 + (text-fig. 45) connected at its lower end with the main mass of hypoblast but its attachment is obviously becoming constricted. Further, on examining the series of sections it is seen that the stalk of attachment has become solid, thus isolating the cavity of the outgrowth from the main gut cavity. In an embryo of stage 25 the diverticulum has become completely isolated from the main mass of hypoblast, forming a rounded sac with thick yolky walls, contained within the prominent projection (Pl. XIV. fig. 25 *l, c.o.*) which forms so conspicuous a feature in the external characters of the young Polypterus of this stage. The sac is separated from the main mass of hypoblast by a considerable distance, filled with mesenchyme.   On its ventral aspect the wall of the sac is found to be completely fused with the ectoderm.   The wall formed by this fusion has become noticeably thin, so that the lumen of the sac approaches in close proximity to the outer surface.

I should mention in passing a condition met with in certain specimens of stage 25.   Six specimens of this stage were examined.   In one of these (9 C) the cement organ rudiments correspond with the description just given, but they differ from the normal in the fact that the two have approached close to one another. In another specimen of the same age (9 A) the two sacs are completely fused together across the mesial plane to form a single sac elongated in transverse section (text-fig. 46, *c.o.*) and betraying traces of its double origin in a slight backward bulging of its lumen on each side.   In the remaining four specimens the organs were paired and in the normal lateral position.

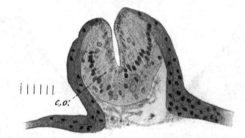

FIG. 47.   Horizontal section through cement organ of stage 26.   *c.o.*, secretory epithelium.

In stage 26 (text-fig. 47) the thin outer wall of the sac has become perforated so that its lumen, originally a part of the gut lumen, opens to the exterior as the cavity of the cement organ.   The numerous yolk granules in the tall columnar cells which form the wall of the cavity betray their hypoblastic origin.   Apart from this they have every appearance of being merely an involuted portion of the deep layer of the ectoderm into which they are continued round the margin of the cavity (text-fig. 47).   The ends of these columnar cells next the lumen of the gland are comparatively free from yolk and have a clear appearance.   In sections there are fine particles of pigment visible—possibly of artificial origin.   The superficial layer of the ectoderm is continued on to the margin of the cup but there ceases.   During the next stages the cement glands are actively functional and during their activity the reserve food in the form of yolk granules becomes gradually used up, but is apparently just sufficient to last through the period of activity.   By stage 30 (larva 16) many of the cells are completely denuded of yolk.   In stage 31 (larva 17) there are only a few large yolk granules left (text-fig. 48) and signs of degeneration are beginning to appear.   The gland cells are becoming more slender, causing the cup to shrivel somewhat, and within the cell bodies there is a considerable deposition of black pigment.   In the next stage (stage 32, larva 18) the yolk is reduced to an occasional odd granule, the shrivelling of the whole organ is more pronounced, blood spaces have increased in abundance in the neighbourhood of the secretory epithelium, and the latter has become vascularized by prolongations of the blood spaces passing into it (text-fig. 49). In stage 30 the process of degeneration has proceeded further and the cell boundaries within the secretory epithelium have broken down, and finally in the last specimen (larva 20, stage 36) of the series the cement organ seems to have completely disappeared as such. The final stages in its atrophy will have to be studied when further material is available.

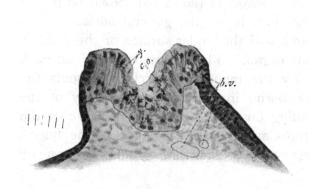
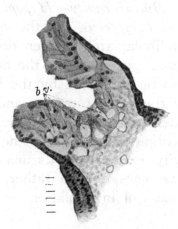

FIG. 48.   Cement organ as seen in transverse section of stage 31. *b.v.*, blood vessels; *c.o.*, secretory epithelium; *y.*, yolk granules.

FIG. 49.   Cement organ as seen in longitudinal vertical section of stage 32. *b.v.*, blood vessels.

It need hardly be pointed out that the remarkable method of development of the cement organs of *Polypterus* and of at least *Amia*[1] amongst the actinopterygian Ganoids raises these organs to a very different level of morphological importance from that which they would have were they mere local concentrations of the glandular elements of the ectoderm. While their morphological importance is quite clear from their mode of development in the relatively archaic Ganoids *Polypterus* and *Amia* it is far from clear what is precisely the morphological significance to be attached to them. They may be interpreted in three different ways: (1) they may be homodynamous with the visceral clefts, (2) they may represent a pair of enterocoelic pouches homodynamous with the coelomic segments, or (3) they may be organs simply *sui generis*—at least amongst craniate Vertebrates. It is clear that the last-mentioned view need only be considered in the event of the first two proving to be untenable as working hypotheses. I do not propose to attempt to discuss for the present this extraordinarily difficult and obscure morphological problem. I would only remark that upon the whole (1) the anatomical features in the development of these organs suggest more strongly their homology with gill pouches, while (2) their physiological features—their formation of excretory material to be used as cement is rather suggestive of enterocoelic homologies. There is perhaps the possibility of the gill pouches and enterocoelic pouches being homo-

---

[1] Phelps, *Science*, N.S. IX. 1899, p. 366; Eycleshymer and Wilson, *Amer. Journ. Anat.* V. 1906, p. 154.

dynamous. This seems on the face of it highly improbable, but it would in my opinion be rash in the present state of our knowledge of the development of the mesoderm of the head to assert definitely that such homodynamy is impossible. As regards the occurrence of homologues of these structures in other forms I need for the present merely indicate the probability that they correspond with the "praemandibular head-cavities" found in various vertebrates.

3. *Buccal cavity—Hypophysis.* The buccal cavity develops in very simple fashion in *Polypterus.* In the sagittal section of stage 32 (larva 18) shown on p. 224 the mandibular arch is seen distinctly projecting above the general surface. The wide space lying between the mandibular arch and the under surface of the head is walled in on each side by the large cement organ. This space, bounded on each side by the cement organs and ventrally by the mandibular arch, represents the still widely open buccal cavity. It will be borne in mind that the cavity of the cement organ which bounds the buccal opening on each side is a derivative of the gut cavity, is in fact in its anatomical features an enterocoelic pouch. The buccal cavity becomes defined rather by the upgrowths of structures round it than by any process of invagination.

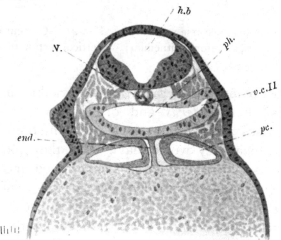

FIG. 50.   Transverse section through spiracle rudiment (*v.c. II*): stage 25.
end., endocardium rudiment; *h.b.*, hindbrain; *N.*, notochord; *pc.*, pericardial cavity; *ph.*, pharynx.

Hypophysis. The pituitary body is first seen in larva 17 (stage 31) as a thickening just at the junction of the entoderm with the outer skin, and agreeing with the entoderm in being rich in yolk. In larva 18 (stage 32) a lumen has appeared, and it forms a pouch-like structure opening close to the primitive mouth opening, i.e. the anterior limit of the entodermal part of the alimentary canal which has also developed its lumen since the preceding stage. The hypophysis thus opens just at the hind end of what will become the stomodaeum. The organ is in the form of a short blindly ending tube which passes backwards from its opening between infundibulum and the entoderm of the pharynx. In its posterior portion it is composed of columnar cells : anteriorly its cells are still laden with yolk.   In larva 19 the aperture of the hypophysis is not visible in the series of 5μ sections, though it is quite clear in the following stage.   Whether there is actually a temporary closure of the pituitary opening cannot be decided till further material is available.

4. *Pharynx and branchial clefts.   Polypterus* does not afford favourable material for the study of the minute detail of gill cleft formation and I will therefore deal with their development very shortly.

The spiracle is the first of the definitive clefts to make its appearance in rudiment in the form of a yolk-laden projection of hypoblast towards the skin.   Text-fig. 50 shows a transverse section through the spiracle rudiment of stage 25.

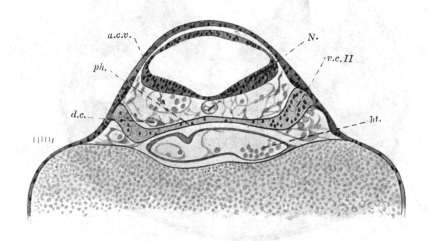

FIG. 51.   Transverse section through spiracle rudiment.   Stage 27—29.
a.c.v., anterior cardinal vein ; *d.c.*, dorsal carotid artery ; *ht.*, heart ; other letters as in Fig. 50.

In stage 27—29 (text-fig. 51) the upward tilt of the cleft rudiment is becoming more marked ; the lumen of the cleft rudiment—as of the pharynx at this level—has disappeared, while its yolky substance (*v.c. II*) is seen to be continuous at its outer end with the deep layer of the ectoderm.

In stage 31 (text-fig. 52) the cleft rudiment (*v.c. II*) forms a thin nearly vertical solid lamina.   At its lower end it bulges out somewhat to join on to the pharynx, which has now developed a secondary cavity (*ph.*) in its lateral portions.

In stage 33 (text-fig. 53) the pharynx has developed practically its definitive relations.   The spiracle now forms an extremely thin nucleated protoplasmic strand (*v.c. II*) passing upwards and outwards from the pharyngeal roof at its outer edge just over the hyobranchial cleft, and passing at its outer end into the deep layer of the ectoderm.   There is no trace of lumen.

Finally in larva 20 (stage 36), the same conditions are found only, a lumen has developed secondarily throughout the spiracle rudiment except at its outer end, which therefore still remains closed.

As regards the clefts posterior to the hyomandibular there is little to be said. The spiracle was noteworthy by its early appearance: the remaining clefts on the contrary, with the exception of the hyobranchial, are very late in making their appearance.

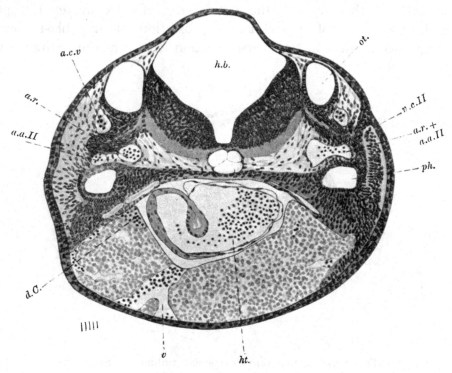

FIG. 52.   Transverse section, stage 31, passing through spiracular rudiment (*v.c. II*).
a.a. II, aortic arch II; *a.r.*, aortic root; *d.C.*, ductus Cuvieri; *ot.*, otocyst; *v.* vein; other letters as in preceding Figures.

In stage 31 the pharynx is completely solid except in the branchial region, where along each side a wide lumen is present (text-fig. 52 *ph.*). This passes out into the opercular cavity by what must be the precociously developed hyobranchial cleft. Why this cleft should develop before the others is not clear, for anterior to it the pharynx is quite solid and there cannot therefore be any passage of water through it. The next cleft is indicated merely by a double row of nuclei marking out the position of its walls, between which a split will appear later. Here and there a yolk granule in amongst the nuclei indicates their hypoblastic character. The next cleft is marked out similarly but less distinctly.

In larva 18 all the clefts are still without lumina except the hyobranchial.

In larva 19 four clefts behind the spiracle are all perforated.

Lungs. The first trace of lung rudiment occurs in stage 33, where there is found on the ventral side of the pharynx a widely open longitudinal groove formed by the dipping down of the slightly thickened pharyngeal epithelium—(text-fig. 54). This groove-like modification of the pharyngeal wall is continued back through one or two sections as a solid ridge of the pharyngeal epithelium.

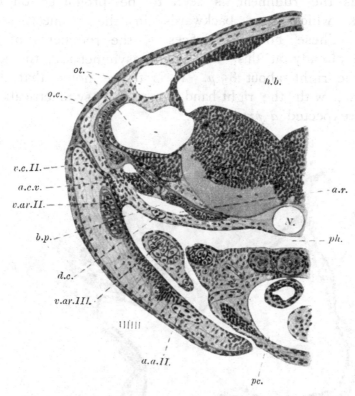

FIG. 53. Transverse section, stage 33, showing spiracular rudiment (*v.c. II.*).
b.p., basal plate; *o.c.*, otic capsule; *v.ar.*, visceral arch; other letters as in the preceding figures.

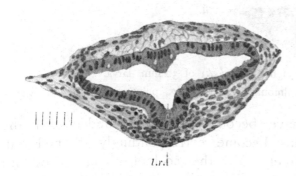

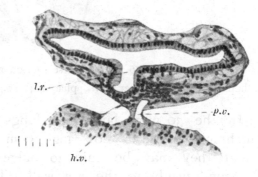

FIG. 54. Section through lung rudiment (*l.r.*), stage 33.

FIG. 55. Transverse section through lung rudiment, stage 36.

h.v., hepatic vein, with liver beneath; *l.r.*, lung rudiment; *p.v.*, pulmonary vein.

In stage 36 (text-fig. 55 *l.r.*) the mid-ventral groove is much deeper and is not so widely open to the pharyngeal cavity. The groove extends through a length of about 75 μ. Posteriorly the bottom of the groove spreads out laterally, so that in transverse section it has the form of an inverted T. The ⊥ is not symmetrical—the left hand limb being considerably longer than the right. Traced backwards the rudiment is seen to be prolonged on each side as a horn-like process which runs backwards in the connective tissue of the splanchnopleure. These horn-like processes, the rudiments of the two lungs of the adult, are already at this early stage asymmetrical, the left being about 20 μ in length, the right about 85 μ. It is to be noted that the glottis is in line, or nearly so, with the right-hand lung—not symmetrical to the two as might have been expected *à priori*.

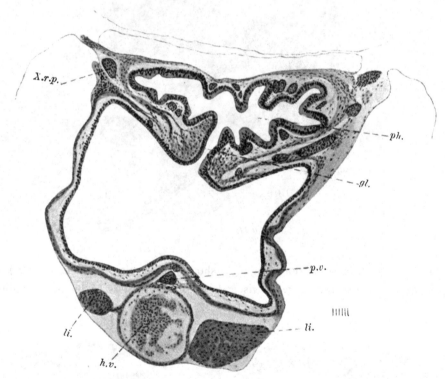

FIG. 56.   Transverse section through glottis (*gl.*) of 30 mm. larva.
h.v., hepatic vein; *li.*, liver; *ph.*, pharynx; *p.v.*, pulmonary vein; *X.r.p.*, pulmonary branch of vagus.

In the 30 mm. larva the lungs have become greatly expanded and the splanchnic mesoderm covering them has become correspondingly thinned out, so that they may be said to bulge freely into the coelom, though they are still bound firmly to the gut wall. The right lung has grown back as far as the cloaca while the left projects only about 3 mm. behind the glottis. The glottis forms a slit about ·2 mm. in length. In the region of the glottis (text-fig. 56), the two lungs are continuous across the middle line. In fact it would be correct to describe the lungs as an unpaired sac which projects back into two horns of very unequal length. Anteriorly also the median lung chamber extends forwards in the form of two horn-like processes. The glottis extends forwards beyond the point where these two horns diverge, and it is to be particularly noted that it extends along the roof of the right hand horn. This brings out strongly the essentially asymmetrical character of the lungs, which was indicated in the preceding stage.

5. *Intestine and spiral valve.* As in the case of Elasmobranchs and Dipnoans the spiral valve is foreshadowed by the gut wall forming a spirally coiled tube as is shown in text-fig. 57, which shows a reconstruction of the hind end of the gut of stage 32.

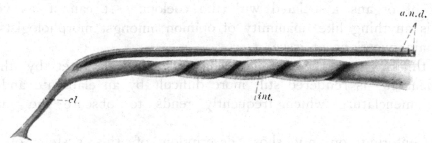

FIG. 57. Reconstruction of intestine (*int.*) of stage 32.
a.n.d., archinephric ducts; *cl.*, cloaca.

6. *Cloaca.* As is clear from the series of sagittal sections the cloacal cavity is a persisting portion of the original archenteric cavity.

7. *Liver and Pancreas.* What appears to be the liver in a dissection of *Polypterus*[1] is really a Hepatopancreas, liver and pancreas being completely fused together. The study of sections shows that the pancreas forms a widely extending sheet of tissue covering over a considerable area of the ventral side of the liver. The material is unfortunately not such as to make it possible to work out the details of the development of the liver and pancreas. I shall therefore mention merely such of the more important developmental features as can be certainly made out from the four oldest specimens of the series.

In stage 32 (text-fig. 40, p. 224) the liver rudiment forms a compact somewhat pear-shaped mass projecting forwards beneath the oesophagus and possessing a wide flattened lumen only in its posterior portion, close to its junction with the gut wall. The pancreas is in the form of three distinct rudiments, each forming a branched projection of the gut wall, the branches being in close apposition so as to form a compact mass in each case. The two ventral rudiments are in close contact with one another, the dorsal rudiment is widely separated from them. The branches of the anterior rudiment are closely united with the liver rudiment.

[1] Cf. Gegenbaur, *Vergl. Anat. der Wirbelthiere*, Bd. ii. Fig. 131. The structures labelled *h'* and *h''* are fatty ridges, not lobes of the liver.

In stage 33 a single large compact pancreatic mass passes forwards beneath the stomach and becomes fused at its anterior end with the liver substance. Apparently this has arisen by the fusion of the three original rudiments. In the 30 mm. larva the pancreas has undergone a complete fusion with the liver, forming a thick layer on its ventral side in the region near the opening of the bile duct. There are two distinct pancreatic ducts which open into the gut close to the opening of the bile duct.

IX. The Coelomic Organs.

Probably no system of organs in the Vertebrate body has been the subject of so much discussion in recent years regarding its general morphology as has the system of organs associated with the coelom. It cannot as yet be said that there is anything like unanimity of opinion amongst morphologists regarding even the main questions at issue.

Clear thinking on the subject—which is already beset by the greatest natural difficulties—is rendered still more difficult by an elaborate and extremely confusing nomenclature which frequently tends to obscure the main points at issue.

Before entering on my short description of this system of organs as occurring in the developing *Polypterus*, it seems essential to define precisely the sense in which various terms are used in this paper, and to state which out of the various divergent views on the general morphology of the coelomic organs I accept as being in my opinion in best accord with our present knowledge of facts.

General Morphology of the Coelom and its Derivatives. As regards the origin of the coelom of the Coelomata, the facts of Vertebrate embryology appear to me to fit in best with Lankester's Enterocoele hypothesis, according to which the coelom arises in the form of diverticula of the primitive gut wall. On this view the primitive rudiments of the coelom would function originally as mere extensions of the digestive cavity, though no doubt the genital cells would be developed from patches of their lining as in Actinozoa amongst existing Coelenterates.

Renal Organs. In 1877, in his *Notes on Embryology and Classification*, Lankester proposed the word nephridium for the excretory tubes found in many of the subdivisions of the animal kingdom. He laid stress on the fact that they usually opened into the coelom internally by a ciliated funnel, and he included in the term the excretory tubes of Rotifers, Turbellarians, Trematodes and larval Gastropods.

Later research showed that in the last mentioned cases the inner end of the organ had no open funnel, but on the other hand was provided with flame cells. Such organs were therefore separated off from those with an open funnel to which alone the term nephridium became restricted.

Later still, Goodrich has brought out the fact that flame cells and funnels opening into the coelom may occur on the same nephridium, and he has attempted—in the eyes of many, successfully—to prove that the open funnels in such cases are not all strictly homologous — the funnel in certain cases, e.g. certain Polychaetes, being a structure, which originally quite independent and serving for the exit of the genital products, has only secondarily become grafted on to the nephridium.

In considering the general morphology of nephridial organs it may probably be safely conceded (1) that it is *à priori* probable that the original excretory organs were ectodermal so that the waste products could readily be got rid of, and (2) that it is similarly probable that the original genital ducts were coelomic in origin.

We may take it that renal organs in general are characterised by the presence of one or both of the following structures:

(1) Flame organs or solenocytes, and (2) nephrostomes. It appears to me that a rough distinction can be drawn between these two sets of renal structures.

The solenocyte is primarily the excretory organ of the sponge-work of the spaces in the mesenchyme: the nephrostome, on the other hand, is associated specially with the coelom. The solenocyte[1] is a filtering apparatus by which water with certain salts in solution is filtered off from the highly valuable fluid which bathes the living protoplasm of a large part of the body: the nephrostome serves for getting rid bodily of the surplus of the relatively less precious coelomic fluid. As to which of these types of excretory organ is the more primitive, it is difficult to decide, though on the whole, perhaps, the probability is in favour of the ectodermal type with solenocytes being so[2].

We may assume that later on the coelom developed a communication with the lumen of the protonephridium, probably at first in the form of a rupture at this weak point caused by the pressure of the ripe genital cells, but later becoming permanent. As the coelom became more and more important, the system of chinks of the mesenchyme became less so. Further, with the develop-

---

[1] The functional activity of the flame cell or solenocyte stands in close relation to the pressure of fluid in the tissues bathing it, as may be seen clearly in a Rotifer where the formation of a minute perforation of the body wall so as to diminish the internal pressure causes instant cessation of movement of the flame apparatus.

[2] In the lower Actinozoa (Alcyonaria) we see very lowly organised animals in which the enterocoelic pouches are not yet closed and which already possess a well-developed mesenchyme. It is clear that a slight advance in development might lead to the formation of an excretory system in connexion with this mesenchyme at a stage in evolution when the enterocoelic pouches were still open, and when therefore the genital products were still able to reach the exterior without the formation of special genital funnels or temporary special openings. For an instructive diagram illustrating the evolution of the type of excretory system under consideration see Lang, *Jen. Zeitschr.* XXXVIII. S. 110.

ment of a large mass of movable mesenchyme—the blood—and the establishment of an efficient circulation, metabolic activity became accentuated in this movable tissue.

The products of catabolic change had in the fixed mesenchyme to be drained from its meshes by a diffused flame cell apparatus, but with the development of the floating mesenchyme there came the possibility of concentrating the excretory function at fixed points, over which the floating mesenchyme was borne by the blood stream. With greater perfection of the circulation this concentration of excretory processes would become more pronounced.

The coelomic funnel very probably served originally for the transmission of the genital cells, but it is easy to see how once in existence it would serve as a safety valve for the coelomic fluid, and how this passing out of excess of coelomic fluid would serve to flush the lumen of the nephridium and thus serving a useful purpose would tend to become accentuated so as to form a slow but continuous outward stream. With the development of the nephrostome —more efficient for the draining away of surplus fluid—and with the concentration of excretory activity in the nephridial wall the flame cell apparatus would gradually disappear.

The nephridial lumen would now serve to conduct to the exterior (1) the excretory products produced by its wall, and (2) the genital cells produced from the coelomic lining. When one considers (1) the paramount importance of the genital cells reaching the exterior in as healthy a condition as possible, and (2) the highly poisonous nature of the waste products of metabolism, it seems only reasonable to expect that the subsequent course of evolution would tend to bring about an arrangement whereby the germ cells would reach the outside without being subjected to a bath of excretory poison *en route*.

Even without any direct evidence I think we might safely assert that there *must* be such a tendency. That such a tendency does exist, however, is surely demonstrated by e.g. the conditions of the sperm ducts in male Anurous Amphibians (*Rana, Bufo, Alytes, Discoglossus*) or by the separation of renal and genital parts of the holonephros of Amniota, even if we ignore such highly probable cases as the evolution of the male duct in *Polypterus* and in Teleosts, or the splitting off of the Müllerian duct from the archinephric duct.

It seems to me, in spite of the arguments of Goodrich in the contrary sense, most reasonable to interpret the series of forms of excretory organ in Polychaetes in this way. I find it very difficult indeed to imagine what possible advantage would result from the mode of exit of genital products becoming more complicated and difficult, while the advantages of the opposite course of evolution are quite obvious.

I. It is fairly certain that Vertebrates are descended from ancestral forms in which the coelom was completely segmented. This is indicated by its segmentation in the young Amphioxus and by the still persisting segmentation of the more dorsal portions in Craniate forms.

II. The original division of the coelomic space into segmentally arranged compartments involves with practical certainty the provision of each of these segments with a pair of nephridial tubes.

III. These nephridial tubes forming the Archinephros (Lankester) or Holonephros (Price) came to open into a longitudinal duct—the archinephric duct on each side. Of the various modes of origin suggested for this duct the most probably correct seems to be that the tubules came to open into one another by the backward migration of their external openings : the opening of each tubule moving back till it (1) became coincident with the external opening of the next tubule behind, and (2) came to open into the lumen of that tubule. Examples of the same type of migration of the external opening are seen in the collecting tubes of the posterior portion of the kidney in Elasmobranchs and in the gill tubes of *Myxine*.

IV. Originally the renal tubules like the other segmented organs of the body developed in regular sequence from before backwards, but a great increase in the size of certain of the anterior tubules enabled them to cope with the excretory needs of the young animal during a considerable period of growth without their needing to be reinforced by the next tubules of the original series. This led to these tubules lagging behind or even entirely dropping out so as to form a gap dividing the originally continuous archinephros into pronephros and mesonephros.

Nomenclature of the Renal Organs. The naturally great difficulties in the way of understanding the morphology of the renal organs are greatly increased by the difficulties of nomenclature. Special names have been coined, frequently without any adequate morphological reason, and terms which had a definite accepted morphological significance have undergone redefinition and been made to mean something quite different. It is advisable then to define precisely the sense in which one or two terms are used in this paper.

The Nephridium is an excretory tube carrying at its inner end either solenocytes (flame cells) or an open funnel communicating with the coelom. The series of nephridial tubes opening into a longitudinal duct and extending the whole length of the coelom is called either by Lankester's name Archinephros or by Price's name Holonephros. Its duct is the archinephric duct.

The opening from coelom—whether general or Malpighian—into nephridium is a nephrostome. It is misleading to apply this term in any other sense in connexion with Vertebrate morphology, seeing that it has a perfectly definite connotation in regard to other groups of Coelomata.

The portion of coelom in the neighbourhood of the glomerulus or nephrostome which may be cut off partially or completely from the general coelom to form the cavity of a pronephric chamber or Malpighian body, may be called Malpighian coelom, or on the other hand, there seems no strong objection to allowing it to retain the name nephrocoele by which it is known in early stages of its development.

The open communication which may persist between nephrocoele and splanchnocoele I call with Brauer the peritoneal funnel, or—when lengthened out into a tube—peritoneal canal.

Coelom generally. The coelom first becomes patent in the lower portions of the protovertebrae, in the regions corresponding to the nephrocoeles and to the lower ends of the myocoeles. Already in embryo 4 F—between stages 20 and 21—a slight cavity has made its appearance in the lower ends of protovertebrae I.—VI. This coelomic cavity is at first irregular, with ill-defined boundaries. It arises apparently like the myocoelic cavities of *Lepidosiren* by the breaking down of the protoplasm and its replacement by fluid. As development goes on the cavity extends up into the myotom, and at the same time its limit becomes sharp and well defined (text-fig. 59, p. 245).

While the nephrocoele remains a patent cavity, as will be described later, the myocoele almost immediately becomes reduced to a narrow fissure as its inner wall thickens to form the main mass of the myotom. The cavity of the nephrocoele becomes extended outwards as the lateral mesoderm splits to form the splanchnocoele. This development of the splanchnocoele takes place very slowly. It is only in stage 33 that it has assumed its adult relations. In stage 32 it has not yet developed over the main mass of yolk nor in the region adjoining the cloaca, although it forms a definite cavity in the intermediate region. In earlier stages the splanchnocoele is a patent cavity only in the region adjoining the pronephros—where the space has spread from the nephrocoele round the liver (stage 31)—and in the pericardiac region. The pericardiac portion of the splanchnocoele is in fact one of the earliest parts of the coelom to become patent. It is first visible in an embryo of stage 25 (cf. text-fig. 50, p. 232). At this period the pharyngeal region of the gut is being nipped off from the main mass of yolk by the active backgrowth of the mesoderm underneath it. The pericardiac cavity arises by the hollowing out of this backgrowing mass of mesoderm. In embryo 7 A the cavity is somewhat ∩-shaped in ground plan, the two halves of the cavity being continuous with one another anteriorly. Round the cavity the mesoderm is somewhat condensed so as to form a definite and fairly thick wall. Traced out over the yolk this wall gradually merges into the superficial layer of the yolk just as does the mesoderm rudiment elsewhere. The part of the mesoderm lying in the sagittal plane between the two limbs of the pericardium forms a vertical partition which is destined later to give rise to the endocardium.

The two pericardiac cavities when followed in a dorsalward and backward direction can in several embryos of about stages 25 and 26 be traced very nearly into continuity with the pronephric chamber of nephrostome A. Whether, however, there is actual continuity I am not prepared to assert definitely without having access to further material. With the development of the heart the two pericardiac chambers become continuous throughout across the mid-ventral line. The pericardiac coelom remains continuous with the rest of the splanchnocoele as late as stage 33. By stage 36 it has become closed: there is no material to show the details of the process of closure.

DEVELOPMENT OF THE NEPHRIDIAL SYSTEM.

Pronephros. The excretory organ of the young Polypterus during a great part of its larval existence consists of a pronephros with two tubules as in Urodeles and Dipnoans. As will be seen from the history of the development these are the tubules belonging to the second and fifth metotic primitive segments which have become enlarged and persisted as functional structures after the tubules of segments three and four have atrophied.

The first trace of the Kidney system which I have been able to detect occurs in 3 A (Pl. XIII. fig. 16 *m*), an embryo with still open medullary groove. Here there arises from the outer side of one of the mesoderm segments, close to its lower end, a solid process which projects backwards for some distance.

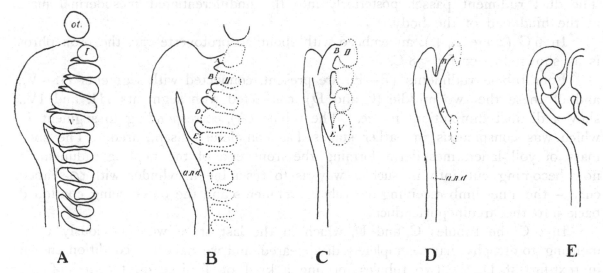

FIG. 58 A—E.   View of pronephros of 4 G (stage 20), 5 D (stage 23), 6 G (stage 24 +), 7 C (stage 25), and 13 B (*about* stage 28).

a.n.d., archinephric duct; *ot.*, otocyst; *I, II*, etc., mesoderm segments; *A, B*, etc., tubule rudiments.

The next stage (4 F) is shown by a series of horizontal sections. Here mesoderm segments II., III., IV., V., VI. send outwards projections from their ventral ends.   These pass into a common mass of heavily-yolked protoplasm, which is continued backwards as far as the hinder limit of the mesodermal segments as the obvious rudiment of the archinephric duct.   The duct rudiment lies between the mass of yolk—and the lateral mesoderm which is in process of being delaminated from it—and the ectoderm.   At its hinder end it is continuous with the yolky mass forming the gut rudiment.

In a slightly older embryo (4 G, text-fig. 58 A) the nephros is in very much the same condition, but there are now rudiments of seven tubules—an additional one (A) having arisen in front from the first mesoderm segment and another (G) posteriorly from the seventh.

In 5 F, a slightly more advanced embryo in its general features, the nephric rudiment is connected up with primitive segments I.—V.

In 6 C apparently there is connexion with primitive segments I.—IX. inclusive.   At the same time it is quite possible that this appearance is illusory and due to the boundary between duct rudiment and primitive segments being obliterated by yolk granules catching on the edge of the knife during the process of section cutting.

In 5 D, an embryo with 15 complete protovertebrae and a 16th incompletely cut off, the connexions between protovertebrae and duct rudiment are seen to be now in the form of distinct tubule rudiments as shown in text-fig. 58 B.

There is a series of five tubules (A—E) connected with mesoderm segments I.—V., not varying greatly in size though the anterior ones are rather the longer. The duct rudiment passes posteriorly into the undifferentiated mesodermal mass at the hind end of the body.

In 6 G (stage 24 +), an embryo with about 20 protovertebrae, the pronephros is as shown in text-fig. 58 C.

Four tubule rudiments (B—E) are present, connected with segments II.—V., and of these the two middle (C and D), connected with segments III. and IV., show a distinct diminution in size.   The tubule (A), corresponding to segment I., which was conspicuous in earlier stages, has completely disappeared.   The solid mass of yolk-laden mesoderm forming the front end of the nephric rudiment is now becoming cut into in such a way as to resemble a cylinder with ∽-shaped curve—the inner limb receiving the tubule rudiments and the outer being continued back into the archinephric duct.

In 7 C the tubules C and D, which in the last stage were obviously commencing to atrophy, have completely disappeared, and we have the condition shown in text-fig. 58 D, the two tubules forming a kind of **T** piece on the end of the archinephric duct.

The pronephros has now assumed the general features characteristic of it throughout its actively functional period, subsequent changes involving mainly the increase in length—and later in diameter as well—of the anterior end of the archinephric duct whereby it is thrown into the complex turns and twists characteristic of its later stages (text-fig. 58 E : see also text-fig. 17, p. 170). In its later functional stages, as seen in larvae 17, 18, 19 and 20 the pronephros reaches a relatively enormous size, occupying the whole thickness of the body wall. In these later stages the tubules as well as the archinephric duct have become much elongated and coiled.

Evolution of the Pronephric Segment. I have described above the development of the pronephros as seen in its broad topographical features. It will be useful now to summarise the history of the segmental unit of the pronephros at its various stages of development.

As already stated, the first stage in the development of the pronephric segment consists in the formation of a blunt backwardly-projecting process from the primitive segment near its lower end. This earliest stage was found only in one embryo of the series, the process in question extending back along the outer side of the protovertebral stalks and between them and the ectoderm. There is no trace of coelomic cavity, either myocoelic or nephrocoelic.

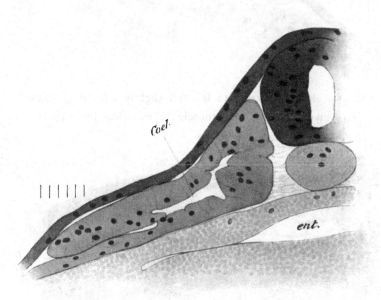

FIG. 59. Part of transverse section (stage 20), showing rudiment of pronephric tubule.
Coel., coelomic space ; *ent.*, enteric cavity.

In the various embryos of batch 4 belonging to stages 20 and 21 (cf. text-fig. 59) myocoelic cavities (*Coel.*) have appeared in the anterior segments, and these are beginning to extend out into the nephric rudiment. In 5 F quite considerable spaces of irregular form have developed within the nephric rudiment, mostly in continuity with the myocoelic cavities. In 5 A (stage 23, text-fig. 60) the examination of transverse sections shows that the cavity at the lower end of myotome I. is developing the characteristic features of a pronephric chamber (*nc.*), continued by a split dorsally into the myocoele (*mc.*) and ventrally into the splanchnocoele (*splc.*)—the latter being however merely a virtual cavity with its inner and outer walls still in close apposition.

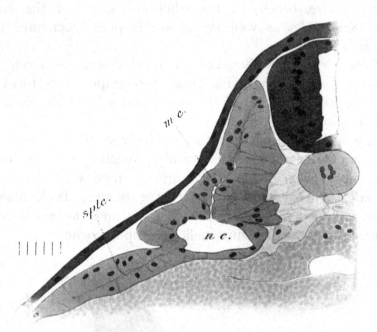

FIG. 60.   Similar section to last but slightly later stage (23).
mc., myocoele; *nc.*, nephrocoele; *splc.*, splanchnocoele.

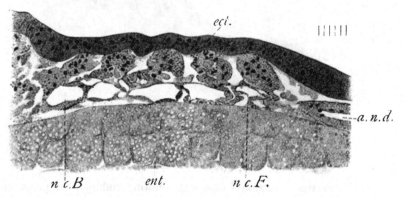

FIG. 61.   Part of a longitudinal vertical section of stage 24 +.
a.n.d., archinephric duct; *ect.*, ectoderm; *ent.*, endoderm; *nc.B* and *F*, nephrocoeles.

In 6 G (stage 24 +), as is well seen in text-fig. 61, there have developed a series of these pronephric chambers. Of these B, C, D and E are large and about the same size, A has apparently disappeared, F is distinct though considerably smaller than those in front of it, while there are none obvious behind this. The four large pronephric chambers, i.e. B, C, D and E are each continued outwards as a narrow space into the cavity of the inner limb of the ∽-shaped pronephric duct.

In the fully-developed pronephros only two pronephric chambers are found on each side, corresponding with the two persisting tubules. This condition is brought about by a gradual diminution in size and eventual atrophy of pronephric chambers C and D.

It will now be convenient to follow out particularly the evolution of pronephric chamber B.

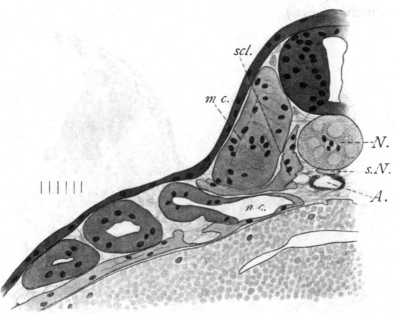

FIG. 62. Part of transverse section, stage *c*. 25.

A., aorta; *mc.*, myocoele; *N.*, notochord; *nc.*, nephrocoele; *s.N.*, hypochorda; *scl.* sclerotome.

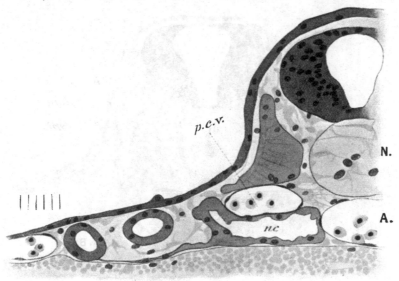

FIG. 63. Part of similar section to last, of about stage 28.
p.c.v., posterior cardinal vein; other letters as in Fig. 62.

Text-fig. 62 shows a transverse section through the first pronephric chamber of 8 A (*about* stage 25). The wall of the chamber (*nc.*) is seen to be much thinner than in the preceding stage, being composed of a single layer of flattened endothelium cells—swollen out in the middle owing to the presence of the nucleus.

In the section of 11 A (*about* stage 28), shown in text-fig. 63, the chief change lies in the appearance of the posterior cardinal vein (*p.c.v.*) overlying the pronephric chamber (*nc.*).

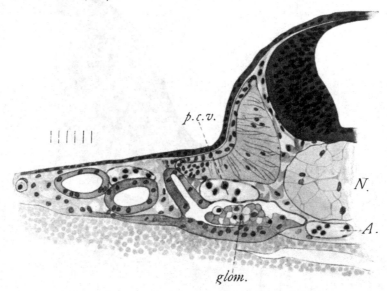

FIG. 64. Similar section, slightly more advanced.
glom., glomerulus; other letters as before.

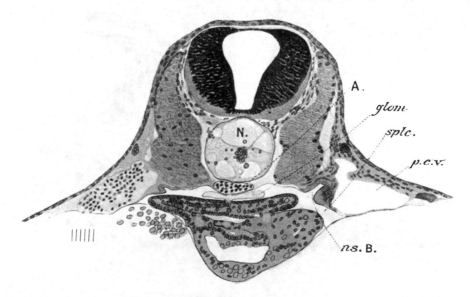

FIG. 65. Transverse section, stage 31.
ns.B., nephrostome B; other letters as before.

In 13 C a slight thickening of the mesial wall of the chamber probably fore-shadows the development of the glomerulus. In the preceding stage a sparse mesenchymatous spongework was seen in the intertubular spaces of the pronephros: in the present stage this spongework has been replaced by a wide lacunar blood space in communication with the anterior cardinal vein.

In 13 F (text-fig. 64), a slightly older embryo, the glomerulus has developed considerably and forms a large vascularized projection into the pronephric chamber from its ventral wall, and filling up most of the cavity of the chamber.

Larva 17 (stage 31), which forms the next series of transverse sections available is separated by a considerable break from that last described. The pronephric chamber is now seen to be continued (text-fig. 65) in a ventral direction into the splanchnocoele (*splc.*), which has become patent in the region of the oesophagus and stomach. The glomerulus (*glom.*) has altered consider-ably in its relations as seen in transverse section, its base of attachment is greatly narrowed, and it has been carried by differential growth on to the mesial wall of the coelomic cavity. This description applies only to the anterior (B) of the two pronephric chambers. The posterior one (E) lies dorsal to the splanchno-coele, and isolated completely from it except at its front end, where it is continued into irregular chinks round the glomerulus, which open anteriorly into the splanchnocoele. The glomerulus as seen in transverse section also retains its original ventral position in this posterior pronephric chamber.

Larva 19 shows little change as regards the relations of nephrostome B. The splanchnocoele with which it communicates has now spread considerably ventralwards on the right side of the body. As regards nephrostome E, it still communicates with a distinct pronephric chamber, having the glomerulus on its ventral side. Traced forwards the posterior pronephric chamber as it tapers off becomes completely occluded by the glomerulus which touches the wall of the chamber all round. There is consequently no patent communication between this chamber and the splanchnocoele in this particular specimen. Whether this is the normal condition or whether it is more usual for the cavity, which is here only virtual, to be patent, can only be settled by the examination of other specimens of this stage. The walls of the tubule for some distance from the nephrostome in the case of both B and E show a deposition of dark pigment in their protoplasm as in the case of various amphibians.

In larva 20 the anterior nephrostome (B) shows the same relations to the splanchnocoele as before, except that the splanchnocoele is now a wide cavity. The glomerulus is relatively further back, its anterior end being some distance posterior to the nephrostome. It forms a flattened horizontally-placed structure, which when traced backwards is found soon to become fused by its outer edge with the somatopleure so as to separate off the portion of coelom lying above it from the general splanchnocoele. Tracing this special portion of coelom back-wards, it is found to become very narrow and then to expand again to form the pronephric chamber from which the hinder nephrostome (E) opens. In other words, at this stage pronephric chamber B has become completely merged in the general splanchnocoele, while pronephric chamber E communicates with the splanchnocoele by the opening which formed in earlier stages a communication between it and the anterior chamber.

In the 30 mm. larva Budgett (p. 168) discovered the still persistent remains of the pronephros, the archinephric duct forming anteriorly a large coiled mass lying close under the skin and communicating with (1) a pronephric chamber containing a large glomus, and (2) with the general coelom by a peritoneal funnel. Unfortunately the sections of this region are not arranged in sequence on the two slides concerned, and it is therefore impossible, without repeating the elaborate reconstruction of this region which Budgett made, to do more than confirm his results in a general way. A search through the sections passing through the anterior portion of the pronephric vestige (these had not been disarranged) disclosed the interesting fact that another pronephric nephrostome existed on each side in front of that described by Budgett. And further this nephrostome showed the peculiarity that it opened not into a special pronephric chamber but into a pocket-like portion of the general coelom extending forwards along the dorsal side of the pharynx in the region of the glottis. In these two nephrostomes on each side there is no difficulty in recognizing the persisting two pronephric nephrostomes (B and E) of the preceding stage.

Mesonephros. In larva 17 no traces were observed of mesonephric tubules.

In larva 18, on the other hand, such rudiments were observed corresponding in position with the undermentioned myotomes:—Right side 13, 16, 18, 19, 21 and 22; left side 10, 11, 12, 14, 15, 16, 17, 18 and 19; i.e. on the right side there are rudiments of six tubules, on the left side nine. Histologically each rudiment is seen to consist of an aggregation of nuclei embedded in protoplasm, rounded in shape in the least developed, and drawn out into a pear shape with the point posterior in the more advanced. The latter also show a cavity in their interior.

In larva 19 the rudiments are on the right side 11 in number, on the left 9 or 10. The various rudiments have now also become joined on to the duct at their lower ends.

In larva 20 but little advance appears to have taken place, but the state of preservation of the kidney is so bad that it is impossible to make out the finer details.

Theory of the Archinephros or Holonephros.

Although the material has yielded only very scanty information regarding the mode of development of the mesonephros it is quite clear that the mesonephric tubules develop independently amongst the mesenchyme, i.e. that their mode of development is quite different from that of the pronephric tubules. But I should not adduce this difference as evidence against the homodynamy of the two sets of tubules. In fact I find it difficult to understand how any morphologist can at the present day—more especially since Brauer's most admirable work on the development of the kidney in Gymnophiona—any longer refuse to accept the idea of the serial homology of the primary tubules throughout pronephros and mesonephros. I do not propose to enter here into a discussion of the matter, and need only say that my own investigations on the development of the kidney system in various vertebrates, including Dipnoans, lead me to almost absolute agreement with Brauer's theoretical views[1].

Almost the only point in which I feel inclined to differ from him is brought out in the following sentence in his paper[2]. He refers to the originally separate tubules of the holonephros becoming fused at their openings to form a longitudinal canal which later, becoming covered in, formed the longitudinal duct—"An dessen Bildung sind Anfangs alle Canälchen betheiligt gewesen, so wie es jetzt noch die vordersten zeigen; je mehr der vordere Theil des Holonephros zum larvalen Organ sich umbildete, um so schneller wurde der Gang entwickelt und dadurch die hintern Abschnitte mehr und mehr von einer Betheiligung an seiner Bildung ausgeschlossen." It appears to me more reasonable to look at matters in a slightly different light and to suppose that the excretory needs of the growing larva are

[1] To argue as various morphologists do for a former extension of the "pronephros" backwards through the body involves a usage of the term quite at variance with the meaning of the word as defined by its inventor Lankester. "The more anterior nephridia form the pronephron,...the posterior nephridia forming the mesonephron," *Q. J. Micr. Sci.*, N.S. XVII. 1877, p. 429.

Evidence of this supposed extension backwards of the "pronephros" plays the chief part amongst the objections brought forward to the idea of the Holonephros, e.g. Felix (*Hertwig's Handbuch*, III. 1, p. 416): "In dem Moment, wo wir den Nachweis liefern können, dass die Vorniere sich ursprünglich in ganzer Ausdehnung der Leibeshöhle erstreckt, haben wir festgestellt, dass Vorniere und Urniere in ihrer ganzen Ausdehnung nebeneinander vorkommen; die Theorie des Holonephros, der sich vorn zur Vorniere, hinten zur Urniere entwickelt, wäre damit unmöglich geworden. Nun haben wir aber im ersten Abschnitt über die theoretische Auffassung der Vorniere die Hypothese zu begründen gesucht, dass sich die Vorniere des Vertebratenvorfahren über die ganze Ausdehnung der Leibeshöhle erstreckt haben muss. Stützt sich unsere Hypothese auf ein genügendes Thatsachenmaterial, so wäre durch sie die Sedgwick-Field'sche Hypothese von der Abstammung der Vorniere und der Urniere vom gleichen Ahnenorgan widerlegt. Wir brauchen aber gar nicht auf diese Hypothese zurückzu-greifen, schon der Nachweis, dass Vornierenkanälchen und Urnierenkanälchen im gleichen Segment nebeneinander vorkommen, würde gegen den Holonephros sprechen; und dieser Nachweis ist erbracht." It will be interesting to see whether future research confirms the presence of pronephric and mesonephric tubules in the same segments.

[2] *Zool. Jahrb. Anat.* Bd. XVI. S. 147.

fully met by the great increase in size of the pronephros—for its relatively enormous size can hardly be considered as probably primitive—and that correlated with this hypertrophy of the pronephros the development of the mesonephric tubules, so rendered unnecessary functionally, becomes postponed, the necessary duct however still retaining its early development.

ORIGIN AND DEVELOPMENT OF THE ARCHINEPHRIC DUCT.

Material is wanting to show in detail the first stages in the development of the archinephric duct. As has already been indicated the very earliest stage is shown in embryo 3 A (stage 19), where there exists a backwardly projecting process from the lower end of one of the anterior mesodermal segments. In the next available stage five such projections are present, and they are fused together at their outer ends into a solid mass of heavily yolked protoplasm, which forms the rudiment of the archinephric duct, and which already extends back to the hinder limit of the segmented mesoderm. There is no material to show whether the anterior portion of this is actually formed in development by a process of fusion of the originally separate nephrotomal outgrowths, but appearances fit in quite well with this idea, and we may probably assume with safety that the front end of the archinephric duct does arise actually in this probably primitive fashion just as it does in various Amphibians and Dipnoans.

The hinder end of the duct arises otherwise.

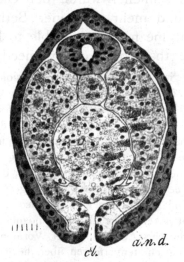

FIG. 66.   Transverse section through cloaca in stage 23.
a.n.d., archinephric duct opening into cloaca (*cl.*).

In text-fig. 66 is shown a transverse section through a specimen of stage 23 at the level of the cloaca. The mesoderm rudiment is seen on each side as a typical enterocoelic "outgrowth" except that it is solid. Into this mesoderm on each side there passes a prolongation of the cloacal cavity which represents the hinder end of the archinephric lumen. From the outer or dorsal end of this cavity the archinephric rudiment can be traced forwards in a nearly horizontal direction, i.e. the archinephric duct in this region is formed directly out of the entire thickness of the lower end of the mesoderm segments. In other words, the hinder part of the archinephric duct is formed by the fusion of entire nephrotoms, not merely of outgrowths from nephrotoms, as appears to be the primitive mode of formation of this duct. Further, such a section as that figured suggests the possibility that the opening of the archinephric duct into the cloaca is morphologically the original communication between enteron and an enterocoelic pocket, which alone of the series retains in the adult its communication with the gut owing to its having to carry out a renal function. Should this speculation be well grounded we should still regard this method of opening of archinephric duct into gut as a secondarily acquired arrangement, just as we should in the case of the origin of the archinephric duct from a series of fused nephrotoms. Considerations much too weighty to be ignored support the belief that the archinephric duct arose in phylogeny by the fusion of the outer ends of tubules which originally opened directly to the exterior.

The lumen of the archinephric duct which appears as a secondary excavation of the originally solid rudiment develops very rapidly during stage 23, some specimens of which show the rudiment still solid, while in others the lumen is completed and opens into the cloaca.

X. Vascular System.

Before describing the sequence of developmental features shown by the vascular system it will be convenient to give a short account of the main features of the vascular system in an embryo in which the system has been blocked out as regards its main features. Such a stage is afforded by the larvae belonging to batch 11 (*about* stage 28).

Heart and Arterial System. The heart is in the form of a bent tube. Its topographical relations are puzzling until a reconstruction is made, when it is seen (text-fig. 67) that the heart has—owing to the hypertrophy of the yolk-laden gut wall—had its ventricular portion (*v.*) pushed forwards so as to bring about an inversion[1]. The ventral aorta (*v.a.*) on leaving the heart runs not forwards but backwards. It soon bifurcates to give rise to the only two aortic arches which are present—the hyoidean (*a.a.II.*). Each of these passes outwards as the afferent vessel of the external gill. In sections through the external gill the afferent vessel is found to be dorsal instead of ventral. Were this arrangement primitive it would be a strong argument against the homology of the external gills in the different vertebrate groups in which they occur. Inspection of a reconstruction of this stage—as for example that shown in text-fig. 67—at once demonstrates however that the unusual arrangement is a secondary one: the course of the vessels at the base of the gill showing that it has undergone a rotation in the direction in which the arm is rotated from the pronated to the

---

[1] As in various Teleosts.

supinated position. The afferent and efferent limbs of the vessel are joined by loops belonging to the pinnae which are just beginning to develop. The efferent vessel on leaving the external gill at once joins the aortic root (*a.r.*)—which is continued forwards as the dorsal carotid (*d.c.*), to pass down the side of the thalamencephalon, gradually tapering away to its front end,—and backwards to unite with its fellow and form the very large dorsal aorta (*d.a.*).

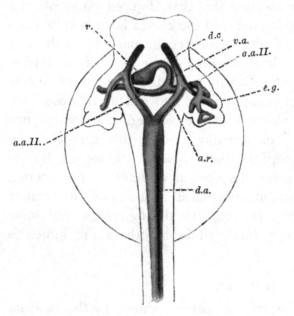 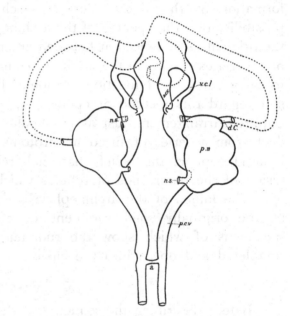

FIG. 67. Dorsal view of main arterial trunks in larva 11 *c.*, between stages 27 and 29.

a.a.II., aortic arch II.; *a.r.*, aortic root; *d.a.*, dorsal aorta; *d.c.*, dorsal carotid; *e.g.*, external gill; *v.*, ventricular portion of heart; *v.a.*, ventral aorta.

FIG. 68. View of main venous trunks at stages 27—29; the heart and the ductus Cuvieri are shown by dotted lines.

a., anastomotic vein; *a.c.v.*, anterior cardinal; *d.C.*, ductus Cuvieri; *n.s.*, nephrostome; *p.c.v.*, posterior cardinal; *p.n.*, pronephros; *v.c.l.*, lateral vein of head.

Venous system (text-fig. 68).   The most conspicuous part of the venous system is a great irregular blood sinus which bathes the surface of the pronephric tubules (*p.n.*).   This sinus or system of sinuses is continued backwards lateral and slightly ventral to the dorsal aorta as the posterior cardinal vein (*p.c.v.*). These pass backwards as separate vessels, but it is to be noted that the ultimate fusion of the two is already foreshadowed at this stage by the existence of anastomotic connexions as in the specimen figured, where one such bridge-like connecting vessel (*a.*) is seen.   Anteriorly the pronephric system of sinuses is continued forwards dorsally into the venous sinuses connected with the head. These show important variations in the larvae of the age under consideration. In the less advanced ones a typical anterior cardinal vein passes forwards in the ventral angle between the otocyst and the brain wall.   At its front end this anterior cardinal vein becomes dilated to form a great sinus which gives off irregular branches into the mesoderm of the front end of the head.   In other specimens a new anastomotic channel has developed lying ventral to the otocyst and connecting that part of the anterior cardinal vein lying posterior to the otocyst with its dilated front portion lying anterior to the otocyst. In still others, as in the specimen figured, the anterior cardinal (*a.c.v.*) has become constricted so as to form a narrow neck behind the level of the otocyst. Finally in others this narrow neck has become completely obliterated so that the whole of the blood stream passes back by the anastomotic vein (*v.c.l.*) lying external to the otocyst, and the part of the anterior cardinal vein lying on the mesial side of the otocyst forms merely a backwardly directed branch of what is now the main vein of the head. In the above-described varieties of venous arrangements in the head we have illustrated the steps by which in *Polypterus*, as in many other vertebrates, a large portion of the original anterior cardinal, lying on the mediad side of the otocyst, becomes replaced functionally by a secondarily developed lateral cephalic vein[1] (*vena capitis lateralis*) lying in a more superficial position. At about the anterior limit of the rhombencephalon the anterior cardinal vein gives off dorsally another branch rather smaller than the backwardly directed one just described. This runs dorsad in the deep groove between rhombencephalon and mesencephalon.

Besides the veins mentioned an important set of veins is found on the surface of the gut rudiment. In the posterior portion of the body behind the main mass of yolk a large subintestinal vein runs along beneath the gut rudiment. As the swollen out yolk mass is reached the subintestinal vein breaks up into a network of vessels which lies over the surface of the yolk. Anteriorly the vessels forming this network pass into a main vitelline vein on each side—the two vitelline veins eventually uniting to form the morphologically posterior end of the heart.

From the pronephric sinus connexions pass outwards into the vitelline network. The most important of these (*d.C.*) passes forwards and outwards, and making a wide sweep passes eventually into the main vitelline vein. It is to be looked on as corresponding to the *ductus Cuvieri*, although at this stage it is merely a part of the general vitelline network. A similar condition of the ductus Cuvieri is well known to occur in some Teleostean fishes (Ziegler, "Randvenen" of Wenckebach).

Vessels of hind end of body. The relations of the main vascular trunks towards the hinder end of the embryo deserve special mention. In this portion of the body—behind the main mass of yolk—transverse sections show four main vessels arranged round the cylindrical gut—dorsally the dorsal aorta, ventrally the large subintestinal vein, and dorso-laterally on each side the posterior cardinal vein. Of these the posterior cardinal veins pass backwards just dorsal to the archinephric ducts, and as they approach the neighbourhood of the cloaca become gradually reduced in size and at the same time show irregular constrictions. Wide communications now occur—arranged apparently irregularly between posterior cardinal veins and dorsal aorta. Wide communicating vessels also arise from the posterior cardinal veins, which pass ventralwards between gut wall and archinephric duct and open into

[1] Rabl, *Leuckart's Festschrift*, 1892.

the subintestinal vein. Finally just in front of the cloaca the subintestinal vein gives off on each side a wide vessel which passes up into the posterior cardinal vein, the latter in turn communicating with the dorsal aorta. These hooplike vessels are probably to be looked on as representing morphologically the bifurcated hind end of the subintestinal vein, as the cavity of this vessel comes to an end almost immediately after giving off the two branches. This view is borne out by the fact that the two paired vessels—formed on each side by the fusion of the lateral branch of the subintestinal vein with the posterior cardinal—are continued back past the region of the cloaca and then unite to form a postanal "subintestinal" vein running backwards beneath the postanal gut and bearing exactly the same relations to it as does the subintestinal vein *sensu strictu* to the rudiment of the functional gut.

Traced further back both dorsal aorta and postanal subintestinal vein gradually thin out and disappear.

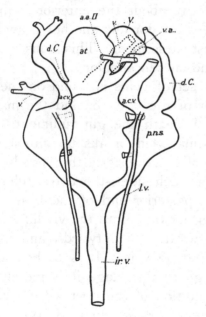

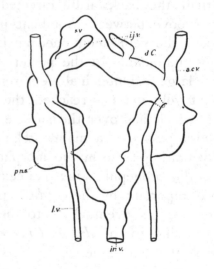

FIG. 69.   Dorsal view of main venous trunks in stage 31.
a.a.II., aortic arch II.; *a.c.v.*, anterior cardinal; *at.*, atrium; *d.C.*, ductus Cuvieri; *ir.v.*, interrenal vein; *l.v.*, lateral cutaneous vein; *p.n.s.*, pronephric sinus; *V.*, ventricle; *v.*, vein; *v.a.*, ventral aorta.

FIG. 70.
Main venous trunks at stage 33.
i.j.v., inferior jugular ;
s.v., sinus venosus ;
other letters as in Fig. 69.

Developmental changes in the main venous trunks. The posterior cardinal veins were left in stages 27—29 (larva 11) as separate paired trunks, but soon after this stage they undergo fusion in the middle line. In larva 17 (stage 31, text-fig. 69) the two vessels have coalesced to form a median "interrenal" vein (*ir.v.*). Anteriorly they separate as two vessels of approximately equal size, each being continued into the system of sinuses of the pronephros (*p.n.s.*), or more accurately the single great sinus on each side in the cavity of which the tubules wind about.

Anteriorly and dorsally the pronephric sinus becomes much narrower and then almost immediately divides into two branches. Of these one runs forwards above the pharynx and is the anterior cardinal sinus (*a.c.v.*)[1], the other, the ductus Cuvieri, which is now clearly marked off from the general vitelline network, turns ventralwards and runs towards the middle line. As it does so it receives tributaries (*v.*) from the yolk, including that part which is developing into the liver rudiment, and eventually unites with its fellow to form the morphologically hind end of the heart. The two ductus Cuvieri have become markedly asymmetrical, the right making a wide sweep over the yolk before running over towards the left side of the body to enter the heart, while the left has become greatly shortened. The liver rudiment at this stage, like any other part of the yolk mass, is supplied with blood simply by branches of the general vitelline network. Near the anterior end of the pronephric sinus on each side there passes into it a small vein which may be traced dorsalwards and then backwards beneath the lateral branch of the vagus nerve as a typical "lateral cutaneous vein" like that seen in Urodela, etc. The postanal "subintestinal" vein has increased in size considerably and forms the caudal vein—traces of the postanal gut being still visible here and there between it and the dorsal aorta to which it is closely apposed. At the point where it originally bifurcated to pass on each side of the cloacal rudiment irregularities are seen in the walls of the vessel and a tendency to give off branches, but the main cavity is continued straight on into the interrenal vein formed by the fused posterior cardinals.

Larva 19, *stage* 33 (text-fig. 70). The right posterior cardinal is now seen to be distinctly larger than the left. Each passes forwards, dilates to form the pronephric sinus, and is then continued forwards to the ductus Cuvieri. The study of sections shows that on the right side there exists a large direct channel lying ventral to the pronephric tubules through which the blood stream can make its way direct to the ductus Cuvieri without filtering through between the tubules of the pronephros. On the left side this direct channel is not developed. As will be seen from the figure the right ductus Cuvieri is much shorter than in the preceding stage, and it now pursues a straight course to the heart. In connexion with the liver a definite portal vein is now developed in the form of a specially enlarged vessel, which, collecting blood from the general enteric network, runs forwards on the left side and to the left of the pancreas and enters the liver rudiment. It can be traced into the liver rudiment for some distance running parallel to the bile duct and slightly to the left and dorsal side of it. It gradually disappears, giving off branches to form the capillary network of the liver. Paired rudiments (*i.j.v.*) of inferior jugular veins may be traced for a short distance on each side. The lateral veins are very conspicuous.

Larva 20. This 12-day larva (stage 36) is unfortunately in very poor histological condition. It further shows remarkable features in its venous system which

---

[1] With, of course, its intercalated portion of lateral cephalic vein as already described.

give the impression of being abnormal, e.g. the portion of the left ductus Cuvieri lying between the inferior jugular vein and the heart has completely disappeared[1]. The whole of the blood stream from the liver passes into the remaining right ductus Cuvieri. Whether this also is an abnormal arrangement is impossible to decide without further material.

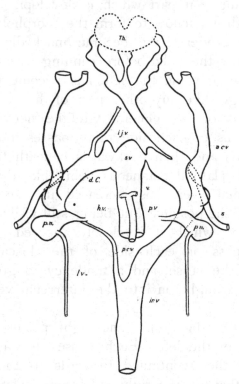

FIG. 71.   Main venous trunks of 30 mm. larva.

h.v., hepatic; *p.v.*, pulmonary; *s.*, subclavian; *Th.*, thyroid; other letters as in preceding figures.

30 mm. Larva (text-fig. 71). The most striking difference from stage 33 is seen in the increased asymmetry of the venous system. The portion of the left posterior cardinal vein lying just behind the pronephros was seen to be distinctly smaller than its fellow: it has now become reduced to an insignificant vestige so that practically the whole of the blood from the interrenal vein is drained forwards into the right ductus Cuvieri. The main blood stream passes forwards by the large sinus-like cavity lying on the ventromedian aspect of the right pronephros. The left pronephros has its tubules surrounded by a system of sinuses just as on the right side and anteriorly a channel (*) representing the anterior end of the posterior cardinal vein leads from this into the ductus Cuvieri, which latter is considerably smaller than that of the right side. From the pronephric sinus of each side the lateral cutaneous vein (*l.v.*) passes tailwards, reduced to an almost vestigial condition.

[1] Cf. with normal condition in *Salmo*. Ziegenhagen, *Verh. anat. Ges.*, Berlin, 1896.

Each ductus Cuvieri gives off ventrally a well developed subclavian vein (*s.*) which passes backwards and then outwards to the pectoral ·fin. On its headward side the ductus Cuvieri gives off the inferior jugulars which pass forwards on the dorsolateral sides of the pericardium. These (*i.j.v.*) are now strikingly asymmetrical, the left having become relatively reduced, the right having taken on its function. The right forms a large vessel which passes forwards and inwards until it reaches the middle line. It then gives off a large lateral branch on each side, and the trident-like arrangement of vessels so formed serves mainly to drain away the blood from the thyroid. The left inferior jugular vein breaks up a short distance behind the point of trifurcation of the right. It remains to be said that the sinus venosus is continued back as the large—now median—hepatic vein. The pulmonary vein, formed by the fusion of the veins from the two lungs, runs forwards immediately dorsal to the hepatic vein and eventually opens into it a short distance behind the sinus venosus.

Subintestinal vein. In larva 18 the praecloacal part of the subintestinal vein still exists as a single vessel as far forwards as the point where the gut rudiment swells out into the main mass of yolk. Here it breaks up into the vitelline network.

In larva 19 the subintestinal vein has become replaced by a network throughout its extent and no longer exists as a discrete vessel.

Pulmonary veins. In larva 20 the anterior portion of the liver is slung up to the ventral side of the alimentary canal by a median mesenteric fold. In the region of the glottis (text-fig. 55, p. 235) the dorsal side of the liver comes into close contact with the ventral wall of the pharynx, and the hepatic vein (*h.v.*), which lies on the dorsal side of the liver nearly in the mesial plane, receives a branch (*p.v.*) on its left side from the mesodermal portion of the pharyngeal wall. A little further forwards another and larger vein opens from the pharyngeal wall into the hepatic vein on its dorsal side. In these, no doubt secondary, venous connexions between hepatic vein and the pharyngeal wall in the neighbourhood of the glottis we have in all probability the foreshadowing of the pulmonary veins.

In the 30 mm. larva the pulmonary veins show the main features of the adult arrangement. The blood from each lung drains into a vein on the medioventral edge of the lung—the right larger than the left in correlation with the larger size of the lung of this side. Anteriorly the two veins unite and run along between the two lung cavities ventrally. In the region of the glottis the pulmonary vein lies just dorsal to the hepatic vein (text-figs. 71 and 56, p. 236) and finally opens into this vessel in the middorsal line. Just in front of its opening a large vein from the pharyngeal wall opens into the hepatic vein and from this pharyngeal vein a small branch passes back on each side on to the root of each lung dorsally.

The fusion of right pulmonary vein with caudal vein described as occurring in the adult has not yet taken place—the lung ending freely at its posterior end.

Arterial System. Unfortunately the deficiencies of material make it impossible to give a proper account of the development of the arterial system, and I will therefore dismiss the subject with a few general remarks. The most striking feature in the development of the arterial system is already apparent in the stage already described (text-fig. 67, p. 254), viz. that in accordance with the fact that the single persisting external gill—that of the hyoidean arch—forms practically the sole special organ of respiration during a prolonged period of larval life, the other aortic arches have their development greatly postponed. In fact it is not until about stage 32 that the other aortic arches make their appearance. In larva 19 (stage 33) III., IV., V. and VI. are developed, though each of them is a very thin vessel, capillary in fact, in size. There are also distinct traces of aortic arch I. In the 30 mm. larva where the general arrangements of the arterial system already resemble those shown in the figure on p. 110 aortic arches III., IV. and V. are still relatively very small. II. is large in correlation with the continued activity of the external gill: VI. is large in its proximal portion in correlation with the development of its large pulmonary branch to the lung and adjacent parts of the pharyngeal wall[1]. It is to be noted that the pulmonary artery is still a branch of the ventral (afferent) portion of the aortic arch. There appears to be a minute forward continuation of the ventral aorta to the region of the mandibular arch, though on account of the thickness of the sections I am not able to be quite certain on this point.

Appearance of first blood vessels. The first blood vessel to become clearly defined is the dorsal aorta and its roots. In larva 7 A (stage 25) the dorsal aorta can be traced as a roughly tubular cavity, interrupted here and there, extending back to nearly the level of the cloaca. At about the level of the front end of the pronephros it bifurcates into the two aortic roots. Each aortic root can be traced forwards on each side over the mass of hypoblast. At the level of the external gill rudiment it turns outwards and then curves ventralwards round the edge of the pharyngeal rudiment. It can be traced for some distance round the side of the pharynx as a fairly regular space but soon narrows down into a mere chink in the mesenchyme and then terminates without its being possible to trace it into continuity with the heart rudiment.

At this stage the dorsal aorta is in most parts quite well defined by an arrangement of mesenchyme cells bounding it. In the preceding stage as shown in larva 6 A there is no arrangement of cells to mark off the dorsal aorta, but it is a remarkable fact that its position is already distinctly marked out as a clear space amongst the delicate reticulum which is visible between the various organ rudiments of the larva.

The natural tendency would be to regard this delicate reticulum as an artificial product brought about by the action of the fixing fluids upon the albuminous

---

[1] Boas has already chronicled the fact that in the adult *Polypterus* the pulmonary artery sends branches to the wall of the pharynx (*Morphol. Jahrb.*, Bd. VI. 1880, S. 344).

substances contained in the fluid which bathes the tissues of the embryo, but the appearance which I have mentioned, that this reticulum is capable of foreshadowing in its own substance the development of a particular organ gives decided cause to reflect whether it is not really a living protoplasmic reticulum.

Mode of origin of blood vessels.   The problem of the mode of origin of the blood vessels is in *Polypterus* as in other forms one very difficult of solution, and I do not feel at present in a position to speak dogmatically on the subject.

The vessel whose mode of development can be studied with least difficulty is the dorsal aorta, owing to the fact that this vessel develops from before backwards, and to the fact that the longitudinal axis of the young embryo is a straight line, unlike those vertebrate embryos where the trunk is curved round the yolk.   These facts make it possible to obtain very instructive horizontal sections which show a long stretch of the dorsal aorta in a single section, the state of development towards the front end of the section being much more advanced than towards the hind end.   The dorsal aorta makes its appearance about stage 24.

On examining horizontal sections through 6 D (stage 24) no cell elements are to be detected in the region where the aorta will develop, i.e. between the subnotochordal rod dorsally and the surface of the hypoblast ventrally.   Only a fine coagulum occurs in this region.

Corresponding sections through 6 C (also stage 24) show a similar condition except that one or two rounded yolk-laden cells are to be seen at long intervals apart lying beneath the subnotochordal rod.

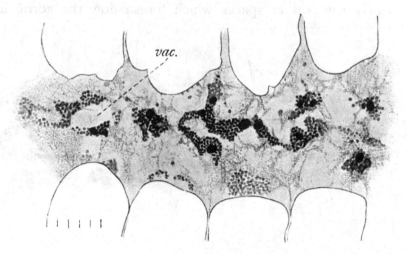

FIG. 72.   Part of horizontal section through aortic rudiment in 6 E (stage 24 +).
vac., vacuole.

Embryo 6 E (stage 24+, text-fig. 72) shows the first obvious rudiment of the dorsal aorta.   Here in the space below the subnotochordal rod irregular masses of yolky protoplasm have arranged themselves in an anteroposterior row.   Here and there clear spaces (*vac.*) are seen, evidently foreshadowing the aortic lumen.

These spaces appear to arise within the protoplasmic masses—to be in fact enormous vacuoles—but careful examination with high powers shows that the protoplasm bounding them is multinuclear, so that there are difficulties in the way of describing them as either intra- or intercellular.

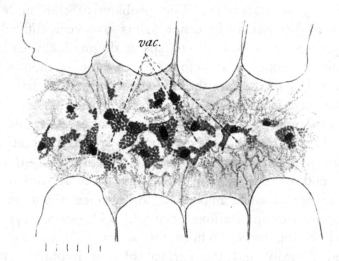

FIG. 73.   Similar section to last, from 6 F (stage 24 +).

Text-fig. 73 shows a high power view of a horizontal section from 6 F belonging also to stage 24 + which is, as regards the aortic rudiment, slightly more advanced than 6 E, the rudiment extending over a greater anteroposterior distance and showing very clearly the row of spaces which foreshadow the aortic lumen.

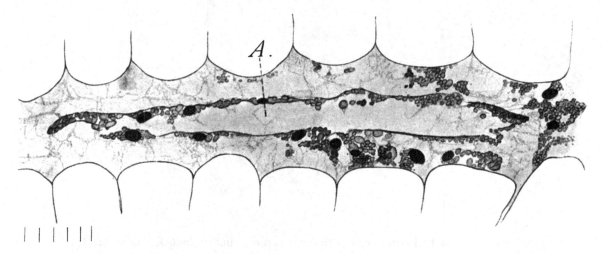

FIG. 74.   Horizontal section through aortic rudiment (A.) in 7 C (stage 25).

A decided advance from the condition just described is seen in 7 C (stage 25). Horizontal sections of this embryo show at about the level of myotome 10 the dorsal aorta in a similar condition to that described in the preceding stage. On tracing the rudiment forwards however the successive cavities are seen (text-fig. 74) to fuse together, the protoplasm separating them becoming apparently withdrawn into the lateral walls. From about myotome 10 forwards there is thus found no longer a row of cavities one behind the other but a single cylindrical cavity, the definitive cavity of the aorta. Posteriorly this continuous cavity stops quite abruptly and its place is filled by a row of still separate cavities such as alone existed in the preceding stage.

I have said nothing as to the origin of the protoplasmic masses which give rise to the rudiment of the dorsal aorta. As regards their broad significance there is no doubt. Their heavily yolked character shows them to be mesenchymatous masses derived from the primitive entoderm. But as to whether they have been derived directly from the enteric wall underlying them or from the sclerotom— that region of the segmented mesoderm where the budding off of mesenchyme elements is particularly active—is a question of which the answer is less obvious, and I remained for some time in doubt between the two possible answers.

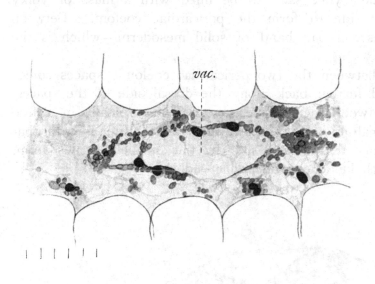

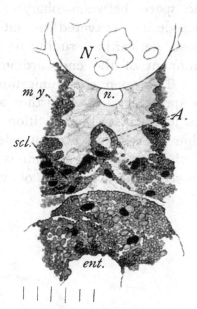

FIG. 75.   Similar section to last but at slightly different level.
vac., vacuole in aortic rudiment.

FIG. 76.   Part of transverse section through 7 G.

A., aortic rudiment; *ent.*, cavity of gut; *my.*, myotom; *N.*, notochord; *n.*, hypochorda; *scl.*, sclerotom.

Fortunately the examination of particularly perfect series of sections has now removed all doubts about the matter. The protoplasmic masses which form the aortic rudiment are derived from the "sclerotom." Put in the ordinary wording of embryology the aortic "cells" (angioblasts) migrate towards the middle line from the sclerotom and then take up their position in a row to form the aorta. A form of wording which however in my opinion presents a more correct picture of the actual facts is to say that coarse heavily yolk-laden protoplasm from the sclerotom streams inwards along the strands of the fine meshwork of living substance which traverses the spaces between notochord, myotoms and enteron, and collects in the neighbourhood of the mesial plane in the form of irregular masses, which form the aortic rudiment. In these more perfect series of sections of this stage (25) the dorsal enteric wall is bounded by a perfectly regular clean cut surface which quite excludes the possibility of its giving rise to the aortic rudiment (cf. text-fig. 76).

Origin of Heart. Early stages in the development of the heart are interesting from their extraordinarily misleading character as observed in sections. Sections of certain isolated stages appear to show most conclusively the directly hypoblastic origin of the endocardium, while the study of a more complete sequence of stages brings out the probability that endocardium like myocardium is of mesoblastic origin.

Transverse sections through 8 A (about stage 25, text-fig. 50, p. 232) show the space between pharynx and "yolk sac" to be filled with a mass of yolky mesoderm excavated on either side to form the pericardiac coelom. Between these pericardiac rudiments lies a broad band of solid mesoderm—which is the rudiment of the endocardium.

The open communication between the two pericardiac coelomic spaces comes to spread gradually further and further back along the dorsal side of the spaces. As it does so the partition between them comes to terminate dorsally in a free edge over which the mesodermal wall of the right space is continuous with that of the left. The septum is covered on each side by this splanchnic mesoderm, between the two layers of which lies the endocardiac rudiment.

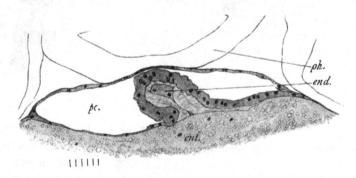

FIG. 77.

ent., endoderm; *end.*, endocardium rudiment; *pc.*, pericardium; *ph.*, pharynx.

In larva 10 A (text-fig. 77) this condition is clearly shown. The septum is also seen at this stage to be leaning over towards the right hand side and an irregular split is appearing within the endocardium, foreshadowing the cavity of the heart. It is this stage which is so misleading in section for the line of demarcation between endocardium and underlying hypoblast is frequently quite indistinguishable.

Origin of Blood Corpuscles. The primitive corpuscles appear suddenly in large numbers in the circulation. Thus larva 9 B (stage 25) shows the dorsal subintestinal vein and other of its main vessels well developed with wide lumen, but in this lumen there is nothing but a fine coagulum representing the fluid contents. Not a single corpuscle is to be seen. Again in larvae 10 B and 10 C (stage 27) the heart has developed its lumen, but there are still no corpuscles. On the other hand larvae 10 D and 10 E of the same age show the heart lumen crowded with corpuscles. The explanation seems to be that corpuscles are suddenly set free in large numbers almost synchronously in certain parts of the peripheral vessels. From the broad morphological point of view[1] blood corpuscles are in all probability to be regarded simply as mesenchyme cells which have lost their protoplasmic connexion with neighbouring cells so as to become freely movable. The appearance of sections of the young Polypterus seems to indicate that here the blood corpuscles actually originate in this manner from mesenchyme cells, and their sudden appearance in large numbers may point possibly to their being set free at the time of an epidemic of mitosis, for the onset of the mitotic process frequently induces a retraction of cell processes and an assumption of a nearly spherical shape.

As regards the precise region or regions in which this freeing of corpuscles takes place I cannot speak with any great confidence. On the whole appearances favour the view that at least one region in which they are formed is that of the pronephros, where in early stages corpuscles may be seen lying in meshes of the mesenchyme and differing from the ordinary mesenchyme elements only in the fact that they are not connected with their neighbours by protoplasmic strands. It is necessary of course to exercise caution in concluding that corpuscles are really still at their point of origin in view of the consideration that circulation of the blood plasma is in all probability fully established before it contains any cell elements.

XI. The Cartilaginous Skeleton.

The cartilaginous skeleton of the 30 mm. larva has already been described by Budgett. Of the series of developmental stages dealt with in this paper it is only the last three (larvae 18, 19 and 20, representing stages 32, 33 and 36) which possess a cartilaginous skeleton, and these will now be briefly described. It will be convenient to take the oldest stage (36) first.

Stage 36. Text-fig. 78 represents a view of the cartilaginous skeleton of the head region of this specimen as seen from the dorsal side. An extensive basal or parachordal plate of cartilage spreads outwards from the notochord on each side. It is to be noticed that the anterior limit of the parachordal cartilages lies a long way—nearly ·25 mm.—behind the tip of the notochord, the anterior end of which· projects freely forwards. Traced outwards the parachordal plate

[1] Cf. Lankester.

is seen to pass without interruption into the cartilaginous capsule which ensheaths the otocyst except on its mesiad face. Posteriorly the parachordal cartilage on each side is continued backwards for a short distance along the notochord: this prolongation agreeing in its general relations to the notochord with the neural arch elements farther back. Anteriorly the cartilage of the parachordal region is continued forwards into the trabecula on each side. The trabeculae are considerably flattened from above downwards, and they are widely separated from one another as in Lampreys and Amphibians and various Selachians. Slightly converging as they pass forwards the trabeculae eventually pass into a broad ethmoid plate of cartilage which underlies the olfactory organs. This plate turns upwards at its outer margin to form a vertical plate of cartilage (not shown in the figure) which rises up on each side external to the olfactory sac. Another vertical upgrowth is found in the mesial plane forming a septum between the olfactory sacs of the two sides and having its dorsal edge in contact with the ventral surface of the brain which extends for some distance forwards dorsal to the olfactory sacs.

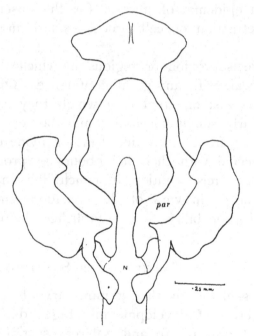

FIG. 78. Dorsal view of chondrocranium of stage 36.
N., notochord; *par.,* parachordal plate.

At the hinder end of the skull the chief feature to be noted is the presence of the pair of large neural arches which become fused with the cranium as already seen by Budgett in the 30 mm. larva.

Meckel's cartilage in this larva forms a long slender rod fused with its fellow anteriorly and swelling out at its hind end to form the large quadrate region. From the latter the long slender palatopterygoid bar runs forwards dorsal to the buccal cavity. This palatopterygoid bar is considerably flattened from within outwards and it passes forwards nearly parallel to the trabecula which lies on its dorsomedian side.

Skeleton of pectoral fin.   Budgett has described the character of the skeleton of the pectoral limb in the 30 mm. larva, and from his description there can remain little doubt that the crossopterygian fin of *Polypterus* is to be interpreted as a uniserial archipterygium which has lost the rays on the inner side of the axis. It was greatly to be hoped that the material now under review would throw further light on this interesting morphological question. The material is however very disappointing in this respect. In the larva now being described the skeleton of the free limb consists simply of a thin uniform plate of cartilage with a few small openings scattered without any apparent regularity. The shoulder girdle consists of a simple curved rod of cartilage on each side which is prolonged forwards for a considerable distance, its anterior portion lying ventral to the fourth branchial arch.

Stage 33. This larva is unfortunately damaged in the head region. It has been possible however by careful reconstructions to make out the main features of the head skeleton, and the differences between it and stage 36 are seen to be comparatively slight. The ethmoid plate is smaller: the cartilaginous roof of the auditory capsule has spread inwards to a much less extent and the direction of the occipital arch on each side is more markedly outwards from its basal attachment to the notochord. The pectoral girdle and the skeleton of the free limb are as in stage 36, only somewhat smaller.

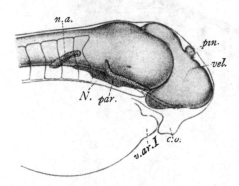

FIG. 79. Side view of brain and skeletal elements in stage 32.

c.o., cement organ; *N.*, notochord; *n.a.*, neural (occipital) arch; *par.*, parachordal; *pin.*, pineal body; *v.ar.I*, mandibular arch; *vel.*, velum. (The dotted outline indicates the rudiment of the trabecula.)

Stage 32 (text-fig. 79). The parachordals are present lying on each side of the notochord but not reaching anywhere near its tip. The anterior half of each parachordal extends outwards into a plate of cartilage lying beneath the otocyst. The trabeculae are marked out merely by a condensation of nuclei. The occipital neural arches have made their appearance, and examination of their outer ends shows that they lie in the septum between myotom 1 and 2 of this stage.

There is no trace of pectoral fin skeleton beyond a slight condensation of nuclei at the base of the limb.

XII. NERVOUS SYSTEM.

On account of the limitations of the material I do not propose to attempt more at the present time than a few notes on certain points of interest in the development of the Brain and of the Olfactory organ.

Brain. The appearances of the medullary plate as a whole have been dealt with in the section on external features. The medullary plate is as usual formed entirely by a thickening of the deep layer of the epiblast (text-fig. 80). A neural tube is formed by overarching of the medullary folds. The infundibular depression of the medullary floor is formed while the medullary groove in this region is still widely open. The optic outgrowths also appear at this early period. After the neural tube is closed its anterior dilated part becomes soon divided into what are probably to be looked on as the two primitive divisions—rhombencephalon and primitive fore brain (text-fig. 81). The marking off of *tectum opticum* from thalamencephalon becomes marked about stage 30. The pineal outgrowth is simple and rounded without trace of division. There is no trace of eye structure at any time.

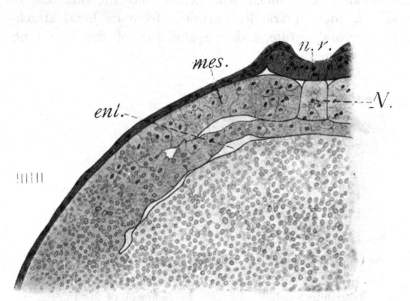

FIG. 80. Part of transverse section through an egg of stage 15.
ent., enteric cavity; *mes.*, mesoderm; *N.*, notochord; *n.r.*, neural rudiment.

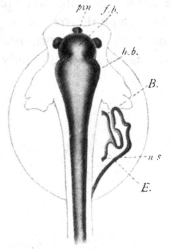

FIG. 81. Reconstruction of stage 27 showing brain and pronephros.

B, E, pronephric tubules; *h.b.*, hind brain; *f.b.*, fore brain; *ns.*, nephrostome; *pin.*, pineal body.

(In the case of the pronephros the reconstruction represents the *lumen*, and does not include the thickness of the walls.)

Brain of Adult. It appears advisable to reproduce a figure (text-fig. 82 A) of a sagittal section through the brain of an adult *Polypterus*, as that recently figured by Burckhardt[1] is apparently immature.   The chief point of interest shown in the mature brain and not indicated in Burckhardt's figure is that the cerebellum projects forwards anteriorly into the cavity of the mesencephalon to form a valvula cerebelli like that of a typical teleostean.   The posterior part of the cerebellum, however, instead of forming a somewhat tongue-like structure dorsal to the medulla oblongata, projects back within the cavity of the fourth ventricle just in the same

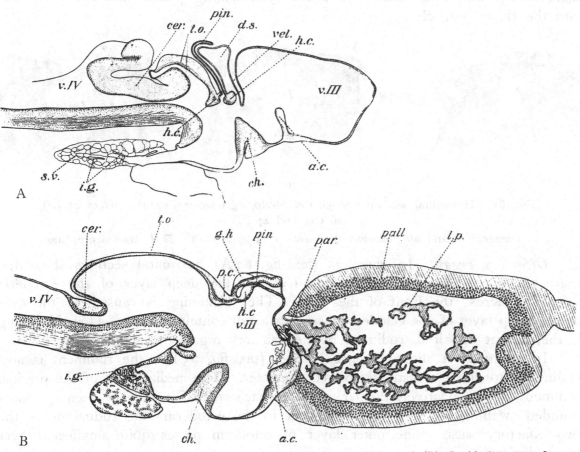

FIG. 82.   Mesial section through the brain of (A) *Polypterus bichir*, and (B) *Lepidosiren paradoxa*.
a.c., anterior commissure; *cer.*, cerebellum; *d.s.*, dorsal sac; *g.h.*, habenular ganglion; *h.c.*, habenular commissure; *i.g.*, infundibular gland; *l.p.*, lateral plexus; *ch.*, optic chiasma; *pall.*, pallium; *par.*, paraphysis; *p.c.*, posterior commissure; *pin.*, pineal body; *s.v.*, saccus vasculosus; *t.o.*, tectum opticum; *v.III*, third ventricle; *v.IV*, fourth ventricle; *vel.*, velum.

way as its front end does into the ventricle of the midbrain[2].   Comparing the cerebellum of the Crossopterygian with that of the Elasmobranch or of one of the higher Vertebrates, such as a bird, we see a good example of the two different ways of attaining to the same increase in area by invagination and evagination respectively.

[1] Semon's *Forschungsreise*, I. s. 571.
[2] Cf. *Acipenser*. Goronowitsch, *Morphol. Jahrb*. XIII. 1888, t. xviii. fig. 17.

The figure also brings out clearly the great anteroposterior elongation of the thalamencephalon in *Polypterus* as compared with e.g. *Lepidosiren* or a Cheiropterygial Vertebrate. Its walls become greatly thickened and there is no bulging outwards to form hemispheres except anteriorly, where on each side the olfactory lobe arises as a small forwardly directed evagination. While there is no outward bulging to form a definite hemisphere the thickened wall of the thalamencephalon finds additional space by becoming pushed somewhat *inwards* into the third ventricle.

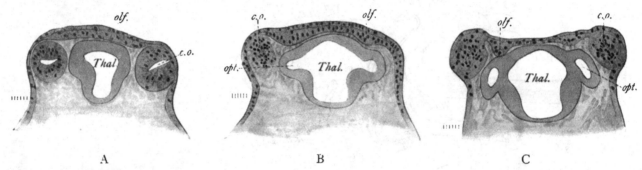

A　　　　　　　　B　　　　　　　　C

FIG. 83.　Horizontal sections through the developing olfactory organ at stages 25 (A), 26 (B), and 27 (C).

c.o., cement organ; *olf.*, olfactory rudiment; *opt.*, optic stalk; *Thal.*, thalamencephalon.

Olfactory organ. In stage 25 (text-fig. 83 A) horizontal sections show that there exists a well-marked thickening (*olf.*) of the deep layer of the ectoderm stretching across the front of the head. The thickening is caused by the cells of the deep layer of the ectoderm assuming a tall columnar form. This thickening is the, as yet unpaired, rudiment of the olfactory organ.

In stage 26 a slight advance is seen (text-fig. 83 B): the rudiment is now beginning to show a distinctly paired character. The median part of the original rudiment is now relatively thinner, while on each side of it is to be seen a distinct rounded swelling (*olf.*) of the deep layer of the ectoderm—the rudiments of the two olfactory sacs. The outer layer of ectoderm passes quite unaffected over the rudiments.

In stage 27 the two rudiments—seen in horizontal sections (text-fig. 83 C) just in front of the origin of the optic stalk (*opt.*) and just mesiad of the cement organ (*c.o.*)—are seen to have become decidedly thicker, and they show in places two layers of nuclei. The external layer of the ectoderm passes as before quite unaffected over the two olfactory rudiments.

In stage 28—29 (text-fig. 84) the ectodermal thickening has greatly increased in depth. In the middle of it a distinct cavity is visible with smooth outline. The cells lining it have developed a tall columnar shape. The cavity is completely closed, although the rudiment is quite continuous with the general ectoderm. The superficial ectoderm whose cells now show melanin granules still passes uninterruptedly and without modification over the olfactory rudiment.

Communication with the exterior is established about stage 30. Thereafter the chief changes consist in the sac growing backwards to form the projections which give to the organ its characteristic complicated form ·in the adult.

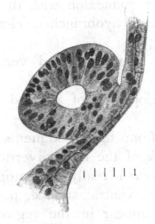

FIG. 84. Longitudinal vertical section through olfactory rudiment.

XIII. SUMMARY AND GENERAL REMARKS.

Summary of Chief Results.

1. The eggs of *Polypterus* are believed to be deposited in the lateral lagoons, where they adhere to the sticks, water-plants etc., pp. 203, 207.

2. The erectile anal fin suggests internal fertilization but there is no evidence to show definitely whether or not this takes place, p. 199.

3. Segmentation is complete and in its earliest stages nearly equal, p. 211.

4. The invagination groove is at first nearly equatorial in position, p. 212.

5. As the curve described by the groove becomes closed an enormous "yolk plug" is formed, p. 212.

6. There is no obvious protostomal seam, p. 212.

7. The neural tube is formed by overarching of the medullary folds, p. 213.

8. Rudiments of external gills and cement organs appear at an early stage, pp. 213, 214.

9. The buccal cavity is for a while a widely-open space bounded by the cement organs, the lower side of the head and the cardiac region, p. 216.

10. The mesoderm of the trunk region arises as it does in *Lepidosiren*, *Protopterus* and *Petromyzon* by "delamination," p. 225.

11. The notochord has a well marked groove along its ventral side in early stages, p. 225.

12. The Hypochorda develops a cuticular sheath and its cells show vacuolation, p. 226.

13.   A well developed solid postanal gut is present, which eventually breaks up and disappears, p. 228.

14.   The secretory epithelium of the cement organs is endodermal in origin, arising as a pair of hollow enteric diverticula which become cut off from the rest of the endoderm and establish a connexion with the outer surface, p. 229.

15.   The hyomandibular and hyobranchial clefts are laid down early; the others develop very late, p. 233.

16.   The lung rudiment is median and ventral and very soon develops asymmetry, p. 235.

17.   The spiral valve develops out of a coiled condition of the gut rudiment as in other forms, p. 237.

18.   The pancreas arises from three rudiments, one dorsal and two ventral, the latter arising from the stalk of the liver diverticulum, p. 237.

19.   The "liver" of *Polypterus* is really a hepatopancreas—the pancreatic tissue being spread out over part of its ventral surface, p. 237.

20.   Coelomic spaces first appear in the region of the nephrocoeles, i.e. in the lower ends of the mesoderm segments, and next in the pericardium, p. 242.

21.   The pronephros possesses, through the greater part of its existence, two nephrostomes (B and E) on each side—those belonging originally to metotic mesoderm segments II. and V. p. 244.

22.   Material does not suffice to make clear the early stages in the development of the archinephric duct, though appearances favour a similar development to that in Gymnophiona except that the tubule rudiments are solid, p. 245.

23.   In early stages distinct tubule rudiments (A—E) occur in metotic segments I.—V. with doubtful rudiments in several succeeding segments, p. 244.

24.   Tubule A soon disappears; C and D also gradually diminish in size and finally also disappear leaving B and E as the functional tubules, p. 244.

25.   In later stages the pronephros reaches a great size, due to the great growth in length of the anterior part of the archinephric duct and of the tubules, p. 245.

26.   Pronephric chambers B and E each possess a glomerulus. Pronephric chamber B opens out and becomes merged in the general splanchnocoele while E has merely a small opening which leads into pronephric chamber B, p. 249.

27.   In the 30 mm. larva both nephrostomes B and E still persist, p. 250.

28.   The scanty material illustrating early stages in the development of the mesonephros shows that the tubule rudiments are at first independent both of duct and of coelomic lining, p. 250.

29.   The arterial system at an early stage (28) possesses only one pair of aortic arches (II.) which pass out into the external gill, p. 253.

30.   The main venous trunks show during early stages an arrangement departing little from the normal, p. 256.

31. The dorsal aorta arises from cells or protoplasmic masses derived from the sclerotom: its lumen is derived from the fusion of originally separate vacuoles in these masses, p. 261.

32. The endocardium appears to be mesoblastic in origin, p. 264.

33. The blood corpuscles appear suddenly, and it is suggested that they are mesenchyme cells set free by an epidemic of mitosis, p. 265.

34. The posterior cardinals soon fuse to form an interrenal vein, p. 256.

35. The ductus Cuvieri are at first simply portions of the general vitelline venous network, p. 255.

36. The venous system becomes in its later stages strongly asymmetrical, the anterior end of the left posterior cardinal becoming greatly reduced so that the blood-stream from the interrenal vein passes to the heart *via* the enlarged right ductus Cuvieri. The left inferior jugular vein undergoes a similar reduction so that the blood stream from the thyroid region passes back practically entirely by way of the right inferior jugular vein into the right ductus Cuvieri, p. 258.

37. The praecloacal part of the subintestinal vein breaks up into a network and does not persist in the adult as a distinct vessel, p. 259.

38. The chondrocranium is amphibian-like in early stages, p. 265.

39. A single pair of large occipital arches is present, developed between metotic myotoms 1 and 2, p. 267.

40. The skeleton of the pectoral limb in its early stages is formed by a continuous sheet of cartilage, p. 267.

41. The neural tube arises by overarching of the medullary folds, p. 268.

42. Both infundibulum and optic rudiments are clearly recognizable while the medullary groove is still widely open throughout, p. 268.

43. As in *Lepidosiren*, etc., the brain is during the earlier part of its development subdivided into two—not three—regions—the primitive fore brain and the rhombencephalon, p. 268.

44. The pineal outgrowth is single and without any eye-like structure, p. 268.

45. In the adult the cerebellum becomes highly developed and forms anteriorly a valvula cerebelli, while posteriorly it projects back in a quite similar manner into the fourth ventricle, p. 269.

46. The material forming the side walls of the thalamencephalon does not become pushed out to form cerebral hemispheres but is accommodated partly by the great increase in length of the thalamencephalon, partly by its becoming invaginated into the interior of the third ventricle, p. 270.

47. The two olfactory rudiments are apparently connected by an ectodermal thickening across the middle line in early stages, p. 270.

48. The cavity of the olfactory organ is a secondary excavation in the originally solid rudiment, p. 270.

Concluding Remarks.

In the preceding paper I have set down as shortly and concisely as possible the more important points which have emerged from my investigation of the *Polypterus* embryos and larvae collected by Budgett. I propose to leave on one side for the present any elaborate discussion of these results and also their comparison with the data obtained by other investigators on other forms of Vertebrates. I may here merely indicate that on the whole the general phenomena of development in *Polypterus* show frequent striking resemblances with what occur in Dipnoans and in the lower Amphibia. I believe that these resemblances are sufficient by themselves to indicate the probability that the Teleostomes, the Dipnoans and the Amphibians have arisen in phylogeny from a common stem, which would in turn probably have diverged from the ancestral Selachian stock. The ancestors of the Amniota probably diverged either about one or about several points from the region of the stem common to Dipnoi and Amphibia. While admitting such vague speculative conclusions we are, in my opinion, here, as in other phylogenetic speculations, absolutely debarred from making such statements as that the "Amniota" are derived from the "Amphibia" or the "Dipnoi" from the "Crossopterygii." It is perfectly clear that these classificatory groups are merely expressions of the imperfect nature of our knowledge. They are expressions of our knowledge of the forms of life during one, or more, definite periods of geological time, and if the facts at our disposal were not sharply delimited by the gaps in palaeontological knowledge such sharply-defined classificatory groups would not exist.

External Gills[1].

I have not said anything about the details of development in the external gills. I need only say that they develop here exactly as in *Lepidosiren* and *Protopterus* and in the more primitive Amphibia (Urodela and Gymnophiona), i.e. each one arises as an outgrowth from the outer side of the visceral arch (in this case hyoidean) composed of mesenchymatous core with ectodermal covering. Here, as elsewhere, they develop well before the perforation of the gill clefts: here as elsewhere the aortic arch passes out into the external gill rudiment, i.e. the aortic arch itself is in early stages simply the vessel of the external gill. I need not repeat the reasons—already stated elsewhere[2]—which lead to the conclusion that these true external gills are organs of great antiquity which were probably characteristic of primitive Vertebrates. The tendency has been—quite rightly so long as their development was known only in Amphibia—to regard the external gills as mere larval adaptations

---

[1] It is greatly to be regretted that authors still apply the expression "external gills" to ordinary gill filaments which have become prolonged so as to project in a secondary manner to the exterior.

[2] *Proc. Roy. Phys. Soc. Edin.* XVI. 1905—1906, p. 201.

of no special morphological importance. I have every confidence that this view will be given up as developmental material of Crossopterygians and Dipnoans becomes more available for general study. When a given morphological structure is found only in a single group it is justifiable to regard it as a secondary acquirement: when the same structure turns up in two other equally primitive groups and where the evidence is against its being merely adaptive it is no longer justifiable to do so.

The general position which I believe to be justified by our present knowledge of the development of the gill apparatus of Vertebrates is that (1) the true external gills are structures developed absolutely independently of the gill pouches or clefts, and (2) the respiratory epithelium lining the existing gill clefts has arisen by a spreading inwards from the ectodermal respiratory epithelium of the external gills.

The special morphological interest of the true or primary external gills lies in the fact that if really primitive, if really characteristic of the more ancient forms, they would form a series of potentially motor organs, some of which might well have increased their motor function already potentially present in a rudimentary stage, and become the two pairs of limbs so characteristic of gnathostomatous Vertebrates. I have no intention of wasting time and space in restating the reasons which lead me to believe that the paired limbs may well be homodynamous with the "external gills" but it seems advisable to draw attention to one or two points which seem to require more consideration than they are frequently given by writers on the evolution of the Vertebrate limbs.

The Biserial Archipterygium and its Archaic Character.

The material at my disposal has not enabled me to throw any further light on the question of the morphology of the *Polypterus* limb than had already been done by Budgett. In his paper on the structure of the larval *Polypterus* (this volume, p. 167) Budgett showed that the condition of the fin skeleton in the 30 mm. larva was such as to indicate its close relationship to the type of uniserial fin skeleton occurring in sharks.

As is well known, the tendency amongst morphologists in this country and in America is at present to deny the primitive nature of the biserial archipterygium as compared with the uhiserial type. It would in my opinion be of great advantage to morphology if those who take this standpoint would—leaving entirely on one side speculative views as to the phylogenetic origin of limbs—state clearly and precisely what are the facts which lead them to consider such a series as that represented in text-fig. 85 fallacious as regards the evolutionary history of the various types of limb skeleton.

To many who hold no special views as to the origin of limbs the series will seem a decidedly plausible one—so plausible that it would be advantageous to have it put definitely out of court if it can be shown to be in all probability erroneous. The order of appearance of the various types of fin skeleton in geological time would of course be conclusive if we knew it, but unfortunately we have no

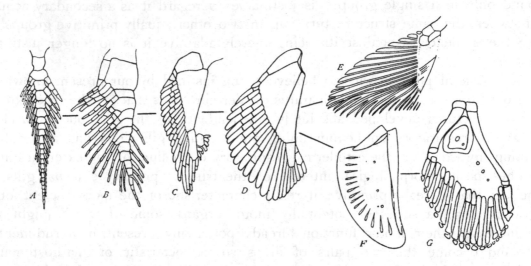

FIG. 85.   Figures of pectoral fin skeletons: *A, Ceratodus* (Semon); *B, Pleuracanthus* (Fritsch); *C, Centrophorus*—embryo (Braus); *D, Acanthias* (Gegenbaur); *E, Cladoselache* (Bashford Dean); *F, Polypterus* larva (Budgett); *G, Polypterus* (Wiedersheim).

The figures are not to the same scale. The figures illustrate the skeleton of the pectoral fin in various of the lower fishes and the arrangement of the figures indicates the way in which each type may be readily referred back to a primitive "biserial archipterygium" like that of *Ceratodus*.

certain knowledge. When it is considered how infinitesimally small are all the excavations and quarryings which have ever been carried out by man as compared with merely the mass of fossiliferous deposits known to exist—without taking into consideration the possibly far greater masses which have been completely destroyed in the course of geological ages—it is impossible not to be impressed by the necessity of the greatest caution in basing conclusions as to the time of appearance of special forms on the mere shreds of evidence available.

Main Lines of Evidence for the Relatively Primitive Nature of the Biserial Archipterygium.

(1) What we do see on taking a broad view of the known data of the palaeontology of fishes is that in tracing backwards the three main groups—Elasmobranchs, Teleostomes and Dipnoans—we find the pointed fleshy fin, like that of *Ceratodus*, becoming in palaeozoic time if not the predominant at least a very conspicuous form of fin. This is in itself evidence—not very strong I admit—of the archaic nature of such fins.

(2)   When we study the clumsy movements of the only existing fish (*Ceratodus*) retaining this type of fin (in *Protopterus* and *Lepidosiren* the limbs are admittedly degenerate) we cannot but admit that as a piece of motor mechanism it is far inferior to the ordinary type of fin in which the muscular apparatus is concentrated at the base, and palaeontology bears out this, for we see how the archipterygial type of fin has undergone almost complete disappearance in the struggle of competition with other forms.

(3)   We know from embryology that in certain Selachians a few postaxial rays occur towards the tip of the fin, as in *Pleuracanthus*—these being best marked during early stages in development.   This embryological fact—meaningless otherwise—is at once explained by the hypothesis that it is a reminiscence of an ancestral stage in which the functional fin was a biserial one.

(4)   Finally, we see by such a series as that of fig. 85 how readily the other main types of fin (including that of *Cladoselache* itself judging by the published descriptions!) can be linked on to the biserial archipterygium.

It would of course be easy to draw up a much stronger case for the relatively primitive character of the biserial archipterygium by going into greater detail. This has however been already done repeatedly by others[1] and I think the few points indicated are quite sufficient to indicate certain lines on which the view will have to be successfully attacked before all morphologists can be expected to agree as to its being discredited.

External Gills and the Origin of Vertebrate Limbs.

I had not intended to touch upon the vexed question of the first origin of the limbs at all in the present paper but the recent appearance of several papers dealing with the subject, particularly one by Goodrich[2], make it seem advisable to define my standpoint in the matter.

There are three distinct views held as to the phylogenetic origin of the paired limbs.

There is the Gegenbaur view according to which the limb has been evolved out of a movable gill septum, there is the Balfour-Thacher-Mivart view that the two limbs on each side are the persisting and enlarged portions of a once continuous fin fold, and there is the third view, held, so far as I am aware, by myself alone, that the paired fins may possibly have been evolved out of the structures which I call true external gills.

Since the days when Gegenbaur and Thacher and Balfour propounded their

---

[1] Cf. Semon, *Forschungsreise*, Bd. I. s. 97.

[2] *Quart. Journ. Micr. Sci.* Vol. L. p. 333. A point of great interest in this paper is the statement that "the continuity of the pectoral and pelvic fin fold is not an essential point" of the lateral fold view. If the little rift which is here suggested in the continuity of the lateral fold only increases in size the acute cause of difference between the two schools may disappear!

views as to the origin of limbs, Vertebrate morphology has not been standing still. Great increases have been made to our knowledge. Now, in considering the working hypotheses of these earlier days of morphology, we should remember that increase of knowledge may greatly alter our point of view, and it seems in my humble opinion that it is conducive to progress, not so much to search for new detailed facts which may bolster up one or other existing hypotheses, as to endeavour to make an impartial survey of the facts as we now know them and then to consider carefully whether the body of facts so surveyed seems to suggest a working hypothesis drawn up on the original lines or one drawn up on somewhat different lines.

Now in regard to the two limb hypotheses I cannot but feel that our standpoint is very considerably altered by what we now know of the external gills. I have been able to make a careful study of the structures in question as they occur in three of the main subdivisions of the Vertebrata—the Amphibia urodela, the Crossopterygii and the Dipnoi—and it is my deliberate opinion—after having carefully gone into the evidence on which the Gegenbaur and the Balfour-Thacher view are based—that these external gills are (1) structures of great antiquity in the Vertebrate phylum, and (2) that they may well have been the forerunners of the paired limbs as we know them. It is of course to be understood that I now, as I have all along, put forward this view as a mere working hypothesis which appears to fit the facts as we now know them, but which may have to be emended or given up entirely with further increase of knowledge[1].

We may however rest assured that sound views on this as on other important problems of Vertebrate morphology will only be arrived at when it is realised that in discussing any such questions it is essential to be acquainted with, and to accord equal consideration to, the facts of development as seen in each one of the main groups of lower Gnathostomata—the Elasmobranchs, the lower Teleostomes, the Dipnoans and the Urodele Amphibians.

The two ordinarily accepted hypotheses are each of them supported by important bodies of facts, as we should expect from the mere names of the distinguished

---

[1] It will be seen that the standpoint taken up in this paper is unhappily much less definite than that commonly assumed by exponents of the lateral fold hypothesis—cf. e.g. Regan (*Proc. Zool. Soc. Lond.* 1906, p. 722). The remarkably definite character of this writer's conclusions is best shown by a short selection of excerpts: "No one who carefully studies them can come to any other conclusion than that the writers had *proved* their case and that the theory of Gegenbaur had been *absolutely* and *finally* disposed of." "The writer of a recent memoir (Kerr) has so little understood Balfour's observations as to offer an absolutely *impossible* explanation of them." "Thacher must be held to have *proved* his case from the facts of comparative anatomy alone." (Amongst the chief points in this case are said to be that "The pectoral and pelvic girdles *must* have been formed by fusion......of the anterior basalia" and that "the type of fin termed archipterygium by Gegenbaur *must* be secondary and the suggested homology of limb girdles with gill-arches *cannot* be seriously entertained.") "Even if this ridge were confined to the Hypotremata it must be *evident* that a very transitory structure, connecting the fin-rudiments only at their first development, *can have no relation* to the secondary extension of the pectoral fins in these Selachians." The italics are not Mr Regan's.

investigators who have accepted them.   It is quite unnecessary to recapitulate these facts—they are so well known[1].

At the same time there appear to me to be very decided weaknesses in the chain of evidence upon which each view is based. In the case of the Gegenbaur view the fatal weaknesses are (1) the total absence of evidence of the existence of transitional stages between the gill septum and the locomotor limb, and (2) the absence of apparent probability that a gill septum such as exists in the lowest fishes should become a locomotor organ.

In regard to the lateral fold view we seem to be met at the outset with the same element of inherent improbability. In discussing any morphological question it is of the utmost importance to keep in constant touch with the physiological side of the question, to bear in mind that structure is merely the handmaid of function.   Now the lateral fold view seems to me to ignore completely physiological considerations of the greatest importance, in particular the consideration that one of the most fundamental features of the Vertebrate body is that it is built for movement by lateral flexure. When we regard the fundamental constitution of the primitive Vertebrate body, with its concentration of skeletal axis and central nervous system close to the mesial plane and its main muscular masses, composed of laterally placed longitudinally arranged fibres, arranged in segments one behind the other, we cannot escape from the conclusion that the creature is built for movement of lateral flexure like the movements of an *Amphioxus*.

How improbable it seems that the primitive Vertebrate possessing this exquisite motor mechanism should start to evolve a new motor apparatus which in its early stages would necessarily be a hindrance rather than a gain! It cannot be seriously suggested that the lateral fold was in its earliest stages merely a *balancing* organ when we recall the very slight difference in specific gravity between the animal and the aquatic medium in which it lives.

By far the most important evidence in favour of the lateral fold view lies in the very strongly marked resemblances between the paired fins and the median fins—which latter we are compelled by overwhelming evidence to regard as having been derived phylogenetically from a continuous fin fold.   I do not feel so impressed by the importance of this resemblance as are many of my colleagues.   I think all who have devoted any considerable time and attention to studying the fin movements of fishes belonging to the various main groups (except Dipnoi) will agree that the movements of paired fins and median fins are alike in kind.   In each case waves of flexure to either side of the plane of rest travel along the fin.   And in the two cases the materials out of which the fin mechanism are built up are the same— segmentally arranged muscular and skeletal elements.   Surely we should expect *à priori* the end result to be remarkably alike in the two cases. What can be more alike in the main features of structure than the radiate eye of an insect,

---

[1] Cf. Summary in *Proc. Camb. Phil. Soc.* Vol. x. p. 227.

and the radiate eye of one of the higher Crustacea? Yet who would seriously try to persuade us that the detailed resemblances are inherited from a common ancestral form? We should surely agree that in this case we had to do with organs evolved out of similar materials to perform a similar function and that *hence* the startling secondary similarity.

It seems to me then that while fully admitting the remarkable structural resemblances between median and paired fins it is quite unnecessary to go farther than the adaptation to the performance of similar functions for an explanation of this resemblance.

It may be well to enumerate Goodrich's objections to what he terms the "gill arch theory."

(1) "It offers no intelligible explanation of the participation of a large number of segments in the formation of the paired limbs."

It appears to some morphologists a perfectly "intelligible" explanation to regard the anterior and posterior myotom buds as representing a kind of trail—a reminiscence of the backward and forward migrations or "transpositions" which all admit the limbs to have undergone. Goodrich admits them as evidence of *concentration*, i.e. of migration of the margins of the fin, but will not admit them as evidence of migration of the fin as a whole. The logic of this is somewhat obscure and difficult to follow.

(2) A sharp distinction is drawn between the origin of the muscles of the limbs from the myotoms and of those of the visceral arches from the splanchnopleure.

It seems to me that too much may be made of this distinction. The inner wall of the myotom is originally perfectly continuous with the mesoderm of the splanchnopleure : the two are parts of the same layer of cells. In the development of certain "visceral" muscles it is perfectly clear that the myotoms play an important part, e.g. the constrictor muscle of the pharynx in *Lepidosiren* and *Protopterus*[1].

(3) Great weight is attached to the fact that the visceral arches lie inside, the limb girdles outside, such important features as the coelom, etc.

Here again when one bears in mind such points as the obscurity in which we still are regarding the primitive relations of mesoderm segments and branchial pouches and when we remember that there are very distinguished morphologists who hold the visceral arches to be of ectodermal origin we cannot but feel the need for caution.

(4) The last objection urged by Goodrich, that of the remarkable resemblances between paired and unpaired fins, has already been alluded to.

I have mentioned these objections raised by Goodrich because they might be applied to the external gill hypothesis as well as to the Gegenbaur view.

In conclusion I must state that I know of no direct evidence whatever

[1] Agar, *Trans. Roy. Soc. Edin.*, Vol. XLV.

that the paired limbs have been evolved out of external gills. In this respect the external gill view is on exactly the same footing as are the other two views. In determining which is to serve as our working hypothesis we have simply to consider carefully on which side lies the balance of probability in the present state of our knowledge. In my own opinion that balance of probability seems to lie on the whole on the side of the external gill hypothesis.

Possible Homology of Visceral Pouches and Coelenteric Pouches.

It would of course be quite ridiculous to suggest that the extraordinary want of agreement between the investigators who have worked at the segmentation of the vertebrate head is due to errors of observation. It can only be due to the fact that the evolution of the head region in the individual Vertebrate does not afford anything like a faithful recapitulation of its evolution in phylogeny. And this great modification in the head region is only what would be expected when it is remembered that intense cephalization is one of the most marked features of the Chordata. (The condition in *Amphioxus* can hardly be regarded as other than secondary in this respect.) Now it is fairly clear from the facts of head development that one of the chief kinds of modification which has taken place has been a varying amount of displacement in an anteroposterior direction of the relative positions of mesoderm segments and visceral pouches. I do not feel convinced personally that it can in the present state of our knowledge be regarded as quite outside the bounds of probability that the enterocoelic pouches were once wholly posterior to the visceral pouches and that the two structures are really homodynamous.

Buccal Cavity.

The buccal cavity of *Polypterus* develops in the more primitive way, i.e. as a simple walling in of a special area of the skin as in Elasmobranchs and Amniota and not in the modified manner seen in Urodela and Dipnoi where the cavity is formed as a secondary excavation in what is to all appearance a mass of yolk laden endoderm. In these forms (at least in *Lepidosiren, Protopterus* and *Triton*) the epithelial layer lining the buccal cavity is developed directly out of the outer layer of the anterior portion of the yolky mass which gives rise to pharynx and other parts of the alimentary canal and which would in the ordinary usage of embryological nomenclature be termed throughout its extent hypoblast. The outer layer in question gradually uses up its yolk and becomes converted by imperceptible gradations into the buccal epithelium, the conversion into definite epithelium spreading from before backwards. In properly prepared celloidin sections, it is quite easy to observe that the buccal epithelium in early stages passes back perfectly continuously into the undifferentiated hypoblast and there can be no possibility of a bodily pushing inwards of an ectodermal layer to line the buccal cavity. I wish to accentuate this point as Greil[1] has recently again asserted that an actual inpushing of ectoderm takes place in the formation of the buccal lining of Urodeles.

The peculiar method of formation of the buccal lining in Urodeles and Dipnoans must of course be looked on as a secondary modification. It is not to be taken as indicative that in the strict morphological sense the buccal cavity of these forms is lined by endoderm, but simply as emphasising the fact that the boundary between the germinal layers is, like other biological boundaries, not to be regarded as a mathematical line but as a broad and variable zone, the special distinguishing features of one cell layer being liable to spread across its original boundary.

Development of Brain.

The development of the brain here as in *Lepidosiren* and *Protopterus* goes to support the view that the most primitive division of the Vertebrate brain is not into three regions lying one behind the other but into two, the primitive fore brain (*Vorhirn*—Kupffer) and the primitive hind brain or rhombencephalon (*Nachhirn*—Kupffer). The separation off of the hinder portion of the primitive fore brain to form the mesencephalon is on this view to be looked on as quite secondary.

General Morphology of the Fore Brain Region of Teleostomes.

As has been pointed out elsewhere[2] I hold to the view of von Baer, Reichert, Goette and Studnička that the true cerebral hemispheres as seen in Vertebrata from the Dipnoi upwards, are to be looked on as primitively paired structures—as lateral evaginations of the wall of the primitive fore brain, developed doubtless in order to give space for the great increase in the mass of nervous matter in this region correlated with the increasing development of the olfactory organ. I find it difficult to realise how anyone can fail to be convinced that this is the correct view to take of the morphology of the hemispheres, looking to their mode of development in the Dipnoi and Amphibia and to their adult relations in the higher forms where that potent disturbing factor in development—the yolk sac— is present. In *Polypterus* a quite similar increase takes place in the mass of nervous matter forming the sides of the primitive fore brain, but in this case there is no evagination of the brain wall to form hemispheres, beyond the pair of small olfactory lobes. Room is found for the nervous mass in other ways : (1) the side wall becomes greatly thickened to form the so-called "basal ganglia," (2) the thalamencephalon increases much in length, and (3) the thickened portion becomes slightly invaginated

---

[1] *Verh. anat. Ges.* (Genf), 1905, s. 27.
[2] *Quart. Jour. Micr. Sci.* XLVI. p. 448.

instead of being evaginated. The nervous material which corresponds with the whole of the hemisphere in the higher forms—including the pallium or mantle—lies in the thickened wall of the thalamencephalon. What is ordinarily termed the pallium in a Crossopterygian is nothing more nor less than the roof of the thalamencephalon, which is of course epithelial here as elsewhere. The conditions in Actinopterygian Ganoids and Teleosts are obviously similar to those in *Polypterus*: what is ordinarily called the pallium in these forms is simply the epithelial roof of the primitive fore brain, while the so-called basal ganglia are the thickened walls including what corresponds to the whole of the hemispheres in higher forms. It remains now to add that although the above results were arrived at by myself quite independently from an examination of sections of *Polypterus* and their comparison with Ganoid (*Amia*) and Teleostean material, Studnička has already given utterance to exactly the same views[1].

Curiously enough they have been to a great extent ignored and we still read in almost every text-book that one of the most important features of the Teleostomi is their membranous pallium.

XIV. DESCRIPTION OF PLATES.

Introductory Note.—There were unfortunately no descriptive labels giving the age of the various eggs and larvae but it has been possible from Budgett's diary to arrive at some rough idea of the age of certain stages. Stage 5 appears to have been reached in the case of certain eggs about 12 hours after fertilization; stage 14, about 17 hours; stage 15, 21½ hours; stages 16—19, all amongst 27 hour eggs; stages 20 and 21, 33 hours; stage 23, 39 hours; stage 24, 45 hours; stage 25, 51 hours; stage 26, 63 hours; stage 27, 89 hours; stage 30, rather over 4 days; stage 31, probably 5 days; stage 32, 6 days; stage 33, 7 days; stage 34, 8 days; stage 35, 10 days; stage 36, 12 days.

These times are subject to considerable possibility of error in correlating the statements of Budgett's diary with particular tubes of specimens—there being no descriptive labels. Even if approximately correct as regards the particular larvae it must be borne in mind that tropical eggs and larvae show as a rule extraordinarily great variations in the rate of development—particularly those developing in the still waters of lagoons or swamps which are liable to very great variations in temperature.

[1] *Sb. Böhmisch. Ges.* 1896 and 1901.

GENERAL LIST OF CONTRACTIONS USED IN PLATES AND TEXT-FIGURES.

The letters affixed to the numbers of the figures signify: *m* view of dorsal aspect, *h* view of posterior aspect, and *l* lateral view. The Scales affixed to certain text-figures show hundredths of a millimeter.

A.	Aorta.		*N.*	Notochord.
a.a.	Aortic arch.		*n.*	Nucleus. Hypochorda.
a.c.	Anterior commissure.		*n.a.*	Neural arch.
a.c.v.	Anterior cardinal vein.		*nc.*	Nephrocoele.
a.n.d.	Archinephric duct.		*n.r.*	Neural rudiment.
a.r.	Aortic root.		*ns.*	Nephrostome.
at.	Atrium.		*o.c.*	Auditory capsule.
bp.	Blastopore.		*olf.*	Olfactory organ.
b.p.	Basal plate.		*op.*	Opercular fold.
b.v.	Blood vessel.		*opt.*	Optic rudiment.
cer.	Cerebellum.		*ot.*	Otocyst.
ch.	Optic chiasma.		*p.*	Melanin pigment.
cl.	Cloaca.		*pa.*	Pancreas.
c.o.	Cement organ.		*p.a.g.*	Postanal gut.
coel.	Coelom.		*pall.*	Pallium.
d.a.	Dorsal aorta.		*par.*	Parachordal. Paraphysis.
d.c.	Dorsal carotid.		*pc.*	Pericardium.
d.C.	Ductus Cuvieri.		*p.c.*	Posterior commissure.
d.s.	Dorsal sac.		*p.c.v.*	Posterior cardinal vein.
e.	Eye.		*p.f.*	Pectoral fin.
ect.	Ectoderm.		*ph.*	Pharynx.
e.g.	External gill.		*pin.*	Pineal outgrowth.
e.n.	Nasal aperture.		*p.n.*	Pronephros.
end.	Endocardium.		*p.n.s.*	Pronephric blood sinus.
ent.	Enteric cavity or Enteron.		*p.v.*	Pulmonary vein.
F.b.	Primitive fore brain.		*r.p.*	Pulmonary branch of Vagus.
g.h.	Habenular ganglion.		*s.*	Subclavian.
gl.	Glottis.		*scl.*	Sclerotom.
glom.	Glomerulus.		*s.i.v.*	Subintestinal vein.
h.b.	Hind brain.		*s.N.*	Hypochorda.
h.c.	Habenular commissure.		*splc.*	Splanchnocoele.
ht.	Heart.		*s.v.*	Sinus venosus. Saccus vasculosus.
h.v.	Hepatic vein.		*Th.*	Thyroid.
hy.	Pituitary body.		*Thal.*	Thalamencephalon.
i.g.	Infundibular gland.		*t.o.*	Tectum opticum
i.j.v.	Inferior jugular vein.		*V.*	Ventricle.
ir.v.	Interrenal vein.		*v.*	Vein.
l.	Lung.		*v.a.*	Ventral aorta.
li.	Liver.		*v.ar.*	Visceral arch.
l.p.	Lateral plexus.		*vac.*	Vacuole.
l.r.	Lung rudiment.		*v.c.*	Visceral cleft.
l.v.	Lateral vein.		*v.c.l.*	Lateral cephalic vein.
m.	Mouth.		*v.IV*	Fourth Ventricle.
m.a.	Mandibular arch.		*vel.*	Velum transversum.
mc.	Myocoele.		*y.p.*	Yolk plug.
mes.	Mesoderm.		*I, II etc.*	Mesoderm segments.
my.	Myotom.			

PLATE XIII.

EXPLANATION OF PLATE XIII.

The figures on this plate represent views from above, i.e. in the later stages from the dorsal side. The number attached to each figure indicates the stage of development. The average magnification is about 30 diameters.

bp.	Blastopore.		*pn.*	Pronephros.
c.o.	Cement organ.		*y.p.*	Yolk plug.
e.g.	External gill.		IV *v.*	Fourth Ventricle.
ot.	Otocyst.			

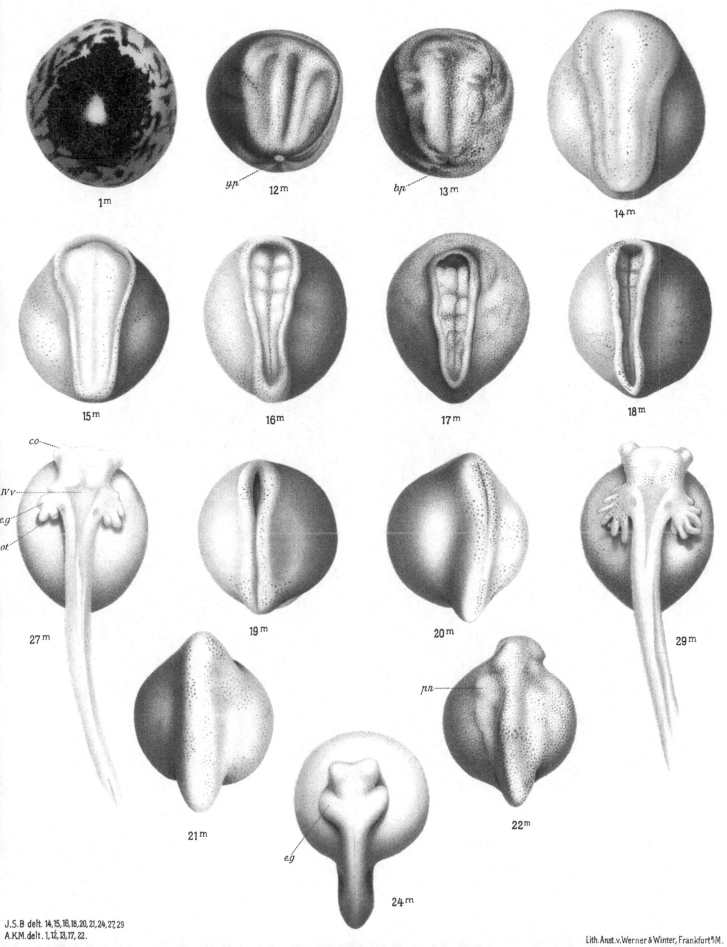

PLATE XIV.

EXPLANATION OF PLATE XIV.

The figures in the top row are views from behind; the remaining figures are from the side. Magnification as before.

c.o.	Cement organ.		*pn.*	Pronephros.
e.g.	External gill.		*y.p.*	Yolk plug.
h.b.	Hind brain.			

PLATE XIV.

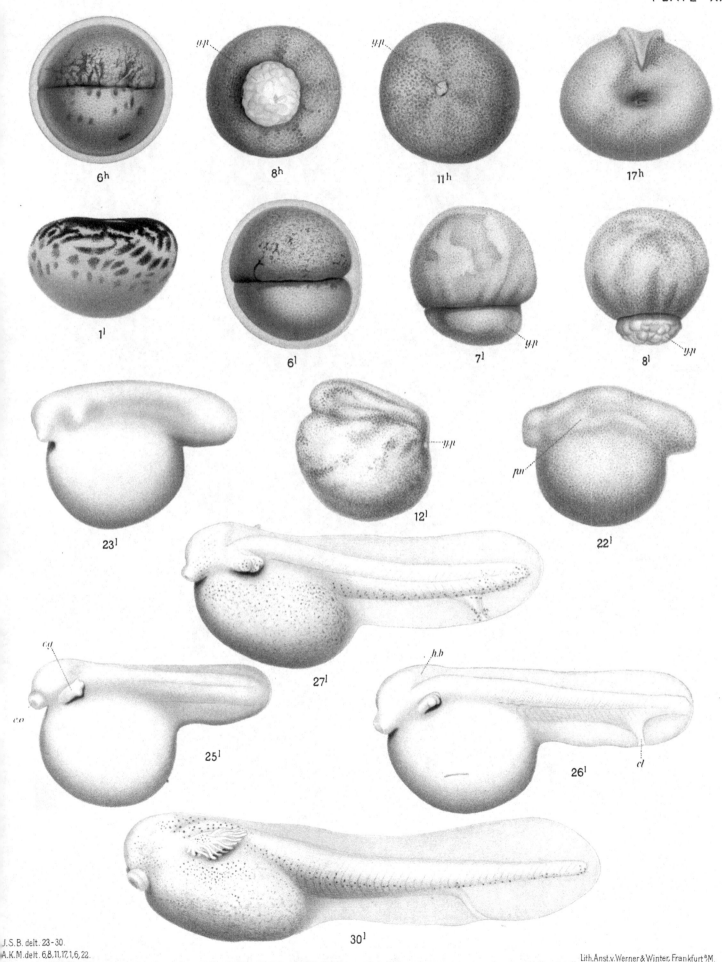

PLATE XV.

EXPLANATION OF PLATE XV.

Views of the latest stages collected by Budgett; the three lower figures drawn by Budgett from the living specimens. Magnification as shown by scale.

c.o.	Cement organ.		*m.*	Mouth.
e.	Eye.		*op.*	Operculum.
e.n.	Nasal opening.		*p.f.*	Pectoral fin.

J.S.B. delt.

Figs.31-33 ├──────┤ 1mm

Figs.34-36 ├──────┤ 1mm

XIII.   NOTE ON HABITS OF POLYPTERUS.
BY J. HERBERT BUDGETT.

[Contributed in the form of a letter to the Editor of *The Field*, dated Dec. 3, 1902, and reprinted with permission.]

"In your issue of Nov. 8th you publish an article upon *Polypterus* by Mr Boulenger, in which he gives certain details of the habits of these interesting fish, after having observed them in captivity for seventeen months. I have kept a pair of *Polypterus senegalus* for more than three years and have watched them throughout with very great interest. They were originally brought by my brother, Mr J. S. Budgett, from the Gambia, and were at first very shy. They will now take their daily meal off the prongs of a fork or from the tip of one's finger. I have noticed from time to time what I take to be a courtship, a description of which may interest your readers.

This courtship may sometimes be induced by raising the temperature of the water, which is usually about 75 deg. Fahr. The phenomena I am about to describe may be divided into phases, although I cannot be sure that they always all occur, or necessarily follow one another in the order in which I have placed them.

Attention is first drawn to the fish by their constantly leaping above the water. The leap usually follows a series of increasingly swift darts to and fro, in the course of which they just break the surface of the water with the top of the head. This appears to have an intoxicating effect.

After the leap they usually descend slowly towards the bottom. I have noticed that these evolutions are rarely performed by both at the same time, each being actuated in turn while the one at rest appears to be quite unconcerned in the other's movements except when disturbed by being brushed past.

Usually following this phase the male swims with the female. They do not now leap above the water or make sudden rushes as in the first described phase. The motion is very rapid, but more sustained, graceful and snake-like, their evolutions being so wonderfully concerted that they appear continually to interlace.

Later, the male may be seen to be in constant close attendance upon the female at whatever speed and however eccentrically she whirls about. So much so, that his snout appears to be literally fastened to her head.

Another phase of this courtship is when the female lies motionless while the male moves quietly forward from behind and as he comes up alongside keeps administering a succession of quick very light taps with his head which is jerked sideways, while his anal fin is repeatedly distended in the form of a cup and brought underneath the female so as to brush her under surface. She meanwhile lies quite still, with the pectoral fin on the side nearest the male pressed closely back against her side or moves forwards slowly for short distances at a time, always to be followed and overtaken by him.

The series of dorsal fins of both fishes is often partially or wholly erected, but this does not appear to depend upon either excitement or motion, and they are raised or depressed in order, sometimes from front to back, sometimes from back to front.

I have often observed that the spiracle momentarily opens, especially when the fish is eating, and I am convinced that it is used when the fish breaks the surface of the water in the way I have described above.

The favourite position of the fish when at rest is to have the body concealed in the Valisneria, which grows in the aquarium, whilst its head and pectoral fins are exposed.

Its liveliest times are morning and evening."

XIV. THE DEVELOPMENT OF GYMNARCHUS NILOTICUS.

BY RICHARD ASSHETON, M.A.,

Trinity College, Cambridge;
Lecturer on Biology in the Medical School of Guy's Hospital in the University of London.

With Plates XVI—XXI. and text-figures 86—165.

CONTENTS.

I. INTRODUCTION.

The Mormyridae are generally held to be a primitive family of the Mala-copterygii, which sub-order itself "is nearest to the Ganoids," Sedgwick (109).

At the same time "owing to their freshwater habit and geographical limitation generally, their evolution has been constrained, and limited in the main to certain parts of their organization, with the result that they exhibit a most curious medley of primitive characters coupled with highly specialized and even degenerate characters," Ridewood (106). Nothing was known of the development of any Mormyrid before Budgett's discovery of the eggs of *Gymnarchus* in the Gambia River in July, 1900, Boulenger (24).

The development of *Gymnarchus niloticus*, a member of this family, is therefore likely to present features of considerable interest. Budgett gave a delightful account of the nest made by the parent fish for the reception of the eggs, and of the habits of the young larvae, which develop with marvellous rapidity, in his paper entitled "On the Breeding-habits of some West-African Fishes, etc.," which was published originally in the *Transactions of the Zoological Society of London*, Vol. XVI. Pt. 2, 1901. The anatomy of the adult animal, or of parts of it, has been described by Erdl (39), in 1847; Förg (43), in 1853; Hyrtl (71), in 1856; Fritsch (44), in 1885, and more recently by Ridewood (106), who in 1904 gave a detailed description of the skulls of *Gymnarchus* and other Mormyroid fishes.

Budgett, as may be seen by reference to his memoir alluded to above, thought that the developing eggs of *Gymnarchus* showed a close resemblance to developing eggs of elasmobranch fishes. He says "the development is exceedingly shark-like" (p. 126). He nowhere states what the resemblances are, and he left no evidence that he had worked extensively at the development of *Gymnarchus*. I believe his remarks were made as a result of a cursory examination, and I feel convinced he would not have maintained this opinion had he lived to complete his research.

A "nest" of eggs of *Gymnarchus* was discovered by Budgett in the Gambia River on the 23rd July, 1900. This discovery must have been made very shortly—perhaps not more than a day—after the laying of the eggs.

Budgett preserved a varying number of eggs every day from, and including, the 23rd July, to the 6th day of August. Three later stages have been preserved, only the oldest of which is dated, namely, 3rd September, 1900. The series is thus in one sense very complete; but owing to the exceedingly rapid development, there are rather wide gaps in the earlier stages which interfere with the complete elucidation of the changes that supervene on the segmentation of the egg. The material is beautifully preserved; the fluid used most frequently was acetic acid and corrosive sublimate. Formalin, perenyi, and picric acid were also used occasionally.

Unfortunately Budgett left practically no record of his observations on this material beyond that which occurs in the memoir already mentioned, although the greater part of the section cutting had been done. But the careful and orderly manner in which he always treated his specimens has greatly facilitated the working out of his invaluable collection.

II. EXTERNAL FORMS.

First Day.

23 July, 1900. C_1. The most striking feature of the egg of *Gymnarchus* is its very large size. The eggs vary, but on an average measure 10 mm. in diameter. The largest measured 10·5 mm. along its equatorial diameter, the smallest about 9 mm.

Roughly speaking, the egg is spherical; but there is a more or less flattened pole on which the segmenting part of the ovum lies.

The egg is meroblastic; the segmenting portion forms a "blastoderm," which is circular, and has a diameter of 2·5 mm. in the smallest to 3 mm. in the larger ones. In the smaller ones the blastoderm is less sharply marked off from the yolk than in the larger ones (Pl. XVI. figs. 1 and 2).

Sections through the smallest show that segmentation is already far advanced (text-fig. 86).

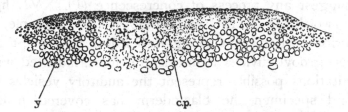

FIG. 86. Section across the centre of the blastoderm on the first day. *y.*, yolk granules; *c.p.*, central plug of fine grained yolk.

The blastoderms with the larger diameters are much more sharply defined; usually a white rim borders the blastoderm. Outwardly there is nothing to suggest which is the anterior or posterior end of the blastoderm, nor is there any trace of an opening into a subgerminal cavity (text-fig. 87).

The egg is surrounded by a tough investment which is perforated by minute pores, as commonly is the case in Teleostean eggs, the so-called chorion of older writers.

FIG. 87. A vertical section through the segmenting pole of the egg on the first day, though a rather later stage than that shown in text-fig. 86.

Second Day.

24 July, 1900. C$_2$. The eggs collected the next day show a still more decided flattening of the blastoderm pole, and a considerably more expanded blastoderm which is on the whole circular and has a diameter of 5 mm. A closer inspection reveals two important changes of appearance. There is a thicker or, at any rate, a darker peripheral rim, and at one spot a beak-like knob projects backwards over the yolk. It projects forwards equally over the thinner central area of the blastoderm, and is the embryonal knob, and marks the head end of the future embryo (Pl. XVI. fig. 3).

Third Day.

25 July, 1900. C$_3$. Of this stage there are only two specimens. A great advance has been made during the 24 hours. In the younger of the two the blastoderm has expanded and crept over rather more than one half of the surface of the egg.

The embryonal knob, which was a mere rudiment before, is now elongated, so as to form a very much attenuated anterior part of the animal, which stretches forwards from the posterior edge of the blastoderm for about 4 mm.

There is no indication of a double origin at the extreme posterior end— and nothing to suggest any process of concrescence (Pl. XVI. figs. 4, 5 and 6).

The hinder two-thirds of the embryonic ridge is simple, and so also is the extreme anterior portion; but along rather less than one-third of the length of the ridge a deep groove divides it into lateral folds, indented near their posterior ends, which indentations possibly represent the auditory vesicles (Pl. XVI. fig. 5).

In the second specimen the blastoderm has covered nearly four-fifths of the whole egg, and the embryonal ridge, which is not very conspicuous, has increased in length to 5 mm., but is extremely narrow. In the anterior region a series of constrictions occur which may indicate a primitive segmentation of the nerve tube, although at this stage the nerve rudiment is still a solid cord.

Fourth Day.

26 July, 1900. C_4. There are two eggs of this date, and one imperfect series of sections. Of these, in one the blastoderm has completely surrounded the ovum, and there is no trace of a yolk blastopore; in the other there is perhaps still a small area uncovered.

In that which I take to be the younger of the two eggs of the fourth day, the most conspicuous structure is the neural tube (Pl. XVI. fig. 8)—which is very much attenuated. This stretches forwards from the dark rim as a narrow straight white line gradually increasing in width until it suddenly widens out close to the anterior end forming the vesicle of the fore brain, and appears to be turned over on its side. The whole length of the embryo is about 8—9 mm., and there are 14 pairs of protovertebrae.

The head region, i.e. the part in front of the protovertebrae, measures 4 mm. The fore brain extends over a fifth of this length. About half-way between the anterior end and the first pair of protovertebrae is a pair of white masses lying on either side of the neural tube. These I take to be the side plates of the original solid nerve rudiment which do not form part of the central nervous system, but give rise to the ganglia of the hindermost cranial nerves.

There seems to be an unsegmented plate of mesoderm lying on each side of the neural tube along the anterior 4 mm. of the embryo. Posteriorly to this point the mesoblastic plates are segmented and form protovertebrae. The first half-dozen of them are fairly distinct; the more posteriorly placed are ill-defined, and gradually merge with the unsegmented plates of mesoderm of the hinder region. They are widely separated. In fact, the whole appearance of the embryo suggests much stretching and attenuation.

Adjoining the hind end of the embryo is a dark elongated area which can without doubt be recognised as the just coalesced rim of the blastoderm. Along the median line is a groove which stops short of each end by about the width of the dark rim. This groove, which in places may be still open into the yolk, marks the final disappearance of the yolk mass from the surface (Pl. XVI. fig. 8).

It is a matter of some interest to notice that the blastoderm has completely surrounded the egg before there is any sign of a tail bud; that is to say, the embryo is still flat upon its ventral surface. It is more flattened than in the trout, Henneguy (63), Kopsch (80), where there is a slight swelling to form a tail bud. When the tail bud forms, it must do so as a tail fold more after the manner of an Amniote.

The growing point of the tail of the embryo is the anterior end of the "primitive streak" like coalesced blastoderm edges, and in this case is obviously part of the embryonic rim; the yolk blastopore being the true blastopore.

This condition resembles the avine or mammalian condition only in so far that the embryo is at first absolutely flat; it resembles the elasmobranch condition in that the "embryo" is obviously in contact with the blastoderm rim, but differs in the fact that the mid-dorsal point of the dorsal lip of the blastopore advances over the yolk mass, whereas in Elasmobranchs it advances but little—if it is not quite stationary—with the result that a prominent tail bud is formed almost immediately; on the other hand, it resembles exactly the Amphibian condition, for the epiblast and hypoblast are continuous with each other throughout the rim, which is not the case with the edge of the blastoderm of Amniotes.

Probably the prime cause of separation of the embryo, together with the primitive streak from the edge of the blastoderm in Mammals and Birds, is the copious secretion of fluid into the archenteron which takes place in all those forms in correlation with the complete absence of open blastopore in the earliest stages. This accumulation of fluid is almost entirely absent in all Ichthyopsida.

The flattening of the embryo, which is so remarkable in Mammalia, Birds, and *Gymnarchus*, and also in a less degree in Teleosteans in general, would seem to be due to different causes. In Mammalia and Birds it is due to the rapid infiltration of the fluid in the archenteron resulting in hydrostatic pressure from within—in *Gymnarchus* it is perhaps due to the presence of a very strong resistant investment together with the rapid travelling backwards of the growing point over the peculiarly large egg.

In the second specimen of the fourth day, which is slightly more advanced than the other, the embryo itself seems to be less well preserved, and the yolk cracked and shrunk somewhat, and the mesoblastic somites are less distinct. I regard it as an older stage, because the cerebral vesicle is more over-turned, and because there is no trace at all of an elongated coalesced blastoderm rim.

The neural tube shows no groove at all; the cerebral vesicles, which are two in number, lie on their left side, and also slightly bent to the left.

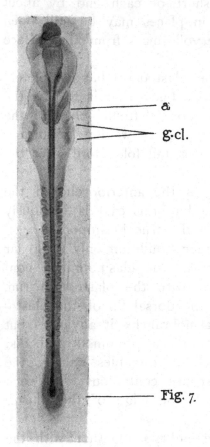

FIG. 88. The embryo of 4½ days. The brain is turned over on its right side. *a.*, rudiment of cranial nerves; *g.cl.*, gill clefts; Fig. 7, the line along which the section shown in the previous figure is taken.

In the first specimen the embryo extended over 90° of the circumference, and the coalesced blastoderm edges over about 60°.

There are two other specimens, one cut, and mounted as a very imperfect series of sections, the other mounted whole, of which a figure is here given (text-fig. 88). These are probably between four and five days old. In them there are about 20 pairs of protovertebrae. The termination is extremely like the primitive streak of an Amniote. The head is turned over on one side, and the brain marked out into optic vesicles, fore brain and mid brain. There are rudiments of two pairs of gill clefts (*g.cl.*). The great length of the hind brain is remarkable.

Fifth Day.

27 July, 1900. C_5. An enormous advance has taken place, and unfortunately there seems to be no intermediate stage.

Whereas on the fourth day there was no trace of tail fold, or lateral fold, and only a slight indication of a head fold, now on the fifth day the embryo is completely marked off and nipped off from the yolk except for the rather extensive yolk stalk (Pl. XVI. fig. 9).

The embryo lies on its side held down by the tough investment, except the tail which is more or less vertical. Optic and auditory organs, gill clefts, and pectoral fins, are now well advanced, although they were quite rudimentary on the previous day.

Blood vessels radiate towards the heart, and posteriorly a large vessel is visible running out from the point where the tail is attached to the yolk.

The advance is so great that one cannot attempt to trace changes in detail between the preceding specimen and this one. The yolk sac has not appreciably altered in size (Pl. XVI. fig. 9).

The total length of the embryo is 11—12 mm. The free anterior end measures about 3 mm., the free tail end 4 mm., and the still attached part of the yolk stalk 5 mm. in length. The head region is large, and shows the ventrally-flexed fore brain. The outline of the head is determined by the large cerebral vesicles, and the beak-like mandibular and hyoid arches. It resembles curiously a bird's head and beak. The beak-like projection (Plate XXI. fig. 54) just shows the division into mandibular and hyoid arches. Behind this is the first branchial cleft, bounded posteriorly by the second branchial arch. These are followed by the third and fourth branchial arches apparently with small clefts between them, and by a less well developed fifth branchial arch. On the external edges of the arches both anteriorly and posteriorly little knobs may be noticed which are the beginnings of long uniramous external gill filaments.

Budgett writes with some ambiguity concerning these gill arches, he says (p. 126), "The larvae soon after hatching develop extremely long gill filaments, which hang down in two blood-red branches from the gill arches, of which there are four."

There are really rudiments of seven visceral arches; there are five clefts, and gill filaments are developed upon four arches only, namely, first, second, third and fourth branchial arches. These filaments form on each side of the embryo a bush or tuft of uniramous vascular loops, "blood-red" no doubt when living, which projects out from the side of the animal.

About the level of the external gill buds, and close behind the branchial region there is on each side an elongated, horizontally placed fold of the integument, the first sign of the pectoral fins. Neither now nor at any other time can I find any trace of pelvic fins.

The brain is large and differentiated into a large thin-roofed hind brain, slightly bent about the middle of its length, a mid brain, and fore brain at right angles to the hind brain. The optic vesicles are very small, and placed far forward (Pl. XXI. fig. 54).

Very little can be seen of the heart in surface view.

Sixth Day.

28 July, 1900. C₇. Up till this moment the embryo, compressed between the stout investing membrane and the yolk, has lain much flattened and distorted, and, as regards its anterior and middle portion, upon the side, after the manner of many Amniote embryos.

And although it lies more frequently upon its right side, it is not invariably upon that side. Of the seventeen specimens which exist of the ages between the commencement of the turning over of the embryo upon the side and the time of the rupture of the investing membrane, thirteen lie upon the right side and four upon the left. It is interesting to note that it is invariably upon the left side that the bird embryo lies. I believe no exception to this has ever been recorded. The determining cause in the case of the avine embryo would appear to be the bending of the heart towards the right. When the general cranial flexure comes about then the embryo tumbles over on to the left, being prevented from falling to the right by the now curved-out heart rudiment, which forms an efficient prop.

There is obviously no such constant determining factor in the case of *Gymnarchus*, though there is probably something which renders the embryo more likely to turn over on to its right rather than its left side.

The eggs which were collected upon the 28th July, that is to say the sixth day, present a totally different appearance, because the tough membrane has been ruptured and lost, and the embryo, freed from its pressure, no longer lies distorted on its side, but has assumed the natural position, and is now quite symmetrical. The yolk, over which the outlines of blood vessels are very clearly visible, is slightly oval (Pl. XVI. fig. 11).

The total length of the embryo is about 8·5 mm. The general shape of the head is much less extraordinary and more like that of other vertebrate embryos of corresponding age than in the previous stages—showing that the peculiarities noticed before were largely due to the great distortion caused by the pressure of the investing membrane. The peculiar beak-like prominence of the mandibular arch seen in Pl. XVI. fig. 9, and Pl. XXI. fig. 54 is quite gone in Pl. XVI. fig. 11.

All the organs of the head, such as eyes, ears, parts of brain, are less prominent, a change which is also due to the absence of compression.

The blood vessels extend over the whole surface of the yolk sac, and the single very prominent yolk sac afferent can be seen on the hinder border of the yolk stalk, and less prominent vitelline veins run to the heart from the anterior lateral region on the two sides. (See also Budgett (26) (Pl. XI. figs. 4, 5).)

Seventh Day.

29 July, 1900. C_8. The length of the embryo proper is now 16 mm., so that it has doubled its length during the 24 hours, and not length only but depth and width as well (Pl. XVII. fig. 12).

The shape of the yolk sac is slightly altered to a more oval form. The gill-filaments are distinct filaments now, and the pectoral fin is spatulate.

The eyes are again conspicuous but small, and just in front and ventral to the eye on each side of the head is a single depression, the olfactory pit.

The stomodaeum is a widely open transverse groove. The pericardial cavity is large, and within it the twisted heart is visible.

Eighth Day.

30 July, 1900. C_9. Again an enormous increase in size has taken place. The yolk sac has become elongated and pyriform, the blood vessels circulating on it have sunk down so as to be no longer visible on the surface in spirit specimens. The embryo appears to be more sharply marked off from the yolk sac, but

the actual diameter of its stalk has not diminished much. The parts in front of the eyes and upper jaw have grown out considerably, and carry away the olfactory pits from the close neighbourhood of the eye. Moreover the olfactory pits are now externally completely doubled (Pl. XVII. fig. 13).

The gill filaments project freely.

There is along the mid-dorsal line a median fin, which starting from about the "level" of the pectoral fin, passes back round the tail and ceases at the anus. A median fin also occurs between the anus and the yolk sac.

These fins are really present on the preceding day, but are not nearly so evident.

The first sign of the operculum can be detected as a fold of the skin growing outwards from the hyoid arch.

Ninth Day.

31 July, 1900. C_{10}. The embryo proper measures 26 mm. All the features mentioned in the last paragraph are now more clearly noticeable, and excepting that the yolk sac has become much elongated I do not think that there is any new feature to be seen in an external view. There is no longer any external trace of the division into the several parts of the brain, as in the previous stages (Pl. XVII. fig. 14).

Tenth Day.

1 August, 1900. C_{11}. The increase in length of the embryo has not been very great; it measures 30 mm. The increase in bulk is considerable (Pl. XVII. fig. 15). The yolk sac has attained its most elongated and curious condition. Hereafter it diminishes rapidly in length and bulk. This elongated condition is far more marked than in other Teleostean fishes though it is characteristic of some (Salmon). The gill filaments are nearly at their maximum length, and now appear all to be coming from underneath the operculum which is well developed.

The change in the shape of the head is very marked, and the animal is more like an adult fish because of the forward growth of the cartilages of the jaws and rostrum.

Eleventh—Sixteenth Days.

During the next few days there is a slight increase in the length of the gill filaments, and the operculum tends to cover a greater extent of them. After the twelfth day the decrease in the length of gill filaments and shortening of the yolk sac are very rapid and marked, so that on the fourteenth day the condition is as seen in text-fig. 89, A.

The head has assumed its characteristic shape, the gill filaments have shortened, and their remains are almost hidden by the operculum; and the yolk sac appears to be much reduced in outward view.

The median fin has become much lower in the caudal region.

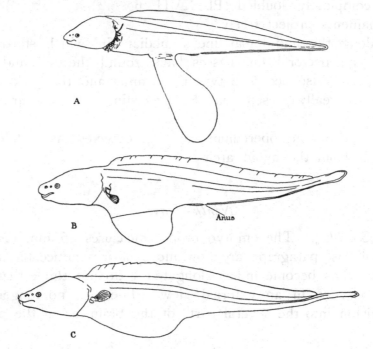

FIG. 89.   A, fourteenth day; B, date unknown; C, forty-third day.

Seventeenth Day.

6 August, 1900. C_{23}. On the seventeenth day the yolk sac has begun to pass within the abdominal cavity. The last two drawings in text-fig. 89, which are from specimens among an undated set, show the subsequent stages up to the inclusion of the yolk sac within the abdominal wall. In the younger (text-fig. 89, B), the pectoral fin has assumed its adult appearance and position, and the dorsal median fin is very prominent, and has lost its continuity with the caudal fin.

Forty-third Day.

3 Sept., 1900. $C_{24,\ 25,\ 26}$. In the oldest specimens the yolk sac has been entirely included within the walls of the abdomen, and is indicated only by a slight bulging. The dorsal fin is high, and the caudal fin is now isolated and has almost gone altogether; the ventral fin has disappeared entirely.

The length of the larva is 65 mm. (text-fig. 89, C).

III. DEVELOPMENT OF THE ORGANS.

1. THE ALIMENTARY CANAL.

In the egg of the first day a blastoderm forms by the segmentation of one pole of the egg. There is a cap of cells, circular in outline, which in surface view shows no distinction into anterior and posterior ends, or any specially thickened areas. It is continuous round its edges with the yolk upon which this whole disc rests. This connection is more particularly evident with reference to the outermost layer of cells of the cap which forms a more distinct membrane. The yolk contains nuclei in every way similar to those in the already segmented cells, and nuclear spindles are present both in the blastomeres and in the unsegmented yolk just beneath it. One cannot as yet speak of a hypoblastic layer. Segments are still being cut off from the yolk mass whose destination one cannot at present foretell.

The part of the yolk mass in which the nuclei alluded to lie consists of a layer of protoplasm crowded with fine grains of food yolk forming the superficial layer of the egg. It extends, however, only just outside the edge of the blastoderm. In the centre it is much deeper than elsewhere and forms a funnel-shaped plug. The protoplasmic layer passes into the yolk mass, becoming more and more broken by the large yolk spheres until deep down one can find no trace of any material between the yolk spheres.

On the second day the blastoderm has spread and with it the sublying protoplasmic layer, which contains the nuclei as before.

This would seem to be, to a large extent, an actual spreading, because the blastoderm has become very much thinner centrally where it is at the most two cells in thickness, whereas on the day before it was five or six cells thick; at the same time cells are being added to the blastoderm from yolk, at any rate, round its advancing rim.

At one side (the embryonic knob seen in the surface view, Pl. XVI. fig. 3) the blastoderm is thickened—9 or 10 cells deep in the median line—and I cannot say if there are any actual connections between the blastoderm of this part and the subjacent protoplasmic mass. There are certain cavities here (text-fig. 90 ves.) which may represent archenteron.

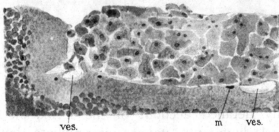

FIG. 90. Sagittal section through the embryonic knob of the second day; i.e. through the dorsal lip of the blastopore; *m.*, yolk nucleus; *ves.*, vesicular spaces perhaps representing rudiments of archenteron.

The nuclei in the subjacent protoplasm are now distinctly larger, and they stain more deeply than the nuclei of the blastomeres. These are the yolk nuclei of later stages, and appear to be directly derived from single ordinary nuclei by hypertrophy.

At the embryonic end no layers can be identified, but at the centre and towards the anterior end there is :—

(1) an outer layer of squamous cells.
(2) an inner layer of cubical cells,
(3) isolated cells lying between this and
(4) the yolk mass with large nuclei.

In specimens of the third day great advance has been made, the nervous system and protovertebrae and "deuterogenetic centre" are all well established, and under the whole of the "embryo" the hypoblast is present as a distinct layer of cells continuous with the underlying yolk by means of connecting strands and continuous with the mesoblast upon each side of the notochord in the same way. This is most obvious in the anterior part of the embryo.

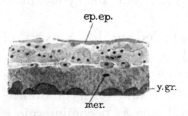

Fig. 91. Section through the more central part of the blastoderm of the second day. *ep.ep.*, epidermic layer of epiblast; *mer.*, yolk nucleus; *y.gr.*, yolk grain.

The notochord is fully formed and is everywhere quite distinct from the epiblast above it, and from the mesoblast and hypoblast also, although the line of demarcation is less easy to see in these cases. At the deuterogenetic centre (text-fig. 92) *all layers merge*, including the hypoblast, and I think on each side there is a region where here and there hypoblast cells may be seen becoming cut off from the yolk mass, though just immediately beneath the streak I do not find connection.

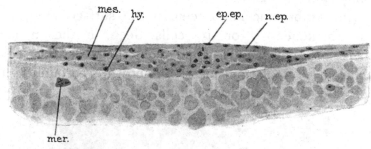

Fig. 92.   Transverse section through the growing point ("deuterogenetic centre") on the third day. *ep.ep.*, epidermic layer of epiblast; *n.ep.*, nervous layer of epiblast; *hy.* hypoblast; *mer.*, yolk nucleus; *mes.*, mesoblast.

I cannot give any detailed account of the formation of the hypoblast or of the notochord as I have no stages intermediate between the two just described.

The following points may be noted :—

(i) The secondary growing point (deuterogenetic centre) has the form of a distinct primitive streak comparable in general appearance with that of an Amniote at the corresponding stage.

(ii) It differs however in being continuous with the hypoblast (and yolk ?), unlike the state of affairs in the Amniota, but like the condition in the Amphibia in that respect.

(iii) There is no suggestion of a double origin to the growing point such as one might expect to find if the growth in length was due to concrescence.

(iv) There is no suggestion of a "liquefaction" of yolk.

It seems probable that the condition of the third day as regards the hypoblast is derived from that of the second day by the conversion of the lower layer cells of the embryonic knob, together with certain other cells budded off from time to time from the sublying yolk mass into a definite layer. Others of these cells which do not take part in the actual layer of hypoblast, become the mesenchyme cells of the anterior region.

The yolk contains, besides the yolk grains and cytoplasm,

(1) enormous very irregular nuclei, which stain deeply,

(2) nuclei with single nucleolus,

(3) nuclei with scattered chromatin,

(4) small ordinary nuclei.

Fourth Day.

There are only three specimens of this stage. One has been cut into series of transverse sections. The two others are intact, of which one is represented in part or whole in Pl. XVI. figs. 7 and 8.

The alimentary canal as such is still very rudimentary, and is extremely difficult to follow. There is nothing comparable to a head fold involving the hypoblast, though the neural tube (and skin) project forwards a little.

The yolk shows yolk nuclei, and beyond the region of the embryo epiblast and hypoblast are clearly distinguishable, where, I think, mesoblast cells may be detected also.

The neural tube is so much distorted and pressed down into the yolk that it is impossible to make out exactly the layer lying immediately beneath it. On either side a hypoblast is always visible, and in places it can be traced underneath the nerve tube. In the anterior region, in front of the notochord, this layer is sometimes visible as a thicker strand, and in places it is continuous with the yolk mass (text-figs. 93 and 94). At any rate at this region there is no *canal*, nor is there any notochord.

In the region of the mesoblastic somites the notochord is very large (text-fig. 95). In this case there is no trace yet of the alimentary canal, but there is a layer of hypoblast cells, or at any rate nuclei on the yolk representing the hypoblast.

At the extreme posterior end this hypoblast layer is distinct and is certainly in connection with the yolk.

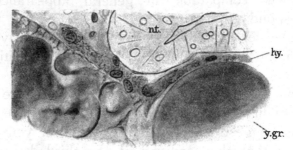

FIG. 93. Transverse section through the anterior region of the embryo of the fourth day. *hy.*, hypoblast; *nt.*, neural tube; *y.gr.*, yolk granule.

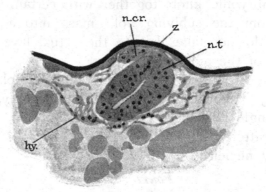

FIG. 94. Transverse section through the cranial region on the fourth day. *hy.*, hypoblast; *n.cr.*, neural crest; *n.t.*, neural tube; *z.*, zona.

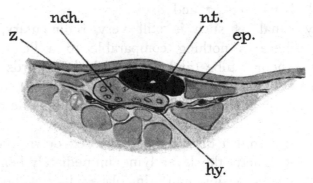

FIG. 95. Transverse section through the hinder part of the trunk of an embryo of the fourth day. *ep.*, the epiblast; *hy.*, hypoblast; *nch.*, notochord; *n.t.*, neural tube; *z.*, zona radiata.

$4\frac{1}{2}$ *Day.*

There are two specimens of this date, one of which had been cut into sections before I received the material, and unfortunately the series is incomplete. It represents a stage intermediate between figs. 8 and 9, Pl. XVI., and so if complete would be of the greatest value. The other specimen is mounted whole (text-fig. 88).   The first part of the series is missing.   The most anterior sections remaining pass through the head in the region just behind the optic cups, but perhaps cutting the optic stalks.

The embryo is greatly distorted and lies crumpled upon its right side. This crumpling brings about a wrinkling of the epiblast of that side.

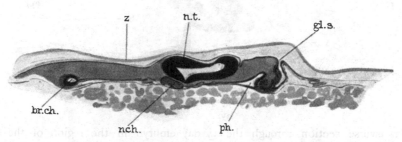

FIG. 96.   Transverse section across the hinder part of the branchial region of an embryo of $4\frac{1}{2}$ days.   *br.ch.*, branchial channel; *gl.s.*, branchial cleft; *nt.*, hind brain; *nch.*, notochord; *ph.*, pharyngeal canal; *z.*, zona radiata.

Owing to this distortion combined with the imperfect condition of the series certain structures are exceedingly difficult to make out.   The above section (text-fig. 96) is clearly through the branchial region.   The cavities *br.ch.* are the branchial canals of the later stages, and are lined by a double layer of epiblast. The pharynx is solid here, and seems to be represented by the nuclei on the surface of the yolk.

Posterior to the branchial region the coelom is well developed and extends dorsally almost to the middle line, leaving quite a small mesentery.   In this region the alimentary canal is distinguishable as a canal lined dorsally and laterally by a single layer of cubical cells.   The floor in the foremost part of this region is formed also by a layer of cubical cells, but further back by the yolk mass (text-fig. 97).

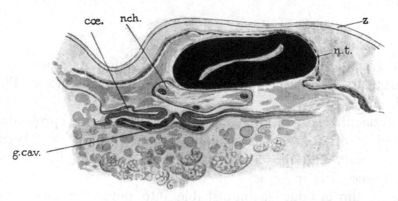

FIG. 97.   Transverse section through the trunk of a $4\frac{1}{2}$ day embryo.   *cœ.*, coelom; *g.cav.*, cavity of the gut bounded above by hypoblast, below by yolk; *nch.*, notochord; *n.t.*, spinal cord; *z.*, zona radiata.

Still further back, i.e., in the pronephric duct region, there is no lumen in the alimentary canal as yet, but a solid cord of cells lying between the notochord and the yolk probably represents the alimentary canal.

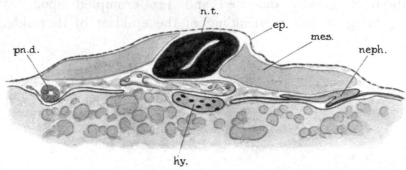

FIG. 98. Transverse section through the 4½ day embryo in the region of the pronephros. The gut is here a solid strand of cells. *ep.*, the two epiblast layers; *hy.*, gut; *mes.*, meso-blastic somite; *neph.*, nephrostome of pronephros; *n.t.*, spinal cord; *pn.d.*, pronephric duct.

Text-fig. 98 is a figure of which the outline was drawn with a camera, but it must be regarded more as an interpretation of the condition judged from this and other sections rather than an accurate reproduction of a single one. The condition of the excretory system, and its connection with the coelom, is seen on the right side, but this is referred to again later.

So far as I can determine it the formation of the gut cavity would appear to be as follows. The lower layers of the segmentation spheres together with nuclei within the yolk form a solid strand in the middle line. This is extremely thin in the anterior region; a cavity arises in the middle region between the segmentation sphere layer and the yolk mass, which extends backwards and forwards, though the extension forwards is considerably delayed. Posteriorly the cavity occurs within a strand of cells, rather than between the cells and yolk mass as in the middle region.

Fifth Day.

Although greatly distorted and pressed over upon its side, the embryo now shows the rudiments of all the important organs quite distinctly (Pl. XVI. fig. 9 and Pl. XXI. fig. 54).

The stomodaeum is a wide cavity overhung by the projecting fore brain and accompanying organs. The hypophysis has, I think, just become detached from the stomodaeum and lies closely applied to the infundibulum as a spherical solid mass of tissue rich in nuclei.

The skin is throughout distinguishable into outer covering layer and inner layer of epiblast, which fact affords a valuable test for the extension inwards of the stomodaeum.

An enormous advance has been made during the preceding 24 hours. The alimentary canal is formed now throughout its whole length, though still solid in the pharyngeal region.

The cavity of the stomodaeum is not continued into that of the mesenteron, and nearly the whole of the pharynx is without a lumen; the pharynx is marked by a flat plate of cells which is of two layers, corresponding probably to its dorsal and ventral walls. Those two layers separate further back at the level of the last pair of gill slits to form the cavity of the mesenteron. In the anterior part they coalesce and form a solid plate which joins along its outer edges with the epiblast at the inner ends of the branchial pits. Here they are nearly always to be found continuous with the inner layer of the skin or true epiblast ("nervous layer").

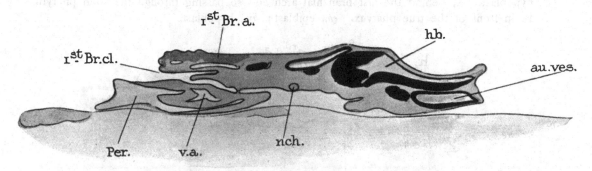

FIG. 99. Transverse section through the region of first branchial cleft in an embryo of the fifth day. *au.ves.*, auditory vesicle; *h.b.*, hind brain; *nch.*, notochord; *Per.*, pericardium; *v.a.*, ventral aorta; *1st Br.a.*, first branchial arch; *1st Br.cl.*, first branchial cleft.

In text-fig. 99 I believe the whole of the lining of the branchial channels and the "pharynx" is of epiblastic origin, it being possible throughout to find the two layers, the so-called epidermic and nervous layers of epiblast. This section is taken just beyond the extreme anterior end of a true hypoblastic pharyngeal plate. In front of this there is complete communication from side to side; posterior to this in constantly increasing degree the centre of the space is occupied by the plate of hypoblast. In this section (text-fig. 99) the two layers of epiblast from the left and right gill pocket invaginations meet as in the diagram B, text-fig. 100. Farther back the hypoblast of the solid pharynx is seen in the median line, diagram A.

The stomodaeal cavity leads into a branchial canal on each side, which perforates the hyoid arch, the first, second, and third branchial arches, and forms wide channels as far as the third gill cleft, narrow chinks after that, and terminates close to the surface near to the spot where the external opening of the fifth branchial canal is forming.

Sections 101 and 102 are taken between the second and third open clefts, i.e. through the second branchial arch on the left side and the second branchial cleft on the right side. The branchial channel, *br.ch.*, is seen on the left side lined by a double layer of "ectoderm" throughout. On the outer wall both layers are thickened, on the inner wall both layers are attenuated, the outer layer of epiblast being the thinner of the two. On the right side the lower wall of the cleft is lined by two layers of cubical cells, the upper by two layers of squamous cells.

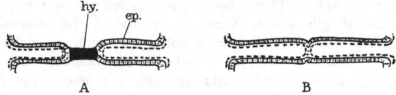

FIG. 100. Diagram to show the relation of the epiblast to the gill cleft and the pharyngeal hypoblast. A, behind the first branchial arch and so passing through the solid pharynx. B, in front of the true pharynx. *ep.*, epiblast; *hy.*, hypoblast.

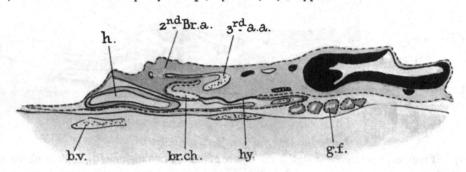

FIG. 101. Transverse section through a fifth day embryo in the region of the second branchial arch. *br.ch.*, branchial channel; *b.v.*, vitelline blood-vessel; *g.f.*, gill filament; *h.*, heart; *hy.*, fused walls of pharynx (hypoblast); *2nd Br.a.*, the second branchial arch; *3rd a.a.*, the third aortic arch.

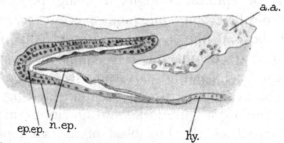

FIG. 102. Section taken through a branchial channel and aortic arch of an embryo of the fifth day. *a.a.*, aortic arch; *ep.ep.*, epidermic layer of epiblast; *n.ep.*, nervous layer of epiblast; *hy.*, hypoblast (solid pharynx).

Immediately behind, and rather ventral to the fifth branchial cleft, the hypoblast of the pharynx thickens and approaches nearer to the actual surface, and touches the skin where there is no, or very little, invagination of the epiblast (text-fig. 103 *g.*). This is in a depression common to it and the invagination of the fifth branchial cleft, but a little more dorsally placed. There are, thus, two passages leading from this depression, one ventro-anterior leading towards, if not actually continuous with, the termination of the branchial canal, the other dorso-posterior leading into a true gill cleft. Text-figure 103 is a diagram supposed to be taken horizontally through this depression, and shows both the true gill cleft and the branchial canal and clefts 3 and 4.

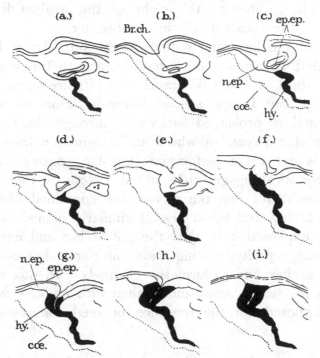

FIG. 103. A consecutive series (a—i) of sections taken transversely through the embryo, showing the region of the common pit into which the true gill cleft and the fifth branchial channel open. a. is the most anterior section. cœ., coelom; Br.ch., fifth branchial channel; ep.ep., epidermic layer of epiblast; hy., hypoblast of true gill cleft; n.ep., nervous layer of epiblast.

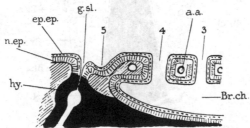

FIG. 104. A diagram (horizontal) to show the relation of the true gill cleft to the pit and to the branchial channels. a.a., aortic vessel of the fifth visceral arch; Br.ch., branchial channel; ep.ep., epidermic layer of epiblast; hy., hypoblast; g.sl., true gill cleft; n.ep., nervous layer of epiblast; 3, 4, 5, branchial channels.

If the clefts I call branchial channels were true "gill clefts," this latter should be regarded as a seventh gill cleft. It has a very narrow chink which almost, if not quite, opens on the surface (text-fig. 103, g, h, i.), and in the other direction it widens into a circular canal, passes backwards, and then inwards to the middle line, where it joins the wide cavity of the pharynx or oesophagus. In front of this spot there is no lumen. Thus it is a true gill cleft, and with its fellow on the other side forms the only pair of true gill clefts. The gill cleft of the right side joins a little further back; and at the same point, but from the dorsal wall, a little to the right of the median line, a diverticulum is given off which is the rudiment of the air bladder.

The condition of the branchial region on the fifth day is in short as follows. A hyomandibular cleft is present and extends nearly to the skin, but is not open. The first, second, third, and fourth branchial clefts are open, but they all lead, *not into the pharynx*, but into a special branchial canal on each side lined by ectoderm. This canal is prolonged backwards through the fourth branchial arch and approaches the skin beyond it where an invagination from the skin meets it, but I think there is not yet a perforation (text-fig. 104, 5). Now all these five clefts are not strictly speaking gill clefts at all. None of them perforate the gut. Yet there can be no doubt that the arches and contained blood vessels are the homologues of the arches and blood vessels similarly named in other Vertebrates.

The probable explanation is that the gill arches and external gill filaments grow out with great rapidity, leaving wide inter-arch spaces which deepen and coalesce under the arches, and so form the channels. In this way a large part of the "pharynx" is really stomodaeum and clothed with epiblast, which fact accounts, as is commonly supposed, for the presence of teeth on the pharyngeal bars of Teleostean fishes.

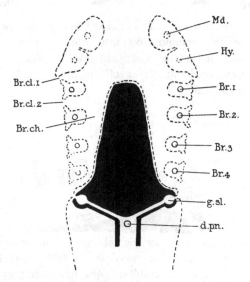

FIG. 105. Diagram to illustrate the branchial channels and their relations with the pharynx and true gill clefts in an embryo of 6—7 days. *Br.* 1, 2, 3, 4, aortic vessels of branchial arches; *Br.cl.* 1, 2, branchial cleft 1, 2, etc.; *Br.ch.*, branchial channel; *d.pn.*, ductus pneumaticus; *Hy.*, aortic vessel of hyoid arch; *g.sl.*, true gill cleft; *Md.*, aortic vessel of mandibular arch.

There is, however, one pair of true gill clefts, each of which is lined throughout by hypoblast, and has the form of a backwardly directed tube (text-fig. 105). This true slit is transitory.

Gill filaments arise as little knobs both on the anterior and posterior edges of the first, second, third and fourth branchial arches. It is quite clear that the external gills are of epiblastic origin and are not homologous with the external gills of Elasmobranchs.

The walls of the oesophagus are composed of a single layer of columnar epithelium, surrounded by a mass of undifferentiated tissue showing a good many blood capillaries. The oesophagus projects into the coelom, and the lumen narrows and soon becomes greater in vertical than in horizontal diameter. As it passes back the floor suddenly deepens into a long gradually shallowing recess, from the base of which several short diverticula arise which lie embedded in highly vascularised tissue in continuity with the yolk mass. The recess (text-fig. 107, *g.bl.*) is the gall bladder. After narrowing, the alimentary canal is continued back as a quite straight and exceedingly fine tube to the anus. At no point is there any connection between the cavity of the alimentary canal and the yolk. Throughout the whole length the intestine is suspended by a dorsal mesentery and a ventral mesentery attached along its ventral edge to the yolk mass, thus completely dividing the coelom. It contains besides the alimentary canal certain masses of mesoblast cells and blood vessels, and also a strand of yolk containing large yolk nuclei.

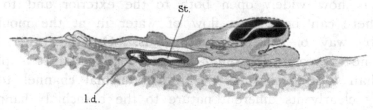

FIG. 106. Transverse section through an embryo of the fifth day. *l.d.*, diverticulum from which the liver is formed; *St.*, stomach.

Text-figure 107 represents diagrammatically the condition of the alimentary canal as seen in sagittal section.

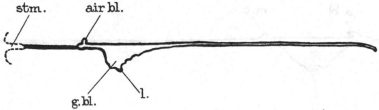

FIG. 107. Diagram of the alimentary canal of the fifth day as seen from the side. *air bl.*, air bladder diverticulum; *g.bl.*, recess which becomes the gall bladder; *l.*, liver diverticulum; *stm.*, stomodaeum.

In a specimen slightly older than the last the branchial canal of the left side is continued to and opens into the invagination of the skin for the fifth branchial cleft. On the right-hand side, although the pit is very wide, there is not yet a communication between the pit and the canal. It is a rather interesting case, because it indicates the method of formation of these canals by pouchings which burrow under the arches from the adjoining pits. The pit of the fifth gill cleft extends inwards and forwards beneath the fourth branchial arch touching, but not actually fusing with, the pouch from the epiblast of the fourth cleft.

In this specimen the stomodaeum, as determined by the presence of the covering layer of epiblast, extends back certainly so far as to line the mandibular, hyoid and first branchial arches.

The true gill cleft is not open to the exterior on either side, though it comes so close to the surface and to the posterior end of the branchial canal as to be separated from them by the covering layer of epiblast only. The hepatic diverticulum shows a tendency to project slightly forwards, but otherwise no change has taken place. Posteriorly the intestine ends in a little dilatation. I think it is not yet open to the exterior, although it receives in its dilated part the common termination to the two pronephric ducts.

Sixth Day.

The embryo is no longer confined by the zona radiata, and has nearly recovered from the distortion of the previous day. There has been no change in the condition of the pharynx and branchial channels, except that the fifth branchial cleft is now widely open both to the exterior and to the branchial channel. So there can be a free flow of water in at the mouth and out by all the clefts by way of the branchial canals. On the other hand the true gill cleft does not perforate the outer layer of epiblast[1]. It is placed distinctly further back than the opening of the fifth branchial channel to the exterior, and marks very clearly its different nature to the branchial channels and clefts anterior to it.

The swimming bladder diverticulum (which arises at the junction of the pair of true gill clefts) is embedded in a very vascular tissue, and bends slightly backwards (text-fig. 108 *a.bl.*).

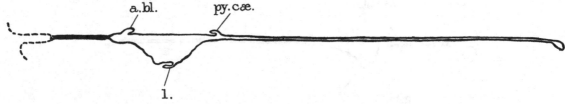

FIG. 108. Diagram of the alimentary canal of the sixth day. *a.bl.*, air bladder diverticulum; *l.*, liver diverticulum; *py.cæ.*, diverticulum from which arise the pyloric caeca?.

[1] v. p. 435—6 this volume. Such a view explains the non-perforation of the outer layer, but renders the outer layer of less value as a test for extension of epiblast.

The lumen of the alimentary canal behind the junction of the two true gill pouches is much wider than before, and the hepatic recess is larger. The epithelium on the dorsal wall is thinner than on the ventral wall. From the ventral and ventro-lateral regions numerous diverticula are forming, and two specially long ones (one of which (*l*) is shown in the figure) arise from the ventral wall and run forwards and end blindly.

A short distance behind the hepatic recess a diverticulum arises from the dorsal wall of the alimentary canal, and bends slightly forwards (text-fig. 108).

Just above this diverticulum there is in the mesentery a solid mass of tissue which stains lightly and is surrounded by blood vessels. This is the beginning of the gland-like tissues known as islands of Langerhans present in many Vertebrates (text-fig. 109 *I.L.*).

The diverticulum does not penetrate the cell mass. The hinder part of this tissue is much attenuated and is enveloped by a net-work of blood capillaries from the coeliaco-mesenteric artery.

Passing backwards the alimentary canal continues as a very fine straight tube, with quite rudimentary lumen. Blood-vessels pass from time to time between the splanchnopleure of the yolk sac and ventral wall of the alimentary canal, thus placing the embryo in connection with the vascular system of the yolk mass. Further back, at the point where the tail end becomes free from the yolk, a strand of yolk passes upwards into the mesentery below the alimentary canal (text-fig. 110).

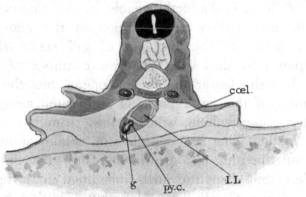

FIG. 109. Transverse section of a sixth day embryo. *cœl.*, coelom; *g.*, cavity of main gut; *I.L.*, Islands of Langerhans; *py.c.*, dorsal diverticulum of gut, probably pyloric caeca.

In the early stages of the segmentation of the egg a part of the yolk may always be found where the protoplasm is more abundant, the large yolk grains absent and where usually many large "yolk nuclei" are present. This patch is still to be found under the hinder end of the trunk, and it is from this patch that the strand of yolk above mentioned passes into the mesentery.

Seventh Day.

The external gills are now well developed, the branchial canals have become enlarged and the area of obliterated gut correspondingly restricted. I cannot say for certain how far this is due to a continued ingrowth of the epiblast,

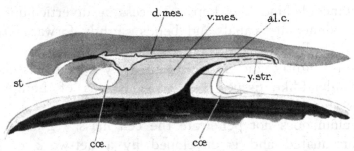

FIG. 110.   Diagram to show the relation of the mesenteries to the alimentary canal and yolk mass. *al.c.*, gut; *cœ.*, coelom; *d.mes.*, dorsal mesentery; *st.*, stomodaeum; *v.mes.*, ventral mesentery.

and how far to a split which may have arisen in the hypoblast lamellae comparable to the formation of the true gill pouch at an earlier stage. I can find the outer layer of the epiblast lining the pharynx as far back as the arch between the second and third open gill clefts. At the level of the centre of the third cleft the hypoblast lamella is separated in the middle line and shows no sign of epiblast, while it is distinctly visible in the channels on each side.

Posterior to this region there is no lumen in the mid-gut line; there are only the two true pouches of the last pair of gill clefts which run backwards and inwards and open into the wide posterior chamber of the pharynx. The one on the left side has the larger lumen and opens into the gut in front of the right cleft. The lumen of the right pouch is obliterated along part of its course.

The air bladder diverticulum, which arises from the dorsal wall of the oesophagus in the median line as regards the whole animal, but to the left of the median line of the gut, has grown out into two branches, one anterior which lies to the left and runs forwards imbedded in a highly vascular tissue as far as the level of the fourth branchial cleft. It is a much dilated vesicle, lined by a single layered epithelium and underlies the left pronephric sac.

The posterior lobe, also lying on the left, passes back in the dorsal mesentery of the oesophagus imbedded in thick vascular tissue, having a columnar epithelium and showing indication all along of the formation of diverticula. It becomes almost separated from the mesentery as it passes backwards, dilates considerably and ends rather abruptly just anterior to the point at which the liver diverticulum arises (Pl. XVIII. fig. 29).

Really the air bladder is a bilobed sac rising to the left of the mid-dorsal line of the oesophagus, of which the left lobe extends in a posterior direction only, while the right lobe extends both posteriorly and anteriorly, though the former extension never nearly equals that of the left lobe.

The lumen of the oesophagus remains large for some time, then as it bends ventralwards it narrows somewhat, all the while embedded in a thick mass of mesoblast which contains many blood vessels.

A great change has occurred during the past day. Apparently the great ventral recess which was noticed as having a different kind of epithelial lining and showing diverticula from all sides has in some way become constricted off from the upper part of the chamber which remains as a small tube lined by tall columnar cells, and is attached to the recess by a small duct only. The recess has become the gall bladder; and the numerous diverticula have increased and subdivided and become folded, while the intervening spaces have been invaded by blood from the vitelline veins, the whole mass now representing the liver.

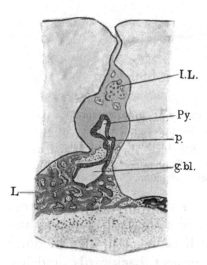

FIG. 111. Transverse section through the liver on the seventh day. *g.bl.*, gall bladder; *I.L.*, Islands of Langerhans; *L.*, liver; *p.*, pancreas?; *Py.*, pyloric caecum?.

As to the little dorsal diverticulum seen on the sixth day (text-fig. 108, *py.cæ.*) I am not quite certain. It is not the pancreas, which may be seen arising from the bile duct (text-fig. 111, *p.*) as several little buds which grow out later at this point into racemose glands (text-fig. 113, *p.d.*) along the mesentery and surrounding the islands of Langerhans.

It is more likely the rudiment of the pyloric caeca, which are now showing as little recesses (text-fig. 111, *Py.*).

In this case the usual dorsal pancreatic rudiment would be absent; unless, as seems possible, the islands of Langerhans tissue is budded off at an earlier time from the dorsal wall of the gut and represents the dorsal rudiment.

The intestine remains small until just before the anus where the lumen swells slightly and is widely connected as before with the two pronephric ducts.

Eighth Day.

The external gills now form great bushes on each side of the head entirely hiding the gill clefts from view. The operculum has just begun to grow back. The first gill cleft is very narrow, being open only through three sections. The stomodaeum extends back as far as the third gill arch. Only quite a narrow strip on the mid-ventral floor is not covered by the ectodermal layer. This narrow strip shows no lumen. The true gill pouches have now completely closed and the whole hinder part of the pharynx and the oesophagus is "solid" up to the point of origin of the swimming bladder where the lumen suddenly appears widely expanded. The left lobe of the swim bladder now extends forwards as far as the second branchial arch, and is itself bilobed, the lobes lying symmetrically.

The left primary lobe is still an undivided sac, though it has grown in length slightly. The oesophagus widens out so as to form quite a capacious tube which passing back gives off a wide but short blind pocket which is the future "stomach" (text-fig. 112). Shortly before the end a narrow tube runs at

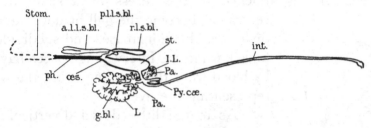

FIG. 112.   Diagram of the alimentary canal of the eighth day.   *a.l.l.s.bl.*, anterior lobe of the right division of the air bladder; *p.l.l.s.bl.*, posterior lobe of the right division of the air bladder; *r.l.s.bl.*, left division of the air bladder; *g.bl.*, gall bladder; *I.L.*, Island of Langerhans tissue; *int.*, intestine; *L.*, liver diverticula; *œs.*, oesophagus; *Pa.*, pancreatic diverticula; *ph.*, pharynx; *Py.cœ.*, Pyloric caeca; *Stom.*, stomodaeum.

right angles to the wide pouch and bends to the left and ventrally and then turns at right angles to be continued into the intestine. At the angle it receives the duct from the gall bladder which has become very much reduced in size by now; and two long narrow caeca, the pyloric caeca, which project backwards, and are covered, like the stomach pouch, by a thick investment of mesoblast. After this point the gut narrows to a fine tube and as such reaches the anus, the lumen however expanding just before that point. There appears to be no proctodaeum, the epithelium of the alimentary canal extends to the surface; on either side of this point the double layered epiblast is seen to end abruptly. There is no longer any suggestion of connection with the pronephric ducts.

The pronephric ducts open by a common aperture posterior to the rectum. There are no abdominal pores.

The "Islands of Langerhans" are present (just above the bile duct) in the mesentery which carries the mesenteric artery from the dorsal aorta and are now much broken by blood-capillaries.

Scattered throughout the mesentery and tissues surrounding the bile duct, and stomach, and extending up to the Islands of Langerhans are little groups of deeply staining cells some of which I can trace to be in connection with small diverticula from the bile duct.

It is extremely probable, that although some other scattered groups may be actually separate at the moment (though of this I am not at all sure), that all have been derived from gut epithelium like those which can now be seen to have connection with the bile duct. These correspond in position and in character with the diffuse gland of later periods, the pancreas. There are certainly three diverticula if not more—hence as many ducts to the pancreas.

One cannot draw any very close distinction between the liver and pancreas as regards development except in time of origin.

Both organs are developed from a series of diverticula of the "recess" noticed in earlier stages; the recess itself constituting the gall bladder and bile duct. The diverticula which perhaps develop earlier and develop in the direction of, and hence in connection with, the blood supply from the vitelline veins, become the liver, and those developing in the other direction towards the arterial blood supply of the mesentery become the pancreas (text-figs. 112, 113).

The doubling of the mesentery is probably the result of the folding of the gut produced by the great growth of the dorsal wall of the post-oesophageal gut to form the blind stomachic pouch, and the reduction of the recess into liver, gall bladder and bile duct.

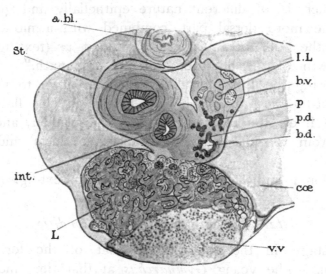

FIG. 113. Transverse section through the liver on the eighth day. *a.bl.*, air bladder; *b.d.* bile duct; *b.v.*, vessels into which the mesenteric artery divides in the islands of Langerhans (*I.L.*); *cœ.*, coelom; *L.*, liver; *int.*, intestine; *p.*, pancreas; *p.d.*, pancreatic duct; *St.*, stomach; *v.v.*, vitelline vein.

Text-fig. 112 is a diagram representing the relation of the several organs to each other: and text-fig. 113 is a transverse section of the eighth day larva showing the liver below the bile duct, where it is swollen at the point from which the pancreatic diverticula arise; the pancreatic tubules (*p.*) which extend into the mesentery towards the islands of Langerhans, as well as ventrally towards the liver; the stomach (*St.*) and commencement of the intestine (*int.*) before its junction with the bile duct.

Owing to the paucity of intermediate stages between that described as a specimen of the sixth day and that of the eighth day one cannot be certain as to the course of development, but I may point out that such a change could be brought about by a constriction occurring between the lower part of the

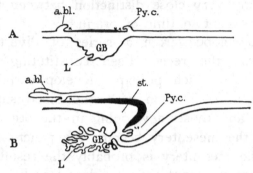

FIG. 114. Diagram to show the conversion of the ventral chamber of the gut (*GB*) in A of the sixth day into the gall bladder (*GB*) in B of the eighth day, and formation of the stomach (*st.*) by greater growth of the dorsal wall between the spots *x* and *xx*.

hepatic recess (which is of different nature epithelially and gives rise to many diverticula) and the more dorsal part, combined with a more rapid growth of the dorsal wall of the gut just anterior to the point *xx* (text-fig. 114, A)—which causes the outgrowth of the blind end of the "stomach."

This would bring the little dorsal diverticulum down to the position of the pyloric caeca (text-fig. 114, B). The diverticula in the floor of the hepatic recess become the liver, the recess itself the gall bladder, and the constriction the bile duct. From the bile duct the pancreatic ducts and tubules develop subsequently.

This explanation is indicated diagrammatically in text-fig. 114.

Attainment of "final larval" condition.

The oldest stage in the collection is that of the forty-third day after laying of the eggs. The young *Gymnarchus* at that time measures 60 mm. in length (nearly $2\frac{1}{2}$ inches) and all trace of the yolk mass is obliterated in an external view. The external gills have gone and the gill clefts are covered by the operculum. In fact both internally and externally most of the organs have acquired the features characteristic of the adult, excepting only the reproductive apparatus which remains in a very rudimentary condition. Figs. on Pl. XVIII. illustrate the condition of the alimentary canal at this time, and comparison of these with the figures given by Förg (43) and Hyrtl (71) will show that the adult condition has nearly been attained.

The mouth leads into a buccal cavity very much compressed dorso-ventrally. Teeth are well developed in both upper and lower jaws, and are large and pointed. All stages are to be found from fully formed teeth with sharp enamel points perforating the stratified epithelium of the jaws down to mere rudiments. The tip of the tooth is a sharp point of clear enamel which is continued only a very short way along the dentine. Young teeth develop from enamel organs which bud off from the enamel organs of an older tooth. Their bases are quite free but lie often surrounded by sockets of bone, but they are not fused with them. Although there is nothing peculiar to record in connection with the development of the teeth, the dates of the different stages may be given. There is no sign of them on the seventh day, but on the eighth day rudiments of the enamel organ are distinctly present as little thickenings of the malpighian layer of the epidermis (which is now many layered), underneath which the pulp rudiments appear as little dense centres of nuclei of mesodermal tissue, in which numerous mitoses indicate an increased activity. These gradually sink deeper into the tissues and on the tenth day the first indication of bone and enamel may be seen. At this stage also the first membrane bone of the jaws is forming, namely the dentary.

The teeth rudiments are arranged in a row on the anterior margins of upper and lower jaws. There is no trace of teeth at any other spot.

By the eleventh day the root has reached down to the level of the developing bone, but does not as yet appear within a socket. The fang has developed considerably, but does not project above the surface. The lengthening of the fang is compensated for at first by the sinking in of the root.

By the forty-third day, however, the rudiment has reached its furthest limit and now additional growth has resulted in the forcing outwards of the tip of the fang through the stratified layers of the epidermis.

Thus the first formed teeth start close to the surface, the rudiments of the later teeth are from the first more deep-seated, arising as they do as buds of the sides of the deeply sunk enamel organ of the first tooth.

There is no perforation of the oesophagus until after the ninth day. On the eleventh it is widely open but the intestine is still a small canal. On the thirteenth day the intestinal canal is large but there is no trace as yet of solid food in the alimentary tract. The forty-third day specimens, however, have been feeding freely.

The whole of the pharynx and the oesophagus as far back as the opening of the ductus pneumaticus is lined by a stratified epithelium. The roof mucous membrane is thrown into fine longitudinal ridges, and on the floor is a protuberant cushion of soft spongy tissue which simulating a tongue in outward appearance, contains, however, no muscle fibre. The surface of this tongue and indeed roof and floor of the whole of the alimentary passage to the oesophagus contains little sense organs probably of the nature of taste buds scattered about it quite irregularly. These are all of the same type. A bundle of columnar cells apparently of the malpighian layer with the nuclei at their bases and their distal ends fused together form a projecting granular mass. Underneath is a cell with clear cytoplasm and an oval nucleus lying at right angles to the bundle.   This cell is accompanied by mesoblast and I think nerve fibres and blood capillaries.

The question how far the stomodaeum extends is a very difficult one to answer.

From the description given some few pages back it would seem that a very great part of the lining of the branchial chamber is derived from epiblast, so that a comparatively narrow tract of hypoblast along the mid-dorsal and mid-ventral line of the hinder part of the pharynx alone can remain.

The Permanent Gills.

The history of the branchial clefts, external gills and permanent gills is simple.

Five pairs of branchial clefts open, which correspond to the branchial clefts 1—5 of other Vertebrates.   I have observed no perforated cleft between the mandibular and hyoid arches.   None of them close.   I have discussed already whether the pair of pouches which develop at the end of the pharynx in connection with the fifth branchial pits may represent a seventh pair of gill clefts or whether it is the sole representative of the true gill clefts of other Vertebrates, in which case it probably represents the fifth branchial, that is to say the sixth visceral cleft.   At an early period, namely on the fifth day, gill filaments grow out from the anterior and posterior edges of the outer surfaces of the gill arches between the 1 and 2, 2 and 3, 3 and 4, 4 and 5 slits.   These are undoubtedly external epiblastic gills. They never branch; each is a simple elongated loop of a capillary with its covering of epiblast and contains one long afferent capillary and one long efferent capillary continuous with each other, at first only at the apex.   They attain their greatest length on the eleventh day.   A kink may appear but never a branch. But by a later development of connecting loops between the afferent and efferent near the base the permanent filaments are formed; and as these develop the long terminal loop shrivels up (text-fig. 125, p. 338).   The efferent filament vessel later becomes divided at its base.

These are attached along the outer edges of each of the arches named.   In the sixteenth day specimens there is still a trace of the external gill filament, but in forty-second day specimens there is none, and only the basal part remains.

From this account it appears that not only the embryonic, i.e. larval external filaments, are epiblastic and external, but that also the permanent gills must be of the same origin.

The Lung.

In the advanced larva the end of the pharyngeal region is marked by a sudden constriction, the oesophagus, which is bounded by a thick muscular layer.

Opposite the anterior end of the notochord the canal widens and from the dorsal wall of this widened region the ductus pneumaticus is given off, as before, a little to the left of the median line of the alimentary canal.   On the fifth day the stratified epithelium does not pass back beyond this point, but later it extends up to the beginning of the stomach.

The ductus pneumaticus is a wide tube leading into the air bladder which is now a large lung-like sac lying along the dorsal wall of the abdominal cavity (Pl. XVIII. figs. 26, 27).   The air bladder in surface view appears conical, with its apex reaching backwards nearly to the posterior end of the abdominal cavity, being between the kidney and peritoneum.   There is a large central lumen which leads on all sides into lateral air chambers with their walls carrying the blood vessels.   The single artery runs along the ventral border near the middle line.   The epithelium of the pharynx and oesophagus contains many mucous cells (goblet cells) and is stratified.   That of the ductus has no goblet cells and is not stratified.   The greater part is squamous, the floor of the duct is columnar (text-fig. 115).

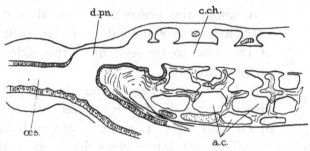

FIG. 115.   Diagrammatical sagittal section of the ductus pneumaticus of the forty-third day larva. *a.c.*, air cells; *c.ch.*, central channel of the lung; *d.pn.*, ductus pneumaticus; *œs.*, oesophagus.

In front, lying at the sides of the head close against part of the auditory labyrinth and covered by the supra temporal bone are two vesicles which have been developed from the air bladder but are now almost separated from it. These, which were described by Erdl (39) as the "schwimmblasenartigen Organe," and recently wrongly described by Allis (5) as "a large balloon-like distension of the non-sensory portion of the sacculus," lie between the sacculus and the supra-temporal bone (Gehüdestal of Erdl), bounded by the semi-circular canals, v. Pl. XVIII. fig. 31, *t.au.*

These conditions have been arrived at by an extension forwards and outwards of the anterior bifurcated lobes as described for the seventh day.

The form of the swimming bladder on the ninth day suggests that it is really a bilobed structure.   At this stage the ductus pneumaticus passes vertically upwards, slightly to the right, and then turns sharply backwards almost in the median line.   Along its right side runs the proximal end of the anterior bifurcated sac which soon joins the ductus but continues backwards a short distance breaking up into many small chambers.   The main duct passes back, giving off diverticula alternately on each side. At the posterior end the whole lobe projects into one side of the divided coelom (text-fig. 116).

Thus, like the lungs of higher Vertebrates, it is a bilobed sac. The left lobe is well developed and entirely posterior, while the right lobe is slightly developed posteriorly but sends forwards a lobe which, bifurcating, ends in a pair of vesicles closely attached to the auditory sac (Pl. XVIII. fig. 28).

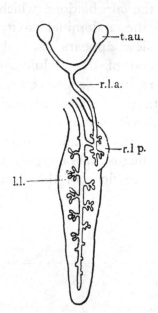

FIG. 116. Diagram of the internal cavities of the air bladder on the seventh day. The left lobe extends backwards only, the right lobe is small but sends along anterior branch forwards which bifurcates. *l.l.*, left lobe of lung; *r.l.a.*, anterior division of right lobe; *r.l.p.*, posterior division of right lobe; *t.au.*, tambour in contact afterwards with auditory labyrinth.

In the older specimens the anterior vesicles of the air bladder retain their position in close contact with the auditory labyrinth covered by the supratemporal membrane bone, but the connecting tubes become more and more drawn out as they pass back between the cartilage of the base of the cranium and the covering membrane bone until in the oldest specimen I am unable to trace any connection at all on the left side, and am not at all sure about the presence of an absolutely unbroken connection on the right side. There is, however, much more of the tube to be seen upon this side. The tube is thus inclosed in bone.

During the tenth and eleventh days the posteriorly directed part of the right lobe of the lung expands very much, and later it is not distinguishable from the left.

The left lobe grows back as a thick walled simple tube on the eighth day and on the ninth day it has given off a number of diverticula, which ramify in this thick wall. On the tenth day it has travelled back further than the posterior end of the stomach, but has not yet reached its final limit. The tissue surrounding the diverticula is becoming looser and more spongy, but the diverticula have not yet begun to expand; they are still growing in length and branching (text-fig. 117 A). The oesophagus remains solid.

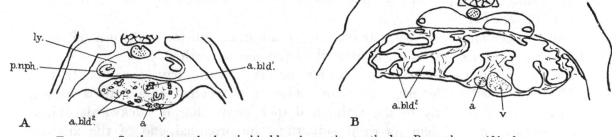

FIG. 117. Section through the air bladder A on the tenth day, B on the twelfth day. *a.*, pulmonary artery; *a.bld.*[1], main channel of air bladder; *a.bld.*[2], lateral chambers or air cells; *ly.*, lymphatic tissue surrounding *p.nph.*, pronephric ducts; *v.*, pulmonary vein.

The eleventh day shows only a greater conversion of the stroma into spongy tissue and more complete vascularisation and ramification of the diverticula. The organ still does not reach as far back as the mesonephros. The germinal ridge at this stage is very evident as an unpaired ridge lying ventral to the air bladder vein. The oesophagus is open.

On the twelfth day a great advance has been made. The organ now reaches back as far as the posterior end of the abdominal cavity, and the whole thing has become enormously swelled out by the dilation of all the diverticula into air chambers (v. text-fig. 117 B).

The stroma is now a gelatinous spongy tissue with minute nuclei and resembles the most advanced condition.

Stomach, Liver, Gall-bladder and Pancreas.

The stomachic pocket grows back slowly, dilating as it does so, and hangs freely in the abdominal cavity unattached by any mesentery. The walls are very muscular, the fibres running chiefly in a circular direction but others are also present. The mucous membrane is so much folded by the forty-third day that it forms in the hinder portions distinct simple tubular glands. The epithelium is a high columnar and single layered epithelium, and the cells are clear with a deep staining nucleus close to the base. At this time the yolk supply is nearly exhausted and the larva is feeding freely. From the contents of the stomach it appears that the food is chiefly vegetable.

As the stomach extends backward it draws with it the duodenal loop and the bile duct, with the result that the gall-bladder is drawn backwards away from the liver so that it comes into sight in a ventral view dissection (Pl. XVIII. fig. 25).

The changes which the liver undergoes are mainly a growth in size and a histological differentiation. The diverticula of the walls of the hepatic recess become more and more branched and tubular, and up to the twelfth day the liver resembles closely the appearance of an ordinary racemose gland, with large blood capillaries and channels between the acini, and with no connective tissue except that belonging to the blood vessels. After this the lumina of the tubules become difficult to distinguish and the cells themselves appear to be honeycombed by vacuoles.

The pancreas, which retains its diffuse character, does not change much. Its cells remain as characteristic secreting cells and the lumina of the tubules usually are distinct. These are easily distinguished from the liver with which some of its branches are in close contact.

It extends ultimately also down the mesentery attached to the pyloric caeca and some of its lobules become embedded even in the splenic tissues at the apex of the stomach.

The pyloric caeca which began as either a single or a pair of simple diverticula of the duodenal loop almost opposite to the opening of the bile duct gradually grow out backwards as blind tubular pockets. They are lined by a simple columnar epithelium, which in the latest stage becomes slightly folded. One caecum is shorter and has greater folds in it than the other.

The intestine, which at first is a straight narrow tube, becomes bent ventrally and to the left at the time of the formation of the liver and stomachic pouch; and if we imagine its attachment to the stomach being drawn back by the growth backwards of the stomachic pouch we see how a loop—the duodenal loop pointing forwards—may be produced. Eventually the distal limb of the duodenal loop turns right back to form the intestine and passes almost straight to the anus (Pl. XVIII. figs. 25 and 28).

The lining between the stomach and distal loop is thrown into folds. In the intestine proper it is simpler. The epithelium is columnar. Thus it passes backwards suspended by a dorsal mesentery and attached to the yolk mass by a ventral mesentery (omentum). It forms a slight bend to the right round the apex of the stomach, expands slightly and runs back to the anus. After the actual inclusion within the abdominal wall of the yolk mass the intestine becomes almost embedded in the mass. The lumen now is large, and since the larva began feeding has no doubt been functional. Until this time the lumen of the intestine is minute.

At an early stage, there is a slight trace of a spiral valve, but this has ceased to be recognisable in the forty-third day specimen.

The Thyroid Gland.

On the fifth day the thyroid gland is well established. It consists of a mass of cells entirely separated from the epithelium and lying in the region of the middle of the first branchial cleft in front of the bifurcating anterior end of the cardiac aorta. In an older specimen, one of $5\frac{1}{4}$ days, there is some trace of a connection with the epithelium of the pharynx, while in the next day's specimen, the gland although not actually in contact, has the appearance of having just separated. The spot at which the contact would seem to have been is where the outer layers of epiblast from the two sides are doubled back, that is to say, just where the endoderm begins, at the hinder border of the first branchial cleft. The gland extends between the bifurcation of the aorta and passes back along the ventral wall of that vessel as far as the heart, which is very far forward.

By the eighth day it has passed back together with the cardiac aorta to the hinder regions of the second gill cleft and is still a spherical mass of cells, among which some blood capillaries ramify. It retains its position as regards the aorta, lying ventral to the bifurcation, but the whole appears to lie further back as the lower jaw and branchial apparatus grow forward.

By the twelfth day the cells have arranged themselves as a typical secreting epithelium. The lumina of the tubes thus formed are filled with a secretion which stains with eosin. Many chinks occur between the tubules and a large central cavity. These are lined by a squamous epithelium. There is no thyroid secretion in these chinks, which are probably lymph spaces. There is a remarkable paucity of blood vessels. In the oldest stage, the gland has a pyriform shape and lies just ventral to the anterior end of the cardiac aorta. It has become highly vascularised while the lymphatics have diminished if they have not disappeared altogether.

Observations on the Organs of the Alimentary Canal.

The features of chief interest in the development of the alimentary canal of this fish are the peculiar development of gill clefts, and the development of the air bladder. Other points to be noted are the relations of the pancreas to the islands of Langerhans, the presence of a spiral valve in the intestine, and the formation of the permanent gill filaments from the external gills which are regarded as epiblastic.

The presence of the single pair of true gill clefts (or at any rate one of a different nature to the anterior clefts), the homologue, apparently, of the sixth or seventh visceral cleft of elasmobranchs requires some explanation.

The question of the origin and homology of the gill clefts and of the gill filaments has been discussed quite recently by Goette (53); Moroff (94); Greil (57).

Goette would assign an epiblastic origin to the lining of the outer part of the gill clefts including the gill filaments, not only in Dipnoi, Ganoids, and Teleosteans, but also in Elasmobranchs with the exception of the pseudobranch of the spiracle.

Moroff, in his description of the development of the gill clefts of the trout, shows that the spiracle is partly lined by ectoderm, but that the pseudobranch is derived from the deeper part of the wall of the cleft and is endodermal—thus agreeing with Goette's account for Torpedo.

The first branchial cleft, however, he believes to be entirely lined by ectoderm.

Greil disagrees with Goette as regards an ectodermal origin to the gill cleft of Elasmobranchs and describes the course of events in Acanthias. Furthermore, he would assign an endodermal origin to the lining of the clefts of Amphibians, describing how the endoderm creeps outward from the clefts in the newt underlying the outer layer of ectoderm and taking the place of the inner layer. It will be remembered that Balfour (8), pp. 210—211, felt some uncertainty on this point. He says "The gill slits arise as outgrowths of the lining of the throat towards the external skin. In the gill slits of Torpedo I have observed a very slight ingrowth of the external skin towards the hypoblastic outgrowth in one single case." Again, "From the mode of development of the gill clefts, it appears that their walls are lined externally by hypoblast, and therefore that the external gills are processes of the walls of the alimentary tract, i.e. are covered by an hypoblastic, and not an epiblastic layer. It should be remembered, however, that after the gill slits become open, the point where the hypoblast joins the epiblast ceases to be determinable, so that some doubt hangs over the above statement."

The question is made no easier when one recollects such processes as that described by Graham Kerr for *Lepidosiren* where the stomodaeal invagination appears to be formed by a sort of conversion *in situ* of endoderm into ectoderm. Possibly one ought to regard these cells as undifferentiated yolk cells, rather than hypoblast.

As regards Elasmobranchs I may add my small quota of evidence, as I have had the privilege of examining the collection of series of sections of Elasmobranch embryos belonging to Mr Adam Sedgwick, and certainly in some of them the distinction between ectoderm and endoderm does seem to be most clear after the opening of the cleft, and in these cases the gill filaments arise certainly from the endoderm and not the ectoderm.

The case of *Gymnarchus* appears to me to be of peculiar interest. It is impossible in this material to study the very earliest stages in the formation of the slits, but this is of less importance, for I think all will agree that in Teleosteans as well as in other Vertebrates the earliest signs of gill clefts are always the pouches of the gut wall (Goette, Moroff, Henneguy, etc.); it is the later development, including the first origin of the gill filaments, that is in dispute.

I have taken in my description above the presence of the outer layer of ectoderm over an inner layer as being evidence of the latter being the inner layer of ectoderm and not endoderm, but if Greil is right, this clearly is by no means a sure test.

The presence in *Gymnarchus* of two quite different types of cleft cannot be doubted; the one type I believe to be lined by ectoderm and to be permanent, the other by endoderm and transitory.

Here they exist in the same animal side by side. On the fifth day one pair of true gill clefts occurs (text-fig. 105); in front of which are the five branchial clefts and their connecting channels (though the fifth at this time is not yet open). The posterior pair (text-fig. 104) is the only true gill cleft—that is to say the only one which at this time can be described as a perforation of the gut and body wall. In the anterior ones the gut wall is not concerned at all—it is not perforated. The branchial channels are to be regarded as due either to in-pittings from without of the ectoderm, or as a result of the growth outwards of the body wall portions of the visceral arches—or more probably as a combination of both processes, and are a later modification of true gill clefts taking the place of the original gill clefts in function, and in a strict sense are not "gill clefts" at all. The first five gill clefts of *Gymnarchus* are no more homologous to the gill clefts of a dog fish (though perhaps Goette would not agree to this) than are the horny coverings of the jaw of *Ornithorhynchus*, a tortoise or a bird homologous to true teeth—in fact not as much so, as they are not even derived from the same germinal layer.   Probably *Gymnarchus* is typical of the Teleosteans in general, and we may look upon the gill filaments of Teleosteans, as being ectodermal structures and of the gill clefts as being respiratory channels having a later origin than true gill clefts.

Another point of interest is with reference to the air bladder, which is lung-like in character.   Erdl described it as a lung and believed that it was divided as in *Lepidosiren*.   He described folds and muscles which enabled the animal to close or open the ductus at will.   Förg and Duvernay and later Hyrtl have all described its character in the adult animal, and corrected Erdl's view that the bladder is double—at least so far as can be seen in an external view of the adult condition. Duvernay strongly protests against its lung-like nature, but Hyrtl (71), drawing attention to Pérottet's account of the habits of the animal and to the extraordinary rich vascular supply and the great distensibility of the organ, considered that it must be used during certain seasons as an organ of respiration.

My description above of the really double nature of the air bladder in development and of the tendency to division of the auricle to be described in the next section, are facts of interest which seem to me to suggest that *Gymnarchus* shows us that Teleosteans may be descended from forms in which a functional lung and perhaps an approach to a double circulation on the Amphibian type were present.

That *Gymnarchus* uses its air bladder as a lung during the larval stage has been proved by Budgett, who found that if the larvae were prevented from coming to the surface they died.

2. THE VASCULAR SYSTEM.

Fifth Day. Series D.

The foundations of the vascular system are laid down during the fifth day. In embryos of the fourth day there are numerous groups of blood corpuscles in the splanchnopleure of the yolk mass, but I am unable to find any trace of vascular system within the actual embryo.   The head end has begun to roll over forwards, so that there is a head fold—and possibly there may be traces of the yolk mass efferents.

In the fifth day specimens a great advance has been made, but as this probably represents the highest point of development of the branchial system of blood vessels it is an interesting and important period.

The five gill clefts of the adult are now present and the external gill filaments are just commencing to show, being clearly discernible on the anterior and posterior outer edges of the first, second, and third branchial arches and perhaps on the fourth. The fifth arch and the hyoid arch do not show any trace of gill filaments.

The whole yolk mass is covered by a net-work of blood channels. These collect into two large yolk mass efferents which converge beneath the "embryo" in the region of the first to third visceral arches. Here the two vessels coalesce and form a single broad channel which runs backwards to the posterior end of the branchial region (text-fig. 118, *Vit.vein*).

This large sinus, formed by the coalescence of the two anterior yolk sac efferent

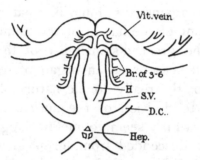

FIG. 118. Diagram of the heart and blood vessels of the fifth day as seen from above. *Br. of* 3—6, branchial afferent vessels to branchial arches 1—4. Mandibular and hyoid afferents are seen arising from a separate vessel in front; *D.C.*, ductus Cuvieri; *H.*, heart; *Hep.*, hepatic vessels; *S.V.*, sinus venosus; *Vit.vein*, vitelline veins.

vessels, may be termed sinus venosus. Into it at some little distance farther back the two ductus Cuvieri open; and behind these, but almost at the same level, and in conjunction with several smaller efferents from the yolk mass, there open channels from round about the developing diverticula from the alimentary canal that form the future liver, which are the first signs of the hepatic veins and sinuses. These two large vitelline veins appear as though they joined the heart anterior to the ductus Cuvieri. This is due to the extreme flattening and stretching which the embryo undergoes at an early stage. Text-fig. 118 represents diagrammatically the relation of these vessels to the heart.

The heart is a long almost straight tube with thick muscular walls and an endothelium of squamous cells. The walls are considerably thicker anteriorly, and gradually become thinner towards the posterior end. The diameter diminishes slightly behind.

It lies quite freely in the anterior part of the coelom, attached only by its extreme anterior end to the anterior wall of the pericardium (as it will be eventually), and posteriorly to the floor of the pericardium.

At its anterior end it breaks up into a pair of vessels which turn backwards and give off the four aortic arches on each side to the first to fourth branchial arches, and a pair which run forwards and gives off the vessels of the mandibular and hyoid arches, v. Pl. XXI. fig. 54.

The vessel of the mandibular arch is large and well defined, and passes up into the aorta of its side without giving off any branches. Close to it, but posterior to it and anterior to the first open gill cleft, is a very much narrower channel which however is quite complete and which is no doubt the hyoid vessel. Near its upper end, just before it joins the aorta, it shows a tendency to become broken up into smaller channels, which reunite before joining with the aorta.

There is certainly at this stage no open hyomandibular cleft, but a cleft or chink from the "branchial channel" is prolonged upwards and outwards between the two vessels.

The posterior common branchial afferent runs back, giving off the four vessels for the four branchial arches as it comes to them.

Certain points of special interest may be noted. It is probable that from the size of the aorta all the blood which enters it by the mandibular and hyoid vessels goes forwards to the head, and that which comes by the vessels in the second, third and fourth arches passes backwards. The blood from the first branchial arch probably goes partly forwards and partly backwards. The second branchial afferent appears to be the largest and most important of the posterior group as suggested by the great expansion of the aortic root after junction with this vessel. This vessel is the more advanced, there being distinct afferent and efferent vessels, which are connected at the ventral end as in the tadpole of the frog. It is curious that the vessel which in the higher Vertebrates is the large systemic arch should be here marked out as of special importance. The condition of the third and fourth branchial arches is less advanced. Gill filaments are certainly present in the third, but in neither case are there as yet distinct afferent and efferent channels.

Tending dorsalwards the third branchial arch runs backwards and unites with the fourth before joining the aortic trunk of their side. Posterior to this point the two aortic trunks gradually converge and join to form the single dorsal aorta just in front of the head kidney.

On tracing the aorta back it is seen to continue as a single median vessel giving off lateral branches (the parietal arteries) to the muscles, spinal cord, and general body wall.

A little way beyond the convergence of the two aortic trunks a pair of small arteries, not much larger than capillaries, is given off from the ventral wall of the aorta which vessels run in the walls of the head kidney (the glomerulus). These fall again into the dorsal aorta (text-fig. 119). A median vessel passes downwards into the mesentery suspending the gut and passes on to the oesophagus, a second, the coeliaco mesenteric, breaks up into capillaries in the island of Langerhans tissue.

A further description of the vessels in connection with this pronephros is given under a later section.

Passing backwards still further into the tail the aorta retains its large size. There is no branch passing off on to the yolk mass, nor is there any direct communication with the sub-intestinal vessel.

FIG. 119. Diagram of the arterial system in the region of the pronephros on the fifth day as seen from the left side. *Cœl.a.*, Coeliaco-mesenteric artery; *d.a.*, dorsal aorta; *Œs.* or *A.bl.a.*, Oesophageal or pulmonary (air bladder) artery; *pr.n.*, pronephric chamber.

Venous System.

The blood which is thus carried to the tail of the dorsal aorta is collected in the large caudal vein which increases rapidly in size as it passes forwards.

At the region of the anus the caudal vein divides into (1) a small dorsal vein, (2) a sub-intestinal vessel which passes ventralwards on one side of the gut and runs forwards at the base of the mesentery, which here divides the coelom into right and left halves, as far as the point where the "embryo" is attached to the yolk mass. Here a branch runs forwards, but gradually dies out, the main trunk turning outwards, i.e. ventralwards, and passes on to the yolk mass where it breaks up into a network of vessels which are collected again into the two large "vitelline veins" which coalesce at the anterior end to form the sinus venosus.

Thus it will be seen that the yolk sac circulation is entirely venous. There are no vitelline arteries; the yolk mass is interposed along the course, so to speak, of the sub-intestinal vein. This is the condition characteristic of the Teleosteans throughout larval life, but is found in Elasmobranchs only in the earliest stage (Rückert in Hertwig (64)).

The blood which passes along the dorsal vein or common renal portal instead of taking the ventral course is as yet very small in quantity; it follows irregular spaces bathing the sinuous pronephric duct, but I cannot trace the channels continuously.

Near the heart a fine straight channel outside the head kidney appears which opens into the ductus Cuvieri. This is very small, but no doubt represents the posterior cardinal vein. It does not receive any blood from the head kidney.

The anterior cardinal vein is well developed and takes the usual course (Pl. XXI. fig. 54). After receiving the small posterior cardinal vein it runs in the body wall as ductus Cuvieri and enters the sinus venosus *apparently* posterior to the vitelline vein.

I have not been able to detect the hepatic portal vein—I think it is not yet developed.

Eighth Day.   Series L.

Three days later some considerable changes have been effected. The heart has been drawn forward by the growth forwards of the whole anterior end of the body away from the yolk so that it assumes its normal position relative to the sinus venosus (text-fig. 120).

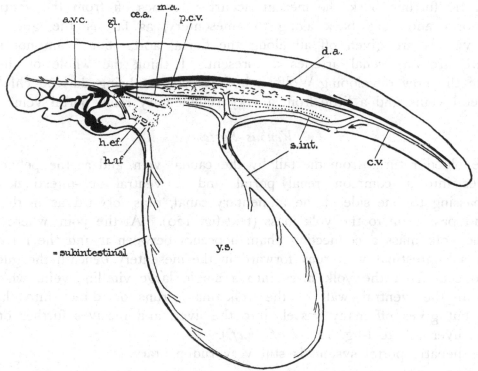

FIG. 120. Diagram of the vascular system as seen from the left side on the eighth day. *a.v.c.*, anterior cardinal vein; *c.v.*, caudal vein; *d.a.*, dorsal aorta; *gl.*, glomerulus of pronephros; *h.af.*, hepatic afferents; *h.ef.*, hepatic efferents; *m.a.*, coeliaco-mesenteric artery; *œ.a.*, pulmonary artery; *p.c.v.*, posterior cardinal vein; *s.int.*, sub-intestinal vein; *y.s.*, yolk sac.

The most noteworthy changes concern the aortic arches. The mandibular and hyoid vessels have disappeared, though a trace of one may be seen. The mandibular arch has grown forwards so that the mouth and jaws are now formed, and in connection with this, the first branchial vessel has been drawn slightly forwards also.

In the figure the left aortic trunk is seen to be continued forward as the common carotid artery supplying head and brain.

All four branchial vessels now have afferent and efferent branches, which are not in any case connected except through the gill filament. These are now long, but quite simple loops. There are no cross loops as yet.

Remnants of the ventral common trunk of the mandibular and hyoid vessels are seen as mandibular vessels.

The junction of the two aortic trunks has crept further forwards, so that the third and fourth branchial efferents join the dorsal aorta itself.

The glomerulus of the head kidney is formed by eight to twelve minute vessels from the ventral wall of the aorta, which after ramifying in the wall of the pronephros join into a single vessel uniting with one from the aorta that passes on to the oesophagus.

A little further back the mesenteric artery comes off from the ventral wall of the aorta and runs back along the mesentery, as far as the gut extends. Parietal vessels are given off all along the dorsal aorta, but I am not satisfied that there are any renal arteries at present. I think the whole of the blood supply of the now developing Wolffian body is venous, being derived chiefly from the parietal veins and a little from a very rudimentary renal portal vein.

Venous System.

The blood returns from the tail by the caudal vein, and at the pelvic region it divides into a common renal portal and a ventral sub-intestinal channel which, passing to one side of the alimentary canal, runs forward as in the earlier stage and passes on to the yolk mass (text-fig. 120). At the point where it turns on to the yolk mass a connecting channel occurs between it and the renal veins. A small sub-intestinal vein runs forward in the mesentery beneath the gut. The blood collects from the yolk mass into a single large vitelline vein, which, embedded in the ventral wall of the yolk mass, runs on direct into the sinus venosus but gives off many vessels into the liver, and receives further on many from the liver (cf. text-fig. 120, *h.af.*, *h.ef.*).

The hepatic portal system is still very rudimentary.

Veins of the Head.

The anterior cardinal sinus is formed by the junction of a large number of venous channels in the head region. It takes the usual course, joining with the posterior cardinal vein (which is now clearly returning blood from the mesonephros) to form the ductus Cuvieri. The ductus running backwards and ventralwards join the sinus venosus one rather behind the other. They form large vessels crossing the pericardium which is still widely open to the abdominal cavity.

A large lateral line sinus is present as a space lying along the lateral line nerve. It increases very much in size and approaches the anterior cardinal sinus very closely just before the junction of that sinus with the ductus Cuvieri, but it does not make any connection therewith.

Eleventh Day. Series J.

The heart consists of sinus venosus (into which open the two large ductus Cuvieri), auricle and ventricle and four aortic diverticula.   Each ductus receives near its entrance into the sinus venosus an inferior jugular which brings back the blood from the lower jaw (text-fig. 121). Posteriorly and ventrally the large single vitelline vein enters the sinus venosus, and dorsally to this opening two small hepatic veins enter together. The opening of the sinus venosus into the auricle is decidedly to the left of the middle line, and is posterior. It is guarded by two large flaps or valves. The ductus Cuvieri are large baggy vessels lying across the cavity of the pericardium and are attached to its walls only at their point of entrance to that cavity. The auricle is dorsal to the ventricle and is a large thin walled sac with a few strands of tissue passing from one part of the wall to another. Its opening into the ventricle is wide and is guarded by a pair of thick valves which have no trace of chordae tendineae. The wall of the ventricle is thin, but its cavity is much broken up by strands of tissue, forming a peripheral spongy portion, while its most central part is quite devoid of interrupting tissue.

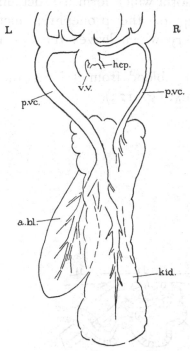

FIG. 121.   Plan of the chief venous trunks on the tenth to the twelfth day as seen from above. *a.bl.*, air bladder; *hep.*, hepatic veins; *kid.*, kidney; *p.vc.*, posterior cardinal vein; *v.v.* vitelline vein.

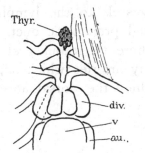

FIG. 122. Ventral view of the cardiac aorta on the eleventh day. *au.*, auricle; *div.*, diverticula on the root of the cardiac aorta; *Thyr.*, thyroid gland; *v.*, ventricle.

At the front end a small opening guarded by two vertical folds leads into the proximal part of the cardiac aorta—the bulbus, whose walls are greatly thickened by what appear to be smooth muscle fibres. Moreover it has two pairs of lateral swellings. Beyond this comes the cardiac aorta proper, which is extremely short, breaking up almost immediately into four pairs of branchial afferents. The trunk for the afferent of the first branchial arch arises from the most ventral part and it alone passes any considerable distance in front of the heart (text-fig. 122).

The branchial efferents of the first and second arches join on each side to form the roots of the dorsal aorta. These unite soon, but form only a small dorsal aorta, for much blood has passed forwards from the dorsal ends of the gill arches along the carotids (text-fig. 123 A).

The third and fourth branchial efferents unite and then join the dorsal aorta a little way in front of the head kidney. A pair of vessels of considerable size leaves the aorta supplying the walls of the pharynx and the thymus glands (text-fig. 123 *Thy.*).

In the region of the head kidney the glomerulus consists of a number of small vessels given off from the ventral surface of the aorta which form a reticulum of blood capillaries projecting into the wide opening of the pronephric duct. The blood is collected into the swimming bladder artery which leaves the dorsal aorta immediately behind the glomerulus capillaries.

Posterior to the glomerulus and not receiving any blood from it is another large median ventral vessel, the coeliaco-mesenteric (text-fig. 123).

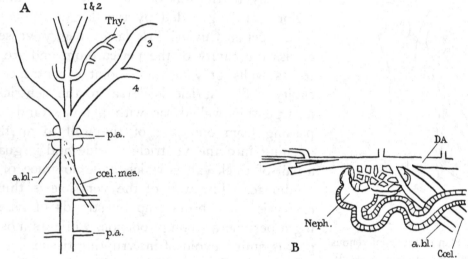

FIG. 123. A. Plan of the arterial system on the eleventh day. B. Side view of the arterial system in the region of the pronephros. *a.bl.*, air bladder or pulmonary artery; *Cœl.*, *cœl.mes.*, coeliaco-mesenteric artery; *D.A.*, dorsal aorta; *Neph.*, pronephric chamber; *p.a.*, parietal arteries; *Thy.*, artery to thymus.

The dorsal aorta gives off parietal arteries but no vessels to the lymphatic tissue which lies below it surrounding the much convoluted pronephric duct, and runs back to the hinder end of the abdominal cavity where the tubules of the mesonephros occur. Here the dorsal aorta gives off many arteries which supply the malpighian bodies of these tubules. Behind the kidney the dorsal aorta is continued as the caudal artery.

Venous System.

The "lymphatic" tissue lying between the aorta and pronephric ducts receives a copious supply of blood, but this is wholly venous, and is derived from dorso-lumbar veins.

There is not much change to record of the veins of the head.

The venous system of the trunk and tail is represented in text-fig. 121, which represents the condition one day later (12 days) though essentially it is the same as at the present stage (11 days). The ductus Cuvieri are formed as usual by the junction of the anterior and posterior cardinal veins, but the right vein is small and brings the blood from the lymphatic tissue surrounding the more anterior part of the pronephric duct.

The left posterior cardinal vein is very large and conveys all the blood from the true kidney and the greater part of the lymphatic tissue band. This vein arises in the ventral wall of the kidney as a median vein gathering blood from left and right. As it runs forwards it bends to the left though still collecting blood from the right lymphatic mass. Near the anterior region it receives a large vein from the air bladder or lung.

About this point, also, the right cardinal may be said to take its origin.

I doubt if there is any renal portal system. If it exists, it is rudimentary. The caudal vein passes at the region of the anus to the right of the rectum and runs forwards as a sub-intestinal vein in the mesentery between the gut and yolk mass. Traces of yolk are still to be found all along this portion of the mesentery (text-fig. 110). The sub-intestinal vein is continued ventralwards over the posterior face of the yolk when it branches over the yolk and constitutes the sole yolk sac afferent vessel as heretofore, and until the end of larval life.

Twelfth Day. Series Q.

The vascular system is essentially as described for the previous stage. There is still only a rudimentary renal portal system.

The caudal vein as it passes ventralwards of the rectum skirts the hinder border of the kidney and one or more vessels connect it with the blood system of that organ. Nevertheless I cannot help thinking that the general arrangement of the vessels suggests a flow of blood from the kidney into the sub-intestinal rather than *vice versa*. There is not the faintest trace as yet of a bifurcation of the caudal vein.

The branchial efferent vessels behave as hitherto, namely, the first and

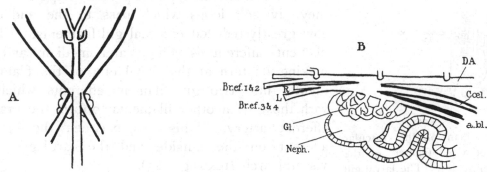

FIG. 124. Diagrams to show the gradual separation of the dorsal aorta from the pulmonary and coeliaco-mesenteric arteries on the twelfth day. A. Plan as seen from above. B. Plan as seen from the left side. *a.bl.*, air bladder or pulmonary artery; *Cœl.*, coeliaco-mesenteric artery; *DA.*, dorsal aorta; *Gl.*, glomus of pronephros; *Neph.*, opening of pronephric duct; *Br.ef.* 1, 2, 3, 4, branchial efferents 1—4.

second fuse, and the third and fourth fuse. There are however certain changes in relation to the dorsal aorta.

The thymus artery is given off as before from the ventral surface of the dorsal aorta; but the common third and fourth efferents unite before they become connected with the dorsal aorta, and, moreover, their junction with the aorta is now very slight. They lie ventral to the aorta (text-fig. 124).

Judging from their relation to the vessels of the glomerulus of the pronephros we may suppose that the ventral part of the aorta from the point of junction of the third and fourth branchial efferents has become constricted off and become part of the third and fourth branchial efferents, so that the afferent vessels of the glomerulus really now come off from the air bladder artery rather than the dorsal aorta (text-fig. 124).

A second constriction occurring still more dorsally nearly effects a separation of the coeliac artery. There is however a connection between all these vessels, which gets slighter as development proceeds, but is never quite lost.

Sixteenth Day. Series MK.

The branchial afferents arise from the heart in the following manner. The bulbus of the heart bears four swellings which form the roots of two main pairs of branchial afferent trunks. Close to the proximal end of the cardiac aorta and just beyond the bulbus, a pair of large vessels is given off. These vessels split almost at once into two—the hinder being the fourth branchial afferent, the anterior the common second and third branchial afferent vessels. This after a short course divides into third and second afferent.

The cardiac aorta continues forward for some distance and then bifurcates at its anterior extremity into the two first branchial afferents, the thyroid gland being ventral to the point of bifurcation.

As the afferents pass up their respective arches they give off loops which pass to the end of the now greatly reduced external gill filaments. The gill filament afferent is single and median, and after making its turn at the distal end of the filament it divides into two gill filament efferents which join with those from other filaments to form the branchial efferent artery. This lies between the branchial afferent on the outside and the cartilages of the visceral arch (text-fig. 125).

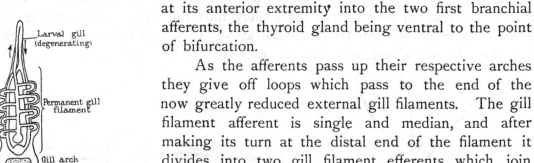

FIG. 125. Plan of the vascular supply of a gill filament on the seventeenth day. The larval gill has almost disappeared.

The branchial efferents 1 and 2 join, and the branchial efferents 3 and 4 join, as in the last stage.

From the dorsal ends of the first branchial efferents the common carotid vessels pass forwards. These divide into external and internal carotids having previously given a branch to the operculum. The internal carotid enters the skull in front of the auditory capsule where it turns downwards and then passes beneath the brain just posterior to the infundibulum and anastomoses with the internal carotid of the other side and gives an anterior and posterior branch supplying the anterior and posterior regions of the brain respectively. The external carotids pass forwards and supply the eyeball and muscles of the jaws.

A considerable change has occurred with reference to the relation the branchial efferents bear to the dorsal aorta. The united first and second branchial efferent of each side joins with its fellow of the other side to form the dorsal aorta.

The united third and fourth branchial efferent may be said to join with its fellow of the other side to form the arterial supply of the lung and alimentary canal, but still it retains connection with the dorsal aorta (text-fig. 126).

The intestine is supplied by a single vessel the coeliaco-mesenteric which arises mainly from the right common third and fourth branchial. There is still a possible communication between all the branchial efferents and all the main arterial vessels—but probably the flow of blood is as follows (text-fig. 126).

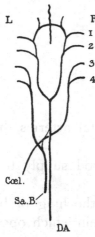

FIG. 126. The branchial efferents as seen from above. *Cœl.*, coeliaco-mesenteric artery; *DA*, dorsal aorta; *Sa.B.*, pulmonary artery; 1, 2, 3, 4, first to fourth branchial efferents.

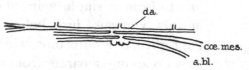

FIG. 127. Plan of the arterial system on the seventeenth day, as seen from the left side. *a.bl.*, air bladder or pulmonary artery; *cœ.mes.*, coeliaco-mesenteric artery; *da.*, dorsal aorta.

The dorsal aorta is probably supplied with blood from the

Right⎫
Left ⎬ first branchial efferents,

Right⎫
Left ⎬ second branchial efferents.

The coeliaco-mesenteric from the

Right⎫
Left ⎬ first branchial efferent,

A small
amount from

Right⎫
Left ⎬ second branchial efferent,

Right third branchial efferent,
Right fourth branchial efferent.

The blood supply of the lung is probably

A small amount from
$\begin{cases}\text{Right}\\\text{Left}\end{cases}$ first branchial efferent,
$\begin{cases}\text{Right}\\\text{Left}\end{cases}$ second branchial efferent.

A small amount from
$\begin{cases}\text{Right} & \text{third branchial efferent,}\\\text{Right} & \text{fourth branchial efferent.}\end{cases}$

The whole of
$\begin{cases}\text{Left} & \text{third branchial efferent,}\\\text{Left} & \text{fourth branchial efferent.}\end{cases}$

The subclavians come off from the dorsal aorta like parietal arteries, but run forwards and ventralwards.

The head kidney has almost disappeared so that the blood supply to it has well-nigh ceased.

The pulmonary artery runs along the mid-ventral line of the lung or bladder, and alongside it the pulmonary vein which opens into the left posterior cardinal.

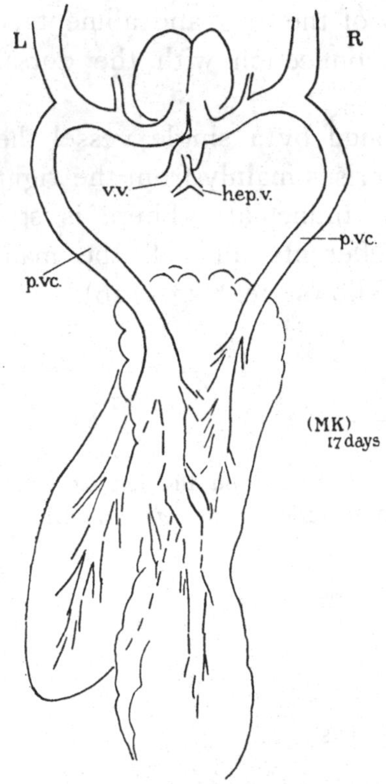

The right ductus Cuvieri is smaller than the left, but the difference is not so great as on the 10—12 days. From the arrangement of the veins it would seem that a large part of the blood from the kidneys still returns to the heart by the left ductus Cuvieri, as well as the greater part of the blood from the swimming bladder. The hepatics now join to form a single vein which unites with the large meatus venosus just before that vessel joins with the right Cuvierian sinus, which then enters the auricle by an opening separate from that of the left Cuvierian sinus. That is to say, whereas in the previously described stage, the twelfth day, the sinus venosus (text-fig. 121) received the two ductus Cuvieri and the meatus venosus and the hepatic veins and opened by a single aperture guarded by valves into the auricle, we now find the left ductus Cuvieri has acquired a separate opening into the auricle, and we can hardly speak of a sinus venosus as any longer existing. The changes are illustrated diagrammatically by the text-figures 121 and 128. As seen from the outside the dorsal chamber of the heart appears to be bilobed. Sections show that the condition which produces the bilobed appearance is the presence of strands of tissue which pass across the cavity of this chamber from the posterior to the anterior wall, from the dorsal to about the middle region of the cavity. These strands in

Fig. 128. Plan of venous trunks on the seventeenth day, as seen from above. *hep.v.*, hepatic vein; *p.vc.*, posterior cardinal vein; *v.v.*, vitelline vein.

some sections completely divide the chamber into two compartments (text-fig. 129 *a—d*). Ventrally, however, there is no trace of such a partition. Into this

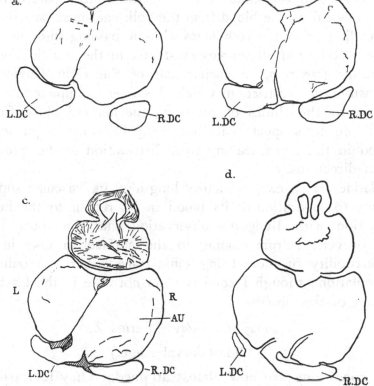

FIG. 129. Four sections taken horizontally through the dorsal region of the heart on the seventeenth day to show the rudimentary auricular septum which tends to separate the inflow of blood from the sinus venosus. *a*, is the most dorsal; *d*, the most ventral sections. *AU*, auricle; *L.DC.*, left ductus Cuvieri; *R.DC.*, right ductus Cuvieri.

chamber open the two ductus Cuvieri. The left opens anterior and dorsal to the right (Pl. XVIII. fig. 17). The left ductus Cuvieri opens dorsally into the left and ventrally into the right auricular chamber. It opens by a slit-like aperture obliquely placed with reference to the rudimentary partition of the auricle, so that blood passes into both sides though the greater length of the slit leads into the left chamber. Its opening is guarded by valves to the right and left, and it is from the tips of these valves, chiefly from the larger right valve, that the dividing strands above mentioned run to be attached to the anterior wall of the auriculo-ventricular opening. This ductus contains blood from the head, the kidneys, and air bladder; the amount from the kidneys is, however, diminishing. The right ductus opens as before stated more ventrally and posteriorly to the opening of the left.

This opening is also guarded by two valves, but they are smaller and no long strands pass from their edges. The right valve is the larger of the two;

at any rate dorsally.   The right ductus joins with the meatus and hepatics before its junction with the heart.   The source of the blood which enters the heart by this aperture is as follows :—The blood entering by the dorsal part of the opening has come from the head and the kidneys, that by the ventral part from the liver, and yolk sac.   A great deal of the blood from the yolk sac must pass direct through the wide channel leading from the yolk mass without passing into the liver capillaries.

Clearly the blood from all sources must mix in the auricle; but the presence of the rudimentary partition, the separation of the right Cuvierian sinus from the sinus venosus, the restriction which becomes complete later of the left posterior cardinal to the duties of returning the aerated blood from the lung, if such it is, lead one to suspect that there may be some slight separation of the streams of blood in the heart, leading to a distribution of the more aerated blood in advantageous directions.

The air bladder is so extraordinarily lung-like, its vascular supply so copious, and the tendency to separation of its blood in its return to the heart so evident, that, in conjunction with Budgett's observation that the young larvae of *Gymnarchus* die if prevented from coming to the surface to take in air, one must consider the possibility of there being some trace either premonitory or vestigial of a double circulation; though I confess I am not able in the least to suggest the actual mechanism of the process.

Forty-third day.   Series Z.

(Oldest larval stage.)

The adult conditions are now almost attained.   Only four pairs of branchial vessels persist.   The external gill filaments have become almost aborted except their proximal ends from which the little loops, visible as soon as the tenth day, arise between the afferent and efferent channels as described for the last stage.

The blood which comes by the third and fourth efferents is almost quite separated from that coming by the first and second branchial efferents; while the connection between the blood-stream of the right third and fourth common efferent is still less intimate with the stream from the common left third and fourth.   In fact this latter connection is of the slightest and only appears in two sections in a transverse series.

The first and second branchials unite to form the dorsal aorta.   The right common third and fourth efferent passes ventral to the dorsal aorta so as to come to lie dorsal to the left common third and fourth efferent.   As it passes under the aorta it makes connection with it and almost immediately afterwards as it passes dorsal to the left common third and fourth efferent it makes connection of the slightest possible character therewith.

Both these communications are very slight and do not give rise to distinct connecting vessels.   It seems probable that the normal flow of blood is as follows, though possibly under certain circumstances an additional supply of blood may still be obtained by one or other of the parts concerned by way of these connecting foramina.

Such blood from the first and second efferents of each side as does not go forwards to the head must pass hindwards by the dorsal aorta to the thymus subclavian, the parietals, and by the caudal artery and caudal vein to the yolk sac.

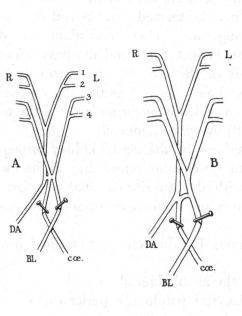

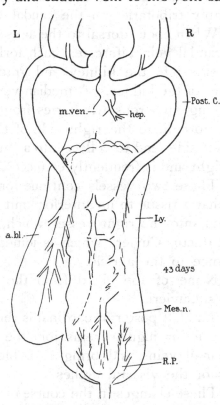

FIG. 130. A. Plan of main arteries in the forty-third day larvae. B. Plan of the main arteries drawn from Hyrtl's description of the adult animal (in both cases as seen from below). *BL.*, pulmonary artery; *DA.*, dorsal aorta; *cœ.*, coeliaco-mesenteric artery.

FIG. 131. Plan of the venous system at the oldest larval stage, forty-three days, as seen from above. *a. bl.*, air bladder and lung; *hep.*, hepatic vein; *Ly.*, lymphatic tisssue; *Mes. n.*, mesonephros; *m. ven.*, vein from yolk sac; *R. P.*, right renal portal vein; *Post. c.*, posterior cardinal vein.

The blood from the right third and fourth efferents passes chiefly to the alimentary canal.

The blood from the left third and fourth efferents passes almost entirely to the air-bladder (text-fig. 130, A).

Changes with reference to the return of blood from the kidneys and air bladder to the heart which had been initiated in the earlier stages are now completed.

The left posterior cardinal no longer receives any blood from the kidney or lymphatic tissues of that region—it conveys solely the blood from the air bladder which is a copious flow. It still remains slightly larger than the right posterior cardinal, which has, however, greatly increased in size as it now returns the whole of the blood from the lymphatic tissues and the permanent kidneys (text-fig. 131).

A renal portal system has become definitely established (*R.P.*).

The caudal vein on approaching the kidney gives off vessels, some into the mass of the hinder part of the kidney, others which do not pass into the kidney are probably tributaries to the caudal vein.

When about dorsal to the anus (the kidney projects backwards beyond the anus) the caudal vein bifurcates, each fork, which may be termed renal portal vein, giving off veins into the kidney and gradually dying out. These run along the dorsal surface of the kidney. A median ventral vessel collects the blood and passes forwards receiving numerous tributaries from each side of the kidney, and gradually passes over more on to the right side of the median line. As it does this the vessels from the left side tend to collect into a parallel vessel which is never so large as the one on the right and is frequently connected therewith by wide channels.

These two vessels continue forwards in front of the actual kidney through the lymphatic tissue to its anterior limit where the left channel ceases and all the blood is carried into the right vessel, which, joining with the anterior cardinal vein forms the right ductus Cuvieri, which is joined by the meatus and hepatic vessel at its point of entrance to the auricle.

None of the blood from the kidney goes by the left post-cardinal into the ductus Cuvieri.

The left posterior cardinal is the vein of the air bladder alone.

Inferior jugular veins join the ductus Cuvieri within the pericardium. These are small veins, and bring back blood from the mandible muscles, and the ventral walls of the visceral arches.

These changes in the course of the arterial and venous systems are shown in the series of diagrams, text-figs. 123, 124, 126, 127, 130, and 121, 128 and 131. Hyrtl (71) says that in the adult fish the whole of the blood of both sides from the third and fourth branchial efferents passes to the air bladder. Sedgwick (109, p. 208) says, "the efferent vessels of the third and fourth branchial arches do not join the dorsal aorta but pass to the air bladder." Hyrtl, however, states that there is an anastomosis between the coeliaco-mesenteric and the air bladder vessel and shows it in a figure of an injection which he made. In this the coeliaco-mesenteric arises from the dorsal aorta and is connected by a short branch with the air bladder artery which is formed by the joined third and fourth efferent branchial arteries of both sides. According to my account of the larva of 43rd day, the coeliaco-mesenteric is a continuation of the joined third and fourth branchial efferents of the right side, and the air bladder is a continuation of the joined third and fourth branchial efferents of the left side. Text-fig. 130 (which should be compared with Hyrtl's fig. 4, Taf. IV.) will show how Hyrtl's condition can be arrived at from mine, by an increase again in the size of the foramen between the coeliaco-mesenteric and the air bladder arteries, and a shifting backwards of the communication between the aorta and coeliaco-mesenteric. Hyrtl accounted for the presence of the communicating loop between the bladder and coeliac arteries by supposing the former organ to vary greatly in its distensibility, so that when widely distended it would receive a large blood supply, when collapsed the blood would pass by the loop into the coeliac system.

There is now no vessel passing from the caudal vein to the yolk mass, nor any one midway as in earlier stages.

The yolk sac, which is within the normal contour of the fish, has lost its embryonic circulation.

There is still a large vitelline trunk which passes on through the liver into the hepatic vein. There are no longer separate openings of the vitelline vein and hepatic vein into the "sinus venosus." They join within the liver tissue.

The surface of the yolk mass is covered still by a network of blood vessels, which has no direct arterial supply—in fact the only supply may be from certain small vessels which pass between it and the liver near where the large vitelline vein joins the liver. In short, the yolk mass has become, as regards its circulation, a part of the liver—and it is to the liver alone that it is in any way attached, and not improbably it may be regarded as having been derived from the liver.

The pericardium is quite shut off from the rest of the coelom. The ventral mesentery has disappeared. The sub-intestinal vein occurs only in the more posterior region.

The Heart.

The heart of *Gymnarchus* is so peculiar that a short account of its development may be of interest. Its chief features are the presence of the large diverticula on the bulbus—a feature common and peculiar to the *Mormyridae* (Hyrtl) and the partial absorption of the sinus venosus into the auricle.

Förg's figures of the heart are poor and his description inaccurate. For instance, in the text he writes of six branchial afferents, and in his figures shows seven in one and eight in the other!

Hyrtl gives figures of the ventral and dorsal view of the heart of the adult fish, which do not show the separate opening of the left ductus Cuvieri as described in the development of the vessels.

The heart on the fifth day is a straight tube with endothelial lining and almost uniformly thick walls, and diminishing slightly in diameter towards the two ends.

At seven days it has become marked by a transverse constriction into ventricle and auricle; and perhaps, as seen in the ventral view (Pl. XVIII. figs. 22, 23), or lateral view, a sinus venosus may be detected. There is also a slight bend to the right. There is no bulbus. At nine days a bulbus is formed and the bend to the side is more marked, and the auricle which from the first tended to be dorsal to the ventricle, has been pushed forward, dragging with it the sinus venosus which opens in its dorsal wall (Pl. XVIII. fig. 20). In front of the ventricle the bulbus is marked off by a sharp bend (Pl. XVIII. figs. 20, 21), and is of an entirely different character to the rest of the heart. Instead of breaking up into cavernous tissue as the ventricle does, the walls of the cardiac aorta remain thick and compact.

Two vertical valves of triangular form guard the opening from the ventricle to the aorta (text-fig. 133). These valves, like those guarding the auriculo-ventricular opening, arise as local thickenings of the sub-endothelial tissue. There appears to be a pair of backwardly projecting pouches, as it were, at the sides of the aortic valves which are the beginnings of the peculiar diverticula so conspicuous later in the "bulbus" of the heart (text-fig. 132).

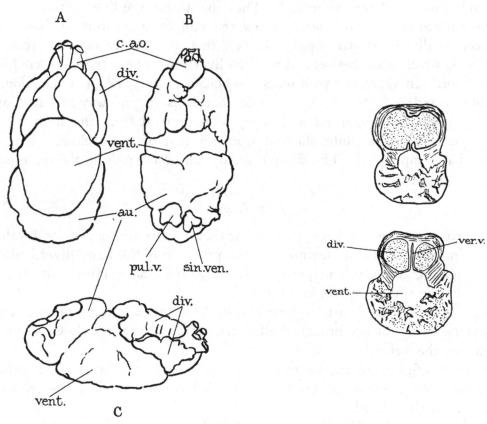

FIG. 132. Heart of adult Gymnarchus. A, ventral view. B, dorsal view. C, lateral view. *au.*, auricle; *c.ao.*, cardiac aorta; *div.*, diverticula on the cardiac aorta; *pul.v.*, "pulmonary vein" or left ductus Cuvieri; *sin.ven.*, sinus venosus without left ductus Cuvieri; *vent.*, ventricle.

FIG. 133. Two sections taken horizontally through the heart at nine days. *div.*, diverticula on the cardiac aorta; *vent.*, cavity of the ventricle; *ver.v.*, two vertical valves.

The character of the changes in the walls of the heart can be followed by an examination of the text-figure 134, *a—d*. On the fifth day the endothelial lining appears to be quite loose. On the sixth day this endothelial lining is attached at certain points to the thickened walls of the heart; where the lining is attached there are little heaps of endothelial cells (text-fig. 134, A, *end.*). On the seventh day these groups are seen to be insinuating themselves in the thick wall of the ventricular portion (text-fig. 134, B, *end.*), and becoming hollowed out gradually convert the thick solid wall into the cavernous and spongy condition typical of the ventricle of the lower Vertebrata, and leaving a fairly regular innermost space, which communicates with the spongy tissue by many openings.

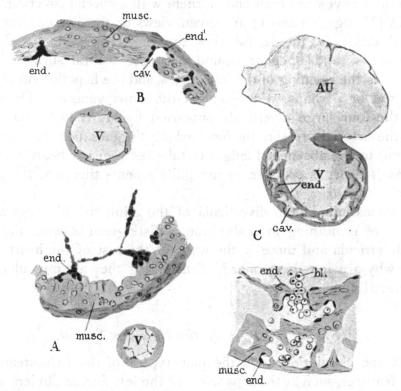

FIG. 134. Three stages in the development of the heart to show the part taken by the endothelial lining in the formation of the spongy tissue of the wall of the ventricle. A, the small figure *V* is a transverse section of the ventricle on the sixth day. The larger figure a small portion of its wall under a higher power. B, the small figure *V* is the ventricle of the seventh day, the upper a portion thereof under a higher power. C, corresponding figure of the eighth day. *AU.*, auricle; *bl.*, blood corpuscles; *cav.*, cavity in wall of ventricle giving rise to "sponginess"; *end.*, endothelial cells; *end'.* in B, shows how a group of endothelial cells has invaded the muscle layer and created a cavity; *musc.*, muscle coat of ventricle; *V.*, ventricle.

This is not the usual way according to descriptions given by other authors. Thus Marshall (90), for the frog, says: "the cavity of the ventricle is much subdivided by muscular trabeculae, which growing inwards from its walls, branch and unite" etc., or for the chick: "the thickening of the ventricular wall…is affected by inwardly projecting ridges of the muscular wall," or for rabbit, p. 415: "just as in frog, by the ingrowth of muscular trabeculae into the cavity." Hertwig (65) also ascribes to the "muscular sac" the initiation in the formation of the spongy tissues, though according to him the "endothelial tube" does take an active part.— Thus (p. 556), "In the ventricle, on the contrary, there occurs a loosening, as it were, of the muscular wall.   There are formed numerous small trabeculae of muscular cells, which project into the previously mentioned space between the two sacs and become united to one another, forming a large meshed network.   The endothelial tube of the heart, by forming out-pocketings soon comes into intimate contact with the trabeculae and envelopes each one of them with a special covering."

In Pl. XVIII. figs. 16 and 17 are given views of the heart in the oldest stage, the forty-third day.   In the lateral view the right baggy ductus Cuvieri is seen passing downwards, and opening behind near the ventral surface.   Above it in the latter figure is the opening of the left ductus, and the hepatic vein is seen entering the right ductus, or perhaps we may call it the sinus venosus.   On the bulbus are seen two of the four large diverticula, supposed by Hyrtl to be the organs for the reduction of the arterial pressure (or for masking the pulsations ?), and are said to be necessary owing to the absence of length of tube between the heart and gill filaments. As *Gymnarchus's* heart is so close to the gill filaments therefore these bulbs are so well developed.

I have not examined the diverticula of the adult fish histologically, but in the larval forms there is nothing to enable one to distinguish between the tissues of the walls of the diverticula and those of the walls of the rest of the heart.   If the latter are muscular why are not the former ?   But even if they are muscular it is not clear what their special function is.

Special Features of the Vascular System.

The yolk sac circulation is on the plan typical of the Teleosteans.   There are many special features, such as the separation of the left ductus Cuvieri, which receives the pulmonary vein, from the sinus venosus and its independent opening into the auricle, which at one time shows signs of a division into right and left sides.

The whole of the kidney blood is carried by the right cardinal sinus, the whole of the lung blood by the left cardinal sinus ; but this is not so at first.

The heart has curious diverticula on the "bulbus"; but as these diverticula appear to be muscular, it may be better to call the proximal part of the cardiac aorta, which by the bye is within the pericardium, a conus arteriosus and not bulbus.   The spongy nature of the ventricle is produced by an ingrowth of endothelial cells and subsequent hollowing out of the muscular wall—not by growth of the walls into the cavity.

There are several features in connection with the development of the vascular system which suggest a more amphibian-like ancestry than any existing fishes.   For instance, the lung-like character of the air bladder ; the large size of the aortic arches of the second branchial arch ; the tendency to a separation of blood from and to the air bladder, and the slight vestige of a separation into right and left auricles.

3. The Excretory Organs.

The first trace I can find of the excretory organs is in the imperfect $4\frac{1}{2}$ day specimen.

The difficulty of interpreting this stage owing to the great compression and distortion of the embryo has been referred to already and adds to the misfortune of lost sections; so that I can give but a most fragmentary account of the early stages of the development of this set of organs.

The pronephric ducts are present as straight tubes with small lumina lying along the sides of the embryo just ventral to the outer edge of the mesoblastic somite (text-fig. 98, *pn.d.*).

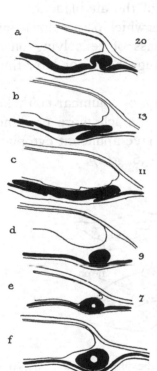

FIG. 135. Transverse sections through the developing pronephros of a four-and-a-half day embryo. The thick black line is the somatopleuric layer of peritoneum. The pronephric duct is seen in the sections *d, e, f.* The numbers 20, 13, etc., indicate the position of each section in the series, counting from behind forwards.

On tracing the duct forwards, it diminishes very much in diameter and the lumen disappears, but the duct runs on as a solid rod which soon becomes continuous with an area of thickened coelomic epithelium of the somatopleure. Within this area of thickened epithelium there are certainly two, perhaps three, folds or peritoneal depressions, of which the most anterior is the deepest and best defined, while the most posterior is shallow but much wider.

The grooves are separated from each other, but all are within the area of thickened coelomic epithelium.

Text-fig. 135 represents certain sections taken transversely through this area. Section (*a*), the most anterior of the series, passes through the most anterior funnel, (*b*) which is the third section succeeding this, passes through the middle funnel, (*c*) two sections further back, through the posterior funnel. Sections (*d*), (*e*) and (*f*) are the 6th, 8th and 16th sections respectively.

At five days the pronephros is firmly established and resembles in all essentials the condition of the pronephros in the Trout after complete formation of the pronephric chamber, as described so thoroughly by Felix (41, 42), though perhaps not yet at its stage of greatest development.

It lies just posterior to the junction of the two roots of the dorsal aorta, as close as possible to the ventral wall of that vessel.

It consists of a pair of sacs, and each sac communicates by an elongated slit-like aperture with the pronephric duct of its side.

The sac is lined by a layer of cells which are best described as spheroidal, which gradually merge into the columnar or cubical cells of the pronephric duct.

The two sacs are quite separated from each other. The walls are invaginated dorsally by small blood channels derived from the dorsal aorta. These run longitudinally and open to the aorta again near the posterior end of the chamber.

About the middle of the length of the pronephros and between the two sacs a vessel comes off from the aorta which passes down on to the dorsal wall of the oesophagus; it divides and runs forwards and backwards, the anterior branch being the longer one. It does not, I think, receive any blood from the glomerulus. There is some indication of a second vessel having the same origin and destination. The former almost without doubt can be identified as the artery of the air bladder.

Posterior to the glomerulus another median vessel arises which passes downwards on to the mesentery and can be traced as far as the mass of cells lying on the mesentery dorsal to the gut which forms the "islands of Langerhans" of later times. This vessel is the coeliaco-mesenteric artery.

Running along the mid-dorsal line of the aorta is a ridge of columnar epithelium which can be traced a considerable distance, but which is much more prominent opposite the glomerulus. The cells composing this ridge have abundant cytoplasm, which stains differently, and have small nuclei (text-fig. 136, *gl.ep.*).

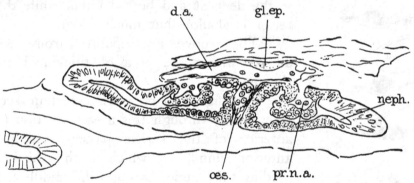

FIG. 136. Transverse section through the pronephric chambers on the fifth day. *d.a.*, dorsal aorta; *gl.ep.*, tract of glandular epithelium with dorsal wall of the aorta; *neph.*, nephrostome; *œs.*, oesophageal (subsequently pulmonary) artery; *pr.n.a.*, primary pronephric artery.

For some distance the pronephric duct runs quite straight, but further back it takes a more sinuous course. There are as yet no diverticula from it, nor is there any sign of bud formation. It is apparently undergoing rapid growth (for mitotic figures are numerous), which, no doubt, gives rise to the sinuosities just mentioned.

Beneath the ducts is a large blood sinus, the cardinal sinus, which communicates with the sub-intestinal vein at one point.

This is the earliest stage at which I can recognise the so-called lymphatic tissue which becomes so prominent a feature later ("Das Pseudolymphoides Gewebe" of Felix). In between the dorsal aorta and posterior cardinal vein lying loosely in the meshes of the mesoblast tissue there are many ovoid cells with deeply staining cytoplasm and small very deeply staining nuclei.   These are the cells which by their multiplication and differentiation give rise to the mass of lymphatic tissue which soon surrounds the pronephric ducts and the posterior cardinal vein (text-fig. 137).

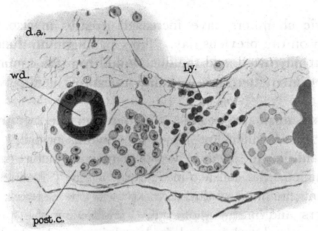

FIG. 137.   Transverse section of seventh day larva behind the pronephros.   *d.a.*, dorsal aorta ; *Ly.*, lymphogenous cell ; *post.c.*, posterior cardinal vein ; *wd.*, pronephric duct.

In a specimen six hours older the two pronephric ducts are seen to join at the posterior end, the one crossing over and opening into a swelling on the other.   This swelling is in contact with a dilatation of the hind gut, and, I think, at one point an actual perforation occurs.

Sixth Day.

On the sixth day the head kidney is at its greatest development.   The character of the cells remains the same, the two longitudinal glomerulus vessels are much as they were before, but more vessels leave the aorta and form a network of blood vessels all round the walls of the pronephros, but do not project into it as the two original glomerulus vessels do.

Most of these collect into the oesophageal vessel (which is single), but some fall into the coeliaco-mesenteric, now a large vessel coming off between the two sacs of the pronephros at their extreme posterior ends.

The coeliaco-mesenteric passes into the mesentery, splits up into capillaries in the "islands of Langerhans," and again collects and passes on as the mesenteric artery.

The pronephric ducts are much convoluted, and the tissue surrounding them and between them is extremely vascular ; in fact the ducts are practically bathed in blood. There is a separation now of the posterior cardinal vein into two sinuses.   The pseudo-lymphatic tissue cells have increased, but are much concentrated as yet just beneath the aorta.

There are no diverticula, or buddings, from the pronephric duct.   The pronephros is still the sole kidney.

Seventh Day.

On the seventh day we find the first indications of the tubules of the permanent kidney.

The pronephric chambers have increased greatly in size, but show no more vascularization than on the previous day. The openings into them of the pronephric ducts are somewhat narrowed, and the ducts take from the commencement a sinuous course. In one specimen there are certainly apertures in the partition between the two chambers.

From the dorsal aorta two vessels pass between the two pronephric chambers, the oesophageal arteries, which are joined by several probably efferent pronephric vessels. A third and larger median vessel comes as before at the posterior end of the chamber which is the coeliaco-mesenteric artery. Besides these, many small vessels pass off from the floor of the aorta to form the network of vessels over the pronephric chambers and in the glomeruli. Following the pronephros is a considerable length of convoluted pronephric duct lying amidst the lymphatic tissue and blood channels of the posterior cardinal vein.

The dorsal aorta gives off no vessels towards the excretory organs until after about three-fifths of the whole length of the pronephric duct, when branches of the parietals begin to pass into the lymphogenous tissue and run towards the coils of the duct.

At this point little proliferations of the walls of the duct occur, quite irregularly, and on all sides (Pl. XXI. fig. 55, *mes. bd.*). Then follows a length of convoluted tube which slightly expanding continues with lessening convolutions to the posterior end. The two ducts unite in a vesicle and open to the exterior, posterior to the gut.

Along the posterior parts of the pronephric ducts, separated by some little distance from the buds alluded to above, are little groups of cells which occur in the mesodermal tissue alongside of but separated from the ducts. These little groups of cells give rise ultimately to the primary tubules of the permanent kidney.

There are thus three distinct and partially separated regions of excretory organs; and the system agrees on the whole, with the account given by Felix for the development of the excretory system in the Trout (Felix (41)).

The early appearance and prominence of the oesophageal arteries, one of which becomes the pulmonary artery, is probably due to the greater importance which the swimming bladder has in the *Gymnarchus* larva.

The absence of arteries from the aorta into the tissues in the neighbourhood of the pronephric ducts along the anterior half is no doubt correlated with absence of tubules in this region.

The Pronephros.

From the seventh day onwards the changes in the pronephros are regressive. It alters its position slightly with reference to the aorta, owing to a contraction of the lower part of that vessel whereby the third and fourth branchial efferents are separated from the dorsal aorta: and with them pass all connections with the glomerulus. (See Section on the blood vessels, pp. 332—7.)

The oesophageal vessels by combination or by development of one of them form a single vessel (the future pulmonary) which withdraws to the posterior border of the pronephric chamber.

As the degeneration of the pronephros proceeds, the single pronephric efferent vessel of each side diminishes, and the glomerulus circulation consists of a number of small apertures in the wall of the pulmonary artery, some of which are no doubt afferent, others efferent.

The regressive changes consist in the decrease in the size of the pronephric chamber, in the extent of the glomus and its concentration to the inner wall of each chamber. This is accompanied by the expansion of the mouth of the pronephric duct (pseudonephrostome of Felix). The changes are illustrated by the accompanying diagram, which represents transverse sections through the middle of the pronephric chamber. The last is really a pair of Malpighian bodies.

In the oldest stage obtained (43 days) the pronephric chambers, although still recognisable, have been reduced to very insignificant proportions.

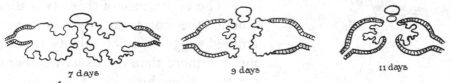

7 days   9 days   11 days

FIG. 138. A series of diagrams to illustrate the degeneration of the pronephros; the conversion of the pronephric chamber into a structure very similar to a Malpighian body.

Accessory "Nephrostomes."

In two specimens I have found what must be regarded as accessory head kidney tubules. In one case, in a specimen of the eighth day, a single diverticulum of the pronephric duct occurs some little distance (25 sections) posterior to the normal pronephros which is in the form of a short straight tubule, whose distal blind end is invaginated by a distinct glomerulus. I cannot determine whether the blood supply is arterial or venous (text-fig. 139).

In another case in a rather older specimen—the ninth day—an accessory pronephric chamber on one side occurs immediately anterior to the normal chamber, which it resembles in general character, but it is smaller, and is continued into a very short blind pocket which is directed forwards and is not in any way connected with the pronephric duct (text-fig. 140).

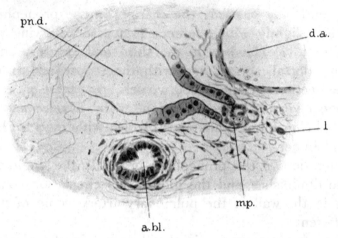

FIG. 139. Transverse section through an accessory tubule with malpighian body. *a.bl.*, air bladder; *d.a.*, dorsal aorta; *l.*, lymphogenous cell; *mp.*, Malpighian body; *pn.d.*, pronephric duct.

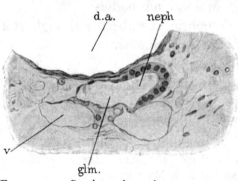

FIG. 140. Section through an accessory nephrostome in the pronephric chamber. *d.a.*, dorsal aorta; *glm.*, glomus continuous with main chamber; *neph.*, nephrostome; *v.*, blood vessel.

The latter instance may possibly have arisen by a later division of a previously single pronephric chamber, or by failure of at first separate funnels to coalesce. In the first case, however, the accessory tubule is more isolated.

The occurrence of these two abnormalities among the comparatively few specimens examined suggests a potentiality for the development of more than the usual number of tubules for the pronephros, or at any rate a not very firmly fixed condition as regards the pronephros, and perhaps indicates a more extended area for this organ in earlier times.

Changes are so rapid in *Gymnarchus* that in the absence of specimens between $4\frac{1}{2}$ and five days I am not able to trace the mode by which the open condition of the former becomes converted into the typical Teleostean pronephros of the later date.

The Pronephric Duct.

The duct which very early becomes convoluted increases its convolutions immediately posterior to the pronephros and remains here fairly conspicuous up to the latest stage, but further back it straightens somewhat—one is inclined to say it becomes stretched out—and undergoes degeneration. For some distance, greater on the right than on the left side—it actually disappears. It reappears further back as a convoluted tube.

The little buds already mentioned are found about the 35th somite and where they occur the duct takes a straighter course; the duct then becomes again convoluted until the region of the mesonephric tubules, when it straightens to the end.

These characters are on the whole retained in the latest stage, but the part in the region of the mesonephros dilates by the 43rd day, and receives the tubules of that part of the kidney at all points.

The two ducts on the fifth day join one another, the right crossing over to open into a dilated terminal part of the left, which opens into a dilated terminal part of the alimentary canal.

Subsequently a separation occurs between the terminal part of the alimentary canal and the ducts, so that the ducts open together to the exterior, posterior to the alimentary canal which ends blindly.

The anus does not open until the twelfth day. On the ninth day the two pronephric ducts after joining turn ventralwards and the single ureter opens to the exterior, posterior to and not attached to the blind end of the gut.

On the tenth day a blind diverticulum has grown forward from the ureter just behind the point of union of the two pronephric ducts. This is the urinary bladder.

On the eleventh day the pronephric ducts are continued back beyond the anus giving rise to the post-anal part of the kidney (text-fig. 141).

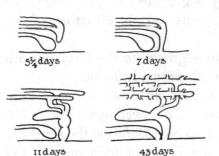

5¼ days   7 days   11 days   43 days

FIG. 141. Diagrams illustrating the changes that occur in the terminations of the pronephric ducts and alimentary canal between the fifth and 43rd days.

In a specimen of the twelfth day the bladder is double, one pronephric duct opening into each. The split is continued down almost to the external opening.

On the forty-third day the same relations obtain, but the bladder is larger and is not double (text-fig. 141).

The length of duct upon which the "buds" above described are to be found is quite short, and does not exceed three or four segments.

These buds are distinctly proliferations of the walls of the pronephric duct, and arise quite irregularly at any point. At first no trace of a cavity can be seen in the buds, but as they grow little pittings into their substances may be detected (Pl. XXI. fig. 56).

These little pittings, however, are really—as the subsequent stage shows—diverticula of the pronephric duct and not excavations into the solid tissue of the bud. There is no doubt that the buds are proliferations of the walls of the duct, see Pl. XXI. fig. 55; but they remain solid, the diverticula being always lined throughout by nuclei which are directly continuous with the lining layer of the duct (Pl. XXI. fig. 56).

These diverticula attain their maximum development on the 11—12 days, after which they diminish and eventually (about the eighteenth day) they have disappeared; but the buds have grown very considerably, and have become quite detached from the duct.

On the oldest day (forty-third) they are in very much the same condition, lying at this time in the thirty-second, thirty-fourth and thirty-fifth segments. At this time they are oval masses of cells showing no hint of a lumen, but some trace of a connective tissue capsule, and of the occurrence of connective tissue strands within them, is present.

The Permanent Kidney.

The permanent kidney arises, as we have seen, from

(1) the posterior ends of the pronephric ducts whose sole duty is the conveyance away of the kidney secretion,

(2) little groups of cells lying in the tissue surrounding the duct, but not in any way at first connected with the duct, which give rise to the kidney tubules and Malpighian bodies.

The first undeniable sign of these kidney tubules occurs upon the seventh day. Possibly little groups of cells rather doubtfully segregated may be identified on the previous day as those which give rise to the more certain structure of the seventh day.

On this, the seventh day, from about the fortieth segment to the forty-seventh a chain of cell groups occurs along the dorsal side of each pronephric duct.

These cell groups are in most cases connected with each other so as to form an almost unbroken though moniliform strand of tissue. The swellings occur rather more frequently than one to each myotome, but taking into consideration the sinuosity of the duct, perhaps one may say that the groups are segmentally arranged and correspond with the segmentation of the muscle tissue of this region. At the thickest part of each swelling the strand is in contact with the dorsal wall of the pronephric duct, but is never fused with it. Elsewhere it is generally separated by blood channels. It is easily distinguishable from the duct by its different colour and also by a slightly different character of the nuclei, which are elongated and on the average are smaller than those of the duct epithelium.

They are clearly in continuity with the general connective tissue of the embryo. They are not derived from the lymphogenous tissue, but from the general connective tissue.

Subsequently the strands break up and each swelling becomes a separate elongated mass of cells in the midst of which a lumen appears and converts the structure into a vesicle, one end of which remains in close contact with the wall of the pronephric duct, the other remains free.

At this stage the difference between the two structures is very marked; the duct has cubical cells with spherical nuclei and stains a pink hue with eosin and haematoxylin, while the vesicle has columnar cells with elongated nuclei and stains a deep blue.

On the ninth day each vesicle has elongated and its distal end has become dilated, and the cells of the distal end of the elongated vesicle which has now come into contact with arterial blood from the dorsal aorta have undergone a peculiar degeneration, losing the greater part of their cytoplasm, with reduction in size of the nucleus. It has very much the appearance of the lining of the pronephric glomus at the time of its greatest development. The arterial vessel then pushes in little tufts and by the eleventh day a typical Malpighian body is produced. Some tubules contain *débris* probably derived from the degenerated cells of the Malpighian capsule.

On the ninth day vesicles have formed at other points apart from the original dorsal strands.

The final form of the kidney is attained by an increase in length and greater convolution of each tubule, by the origin of new tubules on all sides of the pronephric duct and on the tubules already formed.

Until the Malpighian bodies are formed the tubules do not open into the pronephric duct. Each tubule wedges its way into the duct through its wall and then opens. This occurs in the more advanced tubules upon the eleventh day.

In the latest stage, forty-third day, the multiplication of tubules is still in progress.

Pseudolymphoidal Tissue.

This tissue, so characteristic of the Teleostean kidney, is recognisable from a very early date. On the fifth day, when the pronephros has just been formed, the tissue which lies between the two ducts and just ventral to the aorta is composed of a reticulum of mesoblast cells, and many blood channels; and, lying in the meshes of the mesoblast reticulum, a large quantity of cells occurs which are, on the whole, spherical, with cytoplasm and a nucleus that stain deeply. These are the lymphogenous cells (text-fig. 137, *Ly.*).

No change occurs in them until the mesonephros begins to form. They then multiply rapidly and instead of consisting of a few almost isolated units form a more and more compact tissue of large deeply staining cells, broken by innumerable venous channels, and crowding out the ordinary connective tissue cells.

This tissue extends from just behind the pronephros to the end of the body cavity, namely, along the whole length of the pronephric ducts, which soon become embedded in it.

It clearly forms an organ of very considerable physiological importance, from its great size and rich venous blood supply, which is chiefly from the dorso-lumbar veins, but which is in connection with caudal, sub-intestinal and posterior cardinal vessels in the posterior region.   It receives little if any arterial blood.

This tissue is always thicker at its extreme anterior end than elsewhere, but is an important mass throughout.

As the kidney tubules develop and grow into it they break it up so that in the fully formed permanent kidney only a little remains compared with the mass of the kidney tubules.   Anteriorly it is even more hollowed out by blood sinuses.

The histological changes undergone by this tissue from this time are chiefly the disappearance of the large cells with large nuclei and appearance of smaller cells with small very deeply staining nucleus, implying, probably, a conversion of the one into the other.

Comparison.

From the above description it is evident that:

(i) The archinephric duct or pronephric duct is the sole duct of the whole excretory organ, and that the lower end remains as the ureter of the adult.

(ii)   Three regions of tubules are to be distinguished:

(*a*)   The pronephros.

(*b*)   The middle short series of tubules which arises considerably later than the pronephros and which arises as diverticula from the pronephric duct in connection with proliferations of the walls of the duct.   These diverticula are transitory but the proliferations are retained.

(*c*)   A hind kidney which arises at the same time as the middle piece and is formed from the mesoblastic tissue surrounding the duct and not from the duct itself. This is the most important part as it forms the whole of the permanent kidney.

(iii)   A mass of tissue not in any way connected with the kidney though closely contiguous to it forms from certain mesoblastic cells and gives rise to an organ which resembles ductless glands in that it consists of masses of large epithelial-like cells and is honeycombed by blood-vessels.   Possibly on the other hand it may function as an excretory organ in connection with the long convoluted pronephric ducts which it surrounds between the time of the degeneration of the pronephros and development of the hinder part of the kidney system, during which period it is at its greatest development.

At first one is inclined to call the three parts (*a*, *b*, *c*, paragraph ii) pro-meso- and metanephros, but there are certain objections (as Felix has pointed out).   The permanent kidney here develops after the manner of a mesonephros of the Amniote (and Amphibia); that is to say the tubules are developed apart from the archinephric duct, and only make connection later.   So far as I have been able to observe there is no trace of a posteriorly placed series of tubules developed from the archinephric duct itself as in the tubules of the permanent kidney or metanephros of the Amniotes. Therefore on these grounds we must assume the permanent kidney of the Teleostean to be a mesonephros.   Then what name are we to give to the short length in which tubules, transitory though they be, are derived directly from the archinephric duct occurring anteriorly to the mesonephros?

My account of excretory organs differs from that of Felix for the Trout in many details, but in no important point.

I am unable, by lack of intermediate stage, to give so full an account of the development of the pronephros, but undoubtedly there are indications of a rather different condition—a condition which perhaps more closely resembles that of the Amphibians. In the $4\frac{1}{2}$ days specimen there are on each side signs of canals—or nephrostomes—which open into the general body cavity, but end blindly in a mass or cord continuous with the developing pronephric duct which contains a lumen. Of these there are on the right side (which is less injured than the left) three distinct and separate folds of the somatopleure, of which the hindermost one is the widest (text-figs. 98 and 135). Felix says the head kidney is derived from the inner angle of the lateral plate of mesoderm and involves both splanchnopleure and somatopleure.

Although the stage I describe cannot be said to be the earliest stage in the process, yet it certainly does not suggest the implication of more than the somatopleure. There is also a little vesicle closely attached to the somatopleure alongside the pronephric duct posterior to the third nephrostome which may represent a fourth opening.

There is nothing in this specimen that I can identify with the glomerulus, and as this organ is fully formed in the next specimen (five days) and as the pronephric chamber is entirely cut off from the general body cavity, I am unable to give any account of its origin.

Nevertheless it is interesting to find in *Gymnarchus* that a condition exactly like that of the Trout (when the glomerulus is formed and the pronephric chamber is quite cut off from the body cavity) is attained from a condition which is unlike the early condition of the Trout, and is more like that of an Amphibian or Ganoid.

The relation of the vessels connecting the pronephric glomerulus with the aorta is described on page 336 under the section dealing with the vascular system, where it will be noticed that the arrangement differs from the Trout, the chief point of difference being that, after a while, instead of there being only one pair of glomerulus afferents leaving the aorta, there are many channels. The mode of connection of the glomerulus to the lung artery is different in *Gymnarchus*.

Another difference of much greater interest is with reference to the curious buds which grow upon the primary pronephric duct, which Felix described as "Urniere" canals in distinction to the "Nachniere" canals. He believed that these buds became canals with Malpighian bodies like the true canals, but in his later work (42) he recognizes that this is not so. According to his later description these buds which in the Trout arise mingled with the true kidney tubes about the middle third of the primary pronephric duct, become split off and enlarge into spherical and then into oval bodies and later, that is to say some months after hatching, develop into cords.

These cords at the time of maturity resemble a mass of glandular tissue, but never develop lumina or Malpighian bodies, nor do they ever become attached to the primary ureter.

Through lack of material I have not been able to follow their later history in *Gymnarchus*, but their earlier history is rather different.

As in the Trout they arise as proliferations of the walls of the primary ducts, but they arise in *Gymnarchus* quite in front of and quite separate from the true kidney tubules.   They occur about the 37—42 somite, whereas the kidney tubes arise from the 43—56th (in another series, Z, from 32—34, while the kidney lies from 43—53. This latter is a later series so that possibly after separation the buds tend to travel forwards).

Again, they are not strictly segmentally arranged and they may arise from any part of the wall, not from the dorsal surface only.

A more important feature, however, is the fact that they are accompanied by very distinct diverticula from the pronephric duct (Pl. XXI. fig. 56).   These diverticula, however, never attain a great length and never open into any lumen within the bud itself.

They are at their greatest development about the tenth and eleventh days and on the twelfth day have already begun to diminish.

The fact of their presence, although for a short time only, is of great interest, though I am at a loss to make any suggestion as to their import, except such as can be described as hypothetical in the highest degree.

Swaen and Brachet suggest that the body produced by these buds corresponds to the supra-renal bodies of Selachians (Felix (42), p. 221).

A suggestion which is perhaps just worth making, is that the buds represent the part of the mesonephros which in the ancestral form was modified in connection with the conveyance of the sperms as in Elasmobranchs, Amphibians and Ganoids, namely, as vasa efferentia and tubuliferous tissue.

May not the absence of nephrostomes in the Teleostean and the absence of any connection between the renal and reproductive system have been correlated with an enormous development and modification of a lung into a swimming bladder in such a manner as to cut off entirely the renal gland from the peritoneum, and therefore from the body cavity ?

The present absence of air bladder in some families of Teleostean is, of course, to be regarded as secondary.

In Ganoids, in Dipnoi, in Polypterus the air bladder, as in the Amphibians, hangs freely in the body cavity.   May not this be the more primitive condition, sufficient for the function of the air bladder as a breathing organ?   But in such a loose condition the bladder would be in a less advantageous position as a hydrostatic organ than it would be in the fixed position it now normally occupies in the Teleosteans, firmly attached to the most rigid part of the body.   As a fixed organ it would at least save extra muscular and nervous effort necessary to correct the effect of any temporary displacement of a displaceable hydrostatic apparatus.

In the evolution of Teleosteans from Ganoid or even more Amphibian-like ancestors the fixation of the air bladder to the dorsal wall of the coelom may be supposed to have been accompanied by the complete separation of the sexual from the urinary organs.

If, therefore, the Teleosteans are descended from amphibian-like ancestors, in which the "tubuliferous" tissue may have been well developed, it is not extraordinary to find traces of it in the development. May not the curious diverticula and the terminal buds owe their origin to the enlarged tubules of the mesonephros which in Amphibians and Elasmobranchs grow out into the substances of the gonad and give rise to the vasa efferentia in the case of the male, and in Amphibians to great sacs in the case of the female?

When by a development of the lung into a hydrostatic organ and the concomitant necessity of a rigid connection with the main mass of the body, it grew backwards and became firmly attached to the dorsal wall of the body cavity, the old arrangement of conducting away the sexual cells ceased and the new simple ducts were formed.

I cannot say I have any great confidence in the suggestion because, so far as I know, the "tubuliferous" tissue of Amphibians is developed from the kidney tubules and not from the pronephric duct itself, as is the case with the diverticula in question in *Gymnarchus*, so that a strict homology is out of the question.

Still, if the Teleostean ancestry had this tubuliferous tissue (vasa efferentia, etc.), then at the loss of connection between gonad and kidney there would be, so to speak, a liberation of a considerable centre of energy, which might have afforded a stimulus for a growth in the same neighbourhood but resulting in a difference of form and in function from the original structure.

Recent researches have tended to show that originally the renal organ extended from end to end of the abdominal cavity (see Sedgwick (109), Brauer (18), etc.), and it is interesting to note the occasional occurrences of rudimentary kidney tubules with Malpighian bodies as noticed above on page 353, between the pronephros and mesonephros. These accessory ones may be considered to be accidental, but still their occurrence does point to the probability of a continuous series having existed at one time between pronephros and mesonephros (cf. Price (101)).

The buds and their accompanying diverticula are clearly not accidental, but are normal developments with some definite function. They do not develop in any way like the ordinary tubules of the mesonephros, but develop directly from the pronephric duct and so perhaps they should be regarded as belonging to the pronephric part of the series rather than to the mesonephros. They are, however, in close connection with the mesonephros which is a fact the significance of which I am not able to explain, except by the rather heretical suggestion made above.

4. THE REPRODUCTIVE ORGANS.

The oldest developmental stage that I have is too young to enable me to follow to the end the development of the duct of the reproductive gland, although, as regards most of the organs this stage may be said to have approached closely the adult form.

In the forty-third day specimens there is some indication of the structures which become in all probability the duct of the gonad.

This organ, the gonad, is single almost from the earliest times. When, however, the sex cells are first distinguishable they appear near to the peritoneum along the dorsal mesentery, and become arranged along the dorsal wall of the coelom in the usual localities on either side of the mesentery. Thus for a day or two one may describe the reproductive gland as paired, but afterwards all trace of sex cells disappears from the left side.

The earliest stage at which germ cells can be identified with certainty is $5\frac{1}{4}$ days, which is the day before "hatching."

At this time there occur in the hinder part of the animal in the tissue of the splanchnopleure large spherical cells with much cytoplasm, some yolk-like granules and the nucleus typical of a germ cell. These are scattered irregularly, some dorsal some ventral to the gut (text-fig. 142, *g.c.*).

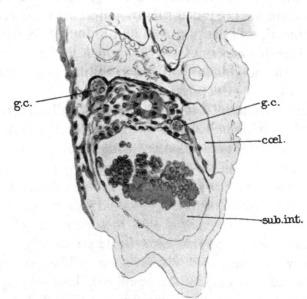

FIG. 142.   Transverse section through the hinder region of the trunk of a five-and-a-quarter day embryo. *cœl.*, coelom ; *g.c.*, germ cell ; *sub.int.*, sub-intestinal vein.

At six days the cells are more readily identified and occur in the corresponding places, not in the peritoneum, but in the mesoderm of the mesentery. In the more anterior part of the embryo they occur more regularly, often in pairs, one on each side of the mesentery, and are situated more dorsally.

The peritoneum is formed of squamous cells throughout and so we cannot speak of the presence of a germinal ridge or epithelium. Sometimes, certainly, the germ cells are so close to the surface as to seem to be superficial, but I think I can always detect the presence of a thin layer over them.

On the seventh day the germ cells are more obviously "in the peritoneum" (text-fig. 143), and mostly occur on the walls of the mesentery. Whether this is due to a migration of the cells, or to a lengthening out of the mesentery, I cannot say.

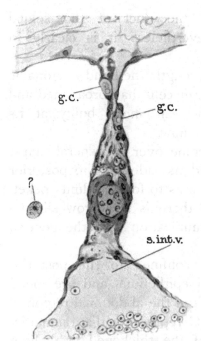

FIG. 143. Section through the gut, mesentery, and sub-intestinal vein of a seventh day embryo in the hinder region. *g.c.*, germ cell; *s.int.v.*, sub-intestinal vein; ?, cell of unknown origin and fate.

On the next day most of the germ cells are found upon one side, the right side, where they are grouped together and are covered by a follicle of peritoneum cells. In the hinder part they project into the body cavity and form almost a continuous ridge.

Isolated germ cells occasionally occur at various points of the body cavity wall. There is no sign of any degenerating cells on the left side, so that it seems probable that the cells migrate from the left to the right side. In some sections a germ cell may be seen between the mesenteries; perhaps this is a germ cell migrating from the left side to the right side.

In the more anterior region they are fewer and more scattered and do not form prominent masses nor a ridge. They occur also more in the dorsal wall of the coelom than in the mesentery.

One day later, on the ninth day, the germ cells have almost disappeared from the left side, and are concentrated on the right side just at the root of the mesentery. The thickening of an external streak of peritoneum is greater, and is often separated from the gonad by a deep groove.

On the left also this band of thickened epithelium is present though, except as a rare occurrence, it is destitute of germ cells. It is not carried so far forwards as on the right side owing to the growth back of the air-bladder which obliterates it by becoming wedged in between it and the actual peritoneum. On the right side it extends as far forwards as about the eleventh somite. The band is not continued backward to the extreme posterior end of the body cavity.

On the tenth day all trace of the left gonad has disappeared, and I find germ cells only on the right ride securely embedded within the thickened epithelium. The gonad forms a conspicuous and continuous ridge projecting into the dorsal part of the coelom. I see no regularity in the distribution of the germ cells or in the swellings which undoubtedly do occur at intervals. There is no greater approach to a segmental arrangement than in the earlier stages. The ridge has been by now removed some distance from the mesentery which lies much to the left, so that the gonad is an almost median structure in the posterior part of the body cavity. Anteriorly it lies more to the right, having been pushed over by the developing air-bladder.

No changes occur for some time.

In the oldest stage (forty-three days) although the duct of the sexual gland is not formed, nevertheless it is clearly in process of formation.

A

Throughout its whole length the band of gonadic tissue with its scattered germ cells has broadened and flattened. It has always been more bulky at its posterior end and so it is now.

A great change has come over its general shape. It has broadened out and its sides at the posterior extremity have folded up so as to form a blind pocket. At the anterior end also there is a narrow slit-like cavity which has no communication with the general abdominal cavity.

B

The germ cells are confined to the posterior portion of the thickened epithelium, and are found only in the hindermost fifth of the abdominal region.

C

FIG. 144. Three transverse sections taken across the peritoneum of the right dorsal wall of the coelom in the region of the germ cells, forty-third day. A, the most anterior about the fifteenth somite; B, about the fortieth somite; C, about the fifty-fifth somite.

Text-fig. 144, A, B, C represents diagrammatically transverse sections of the thickened epithelium through the anterior, middle, and hinder regions respectively. Near the posterior end of the still open part several folds and grooves, and even tubules, occur (which are possibly the commencing tubules of a testis), among which the germ cells may be seen (text-fig. 145).

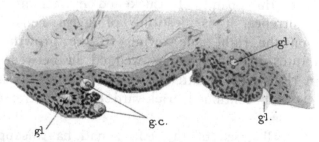

FIG. 145. Section taken through the germinal epithelium on the forty-third day (cf. text-fig. 144, B). *g.c.*, germ cell; *gl.*, gland-like tubules and opening.

Following the blind pocket at the posterior end it is found to run backwards above the peritoneal cavity and to end in the middle of a mass of mesoderm which forms a very wide mesentery between the ureter and the alimentary canal. On the sides are seen the backward running prolongations of the abdominal cavity which reach as far as the epidermis on either side of the opening of the ureter and therefore posterior to the anus, but there is as yet no perforation. No doubt these are the abdominal pores.

In longitudinal section (text-fig. 146) a strand of denser tissue passes from

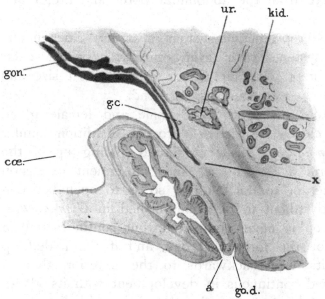

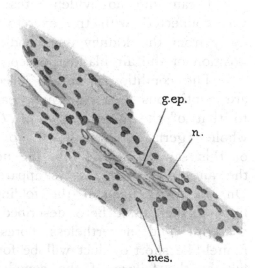

FIG. 146. Sagittal section through the hind end of the gonad on the forty-third day. The section passes through the hind end of the gut as well. *a.*, anus; *cœ.*, coelom; *g.c.*, germ cell; *go.d.*, gonoduct invagination in epiblast; *gon.*, gonad; *kid.*, kidney; *ur.*, ureter; *x*, developing gonoduct.

FIG. 147. The region near the point *x* in text-fig. 146, showing the developing gonoduct. *g.ep.*, cells forming the lining of the duct directly continuous with the periosteum; *mes.*, nuclei of general mesoblast; *n.*, nucleus of peritoneum?.

the extreme end of the gonad towards an ingrowth of the skin (*go.d.*). This lies immediately behind the anus and in front of the ureter, and between the horns of the coelom which probably become the abdominal pores. Into this mass of denser tissue there seems to run a prolongation of the epithelium of the gonad. In text-fig. 147, which is a drawing of the region named *x* in text-fig. 146, the dark nuclei (*mes.*) are those of the mesodermal tissue of the mesentery, and the large clear nuclei (*n.*) which are accompanied by a layer of denser cytoplasm directly continuous with the lining of the gonad, are quite different from the small nuclei (*mes.*) and are probably to be considered as advancing cells of the peritoneum. This is a difficult point to decide. If they are advancing cells of the peritoneum, then the gonoduct is clearly throughout of peritoneal origin and so far resembles a Müllerian duct. If, however, they are metamorphosed cells of the general mesoblast tissue the case would be different. The general appearance of the cells and their neighbours is in favour of the former view.

We may conclude that the probable subsequent history of the gonad and its duct is as follows. The sides of the gonadic plate continue to fold up, as the posterior end has already done, so as to form an enclosed sac, whose cavity is obviously derived from the coelom; its posterior end then bores its way amidst the mesoblast tissue between the ureter and rectum, and so obtains access to the exterior by a channel separate from the abdominal pores and ureter and alimentary canal.

I can find no evidence that the reproductive system is at any time in any way connected with the excretory system. It is from its first appearance far away from the kidney duct, and is kept away by reason of the great size and position of the air bladder except at the extreme posterior end.

The condition of most Teleosteans in which both male and female gland are continuous with the ducts can clearly be attained from a condition similar to that of the oldest stage of *Gymnarchus* by a continued folding up of the whole "germinal epithelium" to form a closed sac, and subsequent extension of this sac backwards into the ureter or to the exterior so as to open between the anus and ureter, as is apparently about to be accomplished in *Gymnarchus*. On the other hand if the folding up is never completed, but continues only a little further than here described for the forty-third day, and if the folded up posterior end nevertheless bores its way backwards to the exterior, then a funnel-like short oviduct will be formed continuous in development with its gland, but by the failure of the complete folding up of the germinal epithelium area the fully developed ovary will not be cut off from the general body cavity as in the female of some Salmonidae.

To what extent can this process be compared with the formation of a Müllerian duct?

If we adopt Balfour's definition of a true Müllerian duct, namely "a product of the splitting of the segmental duct" (Balfour and Parker (9), p. 421) then clearly the gonoducts as developed in *Gymnarchus* cannot be Müllerian ducts so far as their ontogeny is concerned.

They are, however, distinctly developed as folds of the peritoneum and more especially of that part of the peritoneum which harbours the germ cells, but are far removed from the pronephric duct.

The relation of the air bladder to the dorsal wall of the coelom which brings about the separation of the pronephric duct and kidney tubules from the immediate neighbourhood of the peritoneum in the great majority of Teleosteans seems to be a feature of paramount importance in this connection. With it, no doubt, the absence of nephrostomes in the mesonephros has been correlated, and perhaps also the divorce of the kidney tubes from any connection with the testis, and possibly with a disappearance of the true Müllerian ducts (unless vestiges of these old connections still exist, in the curious outgrowths from the pronephric ducts in the anterior part of the mesonephros as suggested in the last section).

The absence of nephrostomes I believe to be complete. I cannot think that the commencing tubules and funnel-like depressions described above (text-fig. 145) as appearing in the walls of the germinal ridge already folding up to form the gonoduct and gonad can be regarded as nephrostomes. They are probably testicular or glandular pits of some sort. They are unlike nephrostomes because they do not show any trace of cilia; and their epithelium is not so markedly columnar, and they appear far later in development than is the case with nephrostomes.

So far as actual development shows us there is some general resemblance to the formation of the oviduct in the Frog according to MacBride (88) who has shown that in that animal there is in ontogeny no connection between the oviduct and the original segmental duct. It is true that a direct participation of the segmental duct in the formation of the oviduct has been described by Fürbringer and Hoffmann in the Salamander and Frog.

If we adopt MacBride's conclusion (in which case neither can the oviduct of Amphibians be regarded as a true Müllerian duct) then although the general resemblance between Amphibians and Teleosteans is established for the formation of the oviduct of the one and gonoduct of the other, still there is this difference, namely, that in the Amphibian it is a part of the peritoneum which is not concerned with the germ cells that forms the duct, whereas in Teleosteans (*Acerina, Zoarces*, Jungersen), and perhaps in a smaller degree in *Lepidosteus*, the germinal ridge itself is involved in the folds and becomes an ovisac, thus bringing about a continuity between gland and duct. This perhaps is not a serious difficulty since the ducts themselves may be homologous and the germ cells, which, as appears to be very probable, travel at any rate from the region of the gut to the dorsal wall of the coelom, may be supposed to take up their position in the duct epithelium as well as anywhere else. (Felix and Bühler, however, maintain that some of the germ cells are derived at a later period from the coelomic epithelium itself (Hertwig (64), pp. 649—50).)

It does not seem to me that the development of *Gymnarchus* offers any decisive evidence with reference to the homology of the ducts of the sexual organs of Teleosteans. If the extension backwards of the gonoducts (text-fig. 146) is really a splitting within the ordinary mesodermic tissue, then that would be strong evidence in favour of the ducts being new structures; but I am inclined to think that the evidence so far as it goes is in favour of the ducts being due to a prolongation backwards of the peritoneal cells. On the other hand there is a distinct involution of ectoderm (text-fig. 146, *go.d.*), which perhaps favours the new formation view.

On the whole I am inclined to agree with those who hold that the gonoducts of Teleosteans male and female are homologous structures, but are not homologous with the gonoducts of other Gnathostomata; unless it be with the oviducts of Amphibians.

They are not true Müllerian ducts, but are probably in part new structures evolved in correlation with the attachment of the air bladder to the dorsal wall of the coelom, thereby cutting off the germinal epithelium from any connection with the renal organs.

The Germ Cells.

As regards the origin of the germ cells I cannot add to what I said a few paragraphs previously. It has been asserted by Beard and lately reasserted by several authors that the germ cells in Elasmobranchs are at first in the yolk sac, and that they then migrate through the stalk into the "embryo" and may be found in all parts of the embryo (Beard (14)), ultimately settling down in the germinal epithelium. The germ cells make their appearance in other parts of the peritoneum as well as the germinal ridge as I have just described to be the case in *Gymnarchus*, and also I believe that they ultimately migrate from the left side to the single streak on the right side. I have, however, not found an undoubted germ cell at any great distance from the peritoneum.

It has been reasserted lately by several authors that the germ cells of Vertebrates do not originate from peritoneal cells, but that they arise elsewhere and migrate thereto. Others have taken equally decidedly the opposite view. Those who have worked upon fishes have more often taken the former view, those on the Amniota the latter view. In the most recent paper, that of Allen, *Anatom. Anzeiger*, Sept. 1906 (to which reference may be made for the literature of the subject), the author, after a study of the Reptile *Chrysemys*, has come to the former conclusion, and lastly, Felix and Bühler (42), maintain that they originate in both ways.

In *Gymnarchus* the germ cells are first obvious on the sixth day. I believe they can be detected—but I cannot be certain—upon the fifth day in the mesentery and round the gut, but I have not been able to trace their previous history.

In specimens of the seventh day cells of large size are found abundantly throughout the tissues—brain, blood vessels, kidneys, coelom, indeed everywhere.

I was at once struck by the resemblance of these to the cells which Beard (14) described as wandering germ cells in Elasmobranchs, especially as regards their extraordinary distribution, and in other respects as well [cf. text-fig. 148, with Beard (14), figs. 4, 7, 52]. After closer examination, however, I see no reason at all to connect them in any way with the germ cells. Their appearance coincides only approximately with the advent of the germ cells. The germ cells come on the sixth day; these aberrant cells upon the seventh day. They differ in being far larger than the germ cells and in having a totally different appearance. If they are cells they contain a small deeply staining nucleus instead of the large vesicle so characteristic of a germ cell (*v.* text-fig. 148, B, C).

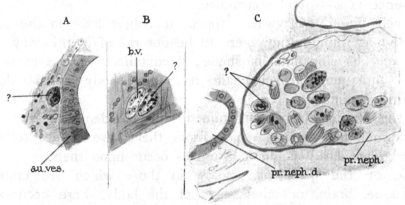

FIG. 148. Showing aberrant cells in embryo of seventh day in A, tissue near the ear; B, in the brain; C, in the pronephric chamber. *au.ves.*, auditory vesicle; *b.v.*, blood-vessel; *pr.neph.*, pronephric chamber; *pr.neph.d.*, pronephric duct; ?, aberrant cells of unknown origin and fate.

If not germ cells, what are they? It may be said they resemble the large nuclei present in the yolk, and chiefly aggregated near the posterior end of the point of attachment to the body of the embryo.

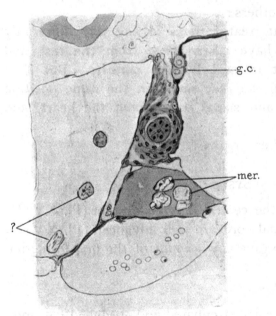

FIG. 149. Transverse section through the hind gut of larva of seventh day. *g.c.*, germ cells; *mer.*, merocytes in strand of yolk (cf. text-fig. 110); ?, aberrant cells of unknown origin and fate.

Although the resemblance here is closer than between these cells and germ cells, I am inclined to regard it as superficial, cf. text-fig. 149.

The yolk nuclei are always crumpled while within the yolk sac. Certainly they have a large nucleolus and resemble in size the cells in question. It is also possible that yolk nuclei though crumpled when in the yolk sac might assume a more shapely form under the altered condition of the embryonic tissues.

The cells have, moreover, a peculiar refrangent character wanting in the periblast nuclei.

Text-fig. 149, which shows some of the cells in question, some yolk nuclei, and germ cells, illustrates the differences fairly satisfactorily.

The third alternative which has suggested itself to me is that they are parasitic protozoa, either sporozoa or infusorians. In the specimens they clearly bring about a certain amount of tissue destruction, which is difficult to explain if their presence is a normal occurrence.

These cells found everywhere appear to collect later in the pronephros and pronephric duct as though they were to be got rid of in this way.

Each "cell" is ellipsoidal in shape; it contains always a mass of chromatin-like material, and generally alongside it another body can be detected, larger and more faintly stained.

When the section has been stained with Heidenhain's iron haematoxylin, connecting strands are seen to pass from these two bodies to the cell wall, and other irregular darkly staining granules occur upon them.

In some of the specimens, mostly in those which are embedded in the connective tissue, brain, or other parts of the body, there occur special bodies which are deeply stained by Heidenhain's stain, but not by ordinary haematoxylin or carmine stains (very much as Beard describes for his wandering "germ" cells), text-fig. 148, B, C.

In these also the outer wall is seen to be marked by slight thickenings or ribs which run from pole to pole, like the markings on certain infusorians.

Can these be infusorians? Why should they occur in all the specimens of six and seven days and in none of the others?

I have no reason to suppose Budgett treated them in any way differently from the others. It is possible he may have taken these from the rest and kept them in a separate vessel which got infested by the parasite. But if it is an accidental parasite why do they occur so very much in the same position in each case, namely, chiefly in the brain and spinal cord, about the heart, and in the pronephros and its duct?

They occur also in less numbers in other organs.

5. THE NERVOUS SYSTEM.

On the second day there is no trace of the central nervous system (Pl. XVIII. fig. 3) and on the third day the brain and spinal cord are well advanced (Pl. XVIII. figs. 4, 5); it is therefore impossible for me to give any account of the first formation of these organs.

Third Day.

On the third day the whole embryo is greatly attenuated and stretched flat upon the yolk in the usual Teleostean manner.

The central nervous system consists of an axial strand of cells continuous with epiblast in front and laterally, and with the secondary growth centre behind.

At this stage it may be said to consist of two parts, an anterior protogenetic portion and a posterior deuterogenetic portion, if I may use the terms suggested a little time ago—[v. Assheton, *Anat. Anz.* 1906]. The former is a thick plate continuous with the inner layer of epiblast, having a strongly developed ventral keel (text-fig. 150, *a, b, c*). At the extreme anterior end it overlies a thick mass of mesoblast which extends some way in front of it. The neural plate, as it passes backwards, thickens into the keel just mentioned, and the mesoblast diminishes, but is distinguishable as a continuous layer between the keel and the hypoblast. A little way back a neural groove appears on the dorsal surface (text-fig. 150, *b*). The groove extends for only a short distance (text-fig. 151).

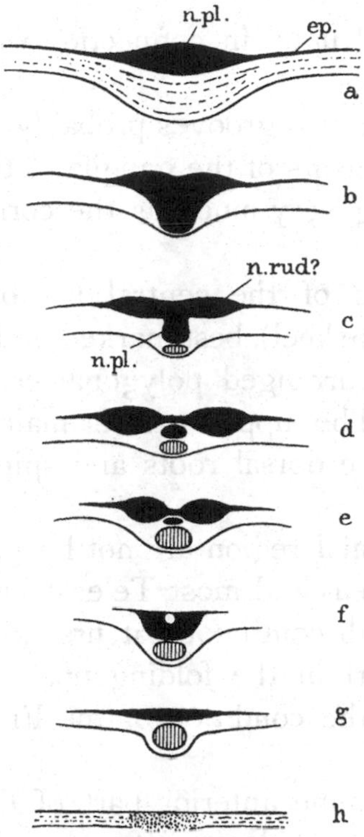

As the groove shallows out the sides of the plate increase in bulk as compared with the central keel, so that the transverse section is shaped like the letter T (text-fig. 150, *c*). Further back the keel diminishes gradually and ultimately dies out, but before it does so it seems to be free from the overlying tissue; but this is a difficult point to determine.

On the other hand the lateral plates retain their thickness and, on the disappearance of the keel, appear as almost isolated bands connected together by quite a thin plate of tissue. Moreover the edges of these thickened bands are distinctly curved inwards, suggesting very strongly a formation of the future cavity of the nervous system by folding rather than by excavation of a solid cord (Pl. XX. fig. 51).

We may describe this part of the central nervous system as consisting of a neural plate with wide neural groove and greatly thickened margins, which are beginning to fold over in order to form a tube. Further back the comparatively wide intervening plate narrows so that the lateral bands come together in the centre, and their incurved edges almost meet; and in fact I believe in places they do meet and enclose the central canal of the future spinal cord (text-fig. 150, *f* and Pl. XX. fig. 52). At any rate there is for a considerable distance a lumen bounded by the neural plate below and laterally, and either by the incurved edges of the neural plate dorsally, or by the outer covering layer alone.

FIG. 150. Sections taken transversely through the neural plate and notochord of third day embryo. Mesoblast tissue is omitted except in the first and last of the series. They are taken at the points *a, b, c,* etc., in text-fig. 151. *ep.*, epiblast; *n.pl.*, keel which probably gives rise to the central nervous system; *n.rud?*, wing-like expansions which probably become the ganglia of cranial nerves.

Near to the posterior end the neural canal is again absent and a slight neural groove takes its place which soon shallows out, and the terminal portion of the neural plate is a flat plate that becomes continuous with the primitive streak (Pl. XX. fig. 51).

There are thus two neural grooves, one occurring anteriorly which belongs wholly to the protogenetic area, the other which belongs to the deuterogenetic area, which is partly curved in to form the canal of the spinal cord; and between the two there is a part which shows no neural groove, and is characterised by the thickness of the lateral margins of the plate. In the region of the anterior neural groove there is a pair of slight bulgings on the sides of the keel which perhaps may be the rudiments of the optic vesicles.

There are several points of interest to be commented upon in connection with this early stage of the central nervous system.

The thickened lateral lobes of the region between the two grooves probably do not take part in the formation of the brain, but are the rudiments of the ganglia of the hinder cranial nerves and of the auditory vesicle, behaving very much as the corresponding parts behave in the Amphibian.

Balfour and Parker (9) note in the early rudiments of the central nervous system of Lepidosteus that the more "superficial part (of the keel), best marked in the region of the brain, is formed of more or less irregularly arranged polygonal cells, and a deeper part of horizontally placed flatter cells. The upper part is mainly concerned in the formation of the cranial nerves and of the dorsal roots and spinal nerves."

Here in *Gymnarchus* the nerve rudiments of the cranial region are not heaped up on to the top of the rest of the rudiments as in Lepidosteus and most Teleosteans, but are lateral expansions as in Amphibians, where although continuous at first with the median part of the neural plate they do not take part in the folding up.

In this respect *Gymnarchus* is intermediate between the condition of the Frog and that of most Teleosteans and Lepidosteus.

There is no evidence as to how the cavity of the extreme anterior part of the neural tube is formed. The neural groove certainly is very wide (text-fig. 150, *b*), but there is no other evidence of a folding up, and the thickness of the median line, the keel, is evidence against it.

Posteriorly, that is to say in the deuterogenetic region, the evidence is in favour of folding, which constitutes an interesting feature. I should be inclined to suggest that in *Gymnarchus* the neural tube of the protogenetic area is formed by excavation of a solid cord and the neural tube of the deuterogenetic region chiefly by folding. At any rate there is a part between the two grooves which certainly shows no indication of a folding.

Another point is the behaviour of the so-called outer layer of epiblast. This takes no part at all in the formation of the central nervous system. It lines the neural groove where it is shallow, but wherever the edges of the plate are to be seen curving inwards, the outer layer takes no part in it at all (Pl. XX. figs. 51—53), and when the canal is fully formed it is quite absent from it. The behaviour of the outer layer of epiblast is here as it is in the case of the formation of the auditory vesicle in the Anura where the nervous layer alone is concerned, and unlike the formation of the central nervous system of the Anura where the epidermic layer is also concerned and gives rise to important parts, namely the neuroglia cells of the brain. Thus it is unlike the epidermic layer of ectoderm of the amphibia.

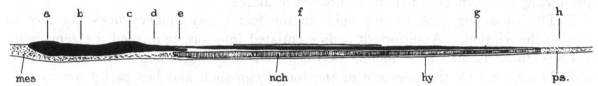

FIG. 151. Sagittal section (semi-diagrammatic) through an embryo of three days. *hy.*, hypoblast; *mes.*, mesoblast; *nch.*, notochord; *ps.*, growing point; *a*, *b*, *c*, etc., indicate the position of the sections in text-fig. 150. The lines *d* and *e* should be moved about 5 mm. further to the right; and perhaps the anterior end of the notochord should be continued forwards a like amount.

Further back, the canal shallows out and the neural plate is reduced to a narrow band, plano-convex in transverse section.

Along a considerable distance of this grooved part a number of large round cells occur just underneath this external ectodermic layer, between it and the inner layer of epiblast. These are probably the cells found at a later stage in a similar position, of whose origin and fate I can say nothing.

Passing to the posterior end of the embryo the neural plate thins out and expands laterally and becomes continuous with the anterior end of the growing point which has, as already described, the appearance of the primitive streak of many Amniota (text-fig. 92, p. 304).

Fourth Day.

The twenty-four hours which separate the stage just described and the next have produced a great change. The nervous system is now a tube throughout almost the whole part of its length, with the characters usual to a craniate Vertebrate (Pl. XVIII. figs. 6 and 7).

The distortion and crushing which render this stage so indescribable in connection with the other organs, do not affect the central nervous system to the same extent.

At the extreme posterior end, immediately in front of the growing point, the nervous strand is solid, but a very short distance anteriorly a small lumen appears, which is continuous with the central canal of the spinal cord. This is, of course, as we would expect to find it; but at a point farther forward there is an obliteration of the lumen. Here for three sections no lumen is perceptible, and the canal in front of this spot lies towards the ventral surface, while the canal posterior to this point lies towards the dorsal surface.

This coincides with the anterior end of the notochord.

Referring for a moment to the previous stage (text-fig. 150), it may be seen that this point corresponds with the part of the neural plate which lies between the two grooves described above, and it is not unreasonable to suppose that the present may have been developed from the previous condition either by the folding up of the two parts of the plate separately, or, more probably I think, by the excavation of the solid keel of the anterior part and the folding up of the neural plate posteriorly, thus producing two canals which have not yet coalesced.

The tube posterior to this point in the fourth day embryo does not require further description. Anteriorly it is differentiated into fore brain and the remainder. I think there is also a distinction between mid and hind brain. The fore brain is already affected by the pressure of the tough egg-shell and lies partly over on its right side. Its extreme anterior part is greatly crushed and difficult to interpret. The optic vesicles are well advanced, and the roof of the tube a little posterior to this region is expanded, and probably represents the roof of the mid brain (Pl. XIX. fig. 40 and text-fig. 88, p. 298).

After this point the neural tube continues as a straight tube, slowly diminishing in size, with narrow lumen, and with sides rather thicker than the roof and floor, and oval in transverse section.

About three-fourths of the distance between the front end and the spot where the lumen is obliterated, a neural crest occurs which is very abundant anteriorily, but it gradually thins out as it passes backwards. The cells composing it are sub-spherical and loosely packed (text-fig. 94, p. 306).

After the point where the obliteration occurs the tube is more upright and alters its character considerably. At first it seems to be compressed, and the lumen remains irregular for some distance and then widens out, so that it is broader horizontally than vertically.

Commencing at the point of obliteration and running back through some 25 sections is a pair of thickenings on the sides of the nerve tube which do not seem to be actually fused with the tube, though in close contact with it and with the epiblast. At first sight they look like mesoblastic somites. They are, more probably, epiblastic in origin, and may give rise, like the similar side plates in Amphibians, to the ganglia of the eighth, ninth, and tenth cranial nerves and lateral line, and possibly to the epithelium of the auditory vesicle as well (text-fig. 88, p. 298).

In a slightly older stage ($4\frac{1}{4}$ days) the distortion has greatly increased, and the series being incomplete, it is not possible to fill in all the details. A considerable advance has been made and the ganglia of the several cranial nerves are separated, but I am unable to add any further evidence as to their origin from the lateral plate.

Fifth Day.

On the fifth day, matters have improved. Distortion is at its height, but the crushing is less, and the object so much bigger that all the parts are easily identified. Unfortunately, however, the advance made has been great, so that we no longer deal with early stages. The optic cup, lens, auditory vesicle, are formed and present no feature of particular interest. The whole "embryo" now lies upon its side. The general characters of the brain are seen in text-fig. 152, and Pl. XIX. fig. 41. The fore brain is sharply bent ventralwards so that the roof of the thalamencephalon is the most anterior part.

The cranial flexure involves the fore brain only; that is to say, the great flexion occurs between the mid brain (*mb.*) and fore brain (*fb.*) (unlike the Elasmobranch). A slight bend also occurs in the hind brain, but so slight as to be hardly perceptible.

Up till now the cells composing the brain and spinal cord have shown no sign of differentiation. Nor has there been any marked thickening or thinning out of certain regions.

On the fifth day, however, nerve fibres are present in the brain, and the fibres of the peripheral nervous system are forming, and marked differences occur in the thickness of the walls and roof of several parts of the brain.

The Fore Brain.

No part of the wall of the fore brain is reduced to less than three or four nuclei in thickness. The thickest part is in the ventral portion (i.e. in the ventral zones of His). Just in front of the optic crossing there is a pair of thickenings, the olfactory lobes, where considerable tracts of fibre occur. Nerves pass between these and the olfactory epithelia which are thickenings of the neighbouring epiblast. Over the greater part of the external surface of the nerve tube a slight development of "ectoglia" (Minot) has begun, "the Randschleier" of His, which is most marked along two bands in the ventral zone. Considerable bands of nerve fibre tissue occur also in these tracts, though they are not well enough developed yet to enable one to determine how the fibres run except in a few places. These tracts of "Randschleier," however, continue backwards, passing to the crura cerebri and on backwards as the lateral columns of the spinal cord. Following the tracts forwards they are seen to pass ventralwards and meet between the optic recess and pituitary body (text-fig. 152, *M.T.*).

Throughout the medulla and spinal cord they are connected ventrally by transverse fibres. Towards the posterior end, all Randschleier tissue and nerve fibre are as yet undeveloped.

The hind brain is already marked by its thin roof, where it is reduced to a single layer of squamous cells (text-fig. 101, p. 310).

The elements are small and difficult to make out. As yet there is no clearly defined epithelial layer. Most of the nuclei are elongated and arranged as usual at right angles to the cavity of the tube. The innermost layers contain large numbers of mitotic figures. Mitotic figures occur here only.

By the level of the root of the fifth nerve the roof of the hind brain has widened greatly. The roots of the eighth and seventh cranial nerves almost coincide, and join the brain at a more dorsal level than the fifth. The ganglia of these three nerves are oval, compact masses of nuclei, and lie as usual close to the brain. The ninth has a ganglionic swelling close to the first branchial gill cleft and is connected by a strand of fibres with the brain.   Just at this point a small collection of nuclei forms an upper ganglion.   At the same point some roots of the tenth nerve leave the brain and run downwards and backwards to an irregular branching ganglionic mass whence strands pass to the hinder gill clefts.   The lateral line ganglion is large and seems to be unconnected with any of the cranial nerves (text-fig. 152, *l.l.*).

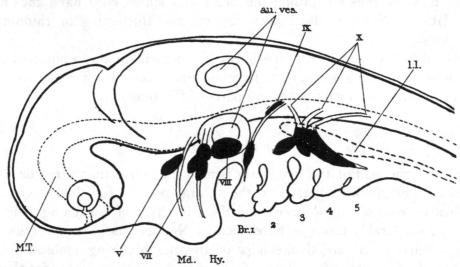

FIG. 152.   Dorso-lateral view of the head region of an embryo of the fifth day to show the chief nerve ganglia and fibre tracts.   *au.ves.*, auditory vesicles; *Br.*1, 2, 3, 4, 5, branchial arches 1—5; *Hy.*, hyoid arch; *l.l.*, lateral line ganglion; *Md.*, mandibular arch; *M.T.*, main tract of ectoglia; *V*, *VII*, etc., ganglia, roots and rudiments of branches of cranial nerves.

Sixth Day.

In the sixth day *Gymnarchus*, immediately after hatching, the embryo has assumed its normal symmetry.

Pl. XIX. fig. 42 is a side view of the brain dissected out, and must be compared with that of the fifth day, fig. 41, which is still greatly compressed. (The latter view is from a specimen cleared and mounted whole and shows the eye which the former does not.)   The fore brain is brought further forwards and the mid brain more dorsal, but the hind brain has a totally different appearance, and is marked by a series of constrictions dividing it into a series of segments of which the anterior one is the most, and the one next behind it the least, prominent.   There can be no doubt that this does resemble on the whole the "neuromeres" of many authors, and more particularly the condition described by Hill of the Trout and Chick (Ch. Hill (66)).   It is also certain, however, that no trace can be seen of these neuromeres in the fifth day specimens, and I feel very doubtful whether to regard them as really neuromeres or as some artificial effect perhaps connected with the pressure of the gill arches.   It is quite conceivable that the divisions may occur in the earlier stage, but that owing to compression by the investing membrane they are masked, and when the pressure is removed they then become markedly visible. I found them in the only two specimens I examined. They are, however, very prominent for so late a stage in development. No trace remains of the primary neuromeres of the mid and fore brain.

The extreme anterior end of the brain at this period is very slightly bilobed, so slightly that it is perceptible only in one or two sections.

The floor of the thalamencephalon is thin and broad at the optic recess.

The mid brain roof is thick and hardly bilobed externally as yet, though it overhangs the rest of the brain at the sides. Its outer surface is covered by spongio-reticulum or "Randschleier." Internally, the division into optic lobes is more marked owing to the groove which occurs along the mid dorsal line, the Deckplatte of His (text-fig. 153, *d.pl.*). This groove as it passes back broadens, and it shows a better marked epithelium than elsewhere Where the "Deckplatte," i.e. the true roof, is broadest it is reduced to a non-nervous layer. There is no trace of the transverse bands of fibres which cross this region later.

The optic recesses are in front of the eyeballs. A well developed transverse band of tissue, continuous as before with the crura cerebri, is found on the ventral surface between the optic recess and the infundibulum. In this are the fibres of the commissurae post-opticae.

The mid brain and hind brain roofs already press against one another, and the mid brain forms two lateral wings between which the solid anterior part of the roof of the hind brain projects, running forwards to a point (text-fig. 153, *cbl.*).

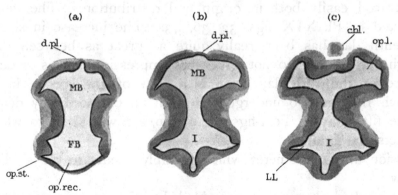

Fig. 153. Transverse sections through the brain of an embryo of the sixth day. (a) through fore and mid brains; (b) through infundibulum and mid brain; (c) through infundibulum and fissura rhomboidalis mesencephalica. *cbl.*, cerebellum; *d.pl.*, Deckplatte; *FB*, fore brain; *I*, infundibulum; *LL*, lobus lateralis; *MB*, mid brain; *op.l.*, optic lobe; *op.st.*, optic stalk; *op.rec.*, optic recess.

The result is that the mid brain cavity is pushed downwards and lateralwards, while the hind brain cavity and thick roof (cerebellum) is pushed forwards and dorsally. The roof of the hinder part of the hind brain is thin and broad. This part of the brain has the form usual in vertebrate embryos of this stage. The spinal cord is of the ordinary type and needs no special notice.

The Nerves.

Of the cranial nerves the fifth is present as a ganglionic mass, one branch of which projects forwards. The root enters the brain very low and can be traced ventralwards. A strong mandibular branch is developed. The ganglia of the seventh cranial nerve are large, and the hyoidean nerve is a fine strand of nerve fibre.

The ninth and tenth arise from a more dorsal position far apart from one another, the ninth by a single stout root which passes to the ganglion belonging to it at a considerable distance from the brain. In this respect it differs from all the others. It passes close behind and almost above the auditory vesicle.

The tenth arises by at least three well defined roots, the ganglion being nearer to the anterior than to the posterior root.

The lateral line nerve is a stout strand of ganglion cells in close connection with thickenings of the epiblast. It lies outside the ganglia of the tenth nerve; and is not connected therewith.

Seventh Day.

A great advance has been made by the seventh day in the nervous system as in all the other organs. The adult characters are foreshadowed and many of the chief tracts of fibres are laid down. All the cranial nerves, except the fourth, can now be traced easily both in origin and distribution. The general external form is indicated in Pl. XIX. figs. 32, 36, 43. The increase in size between the fifth and seventh day has been really quite as great as between the fourth and sixth, although the figures do not give that impression. This is due to the fact that the figure of the fifth day brain is a view of the surface in the plane of flattening which the embryo undergoes to such an extraordinary degree upon the fourth and the fifth days. Text-figs. 96, 97, 99, etc. will show to what extent this flattening affects the brain.

The division into neuromeres which possibly occurs in the sixth day brain is quite lost now.

The most marked change is the decided doubling up transversely of the lateral walls of the hind brain which leads eventually to the formation of the two pairs of swellings on the side of the medulla, the anterior of which comes in close contact with the vesicles of the swimming bladder. This period, sixth to eighth days, is also characterized by the great development of the mid brain, which for the time becomes the most prominent part, causing the fore brain, which up to now has been the largest division, to become pushed backwards and ventralwards (Pl. XIX. figs. 41—43).

The division between the mid and fore brains is much more marked and the latter shows a thinning out of the roof (actually the anterior wall) which extends as a broad transverse band from side to side (Pl. XIX. fig. 43, *vt*). This is indented (as seen in the section, text-fig. 154, *V.T.*) in connection with a small

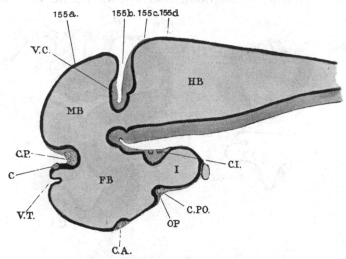

FIG. 154. Sagittal section through the brain of the seventh day larva. *FB*, *MB*, *HB*, fore, mid and hind brain; *I*, infundibulum; *C*, commissura ?; *C.A.*, commissura anterior; *C.I.*, commissura infundibularis; *C.P.*, commissura posterior; *C.PO.*, commissurae post-opticae; *OP*, optic nerve; *V.C.*, valvula cerebelli; *V.T.*, velum transversum; 155 *a*, *b*, *c*, *d*, indicate the planes of sections of text-fig. 155, *a—d*.

blood vessel; and the posterior border of the roof of the thalamencephalon is traversed by a narrow and a very compact band of fine nerve fibre which passes ventralwards and loses itself in the hinder part of the corpus striatum or in the com. post-opticae (text-fig. 154, *C*). Immediately behind this at the bottom of the constriction between mid and fore brains, is a much wider though less purely fibrous tract, the posterior commissure. There is a special bundle of fibres, which at first sight seem to be part of this commissure, but which lie really a little anterior to it and form a well marked strand of stout fibres which pass to the base of the crura cerebri. I feel pretty sure that these fibres have their nuclei in the lateral borders of the roof between the posterior commissure and the pure fibre band noted above (text-fig. 154, *C*), which nuclei are the ganglia habenulae.

These fibres seem to pass to the base of the crura cerebri and do not decussate, but fibres which do cross develop subsequently in connection with the ganglia habenulae.

The sides of the prosencephalon are thickened, forming the corpora striata, and are joined together by a not very compact band of fibres, the anterior commissure, which can be seen in text-fig. 154 on the ventral wall (*C.A.*).

The cavity of the fore brain here is narrow, and a little further back it widens out to form the optic recesses which are prolonged as short horns towards the obliterated cavities of the optic stalks (text-fig. 155, *c*). Immediately behind the optic nerves are the now bulky commissurae post-opticae (*C.P.O.*). In the dorsal wall of the infundibulum are two small commissures, the com. infundibularis and another which may perhaps be part of the former. With the exception of the commissures just mentioned, the median line of the fore brain along the dorsal and anterior surfaces is thin throughout, and shows no nerve cells nor fibres.

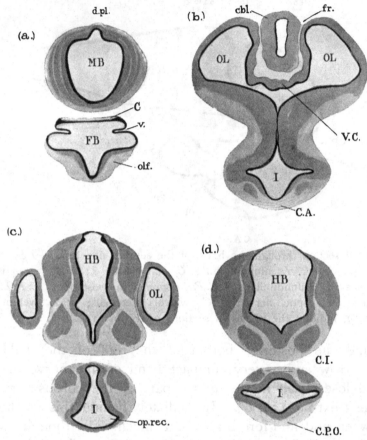

FIG. 155. Transverse sections through the brain of a larva of the seventh day taken approximately along the planes indicated by the 155 *a—d* in text-fig. 154. *C*, com. ?; *C.A.*, com. anterior; *C.I.*, com. infundibularis; *C.P.O.*, com. posterior; *d.pl.*, deckplatte; *fr.*, fissura rhombo-mesencephalica; *olf.*, olfactory lobe; *op. rec.*, optic recess; *v.*, velum transversum. Other letters as in former figures.

The floor of the infundibulum posterior to the com. post-opticae is again reduced to a thin layer in the middle line, seldom more than a single cell in thickness, though ventrally a "Randschleier" is present (text-fig. 154). The dorsal wall of the infundibulum is thickened at one point near its extreme end, where the commissura infundibularis occurs. The text-fig. 156 is a diagram to illustrate the courses of the chief fibre tracts so far as they can be traced in this stage.

The roof of the mid brain in the median dorsal line is a single layer of epithelium with no nervous tissue posterior to the large posterior commissure. This thin part is quite narrow anteriorly, but widens out somewhat farther back. Anteriorly the mid brain shows no sign externally of a bilobed condition, but posteriorly it is slightly separated into two lobes by the thin epithelial strip, which widening separates the more solid lateral areas, until at the extreme posterior region they are produced backwards as very distinct hollow sacs lying upon the

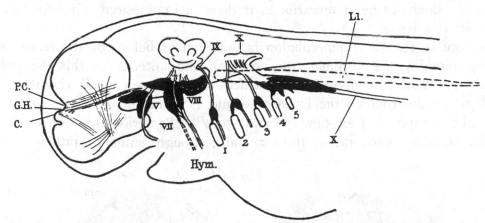

FIG. 156. Side view of the head of the seventh day to show the approximate position of the chief ganglia and nerve tracts on the seventh day. *C.*, com. ?; *G.H.*, ganglion habenulae; *P.C.*, posterior commissure; *Hym.*, hyomandibular cleft rudiment; *Ll.*, lateral line nerve; *V—X*, cranial nerves; 1—5, branchial clefts.

sides of the now prominent median part of the anterior region of the hind brain, which is rapidly developing into the cerebellum (text-figs. 155, *b*, *c*, and 154).

The expansion of the tectum opticum of the mid brain and increase in bulk of the upper part of the anterior end of the hind brain cause the two structures to work against one another in such a way that the mid brain is compressed in the median line, so that its roof becomes a vertical transverse sheet and its lateral regions, as described above, extend backwards on either side.

The cerebellum is at present bilobed.

No fibres occur in it, but some are present running transversely at the base of the bend between the hind and mid brains. The diverticula of the mid brain do not become free from the crura cerebri. Passing backwards, the roof of the hind brain at first is thin but narrow, then by the falling outwards of the sides of the hind brain, it becomes wide and is reduced to a simple epithelium.

The sides and floor of the mid brain and hind brain are, as usual, very similar, and present as yet no peculiar features.

Ninth Day.

The fore brain has made a marked growth, due chiefly to the increase of the corpora striata, while the flexure of the hind brain has become more pronounced.

The net result in the general form is that the fore brain is raised up, whereby the fore brain is seen in a dorsal view, chiefly owing to the increased hind brain flexure, but largely due to the great increase in bulk of the sides and floor of the fore brain itself. The olfactory lobes are clearly visible from below as swellings on the ventral surface of the prosencephalon (Pl. XIX. figs. 33, 37, 44).

There is no sign of a longitudinal division of the prosencephalon whereby two cerebral hemispheres would be formed. The roof of the fore brain remains thin; and the floor does not bend upwards as it does in Lepidosiren (Graham Kerr (77)), and also in the Frog.

The roof of the thalamencephalon is indented as before by the transverse fold containing blood vessels (velum transversum), and anterior to this the pallium is produced into a narrow diverticulum, which runs up to a small rhomboidal piece of cartilage in the roof of the brain case and is firmly united therewith. This is probably the paraphysis (text-figs. 158 and 159, *P*). Posterior to the velum is another larger diverticulum with rather thicker walls, though entirely epithelial, which is

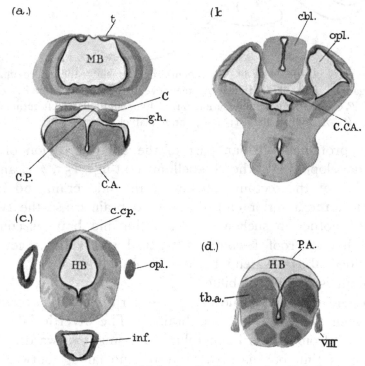

FIG. 157. Transverse sections through the brain of a larva on the ninth day, taken approximately along the planes indicated in text-fig. 158, by the lines 157 *a*, *b*, *c*, *d*. *C*, commissura?; *C.A.*, com. anterior; *C.CA.*, com. cerebellaris anterior; *c.cp.*, com. cerebellaris posterior; *cbl.*, cerebellum; *g.h.*, ganglion habenulae; *inf.*, infundibulum; *tb.a.*, tuberculum acusticum; *t.*, torus longitudinalis; *VIII*, root of eighth nerve. Other letters as in former figures.

attached by its apex by means of a fine band of connective tissue to the same piece of cartilage to which the paraphysis adheres; this is probably the epiphysis. There was no trace of these structures in the earlier stages. The presence of a paraphysis in *Gymnarchus* is of considerable interest, as this organ is not known to occur in other Teleosteans, according to von Kupffer (p. 152, Hertwig (64)).

The mid brain has expanded very considerably. Posteriorly the latero-posterior lobes project further back and are free of the walls and floor. The thin portion of the roof of the mid brain is seen as a dark triangular patch in the dorsal view

(Pl. XIX. fig. 33); the medium-tinted border is due to a thicker margin which appears to be partly the result of the presence of blood vessels (text-fig. 157, *a* (*t*)). The hinder part of the roof, as in the previously described stages, forms a vertical sheet of thickened much folded nervous tissue, joining with the anterior part of the roof of the hind brain, the cerebellum, which is also a vertical sheet. The whole structure forms the so-called valvula cerebelli. A very well marked commissure runs across from side to side, the fibres of which pass backwards along the upper part of the walls of the medulla and are lost somewhere in the neighbourhood of the roots of the fifth nerve (text-figs. 157, *b*; 158, *C.CA.*).

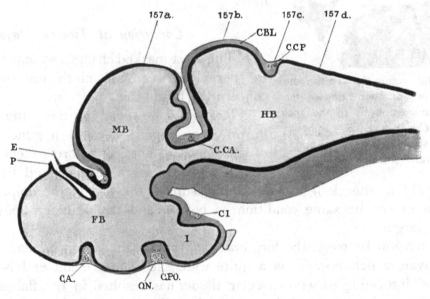

FIG. 158. Sagittal section through the brain of a larva of the ninth day. *P*, a diverticulum attached to a piece of cartilage. This diverticulum is probably the paraphysis. *E*, a strand of connective tissue which runs from the epiphysis to the above-mentioned piece of cartilage; *O.N.*, optic nerves. Other letters as in former figures. 157 *a*, *b*, *c*, *d*, indicate the lines of section of text-fig. 157.

A fine transverse band of fibres also occurs in the cerebellum itself, but in general the cerebellum shows so far very little differentiation into grey and white matter. Posterior to this the roof of the hind brain is as usual a broad sheet of single layered epithelium.

The cerebellar thickening is still bilobed, posteriorly at any rate; that is to say, the median line is as yet thin, but it is covered throughout by several layers of nuclei, while of the part which forms the vertical sheet, I am inclined to say that it shows no trace of bilobed origin. Farther back again, in undoubted hind brain, the roof is even thicker and certainly covered with nervous tissue, and about the level of the hinder extremities of the mid brain pockets, there is a distinct band of fibres (text-fig. 157, *c* (*c.cp.*)). After this the sides of the medulla fall apart and the roof becomes wide and thin, as typical of the hinder part of the fourth ventricle.

A compression of the medulla, probably owing to the upward flexure of the hind brain, shows itself by a great thickening of the walls of the medulla which almost meet in the middle line, a little behind the cerebellar commissure (text-fig. 157, *d*). Further back the walls fall apart again, but the whole cavity of the fourth ventricle becomes much reduced by the enormous thickening the floor of the medulla has undergone, chiefly through the development of nerve fibre.

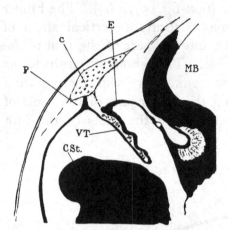

FIG. 159. Sagittal section through the fore brain to show the connection of the paraphysis and epiphysis to the roof of the skull. *C*, cartilage; *E*, epiphysis; *CSt.*, corpus striatum; *MB*, mid brain; *VT*, velum transversum.

Condition at Twelve Days.

The most marked changes which occur during the period between the ninth and twelfth days concern the hind brain.

Referring, however, to the anterior end as in previous stages, we find a general increase of all the parts noticed before except the paraphysis which is much reduced and lies close up against the roof of the skull, and the commissure in the wall of the epiphysis which is present in much the same condition as before, and the pituitary body which is certainly no larger.

A compression between the fore brain and mid brain and an increase in the size of the epiphysis which now forms a quite wide sac, whose distal end is somewhat branched and intermingled with vascular tissue, has resulted in the flattening down and backward pressing of the hinder part of the pallium, so that the cavity of the fore brain projects backwards as a pair of thin walled sacs which hide the ganglia habenulae from a lateral view (Pl. XIX. fig. 45).

With an increase of the cerebral masses the anterior commissure has increased. The post-opticae has not increased to the same extent and the posterior commissure though perhaps more compact than before is no larger. In a transverse section taken through these two commissures, a third narrow band of transversely running fibres may be seen joining the roofs of the optic lobes. This is one of several which run similarly across the thin part of the roof of the mid brain which further back becomes a thin sheet of nerve fibres overlying the epithelial layer. Near the valvula cerebelli the nervous layer becomes very thin, but I think it is never absent from the epithelium (text-fig. 160, *a* (*C.LO.*)).

The valvula cerebelli has its anterior wall, that is to say, its mid brain wall, greatly folded and the commissure at its base very prominent, but the greater thickening is to be seen in its posterior wall, the cerebellum, which now extends back some distance as a thick mass of nervous tissue, containing a prolongation of the cavity of the hind brain. One result of the thickening is that the wide "iter" narrows suddenly into quite a small passage under the valvula cerebelli and cerebellum, which widens out again posterior to the cerebellum. The cerebellum is hollow, but the lumen is small and apparently undergoing diminution by the coalescence of its walls and development from side to side of bands of nerve fibres (text-figs. 161, *a*, 162).

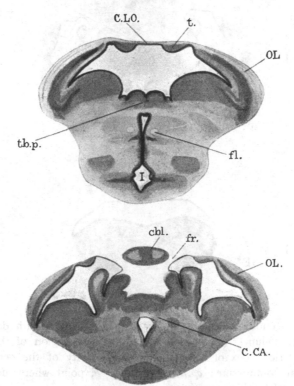

FIG. 160. (*a*) (upper section). A transverse section through the mid brain of larva of twelfth day. (*b*) (lower section), through the fissura rhombo-mesencephalica. *cbl.*, cerebellum anterior end; *C.CA.*, com. cerebellaris anterior; *C.LO.*, one of many transverse commissures between the optic lobes; *fr.*, fissura rhombo-mesencephalica; *OL.*, optic lobe; *tb.p.*, tuberculum posterius superius; *t.*, torus longitudinalis.

Moreover, the cerebellum also projects backwards over the commissura cerebelli a short distance. Posterior to the commissura cerebelli the roof is reduced to a single layer of epithelial cells, and is wide; but the cavity of the fourth ventricle is greatly reduced by the fusion of the sides at the point where close contact occurred in the previous stage—a canal however being left open below. This fusion occurs about the region of the auditory nerve—and separation again comes before the root of the ninth pair of nerves (text-figs. 160, 161). Comp. text-fig. 157, *d* (*tb.a.*) and text-fig. 161, *b* (*F*).

The base of the hinder part of the medulla is enormously thickened and the cavity gradually shallows and narrows by the approximation of the sides but widens again in the hindermost part of the medulla oblongata.

The tenth nerves arise about the end of the brain immediately after which the sides again coalesce and only a small, nearly circular, canal remains, which is continued into the spinal cord as the central canal, representing the ventral portion of the original vertical slit-like lumen. Between the roots of the eighth and tenth nerves in

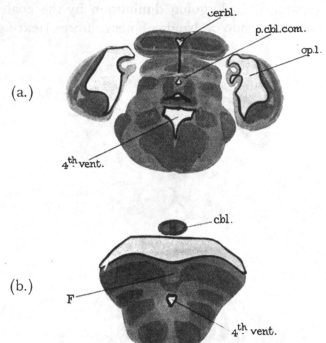

FIG. 161. Transverse sections of the brain of a larva of the twelfth day. (*a*) through the middle of the cerebellum. (*b*) through the hindermost region of the overhanging cerebellum. *cbl.*, posterior apex of cerebellum; *cerbl.*, cavity of the cerebellum; *p.cbl.com.*, posterior cerebellar commissure; *op.l.*, optic lobe; *F*, point where fusion has occurred in the region of the tuberculum acusticum.

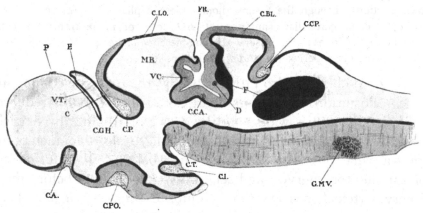

FIG. 162. Sagittal section of the brain of a twelve day larva. *C*, com. ?; *C.A.*, com. anterior; *C.BL.*, cerebellum; *C.GH.*, com. habenularis; *C.I.*, com. infundibularis; *C.LO.*, com. lobi optici; *C.CP.*, com. cerebellaris posterior; *C.CA.*, com. cerebellaris anterior; *C.P.*, com. posterior; *C.T.*, com. tubercularis; *C.PO.*, com. post-opticae; *D*, com. cerebellaris media; *E*, epiphysis; *F*, fusion in middle line of sides of brain; *FR.*, fissura rhombo-mesencephalica; *G.M.V.*, a median group of giant nuclei; *MB.*, mid brain; *P*, paraphysial diverticulum; *VC.*, valvula cerebelli; *V.T.*, velum transversum.

the median line exactly ventral to the central canal there is an elongated group of giant nerve cells, larger than any other in the central nervous system, or indeed in any part of the animal. Beneath them is a group of smaller, though still large, nerve cells. There are about 200 of these large cells (text-fig. 162, *G.M.V.*).

Forty-third Day.

This, the oldest stage, probably closely resembles the condition of the fully developed fish, for such indeed is the case with the other organs (Pl. XIX. figs. 35, 39, 46).

Here again we find that the more marked changes concern the hinder parts of the brain; the fore brain very early attains its final condition so far as its general form is concerned. The olfactory lobes, formerly ventral to the cerebral masses, have become drawn forward and are easily seen in dorsal view (text-fig. 120) and are so much constricted off from the cerebral masses as to form bulbs and stalks. They are connected with the cerebral masses by a curved plate of nervous tissue, containing two distinct tracts of fibres, though the ventral one is insignificant.

The olfactory lobes contain short lateral diverticula of the prosencephalic cavities.

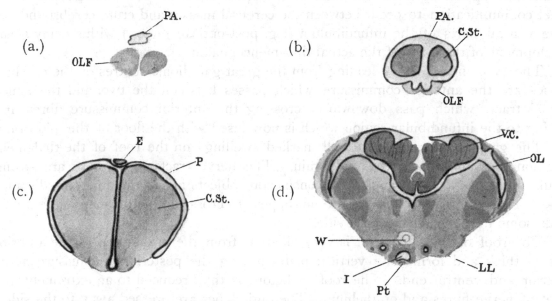

FIG. 163. Sections taken transversely through the brain of a forty-third day larva along the lines indicated approximately in text-fig. 164, by the lines 163 *a, b, c, d.* In (*a*): *PA.*, pallium; *OLF*, olfactory lobe. In (*b*): *PA.*, pallium; *C.St.*, corpus striatum; *OLF*, olfactory lobe. In (*c*): *P*, pallium; *E*, epiphysis; *C.St.*, corpus striatum. In (*d*): *I*, infundibulum; *LL*, lobus lateralis; *OL*, optic lobe; *Pt*, pituitary body; *V.C.*, valvula cerebelli; *W*, basilar artery.

The cerebral masses (corpora striata), which have increased slightly in size in proportion to the optic lobes, are closely pressed to one another so as to obliterate the cavity between them except in the ventral region, where it persists. There is no actual fusion, though so close are they that it would be quite impossible to take a section between them (text-fig. 163, *c*).

The pineal diverticulum, the epiphysis, remains as a hollow epithelial pouch covered at its distal end by blood vessels. Its walls are epithelial, at any rate distally, although of more than one cell in thickness. The posterior surface as it approaches the mid brain becomes covered with nervous tissues and it is traversed by the commissure in its posterior wall, which has, however, undergone no increase in size from quite an early stage (text-fig. 164, *C*). It has, really, slightly diminished in size and is now extremely difficult to trace into the floor of the fore brain.

The paraphysis is still present as a very small hollow diverticulum no longer closely attached to the roof of the skull, but lying along the roof of the brain, projecting backwards (Pl. XIX. figs. 34, 35, and 45, 46).

With a general increase in bulk of the sides and dorsal wall of the infundibular portion, a tendency to a fusion between it and the floor of the mid brain has gone on, being partly an actual fusion and partly an incorporation by general growth in width, not accompanied by growth in length to the same extent. Thus the commissurae post-opticae, which before in a transverse vertical section formed the base of the infundibulum, now appear to be at the base of the crura cerebri. It is thus possible for a direct communication to occur between the cerebral masses and crura cerebri and the more ventral parts of the infundibulum (e.g. post-opticae region) without any great development of the sides of the actual thalamencephalon.

The main nerve tracts leading from the great ganglionic centres of the cerebral masses are the anterior commissure, which passes between the two, and the longitudinal tracts which pass downwards, crossing the anterior commissure fibres and passing to the infundibular region which is now fused with the floor of the mid brain.

The ganglia habenulae are well marked swellings on the roof of the thalamencephalon just anterior to the mid brain. The nerve tracts from these are strong circular bundles of fibres which form conspicuous objects in sections for some distance back. The majority of the fibres do not appear to cross, although each ganglion now sends some fibres to the opposite side.

The roof of the mid brain is very different from the usual type. The anterior part is thick and forms one vertical plate, having the posterior commissure at its anterior and ventral end. The roof as before is thin, reduced to an extremely thin sheet of nerve fibres and epithelium. The optic lobes are pressed away to the sides. The ridges along their edges, bordering the thin sheet, are formed of ganglionic tissue, the tori longitudinales.

A great increase in complexity has occurred in the character of the valvula cerebelli. The transverse commissure is of moderate size, and the mid brain half of the valvula is thrown into numerous and labyrinthine folds. The hinder or hind brain half is much thicker, but without the complexity of foldings. Its sides fuse in the middle line (text-fig. 163, *d* (*V.C.*)).

The cerebellum proper is not large but forms a slightly bilobed hollow outgrowth of the dorsal wall of the hind brain which overlaps the rest of the brain both in front and behind to a small extent.   In general form it much resembles the cerebellum of an Elasmobranch.   The cerebellar commissure is enormous (text-fig. 164, *C, CP.*). Immediately posterior to this, the roof widens and is reduced to a single layer of cubical cells, which further back become squamous.

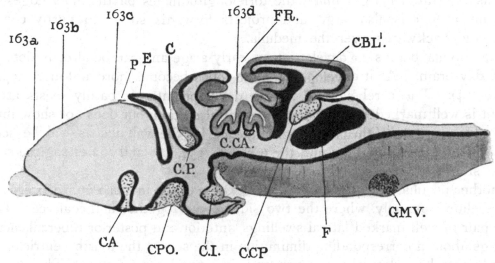

Fig. 164.   Sagittal section of the brain of a forty-third day larva.   The lines 163 *a, b, c, d* indicate the planes along which the sections of text-fig. 163 are taken.   The line *c* should however be farther to the right so as to cut the epiphysis and pass in front of the anterior commissure.   Other letters as in text-fig. 162.

The hind brain is perhaps reduced in length, but its walls are enormously thickened at four points to form the great swellings (Pl. XIX. figs. 35, 39, 46 *t.a.p.t.a.a.*), which, as shown in the previous stage, fuse together by their inner surfaces and so obliterate a large portion of the cavity of the fourth ventricle.   At the extreme posterior end the cavity reappears and shallows slightly as it passes backwards.   Then the walls come together and obliterate the dorsal part of the lumen and leave only the small circular cavity of the central canal of the spinal cord.   Histologically by far the most conspicuous object is the great group of giant nerve nuclei, which occupy the mid ventral line of the medulla about the region of the roots of the vagus nerve, posterior to the eighth nerve referred to in the previous stage (text-figs. 162, 164, *G.M.V.*).

General characters of the Brain.

Plate XIX. contains figures of dorsal views of the brain on the seventh, ninth, twelfth and forty-second days, figs. 32—35; of ventral views of the same dates, figs. 36—39; of side views of the same dates, figs. 43—46; and of side views of three earlier stages, namely four days, fig. 40; five days, fig. 41; and six days, fig. 42.

The oldest stage (figs. 35, 39, and 46) closely resembles the adult brain.

Figs. 34 and 35 have lost the thin pallium forming the roof of the fore brain together with the epiphysis.   Its extension is indicated by the dotted line.

The most peculiar features are the enormous development of the valvula cerebelli which is no doubt the cause of the attenuation of the tectum opticum and displacement of the optic lobes to the sides.   This organ is seen in fig. 35 through the thin tectum as a serrated margin lining the tori longitudinales on the inner edges.   The cerebellum proper is also large, and projects forwards so as to partly cover the valvula, and backwards over the medulla.

The valvula begins to develop at an early stage and can be already detected in the fifth day brain.   As it develops the optic lobes become more and more separated (figs. 33, 34).   The cerebellum is a later development.   It hardly exists at seven days, but is well marked at nine days in sagittal section, but does not show much in surface view until it is thrust into prominence by the valvula, as will be seen by comparison of figs. 34 and 35 and the text-figs. 154, 158 and 162 of sagittal sections of these stages.

Another peculiarity is the enormous thickness of the medulla immediately behind the cerebellum—namely, where the two sides come together and coalesce.   Here it forms a pair of well marked lateral swellings, anterior and posterior tubercula acustica, and brings about a corresponding diminution in the size of the fourth ventricle.

The ventral and lateral views present less unusual appearances.   The absorption of the infundibular region into the crura gives a very compact look to the forty-third day and adult brains.

The whole brain is extremely bulky and concentrated and entirely fills the cavity of the cranium in the larval stages.

The two nerves with ganglionic swellings seen in fig. 35 are the lateral line nerves which are unconnected with the vagus and join the brain close to the roots of the facial nerve.

As compared with the brains of other *Mormyridae* the brain of *Gymnarchus* shows less pronouncedly the special features of that family.

There is a great development of the valvula cerebelli, identified as such by Sanders (108) ("eigenthümliches Organ" of Marcusen), but not to so great a degree as in *Mormyrus longipinnis* (Marcusen (89)) or *Hyperopisus dorsalis* (Sanders (108)) or even of *Petrocephalus isidori* (Marcusen (89)).   In the two former species the valvula appears to be enormous, projecting so far forwards as to hide from view dorsally the whole of the prosencephalon.   In *Petrocephalus*, although the valvula projects forwards so as to cover the optic lobes, it stops short of the anterior part of the prosencephalon.   In *Gymnarchus* the valvula is seen shining through the thin and greatly widened tecta lobi optici, or more exactly the Deckplatte of the mid brain, and although it pushes the optic lobes to the sides it does not rise up above them, nor does the cerebellum project much further backwards or forwards than it does in, for instance, the Salmon.

Reference to my figures (Pl. XIX. fig. 43, text-figs. 163, *d*, and 164, *V.C.*) illustrates sufficiently well the relationship of the valvula to the optic lobes. By a still greater development of this organ the condition of *Mormyrus* might be obtained by its rising up through the fissura rhombo-mesencephalica and spreading out forwards and laterally.

It is not possible from Marcusen's and Sanders' figures of longitudinal sections of the adult brain to make out exactly the cavities of the brain and the thin parts of the roof, but it is probably not correct to speak as Sanders does of the valvula breaking through the tecta lobi optici. It is far more probable that it should rise up as here suggested by way of the fissura. The tecta lobi optici or Deckplatte of the mid brain is in *Gymnarchus* reduced to a single layer of ependymal cells during the early stages of the separation of the optic lobes. If one had a sufficiently complete series of stages it would be an interesting region in which to watch the growth of nerve fibres which develop across it at a later stage.

The fusion of the sides of the medulla in the region of the tubercula acustica described above gives rise to the medium tuberculum impar, which seems to be a character of the *Mormyridae* as of some other fishes, *Cyprinus* (Sanders). There is only a short fusion of the walls of the optic lobes and this occurs at the extreme anterior end.

The thinning out of the Deckplatte of the mid brain gives to the sagittal section of an advanced stage (text-fig. 162) an appearance which at once recalls that of a section of the brain of *Ammocoetes* (*v.* K. von Kupffer (85), p. 52) ; but this is of course a chance resemblance since the thin roof of the mid brain of the *Ammocoete* is a choroid plexus and quite devoid of nerve tissue, while that of *Gymnarchus* has nerve fibres eventually and is not vascular. On the other hand, a certain resemblance occurs in the presence in *Ammocoetes* of the very large commissure at the extreme posterior margin of the cerebellar tissue, and of the bundles of fibres in the roof of the fore brain anterior to the posterior commissure, and to the commissure of the ganglia habenulae, and posterior to the epiphysis. These correspond roughly in position to the commissure described previously in the posterior wall of the epiphysis of *Gymnarchus* (text-fig. 162, *C*, cf. p. 52, v. Kupffer (85)), but I do not know what centres of the brain these commissures in *Ammocoetes* connect.

This commissure in the posterior wall of *Gymnarchus* is interesting for several reasons. It is not easy to identify it with any commissure in other Vertebrates (unless the above suggestion with reference to *Ammocoetes* has any foundation); for it is certainly not part of the posterior commissure because its fibres arise in the walls or floor of the fore brain. It skirts close by the ganglia habenulae but is in no way connected with them.

Minot (92) lays great stress on the importance of the superior commissure, which he says persists in all vertebrate types. It does not seem to be present in *Gymnarchus* in the usual position. The commissure under discussion cannot be the superior commissure, since the latter is always between the epiphysis and the velum transversum, whereas the *Gymnarchus* commissure is posterior to the epiphysis.

The superior commissure of Osborne and Minot is the same as the commissure called by Kupffer the commissura habenularis. The ganglia habenulae are present, but the decussating fibres do not appear till late, and then they pass posterior to the epiphysis instead of anterior to it, and do not form a strong bundle.

The commissure in the posterior wall of the epiphysis is interesting, also, because it develops very early (unlike the superior commissure of other types) and attains a considerable size and then increases no more, and even decreases by the forty-third day.   Its approximate course is shown in text-fig. 156, C.

The commissures of the cerebellar region are also difficult to identify.   The most posterior, which I have called commissura cerebellaris posterior (text-figs. 162 and 164, C.CP.), is a large bundle whose fibres quickly gather together from the substance of the sides of the hinder part of the cerebellum, some coming more ventrally and from farther forwards.   They are clearly fibres which cross from side to side from the lateral regions of the cerebellum and are intra-cerebellar.   The middle one, D, commissura cerebellaris media, is a group of bundles of fibres which connect the anterior regions of the cerebellum.

The commissura cerebellaris anterior is a compact bundle of fibres, the bulk of which can be traced downwards and backwards into the lateral and ventral walls of the medulla oblongata.

This latter commissure (C.CA. in the figures) seems to occupy the position at the base of the valvula corresponding to that of the decussatio trochlearis.   I cannot make out that there is any connection at all with this and the fourth nerve, which can be seen to leave the brain in the angle just in front of the commissura cerebellaris media.   The nerve is very small and its fibres can be traced downwards and slightly forwards in the direction of the ganglion interpedunculare.

From the course taken by its fibres this commissure seems to correspond with the chief cerebellar commissure of amphibia.

The fact of its fibres passing back as thick bundles to the base of the medulla seems to indicate that it is an old and fundamental tract.   The other two, whether both or either correspond to the commissura cerebellaris of Kupffer's description or not, are more local and so very likely may be special developments.

Cranial Nerves.

The cranial nerves develop in the normal way, namely, the olfactory in connection with the olfactory epithelial patches, the optic in connection with the optic cup and stalk, and the fifth and seventh probably from a cranial neural crest, which is present on the fourth day (text-fig. 94, p. 306) as a heap of large, loosely-packed cells lying between the central nervous system and the skin.

This part is in front of the notochord.

The two wing-like expansions (text-fig. 88, a, p. 298) may possibly be the origin of the eighth, ninth, and tenth cranial nerves.   This, however, is little better than guess-work.

In the four-and-a-quarter day specimen a great advance has been made. In the short space of six hours all the neural crest has been broken up and transformed into the ganglia of the several nerves. The series is unfortunately imperfect, so that it is not possible to state the exact conditions.

On the fifth day the position and relation of most of the nerves and nerve ganglia can be determined; and the brain and spinal cord are differentiated in part into grey and white matter (text-fig. 152, p. 376).

First Pair of Cranial Nerves.

The olfactory nerve is not for certain distinguishable on the fifth day, though the olfactory epithelia are present as thickenings of the nervous layer of epiblast on the sides of the head. Immediately opposite, but, as far as I can see, not actually in continuity with it, is a thickening of the fore brain and special development of "white" matter, or at any rate of "Randschleier."

In the five-and-a-half day specimen the olfactory nerve is present as a bundle of fibres passing part way to the brain from the epithelium, a shrinkage cavity, however, occurs between the epithelium and the olfactory end of the bundle so that only one fibre can actually be traced across it. It runs towards the ventral end of the thickening of the fore brain mentioned above, but does not reach it. It is present also on the other side, though its cerebral termination is less easy to see; but the olfactory epithelium shows more distinctly the origin of the fibres.

As development proceeds the nasal epithelium comes to lie further back than at first, so that the olfactory nerve instead of lying transversely to the main axis of the animal, lies from its point of attachment to the brain backwards towards the olfactory epithelium.

On the eighth day the points of origin of the olfactory nerve in the brain are two swellings at the base and extreme anterior end of the prosencephalon. These are the olfactory lobes, and are really in front and ventral to the corpora striata. Near their anterior ends they lie in close contact with two other little thickenings of the epiblast, which are the anterior ends of the skin sense organ system. The olfactory nerve is single at its cerebral end, but splits into several branches before joining the epithelium, two of which may be regarded as more important than the others. On the ninth day the olfactory opening becomes divided so that henceforward there are two external nares on each side. The olfactory epithelium occupies a position about mid-way between the two nasal openings.

On the formation of the jaws and mouth the position of the nasal pits with regard to the brain changes considerably, so that by the eleventh day these organs no longer lie posterior to the lamina terminalis, but have become turned forwards and upwards so as to assume their ultimate position anterior to and slightly dorsal to the fore brain. In consequence the nerves also now run forwards instead of backwards, the pivot of the movement being approximately the wall of the skull.

Second Cranial Nerve.

The optic nerve fibres first appear, I think, a little later than the time of appearance of the olfactory nerves. They are not easy to trace. Possibly they may be present at the optic vesicle end of the stalk at five-and-a-half days, but I think they are not present at the brain end as yet, but it is extremely difficult to say for certain.

On the seventh day the fibres extend in large numbers along the whole length of the stalk between the eye and the brain. Close to the eye the lumen of the optic stalk is still widely open and the fibres run along the ventral and lateral but not along the dorsal surfaces of the stalk. In a way they may be said to run between the cells of the optic stalk. The appearance resembles very closely that in the Frog and Elasmobranch, where the optic nerve has been said to be developed not from the stalk itself but external to it. *Gymnarchus* is not a favourable animal for determining the actual method of fibril development.

Nearer to the brain the lumen of the stalk disappears and the cells of the wall of the stalk seem to be getting pulled apart as in other Vertebrata.

Mr J. T. Gradon in the *Quarterly Journal of Microscopical Science*, Vol. 50, p. 479, writes: "Just over a year ago I began to feel dissatisfied with Assheton's conclusions that the cells of the optic stalk do nothing more than serve as a conductor for the fibres of the optic nerve," and a little farther on he gives a quotation from Robinson (107) in which it seems to be suggested that my conclusions point to the possibility of the framework of the optic nerve being mesoblastic in origin and not part of the spongioblastic tissue of the brain.

Now that was not my contention at all. I did not go very fully into the fate of the optic stalk cells because at that time I was concerned chiefly in attempting to prove:

(i)   that the optic nerve fibres are not developed from the optic stalk,

(ii)   that most of the fibres "grow" from the retina towards the brain,

(iii)   that they pass over the edge of the optic cup and do not pierce its wall,

(iv)   that the choroidal fissure is a necessary consequence of the myelonic origin of the vertebrate eye.

With these hypotheses, however, Mr Gradon does not quarrel. Nor do I see why he should object to the statement that the stalk serves as a conductor of the nerve fibres, for it certainly does so.

I never said that the optic stalk does "nothing more than serve as a conductor for the fibres of the optic nerve." What I did say was "the optic stalk takes no part in the formation of the nervous parts of the organ of sight" and again, p. 1, "As the nerve fibres increase in number they seem to tend to grow in between the cells of the walls of the stalk, and eventually the walls of the stalk become completely broken up, and the cells remain separated from one another, and lie among the fibres of the optic nerve" and I show them as spongioblastic cells in my figure 10 (6).

If, as I believe, the separation of the two layers of epiblast in the Frog is a separation into spongioblastic and neuroblastic elements then one would expect the stalk cells which are continuous with the lining cells of the general brain cavity to assume the same *rôle* in the stalk as elsewhere.

Mr Gradon has, however, followed their history much more carefully and completely than I did and his account of the effect of the development of the lymph system is interesting, although I do not understand whether the lymph spaces are due to ingrowth of mesoblast, or whether they are chinks in between and among the nervous and spongioblastic elements. With reference to one point, namely, whether the first formed nerve fibres are actually outside of or in between the stalk cells—that is to say outside or within the "membrana limitans externa" [ought not this to be called "membrana limitans interna," as it is continuous with the interna—not the externa of the retina?] I quite agree that he is probably right and my figure probably wrong. This would be in accordance with the account given by Froriep in Hertwig's *Handbuch*, and figures of Froriep and Grönroos, Studnicka and Muthmann for the Duck, Grass Snake, Torpedo, Rabbit and the Frog itself, to which I may probably add *Gymnarchus*, although as I explained above, the early stages are not favourable, and so I cannot write with certainty.

Third Cranial Nerve.

On the seventh day the fibre of the third cranial nerve can be found arising from nuclei close to the epithelium of the central canal in the floor of the mid brain (just below where the roof begins to thin out). The nerve can be traced as far as one of the eyeball muscles close to the eyeball without branching, after which it is too faint to be followed. There is no ganglionic swelling. A nucleate sheath surrounds the whole.

There is nothing particular to be noted about its future history.

Fourth Cranial Nerve.

The pathetic nerve is very small. It is seen easily on the twelfth day arising from the angle of the valvula cerebelli.

Fifth Cranial Nerve.

The trigeminal is connected with the brain close to the ventral surface ; it has a large ganglion as usual. The earlier stages have been dealt with already, p. 375. On the seventh day the nerve strands are still difficult to trace, excepting the mandibular, but later the three usual branches take a normal course.

Sixth Cranial Nerve.

This can be detected on the seventh day arising from the floor of the medulla oblongata.

Seventh, Ninth and Tenth Cranial Nerves.

The facial nerve is first indicated by several ganglionic masses united (text-fig. 152, p. 376).   One mass is in contact with the dorsal end of the rudimentary hyo-mandibular cleft.   There is what appears to be a strong ophthalmic branch which is connected by fine strands with the skin sense canals of the head.   The root of this branch passes backwards and does not fuse in the same way as the other branches to form a mass directly connected with the brain by a strong band of nerve fibres, but seems to keep apart outside the general ganglionic mass of the facial and mingles with the roots of the auditory nerve and nerve of the lateral line.

At a later stage, sixteen days, the fifth, seventh, eighth, and lateral line nerves, though coming off from the brain close together have quite separate roots.

Of these the fifth is most ventral and anterior, arising from a swelling just in front of a larger swelling, the tuberculum acusticum, which is so large as to meet with its fellow and obliterate part of the cavity of the fourth ventricle (text-fig. 157 *d*). From this, the tuberculum acusticum (against which the vesicle of the air bladder lies at an early stage), there arise posteriorly and dorsally the lateral line nerve, more ventrally and anteriorly the facial nerve, and the auditory nerve immediately below.   The ninth arises close to the lateral line nerve and its fibres run along with the lateral line until they pass together out of the skull.   There is no connection so far as I can see between the vagus and lateral line nerves, although they cross as they pass out from the skull.   The roots of the facial and trigeminal are quite separate, but many fibres pass from their ganglia to the several branches of each.

Spinal Nerves.

Between the skull and the first vertebra three pairs of nerves pass out.   The first two are of nearly equal size and are larger than the third.   Each nerve has a dorsal root with ganglion and a ventral root.   These roots come off one above the other, those of the first two pairs arise some distance within the skull, and running backwards all three pass out together from the foramen magnum.   The first is the hypoglossal and runs ventralwards and forwards to the tongue.   The second runs some distance alongside the hypoglossal, then turns backwards and, together with the third, supplies the pectoral girdle and fin.

The fourth spinal nerve leaves the cord by two roots as usual, both of which pass out between the neural arches of the first and second vertebrae.   This also supplies the pectoral girdle and fin.   The hypoglossal and second spinal nerves are anterior to the myotomes.   The third runs just in front of the septum dividing the first from the second myomeres.   The third, fourth, etc., occupy corresponding positions with reference to the succeeding myomeres.

The Organs of Special Sense.

The olfactory organs present no peculiar features. The development of the nerves has already been discussed. The thickened epithelium becomes invaginated so that a deep pit is formed, and by the approximation of its lips it acquires a double opening to the exterior. The sac then elongates so that a canal is formed through which presumably a current of water flows. As the epithelium of the surface turns into the pit, it loses its characteristic stratified appearance. The normal stratified epithelium, consisting of a basal columnar layer with superjacent more and more squamous cells, is replaced by a stratified layer of about the same thickness in which the outer layer is columnar.

The middle part of the canal is the sensitive region. Here the sac widens and is divided into two channels or grooves by a median ridge. Only the grooves are apparently of olfactory epithelium. The deeper layers of these groove regions form the nervous elements which are continuous with the olfactory nerve.

Optic Organ.

The eye needs no long description as it develops according to the ordinary mode. It is, comparatively speaking, a small organ; and the choroidal fissure closes very early after the establishment of the optic nerve and formation of the retinal artery.

The lens, after the inner wall has become greatly thickened by the elongation of its cells in the ordinary way, appears to become invaginated, the cavity of invagination, however, being a narrow pit.

Ultimately the nuclei disappear from the elongated cells and only the anterior wall retains its cellular structure.

In the oldest stage the retina is practically fully developed. The posterior wall of the optic cup by that time consists of a single layer of cells whose sides next the retina are produced into long processes filled with fine grains of pigment. The other side of the cell is devoid of pigment and here the cells seem to be continuous with one another. The nucleus lies just within or on the border of the pigmental region.

Outside the layers derived from the optic vesicle, there is a mesoblastic layer of a rather loose character, the inner part of which is closely applied to the pigment layer of the retina, which is highly vascular and pigmented; the outer part, which is continuous in front of the eye with a thin sheet of tissue, the developing cornea, is of some tenuity, but represents the future sclerotic. The skin still forms a thick layer over the eye, not in any way different from the skin elsewhere.

The visual cells tend to form bundles and run together between the larger processes of the pigment cells.

Pigment first appears on the seventh day. On the sixth day there is no trace of it. The space between the lens and the retina is entirely filled by a blood vessel.

The condition histologically of the retina is much the same as in the brain, that is to say, white matter (fibre) is developed on the outer surface, but there is as yet no appearance of fibre among the mass of nuclei.   Just as in other parts of the brain the main tracts of fibres are in process of being laid down, so here the main tract from retina to the rest of the brain is partially developed.   On the seventh day differentiation begins to appear in the nuclear mass.   The outer nuclear layer may be seen to be composed of more deeply staining and more elongated nuclei than elsewhere, and a considerable mass of fibrillar tissue now marks the area of the future inner molecular layer, differentiation as usual occurring in the central regions of the retina before the edges.

On the eighth day all the main regions are with difficulty distinguishable, except the visual cells.   These make their first appearance beyond the limiting membrane on the ninth day, and on the same day the pigment layer shows signs of the growth of its pigment-bearing processes.   The developing rods and cones are of small size and difficult to make out.

The Auditory Organs.

On the fifth day the auditory organ is the usual simple almost spherical sac lying on each side of the head, close to the medulla oblongata and attached to the auditory ganglion.   It subsequently undergoes the usual conversion into sacculus, and utriculus and three semicircular canals.

On the eighth day the labyrinth consists of a ventral posterior chamber, the sacculus, slightly constricted off from the more anterior and dorsal utriculus.   From this utriculus the anterior vertical semicircular canal arises at its anterior end, and a horizontal semicircular canal, which is throughout much narrower than the vertical canals.   The anterior vertical semicircular canal joins high up with the posterior vertical ; the horizontal canal opens behind into the utriculus.

The inner wall of the sacculus, the ventral wall of the utriculus and the inner wall of the horizontal canal are of thickened epithelium.   The rest is thin.

On the tenth day (cartilage having been formed by this time) the auditory labyrinth is partly embedded in the cartilages of the auditory capsules.

The sacculus, which in the previously described stage was only a constricted recess of the utriculus, has now become constricted so much that it remains connected with it only by a very narrow canalis reuniens which is produced upwards as a ductus and saccus endolymphaticus.   This is an extremely narrow tube running up alongside the sinus utriculi superior and ends in a slight dilation within the skull between the anterior semicircular canal and the brain.   The utriculus is wide in front containing a patch of thickened epithelium as its ventral wall, which is continued into the beginning of the anterior semicircular and horizontal canal ; in the latter it forms a high ridge to which a branch of the eighth nerve runs.   The utricle lies up against a part of the ganglion of the auditory nerve, while the sacculus lies against another part.   The wall of the sacculus next the ganglion is much thickened. Against the opposite wall of the sacculus is the terminal bulb of the anterior prolongation of the horn of the swimming bladder which lies between the sacculus and the horizontal semicircular canal.

The floor of the utricle at the posterior end, where the sinus utriculi superior arises, is thickened. Nearly the whole of the horizontal semicircular canal is embedded in cartilage, and so is most of the posterior vertical canal, but the anterior canal which is by far the largest of the three (the posterior is the smallest) although protected on the outside by cartilage is embedded only for a short distance at its anterior end.

By the oldest stage, forty-three days, no further changes of great moment take place. The relation of the membranous labyrinth to the cartilaginous part of the skull remains unchanged. The canals, where they lie within the skull, are somewhat more roomy and the two anterior canals lie on the roof of the cerebellum and almost meet.

The swimming bladder bulb has grown considerably and now lies higher up above the cartilage of the horizontal canal, and no doubt forms a kind of tambour by means of which vibrations are perceptible from the external world by way of the skin, the supra temporal membrane bone, the bulb in question, and the sacculus, which is provided with a rich nervous supply (Pl. XVIII. fig. 31).

The lagena is very long and is embedded in bone, and has its special nerve supply; on its inner wall it has thickened epithelium which bears stiff hairs.

The sacculus and lagena are, I believe, quite separated from the utriculus, which lies wholly within the skull.

Sense Canals and Lateral Line Organs.

The lateral line system consists, when fully formed, of a long closed canal which extends from the tail forwards on each side of the animal about the level of the notochord. In the head region, about the level of the ear it branches, the main stem continuing over the eye and curling round in front and ventral to the olfactory organ. Branches pass off ventralwards between the ear and orbit. Histologically it resembles very closely the auditory labyrinth. The greater part of its walls are thin, but here and there occur patches of sensitive epithelium as in the membranous labyrinths. Many of the smaller canals of the head region are embedded in the membrane bone of the skull. The canal is apparently developed as a thickening of the nervous layer of epiblast which subsequently becomes either folded up or hollowed out. It forms first in the region of the auditory vesicle and extends forwards and backwards.

On the eighth day it is free from the skin and present as a wide tubular canal with uniform columnar epithelium in the region of the ear, but anteriorly it is in many places still solid and attached to the skin. On each side the most anterior point of attachment is slightly in front of and ventral to the nasal epithelium. The second is nearly above it, the third more dorsal but also further back, the fourth posterior to the third, the fifth about the level of the hinder part of the mid brain.   Between these points of fusion with the epiblast lie the canals.

So, also, passing backwards the canal becomes continuous with a thickened ridge of epiblast.

Skin Sense Organs.

In addition to the lateral line canal proper and its branches, many small sense organs occur scattered over the skin, chiefly over the head region, and more especially the front.

There are two kinds, one of which is far more numerous than the other, perhaps in the proportion of twenty to one.   The more numerous one is more deeply situated than the other.   The innermost layer of the stratified epidermis projects inwards and gives rise to the organ.   There is a mushroom-shaped cell with a large nucleus. From its concave surface a clear rod, perhaps of mucus, arises and penetrates the main layer of epithelium.   The convex surface of the cell is clothed by a layer of cubical cells, which are continuous with the epithelium.   Each organ has a blood capillary in connection with it and a nerve.   The less numerous organ is more superficial, and consists of a bundle of tall columnar cells, which are cells of the innermost layer of the epithelium, greatly elongated.   Of these a little central column of cells appears to represent the real sense cell.   To these the nerve passes.

6.   THE SKELETON.
The Vertebral Column.

The skeleton in the oldest stage available is in a less advanced condition than the other organs.   The notochord is remarkable for the long distance that its anterior end is from the anterior end of the nerve tube.   At no time does it extend further forward than the hinder part of the hind brain.   At an early date, before the fifth day, the characteristic vacuolation of the central cells takes place.   The notochord on the seventh day, which is the day before the first appearance of cartilage, extends from the hind brain to the growing point behind. Its front end is free, and shows certainly one mitotic figure.   It possibly extends forward a little way by its own growth but probably not to any great extent.

There is no sheath or any sign of a concentration of tissue suggestive of cartilage formation.   The mesoblast surrounding the notochord is in no wise segmented, except so far as segmentation is suggested by the passage through it of blood vessels and nerves, which are as usual arranged segmentally.   The limiting layer of cells of the notochord is formed by its own peripheral cells, which, being not vacuolated, are capable of forming an epithelium-like superficial layer.   This condition is retained up to the oldest stage when cartilage and bone are formed, and even at this time nuclei are abundant throughout the chorda, and are not restricted as in Dog-fishes to the epithelial layer.   On the eighth day cartilage or a concentration of mesoblast occurs in both the head and trunk region. One day later the foundation of the cartilage is formed throughout, wherever it is going to form. Round the notochord there is a thin hyaline sheath—the notochordal sheath—a little later a second sheath forms. The notochord is quite unconstricted; and attached to its sheath but developing outside it are series of cartilages, two series along the base and two along the upper part. These latter are larger and longer and project upwards, embracing the nerve tube. These are the first pieces of cartilage to be formed in the trunk region and are the basi-dorsals and basi-ventrals of Gadow (47, 48). They are arranged so as to correspond with the septa between the myotomes. The two dorsal pieces grow upwards and meet above the neural tube and form a neural arch, and in the older stages point backwards. Attached to each of the two lower pieces a separate piece of cartilage, the rib cartilage, develops. In the caudal region the basi-ventrals are produced downwards so as to form a haemal arch.

The cartilaginous investment is not completed up to the forty-third day and probably it never does form a complete investment. The area of the cord covered by the cartilaginous rib-pieces is increased somewhat by the growth in bulk of each piece, but the pieces do not join. The centrum of the vertebra is formed by a sheath of bone which develops from the skeletogenous sheath that surrounds the hyaline membrane of the notochord. This skeletogenous sheath appears at the same time as the hyaline sheath and cartilage, as a single layer of cells surrounding the hyaline sheath. This skeletogenous sheath is outside the cartilaginous basi-dorsals and basi-ventrals, and bone is formed upon the notochordal sheath everywhere except where the cartilages cover it, and round a ring between each set of cartilages, the future intervertebral ligament. At first the bone only ensheathes the bases of the cartilages. About the same time bone is being formed upon the apices of the basi-dorsals. It increases its extension so as to cover the cartilages which are thus invested with a coat of bone. In an actual sagittal section of the centrum one sees no cartilages at all with the exception of the apices of the neural spines now grown out as long backwardly projecting spines. The mid-ventral and dorsal lines of the notochord are ensheathed in bone and intervertebral connective tissue.

In the dorsal fin there are two somactids to every segment, i.e. to every neural spine (with which they have no connection).

There is a long axonost and short baseost. The axonosts are entirely cartilaginous, but a cap of bone is to be seen on the baseosts, and the lepidotrichia are bony from the beginning. In the fin itself are fibrous bands, the dermotrichia, and covering these are long blades of bone with expanded bases the lepidotrichia, to which the muscles are attached, which have their origin in the connective tissue and basal portion of the axonost.

To sum up—the vertebral column is formed entirely from the skeletogenous layer surrounding the notochord which retains its cellular character (as in the Urodela among Amphibians) and is surrounded by an inner and outer chordal sheath.

The vertebrae are formed upon the arco-centrous type, and are laid down partly in cartilage, but partly in bone from the beginning.

The cartilaginous elements are restricted to the basi-dorsals and corresponding neural spines, and to the basi-ventrals with their ribs.

The bases of these, although broad, do not coalesce, nor is any cartilage formed in the intervening spaces. The greater part of the centrum of the vertebra is formed from the first in bone which arises from four separated centres of ossification between the bases of the arcualia partially clothing them with bone. An investment of bone very soon covers the dorsal ends of the basi-dorsals (where they pass over the dorsal longitudinal ligament) and the neural spines and the distal ends of the ribs, but the somactids remain uncovered for a longer time, and the lepidotrichia are, as always, bone from the first (v. Pl. XXI. fig. 57).

In addition to these there are ossifications of some importance formed in the outer skeletogenous layer about the level of the dorsal surface of the spinal cord outside the basi-dorsal cartilages (v. fig. 57, z).

These are quite separate from the neural arches of the forty-third day, but probably fuse later and form the zygapophyses-like processes of the neural arches of later stages (v. Hyrtl, Pl. I., and Erdl, Pl. V. fig. 42 A).

The Skull and Visceral Arches.

Sixth Day.

Skull.

The earliest trace of the formations which lead to skeleton structure occurs on the sixth day, when concentrations of tissue are to be seen under the auditory vesicle. The notochord is embedded in the loosest of tissues; there is no trace of parachordals.

On the next day, in addition to the concentrations beneath the auditory capsules, there are very decided strips along the ventro-lateral borders of the brain, the trabeculae, perforated, or divided by the internal carotid artery.

Above the anterior region, concentrations occur at the sides of the mid brain, where at a little later period they form the thick supports to the optic lobes (text-fig. 165). These develop quite independently of the trabeculae which are present further back and ventral to the thalamencephalon. They arise just inside of the lateral line sense tubes.

The auditory capsule concentrations have extended upwards outside the vertical canals.

The trabeculae are now continuous behind with thicker concentrations—the parachordals, which lie alongside of the anterior end of the notochord.

The visceral skeleton is in a more advanced state than the skull. The concentrations more nearly simulate cartilage. The palato-pterygo-quadrate bar is a strand of concentrated tissue lying close to the stomodaeal epithelium and converging, as it passes forwards, with its fellow and lying in close approximation to, but, I think, not fused with, the trabecular concentration.

As it passes backwards it broadens into a thick nearly vertical plate, and in the region of the auditory capsule and point of origin of the ninth nerve, it is separated from another concentration, the hyo-mandibular, which is a massive centre of growth and far removed from the auditory capsule. The arches are difficult to distinguish at this early stage.

Eighth Day.

Skull.

On the eighth day there is no trace of paraphysial cartilages—nor of any part of the skull anterior to the hinder part of the mid brain.

Here we find on each side two plates of denser tissue—hardly amounting to "pro-cartilage"—the one more dorsal, along the sides of the crura cerebri and passing from the sense canal obliquely downwards and inwards, the other more ventral, from the anterior cardinal sinus downwards and inwards alongside of the infundibulum.

The former thicken further back, but die out at the posterior border of the eyeballs; the latter connect with the anterior ends of the trabeculae cranii, which are broad plates, closely approximated except in the region of the infundibulum. Passing backwards the trabeculae thicken and then, joining, become at the point of junction continuous with the parachordal plate at the anterior end of the notochord, whose tip is completely embedded in the plate.

In the otic region the auditory capsule appears as a pro-cartilaginous concentration which underlies the vestibule and horizontal canal, and is continued up outside the anterior vertical canal.

The parachordal plate, which is flattened in front, lies up against the posterior part of the otic vesicle. Farther back an uprising plate passes between the vestibule and the brain and behind it, and forms an investment to the posterior region of the vestibules extending high up, but not so far as to constitute a roof to the hind brain.

Thus the hinder part of the medulla lies in a trough of thick cartilage, in which a foramen occurs on each side for the exit of the vagus nerve.

Posteriorly the cartilage of the parachordals does not enclose the notochord completely.

The parachordals gradually die out behind into the undifferentiated tissue surrounding the notochord.

Visceral Skeleton.

The palato-pterygo-quadrate bar is represented by a dense tissue attached to the trabeculae cranii just posterior to where they separate after forming the rostral cartilage (which is about the level of the ganglion of the trigeminal nerve). Passing backwards, the bars run towards the skin and are at least pro-cartilaginous. This is anterior to the mouth, i.e. to the lower jaw. The bar widens into a nearly vertical plate and articulates with the hyo-mandibular and Meckel's cartilage. Meckel's cartilages are placed almost transversely.

Ninth Day.

Skull.

On the ninth day the most anterior piece of cartilage is a dorsal piece which is a median rhombic plate; attached to it is the paraphysis. It is continued back as far as the mid brain as a band of connective tissue (Pl. XX. figs. 47, 48, *P.C.*).

The substance of the cranium is now distinctly cartilage, which forms a slipper-like investment to the floor and sides of the hind and mid brain (Pl. XX. fig. 47).

In the region of the mid brain, the roof of that organ, the optic lobes, seems to rest upon the thick edges of these side plates (text-fig. 165).

FIG. 165. Transverse section through fore and mid brain on the ninth day, to show the mid brain supported by the sides of the cartilaginous cranium.

In the hinder region of the mid brain the side plates bend outwards somewhat, so that the brain no longer rests on the edges of the side plates, but is supported by them in the ordinary egg-cup fashion.

In front of this region there is a foramen on each side in the floor which is continued backwards almost as far as the hind brain. The olfactory nerve passes out by this foramen.

There is a large pituitary foramen (*pt. f.*), small apertures for the ingress of the carotid arteries (*c. a.*) and a long slit-like foramen. From the posterior end of the long slit-like foramen the fifth and part of the seventh nerves pass out, but part of the seventh passes out by a separate notch.

This long foramen passes underneath the anterior semicircular canal as seen in the figure (Pl. XX. fig. 47).

A large vacuity occurs in the sides of the cranium in the auditory region.

The auditory capsule is thus very imperfect. The sacculus and utriculus are practically not encased in cartilage at all; the anterior and posterior vertical semi-circular canals are protected on the outside by cartilage, but are only completely encased for a short distance at their extreme anterior and posterior extensions respectively. The horizontal canal is encased for nearly the whole of its length.

From this large vacuity, which is ventral to the horizontal semicircular canal, the eighth and ninth nerves pass out, and the auditory diverticulum of the air bladder passes in. The tenth nerve passes out by a single special foramen posterior to the auditory capsule.

There is no cartilage in the roof with the exception of the small piece at the anterior end attached to the paraphysis.

Visceral Skeleton.

There is a piece of cartilage attached to, but not actually fused with, the anterior end of the trabeculae with which it is continuous through pro-cartilage tissue. This runs backwards quite free from the skull proper and is similarly connected by pro-cartilage tissue to a long stout vertical piece which articulates with the auditory capsule (Pl. XX. fig. 48, *P.Pt.Q.* and *Hy.*). The former is the palato-pterygo-quadrate bar, the latter the hyo-mandibular cartilage. Where they unite they articulate also with the Meckel's cartilage, and with pieces of cartilage which pass ventralwards in the hyoid arch, and almost meet ventrally, being divided by a piece of tissue in which bone subsequently develops, but no cartilage. Meckel's cartilages meet ventrally. Posteriorly each is continued back and downwards in a long thin process which is attached to a band of transverse muscle fibre.

The first branchial arch consists of a half hoop of dense tissue extending from near the mid-ventral line to just above the gill cleft in which are developed three masses of cartilage—a large cerato-branchial, an epi-branchial and a pharyngo-branchial. The second is a similar half hoop, but with two cartilaginous centres, which probably are the pharyngo-branchial and the cerato-branchial. The third, a half hoop with a single centre of cartilage, which is that of the cerato-branchial cartilage.

In the fifth there is no cartilage.

There are no bones formed in connection with the skull proper as yet.

There is one, however, formed at this early period in connection with the visceral skeleton, namely, that one outside the hyo-mandibular which runs downwards and backwards far beyond the hyo-mandibular cartilage; it is closely applied to that cartilage at one spot, but does not seem to be formed especially in reference to it, as it far exceeds it in extent; this is the opercular bone.

Traces also occur in the branchiostegal ray condensations.

The pectoral girdle is already indicated by development of cartilage and of bony tissue.

The cartilaginous portion of the pectoral girdle consists of a basal piece which is perforated by the subclavian artery and nerve, and is arranged vertically; passing backwards and outwards into the fin, it turns so as to lie horizontally, and at the same time the cartilage becomes divided into two small processes which are continuous with dense tissue that runs out to the skin.

Eleventh Day.

Skull.

The thickened anterior end of the brain case has grown out to form the peculiar rostrum. The sides have increased over the olfactory region and nearly coalesce with the paraphysial cartilage.

In the orbital region they have quite grown up so as to complete the roof, fusing with the paraphysial cartilage, but in the region of the mid brain there is no cartilaginous roof. The paraphysial cartilage extends backwards as a very thin band of cartilage as far as the cerebellum. The sides of the fore part of the cranium no longer support the optic lobes. Probably the whole anterior part of the skull has extended forwards, bringing the thick sides which at an earlier period supported the mid brain forwards over the thalamencephalon where they have coalesced with the paraphysial cartilage.

In the auditory region the same large vacuities exist at the sides; perhaps the horizontal canal cartilage projects a little further out. Here again the roof has been closed in for a very short distance, but remains open along a narrow slit until the level of the vagus foramen, where also it is closed. The cartilage overhanging the foramen forms a massive ridge, and posterior to this point the cartilage of the occipital ring is thickened, and forms a slight supra-occipital crest.

The foramina are much as before.

Visceral Skeleton.

The palato-pterygo-quadrate.is more slender and is fused with the cranium in front. Its articulation with the Meckel's cartilage is by a glenoid cavity.

Meckel's cartilage has of course greatly lengthened. Its hinder protuberance is prolonged backwards and downwards as before and is attached to a transverse band of muscle fibre.

The hyo-mandibular is no longer distinct from the palato-pterygo-quadrate bar as they lie closely attached to one another; the palato-pterygo-quadrate bar articulates, but is not fused with, the skull in the anterior part of the auditory capsule, just under the horizontal canal.

The cerato-hyals are quite separated from the hyo-mandibulars.

The final Larval Stage.

In the forty-third day specimen the cartilaginous part of the skull is as shown in Pl. XX. figs. 49, 50. The floor of the anterior end of the skull and the paraphysial region have coalesced and grown forward considerably to form the rostrum. The anterior lateral plates of the earliest stage have met the posterior process of the paraphysial cartilage dorsally, and here form a complete roof for a short distance, while posteriorly the occipital region is completely roofed in. The horizontal semicircular canal is also more perfectly encased in cartilage. Foramina exist for the passage of the various nerves and blood vessels, of which one large one at the posterior end of the orbit is for the exit of the fifth, seventh (sixth and third?) nerves. There is also the same large vacuity in the side of the auditory capsule, so that the sacculus is left freely open on the outside, where the bulb of the anterior diverticulum of the air-bladder lies between it—the sacculus—and a membrane bone, the supratemporal (Gehördeckel, operculum pori acustici of Erdl).

The membrane bones are now all formed and have very much the same position and relation that they have in the adult skull.

The earliest to form is, as already described, the bone immediately outside of the hyo-mandibular cartilage, the large opercular bone.

IV. Relation of Teleostean and Amphibian.

No doubt can exist that the development of *Gymnarchus* is essentially Teleostean. There are no features, with the exception of an unusually large yolk mass, which in any way suggest a close relationship to Elasmobranchs. On the other hand there are certain features which do suggest an earlier condition of Teleostean evolution than is the case with other members of the class whose development has been studied hitherto.

Some of these features are such as to have led me to wonder whether they may be explained by the assumption that Teleosteans are descended from a far more amphibian-like stock than is usually supposed, that is to say, from a proto-amphibian rather than any strictly piscine race.

1. The Amphibian-like character of the lips of the blastopore.
2. The vestige of neural tube formation by folding and of wing-like expansions forming the ganglia of certain cranial nerves.
3. The formation of the pronephros as distinct folds of the somatopleure, forming at least three nephrostomes open to the general coelom.
4. Occasional occurrence of supernumerary nephrostomes near the pronephros.
5. Presence of diverticula from the pronephric duct at the anterior end of the mesonephros, which may be remnants of kidney tubules once in connection

with the reproductive system.   (This interpretation obviously is highly speculative.)

6. The lung-like condition of the air-bladder and its vestigial double condition.
7. The suggestion of a double circulation as shown by the separation of the blood that comes from the lung, and its flow into the left side of the auricle.
8. A trace of an auricular septum.
9. The large size of the aortic arch of the fourth visceral arch.
10. The peculiar character of the gill clefts and gill filaments and arches, demonstrating their epiblastic nature and real distinction from true gill clefts, one of which true gill clefts, however, remains, recalling the condition found in Amphiuma, in which the last cleft persists throughout life (Sedgwick, p. 280).

Are Teleosteans descended from a proto-amphibian stock more amphibian than the Dipnoi? Does the race owe the high position it now holds among the inhabitants of the sea to a certain quickening of intelligence or increase of power of adaptability acquired during a period in which its ancestry was making progress under the influence of a better respiratory apparatus and more particularly by living in a less regular and constant environment? Such ancestors we must imagine to have acquired a higher grade of intelligence than that which their forebears possessed when they gave up strictly aquatic habits ; to have had external gills like existing Amphibians, at any rate in the larval stage ; to have had a reduced number of gill clefts in the adult, like Amphiuma ; to have had limbs very little in advance of the piscine paired fin.   They must have had a rudimentary double circulation, with division of the auricle and an enlarged systemic (fourth visceral) arch, and lungs which lay loose in the body cavity after the manner of those of Amphibians, and most other air-breathing Vertebrates.

The mutation, which led directly to the return to aquatic life, may have been the firm connection of the lung to the dorsal wall of the body cavity whereby its use was assured as a hydrostatic organ.   This combined with the greater complexity of brain to give the new race an impetus and greater power of adaptability; and, especially if the new species was for some period a mutable species, it led to the "Teleostean" with its diversity of form, and its success as a race during comparatively recent geological times.

On the final readoption of water habits the external gills of the larval proto-Amphibian became the epiblastic truly external gills of the modern Teleostean.   The disappearing gill clefts were not re-opened, but new clefts were evolved as excavations from without made under the old arches, and so the condition of branchial channels and epiblastic gills seen so plainly in the larval Gymnarchus was attained.   The incipient double circulation was lost, but traces of it appear in the larval Gymnarchus, in which also one of the old reduced number of true gill clefts is retained alongside of the newer form.   The loss of the connection between the reproductive and excretory organs went on *pari passu* with the encroachment of the hydrostatic air bladder along the dorsal wall of the abdominal cavity leading to the extension backwards of the genital gland till it opened to the exterior and created new gonoducts.

These are perhaps idle fancies, which may serve to lighten the labour of the investigator, but he who reads may very justly prefer facts to dreams. Wherefore I will dwell no longer thereon but will conclude with a summary of what I believe to be true facts of the development of *Gymnarchus niloticus*.

V. SUMMARY OF FACTS.

1. The development of *Gymnarchus* is on the whole typically Teleostean.

2. The egg is large and has a tough investment which causes a considerable distortion of the embryo until it is hatched.

3. The development is rapid, so that the larva hatches upon the seventh day whereas the larva of a Trout hatches on the thirty-fifth—one hundredth day acccording to temperature (Henneguy).

4. The embryo is greatly stretched upon the yolk and after closure of the blastopore resembles very closely in form certain amniote embryos with almost typical "primitive streak."

5. The "archenteron" in so far as it occurs, resembles the condition of the archenteron in an early stage of *Hypogeophis* (Brauer) more than it does a "Kupffer vesicle."

6. Pressed down by the tough investing membrane the embryo lies always upon its side from the third day till the time of hatching, its anterior end turning over before the posterior just as in the avine embryo. It lies usually upon its right side.

7. In the region of the primitive streak the hypoblast is continuous with the yolk and the primitive streak as it is in Amphibia, and not separated as it is in Birds and Mammals.

8. The alimentary canal arises as a cleft among the hypoblast cells. At an early stage—or perhaps from the beginning—the whole of the pharyngeal region is without lumen until after hatching.

9. The "gill clefts" of embryonic life are not gill clefts in the true sense since they do not concern the gut. They are invaginations of the ectoderm, which undermine the visceral arches and are lined throughout by ectoderm, and may be termed branchial channels.

10. There are long external uniramous gill filaments upon the first, second, third, and fourth branchial arches, which shrivel after the operculum has grown over them, excepting the proximal ends which give rise to the permanent gills. The whole apparatus is lined by epiblast from first to last.

11.   One pair of true gill clefts however occurs, namely, between the sixth and seventh visceral arches.

12.   There is no spiracle; nor is there a branchial channel to correspond to it, although both mandibular and hyoid aortic arches are present.

13.   The air bladder arises as a single diverticulum of the oesophagus a little to the left of the mid-dorsal line; it has right and left lobes; the left is by far the larger of the two and the right sends forwards a diverticulum which bifurcates and becomes connected with the "auditory" organ.

14.   The air bladder is extremely lung-like; its structure, its vascular supply and the habits of the fish (Budgett, etc.) all point to its use as a lung.

15.   The terminal bulbs of the anterior diverticulum of the right lobe almost lose their connection with the air-bladder and form tambours between the thin supra-temporal bone and sacculus of the auditory organ.

16.   The yolk sac is to be regarded as an appendage of the liver—due to accumulation of yolk in that part of the egg which normally becomes the liver.

17.   The gall-bladder and liver is derived by the constriction off of a large ventral recess of the alimentary canal (just posterior to the oesophagus) from which several diverticula arise in close connection with the yolk sac efferent veins.

18.   The pancreas is developed as diverticula of the bile ducts (the constricted region above mentioned) and these grow backwards along the mesenteries and come later to intermingle with the "islands of Langerhans" tissue and even with the spleen.

19.   The islands of Langerhans arise very early as a solid mass of epithelial tissue which becomes broken up by the splitting of the mesenteric artery.

20.   There is a slight spiral valve in the intestine at an early stage.

21.   The larva begins to feed about the sixteenth day. Its food appears to be chiefly—if not exclusively—vegetable.

22.   The "stomach" is an outgrowth of the dorsal wall of the recess, mentioned in paragraph 17, just anterior to a dorsal diverticulum which, I believe, becomes the two pyloric caeca—but about this I cannot be certain.

23.   The thyroid develops at the extreme anterior end of the true pharynx—but I did not find its earliest stage. It is in connexion with hypoblast when first detected.

24.   The thymus would appear to be epiblastic in origin.

25.   The anus does not form until the twelfth day.

Vascular.

26.   The vascular supply to the yolk sac is typically Teleostean.

27.   The heart is remarkable (1) for the diverticula upon the "bulbus" (cf. other *Mormyridae*) which arise as little spheroidal apparently muscular swellings at the root of the cardiac aorta on the ninth day, (2) for the separation of the left ductus

Cuvieri from the sinus venosus which acquires a separate opening into the auricle about the seventeenth day.

28. There is about the sixteenth day a trace of an auricular septum between the openings of the left ductus Cuvieri (receiving the "pulmonary vein") and the rest of the sinus venosus.

29. The left ductus Cuvieri receives the usual supply of blood from the anterior cardinals, but the left posterior cardinal brings blood from the lung alone.

30. The right ductus Cuvieri brings blood as usual, but with the addition of the whole of the blood from the posterior region of the body which ordinarily is divided between right and left posterior cardinals. It receives no blood from the lung.

31. These changes occur step by step from the ordinary piscine condition between the tenth and seventeenth days.

32. There are the usual anterior six pairs of aortic arches upon the fifth day, though the hyoid arch is small. The second branchial arch is the widest of all at this time.

33. The mandibular and hyoid aortic arches disappear in part at an early date so that when the gill filaments have formed there are four pairs of branchial afferents, and four pairs of branchial efferents.

34. By changes occurring between the fifth day and twelfth day the dorsal aorta receives its supply almost entirely from the first and second branchial efferents, and the lung from the left third and fourth branchial efferents and the alimentary canal from the third and fourth right efferents.

35. The arterial supply to the pronephros is at first by a pair of loops from the dorsal aorta. By constriction of the ventral wall of the aorta, the pronephros afferent and efferent vessels become connected with the air bladder artery.

Excretory.

36. The pronephros arises as a series of open funnels or folds of the somatopleure in connection with the longitudinal or pronephric duct.

37. The part of the coelom into which these open becomes converted in some way not observed into the typical Teleostean pronephric chamber in connection with the glomerulus.

38. Occasionally accessory nephrostomes occur.

39. The mesonephros arises as tubules apart from the pronephric duct. These have their origin in groups of cells which occur in the mesoblast close to the pronephric duct and seem to be arranged segmentally from about the thirty-seventh to the fifty-sixth myotome.

40. In front of the mesonephros buds arise on the pronephric duct itself quite irregularly and are followed by diverticula from the pronephric duct. The buds separate off and remain while the diverticula disappear.

41.   There is never at any time any connection between the reproductive and excretory organs.

42.   A rich lymphoid tissue develops along the whole length of the pronephric duct whose sole vascular supply is venous.

Reproduction.

43.   The germ cells appear first in the mesoderm beneath the peritoneum in the mesentery and the splanchnopleure on both sides.

44.   The germ cells of the left side migrate to the right side where they take up their position under and ultimately among the cells of a thickened strip of peritoneum.

45.   This strip folds up so as to enclose a cavity which is produced backwards to the exterior where it meets an ingrowth of the epiblast between the ureter and anus and forms the gonoduct.

46.   Some remarkable cells occur in all the embryos of a certain age—namely just after hatching—scattered throughout the tissues of the anterior parts of the trunk which apparently ultimately collect in the pronephros and its ducts, and which resemble in some respects the hypertrophied nuclei of the yolk mass, and in other respects certain parasitic protozoa.

Nervous System.

47.   There is some evidence that a portion at any rate of the central nervous system forms by a process of folding up of a thickened plate of epiblast instead of by a hollowing out of a solid ridge as is usual among Teleosteans.

48.   There is also some evidence that the ganglia of the lateral line nerve, the eighth, ninth and tenth nerves arise as wide lateral wings of the central nervous system as they do in Anura.

49.   The lateral line nerve is not connected at any time with the vagus. It arises close to the roots of the seventh and eighth nerves, and is far removed from the vagus, though it leaves the skull by the same foramen as the vagus.

50.   The first three spinal nerves which have dorsal and ventral roots leave the brain within the skull but pass backwards and out by the foramen magnum and out from the vertebral column between the skull and first vertebra. The first is the hypoglossal and passes to the tongue, the second and third pass to the pectoral fin.

51.   The brain is greatly distorted during the fifth day by the pressure of the egg shell, and on the relaxation of this pressure at the moment of hatching it exhibits in the hind brain region a series of constrictions which certainly simulate the neuromeres of some other Vertebrates at an earlier stage in development.

52.   At an early date the preponderance of the tectum opticum shows itself by overhanging the sides of the mid brain and thalamencephalon.

53.   A marked cranial flexure occurs between the mid and fore brains, and but little flexure posterior to that point.

54. The valvula cerebelli attains a high state of development and about the eighth day causes the optic lobes to separate from one another and overhang the sides, where they project backwards on the sides of the medulla. The tectum opticum is intact but where this separation of the optic lobes occurs it is reduced to a single layer of ependyma, though afterwards bands of nerve fibre cross from side to side.

55. The cerebellum is very large and projects forwards over the valvula and here its sides coalesce, at points almost obliterating its cavity; and backwards over the thin roof of the fourth ventricle where it retains its lumen.

56. The anterior commissure is large and is never raised up as in Anura and Dipnoi.

57. The olfactory lobes are early distinguishable from the cerebral hemispheres, from which they become more and more separated as development proceeds.

58. The lobi inferiores become incorporated in the crura cerebri.

59. A commissure occurs in the posterior wall of the epiphysis which connects certain ventral portions of the fore brain, either the corpora striata or some centres nearer the infundibulum. This is a very conspicuous commissure in the earlier stages but diminishes later.

60. A distinct paraphysis is present on the ninth day anterior to the velum transversum and is attached to an isolated piece of cartilage.

61. The anterior commissure, the posterior commissure, the commissurae post-opticae, and the cerebellar and infundibular commissures are all conspicuous objects in a sagittal section.

62. In the region of the hind brain, at about the position of the tuberculum acusticum, the inner walls of the fourth ventricle coalesce and completely fuse.

63. The characteristic peculiarities of the mormyrid brain as seen from above— e.g. including the "eigenthümliches organ" of Marcusen are due to (1) the great development of the cerebellum and valvula cerebelli, which latter is much folded and is seen partly shining through the tectum opticum, though in part it is covered by the anterior portion of the cerebellum, (2) the great lateral displacement of the optic lobes, (3) the great development on the inside of the tuberculum acusticum in combination with the ordinary characters of the Teleostean brain.

64. The ganglia habenulae are conspicuous and I think about equally developed. The greater part of their fibres—the habenulae—pass to their own side and do not cross.

65. A group of giant nerve nuclei occurs in the mid-ventral line of the hinder part of the medulla oblongata and is an absolutely median and unpaired structure.

66. The optic nerve fibres travel back (or develop) along the ventral and lateral parts of the optic stalk separating the cells from each other as they do so.

67. The auditory labyrinth consists of a utriculus and the three usual canals which are only partially embedded in cartilage, of a short ductus and small saccus endolymphaticus, a large sacculus which is quite outside the cartilaginous skull and together with a well developed lagena becomes entirely separated from the utriculus before the forty-third day. It lies closely attached to the bulb of the air bladder.

68. The lateral line sense organs are developed from the nervous layer of epiblast as solid ridges which become hollowed out. The epithelium is thickened at intervals and resembles very closely that of the ampullae of the semicircular canals. The main canals in the head region are soon encased in bone.

69. The spinal canal contains Reissner's fibre on the forty-third day.

The Skeletal System.

70. The cartilaginous cranium is developed from the parachordals, the trabeculae cranii, the side plates and the paraphysial cartilage. The latter forms a cartilaginous roof to the ethmoidal region, and the parachordal by the growth upwards of its outer edges ultimately completes a ring round the occipital region. The sphenoidal region is completed by bone alone.

71. The cartilaginous auditory capsule is very incomplete, the sacculus and lagena being really outside the skull and protected only by the superficial supra-temporal bone.

72. The vertebral column is formed of vertebrae developed on the arcocentrous plan.

73. The chorda retains its cellular character with nuclei scattered throughout as well as superficially. There are both an interna and externa elastica.

74. The cartilage is however restricted to the basi-dorsals and basi-ventrals and ribs and neural spines.

75. The greater part of the body of the vertebra is derived from bone which forms from the inner skeletogenous layer from four centres to each vertebra.

76. Somactids—axonost and baseost—are formed in cartilage in connection with the dorsal fin and bony dermotrichia in the fin itself.

77. In the early larva there is a continuous dorsal caudal and ventral fin. The ventral disappears early and the caudal is separated from the dorsal and eventually the dorsal alone remains and develops to form the chief locomotor organ of the fish.

78. Teeth are formed in the usual way, but only in the upper and lower jaws.

79. Before the forty-third day the larva has attained a sufficient size to enable it to contain the yolk sac within the normal contour of the animal.

80. By this time the yolk sac has lost its afferent blood supply and is now merely an appendage of the liver and receives a small blood supply therefrom.

I am only too well aware of the incompleteness of the work described in the foregoing pages. The section which deals with the skeletal structure is especially meagre and inconclusive ; but it seems undesirable to delay longer the publication of the memorial edition of Budgett's work. No one can regret more than I do that Budgett did not live to give the benefit of his own skill and knowledge to the subject ; but since it has fallen to my lot to endeavour to work out the development of *Gymnarchus*, I can say truthfully that it has been to me a source of unusual interest ; and it is a great privilege to have been given the opportunity of adding my contribution to the memorial volume in token of my esteem for him.

I wish to thank my wife, Frances A. E. Assheton, whose constant help I have had throughout, and also Miss Marie Krull whose assistance in my laboratory—though intermittent—has been very valuable.

To Mr Adam Sedgwick, F.R.S., I am indebted for the use of his series of Elasmobranch embryos, for which I desire to render sincere thanks. Nor can I close without an expression of appreciation and admiration of the work and life of the late John S. Budgett whom I always regarded with real affection. His life must appeal to all who value sincerity and the love of hard work.

VI. Literature.

1. 1878. AGASSIZ, A.: "On the young stages of some Osseous Fishes." Proceedings Amer. Acad. Arts and Science, vol. XIII.

2. 1879. AGASSIZ, A.: "The development of *Lepidosteus*." Proceedings Amer. Acad. Arts and Science, vol. XIV.

3. 1906. AGAR, W. E.: "The spiracular Gill Cleft in *Lepidosiren* and *Protopterus*." Anat. Anz., Bd. XXVIII.

4. 1906. ALLEN: "The Origin of the sex cells of *Chrysemys*." Anat. Anz., vol. XXIX.

5. 1904. ALLIS, E. P.: "The Latero-sensory Canals and related bones in Fishes." Internat. Monatsschrift f. Anat. u. Phys., Bd. XXI.

6. 1892. ASSHETON, R.: "On the Development of the Optic Nerve of Vertebrates, and the Choroidal Fissure of Embryonic Life." Quart. Journ. Micr. Sci., vol. XXXIV.

7. 1905. ASSHETON, R.: "On Growth Centres in Vertebrate Embryos." Anat. Anz., vol. XXVII.

8. 1878. BALFOUR, F. M.: "A Monograph on the Development of Elasmobranch Fishes." London.

9. 1882. BALFOUR, F. M. and PARKER, W. N.: "On the Structures and Development of *Lepidosteus*." The Phil. Trans. Roy. Soc. London, vol. 173.

10. 1885. BALFOUR, F. M.: "A Treatise on Comparative Embryology." 2nd ed. London.

11. 1906. BORCEA, I.: "Recherches sur le Système Urogenital des Elasmobranches." Arch. de Zool. exp. et gén., IV. Série, T. IV.

12. 1889. BEARD, J.: "On the early Development of *Lepidosteus osseus*." Proc. Roy. Soc. Lond. vol. XLVI.

13. 1900. BEARD, J.: "The source of Leucocytes and the true function of the Thymus." Anat. Anz., vol. XVIII.

14. 1902. BEARD, J.: "The Germ Cells." Pt I., Raja batis. Zoolog. Jahrbücher, vol. XVI.

15. 1897. BLES, E. J.: "On the openings in the walls of the body cavity of Vertebrates." Proc. Roy. Soc., vol. LXII.

16. 1897. BRAUER, A.: "Beiträge zur Kenntniss der Entwicklungsgeschichte und der Anatomie der Gymnophionen." I. Zool. Jahrb., Bd. x.

17. 1899. BRAUER, A.: "Beiträge zur Kenntniss der Entwicklung und Anatomie der Gymnophionen." II. Zool. Jahrb., Bd. XII.

18. 1902. BRAUER, A.: "Beiträge zur Kenntniss der Entwicklung und Anatomie der Gymnophionen." III. Zool. Jahrb., Bd. XVI.

19. 1900. BRIDGE, T. W.: "The air bladder and its connection with the Auditory Organ in *Notopterus Borniensis*." Journ. Linn. Soc., vol. XXVII.

20. 1892. BRIDGE, T. W. and HADDON, A. C.: "The air bladder and Weberian Ossicles in the Siluroid Fishes." Proc. Roy. Soc., vol. LII.

21. 1893. BRIDGE, T. W. and HADDON, A. C.: "Contribution to the Anatomy of Fishes. II. The Air-bladder and Weberian Ossicles in the Siluroid Fishes." Phil. Trans. Roy. Soc., vol. 184.

22. 1904. BRIDGE, T. W.: "Fishes." Cambridge Natural History. London.

23. 1901. BOULENGER, G. A.: "Poissons du Bassin du Congo."

24. 1904. BOULENGER, G. A.: "Fishes. Systematic Account of Teleosteans." Cambridge Natural History. London.

25. 1900. BUDGETT, J. S.: "On some points in the Anatomy of *Polypterus*." Trans. Zool. Soc. Lond., vol. XV.

26. 1900. BUDGETT, J. S.: "On the Breeding habits of some West African Fishes, with an Account of the External Features in Development of *Protopterus annectens*, and a Description of the Larva of *Polypterus lapradei*." Trans. Zool. Soc., vol. XVI., Pt. 2.

27. 1902. BUDGETT, J. S.: "On the structure of the Larval *Polypterus*." Trans. Zool. Soc., vol. XVI., Pt. 7.

28. 1895. BURCKHARDT, R.: "Der Bauplan des Wirbelthiergehirns." Morphologische Arbeiten, Bd. IV.

29. 1886. CATTANEO. "Sulla formazione delle cripte intestinale negli embrioni de *Salmo salar*." Rendic. del R. Instit. Lombardo de Scienze e Lettere. Ser. II., vol. XIX.

30. 1900. CATTANEO: "Sul tempo e sul modo di formazione delle appendici piloriche nei Salmonidi." Monit. Zool. Ital. Ann. II., Suppl.

31. 1890. CLAPP, Cornelia M.: "Some points on the Development of the Toad-fish (*Batrachus Tau*)." Journ. of Morphology, vol. V., No. 3.

32. 1898. COLE, F. J.: "Observations on the Structure and Morphology of the Cranial Nerves and Lateral Sense Organs of Fishes; with special reference to the Genus Gadus." Transactions Linnean Soc., vol. VII.

33. 1898. CUNNINGHAM, J. T.: "On young stages of Teleosteans." Proc. R. Irish Acad., Series 3, Vol. V.

34. 1885. CUNNINGHAM, J. T.: "On the significance of Kupffer's Vesicle, etc." Quart. Journ. Micr. Sci., vol. XXV.

35. 1886. CUNNINGHAM, J. T.: "On the Relations of the Yolk to the Gastrula in Teleosteans, and in other Vertebrate Types." Quart. Journ. Micr. Sci., vol. XXVI.

36. 1847. CUVIER ET VALENCIENNES: "Histoire Naturelle des Poissons." T. XX., XXI.

37. 1895. DAVISON, A.: "A contribution to the Anatomy and Phylogeny of *Amphiuma means*." Journal of Morphology, vol. XI.

38. 1853. DUVERNAY, G.: "Note additionelle" to Förg's paper on the pulmonary apparatus of *Gymnarchus*. Ann. d. Sci. Nat. Ser. 2, T. xx.

39. 1847. ERDL, M. P.: "Beschreibung des Skeletes des *Gymnarchus niloticus*." Abhandl. Bayer. Akad. Wiss. München, vol. v.

40. 1897. FELIX, W.: "Die Price'sche Arbeit 'Development of the excretory organs of a Myxinoid (*Bdellostoma sturti* Lochington)' und ihre Bedeutung für die Lehre von der Entwickelung des Harnsystems." Anat. Anz., Bd. xiii.

41. 1897. FELIX, W.: "Beiträge zur Entwickelungsgeschichte der Salmoniden." Anatomische Hefte, Bd. viii.

42. 1904. FELIX, W.: On the Urinary system in Hertwig's Handbuch.

43. 1853. FÖRG: "Remarques sur l'appareil pulmonaire du *Gymnarchus niloticus*." Ann. d. Sc. Nat. (2), T. xx.

44. 1885. FRITSCH, G.: "Zur Organisation des *Gymnarchus niloticus*." Sitz. Akad. Berl. pp. 119—124.

45. 1891. FRITSCH, G.: "Weitere Beiträge zur Kenntniss der schwach elecktrischen Fische." Sitz. Akad. Berl. pp. 439—460.

46. 1905. FRORIEP, A.: "Entwickelung des Auges." Handbuch der vergl. u. exp. Entwickelungslehre der Wirbelthiere. O. Hertwig. Jena.

47. 1894. GADOW, H. and ABOTT, E. C.: "On the evolution of the Vertebral Column of Fishes." Phil. Trans. Lond., vol. 186 B.

48. 1895. GADOW, H.: "On the evolution of the Vertebral Column of Amphibia and Amniota." Phil. Trans. Lond., vol. 187 B.

49. 1901. GADOW, H.: "Amphibia and Reptiles." Cambridge Natural History. London.

50. 1893. GÖPPERT, E.: "Die Entwickelung des Pancreas der Teleostier." Morph. Jahrb, Bd. xx.

51. 1893. GAGE, P. G.: "The Brain of *Diemyctylus viridescens*." The Wildes Quarter Century Book. Ithaca. N.Y.

52. 1878. GOETTE, A.: "Ueber d. Entwickel. d. Central-Nerven Systems d. Teleostier." Archiv für mikr. Anat., vol. xv.

53. 1901. GOETTE, A.: "Ueber die Kiemen der Fische." Zeitsch. f. wissensch. Zool., Bd. lxix.

54. 1906. GOODRICH, E. S.: "Notes on the Development, Structure and Origin of the Median and Paired Fins of Fish." Q. J. M. S., vol. l.

55. 1906. GRADON, J. T.: "Researches on the Origin and Development of the Epiblastic Trabeculae and the Pial Sheath of the Optic Nerve of the Frog, with illustrations of Variations met with in other Vertebrates, and some observations on the Lymphatics of the Optic Nerve." Quart. Journ. Micr. Sci., vol. l.

56. 1903. GREGORY, E. H., Jun.: "Beiträge zur Entwickelungsgeschichte der Knochenfische." Anat. Hefte, Bd. xx.

57. 1906. GREIL, A.: "Ueber die Homologie der Anamnierkiemen." Anat. Anz., Bd. xxviii.

58. 1905. GUDGER, E. W.: "The breeding habits and the segmentation of the egg of the Pipe fish *Siphostoma floridae*." The Proceedings of the United States National Museum, Washington, vol. xxix.

59. 1901. GULLAND: "The Anatomy of the Digestive Tract in the Salmon." Journ. of Anat. and Phys., vol. xv. N. S.

60. 1898. HALLER, B.: "Vom Bau des Wirbelthiergehirns, 1 Theil. *Salmo* u. *Scyllium*." Morphologisches Jahrbuch, Bd. xxvi.

61. 1888. HAY, O. P.: "Observations on *Amphiuma* and its young." American Naturalist.

62. 1889. HAY, O. P.: Journal of Morphology, vol. IV.

63. 1888. HENNEGUY, L. F.: "Recherches sur le Developpement des Poissons Osseux. Embryogenie de la Truite." Journ. de l'Anat. et de la Physiologie, vol. XXIV.

64. 1900—1906. HERTWIG, O.: Handbuch der vergleichenden und experimentellen Entwickelungslehre der Wirbeltiere. Jena.

65. 1892. HERTWIG, O.: "Text-book of the Embryology of Man and Mammals." Trans. 3rd edition. London and New York.

66. 1900. HILL, CH.: "Developmental History of the primary segments of the Vertebrate Head." Zool. Jahrb. Anat. Hefte, Bd. XIII.

67. 1902. HOCHSTETTER: "Die Entwickelung des Blutgefässystems." Hertwig's Handbuch.

68. 1882. HOFFMANN, C. K.: "Zur Ontogenie der Knochenfische." Natur. Verh. der Koninkl. Akademie van Wetens. te Amsterdam.

69. 1887. HOPKINS: "On the Enteron of American Ganoids." Journ. Morph., vol. II.

70. 1854. HYRTL, J.: "Beiträge zur Morphologie des Urinogenitalsystems der Fische." Denkschriften der Kaiserl. Akademie der Wissenschaften, Bd. I. Vienna.

71. 1856. HYRTL, J.: "Anatomische Mittheilungen über *Mormyrus* und *Gymnarchus*." Denkschr. d. Akad. Wiss. Wien. XII.

72. 1871. HUXLEY, T. W.: "A Manual of the Anatomy of Vertebrated Animals." London.

73. 1894. JUNGERSEN, H. F. E.: "Die Embryonalniere von *Amia Calva*." Zool. Anz., vol. XVII.

74. 1899. JUNGERSEN, H. F. E.: "Entwickelung der Geschlechtsorgane bei den Knochenfischen." Arb. Inst. Würzburg, Bd. IX.

75. JUNGERSEN, H. F. E.: "Ueber die Urinogenitalorgane von *Polypterus* und *Amia*." Zool. Anz., Bd. XXIII.

76. 1900. KERR, J. GRAHAM: "The External Features in Development of *Lepidosiren paradoxa*." Phil. Trans. Roy. Soc., B. 182.

77. 1903. KERR, J. GRAHAM: "The Development of *Lepidosiren paradoxa*." Quart. Journ. Micr. Sci., vol. XLVI.

78. 1892. KINGSLEY, J. S.: "The Head of an embryo *Amphiuma*." The American Naturalist, vol. XXVI.

79. 1896. KOPSCH, F.: "Homologie und phylogenetische Bedeutung der Kupffer'schen Blase." Anat. Anz., Bd. XVII.

80. 1898. KOPSCH, F.: "Die Entwickelung der äusseren Form des Forellen-Embryo." Archiv f. mikroscopische Anatomie, Bd. LI.

81. 1901. KOPSCH, F.: "Die Entstehung des Dottersackentoblasts und die Furchung bei Belone acus." Internat. Monatsschrift f. Anat. u. Phys.

82. 1902. KOPSCH, FR.: "Art, Ort und Zeit der Entstehung des Dottersackentoblasts bei verschiedenen Knochenfischarten." Intern. Monatsschrift f. Anat. u. Phys., Bd. XX.

83. 1902. KOPSCH, FR.: "Ueber die künstliche Befruchtung der Eier von *Cristiceps argentatus*." Sitz. Berichte. d. Gesellsch. naturforschenden Freunde, No. 2.

84. 1904. KOPSCH, F.: "Untersuchungen über Gastrulation und Embryobildung bei den Chordaten." Leipzig.

85. 1903. KUPFFER, K. VON: "Die Morphogenie des Centralnervensystems." Hertwig's Handbuch.

86. 1889. LAGUESSE: "Développement du pancréas chez les poissons osseux." Compte rend. hebd. des Séances et Mémoires de la So. de Biol. Sér. IX., T. I.

87. 1894. LAGUESSE: "Développement du pancréas chez les poissons osseux." Journ. de l'Anatomie et de la Phys., T. XXX.

88. 1892. MacBRIDE, E. W.: "The Development of the Oviduct in the Frog." Quart. Journ. Micr. Sc., vol. XXXII.

89. 1864. MARCUSEN, J.: "Die Familie der Mormyren." Mémoires de l'Academie impériale des Sciences de St Pétersbourg. VIIth Ser., Tom. VII., No. 4.

90. 1893. MARSHALL, A. M.: "Vertebrate Embryology." London.

91. 1886. MAURER, F.: "Schilddrüse und Thymus der Teleostier." Morph. Jahrb., Bd. XIII.

92. 1901. MINOT, C. S.: "On the Morphology of the Pineal Region, based upon its Development in Acanthias." The American Journal of Anatomy, vol. I.

93. 1892. MINOT, C. S.: "Human Embryology." New York.

94. 1904. MOROFF: "Ueber die Entwickelung der Kiemen bei Fischen." Arch. für. mikr. Anat., Bd. LXIX.

95. 1904. NICHOLAS, A.: "Recherches sur le développement du pancréas, du foie et de la rate chez le Sterlet." Archives de Biologie, T. XX.

96. 1867. ŒFFINGER, H.: "Neue Untersuchungen über den Bau des Gehirns vom Nilhecht." Arch. f. Anat. u. Physiol.

97. 1898. OGNEFF, J.: "Einige Bermerkungen über den Bau des schwach elektrischen Organs bei den Mormyriden." Zeit. f. wiss. Zoologie, Bd. LXIV.

98. 1873. ŒLLACHER, J.: "Beiträge zur Entwickelungsgeschichte der Knochenfische nach Beobachtungen an Bachforellen." Zeit. f. wiss. Zoologie, Bd. 23.

99. 1866. OWEN, R.: "The Anatomy of Vertebrates." London.

100. 1902. PIPER, H.: "Die Entwickelung von Leber, Pancreas und Milz bei den Vertebraten." Freiburg.

101. 1904. PRICE, G. C.: "A Further Study of the Development of the Excretory Organs in *Bdellostoma Stouti*." The American Journal of Anatomy, vol. IV.

102. 1882. RABL-RÜCKHARD: "Zur Deutung und Entwickelung des Gehirns der Knochenfische." Archiv für Anatomie u. Physiologie.

103. 1898. REINHARD, W.: "Die Bedeutung des Periblastes und der Kupffer'schen Blase in der Entwickelung der Knochenfische." Archiv f. Mikr. Anat., Bd. LII.

104. 1904. RENNIE, J.: "The epithelial islets of the Pancreas in Teleostei." Quart. Journ. Micr. Sc., vol. XLVIII.

105. 1892. RIDEWOOD, W. G.: "The Air bladder and ear of British Clupeoid Fishes." Journ. Anat. and Physiol., vol. XXVI.

106. 1904. RIDEWOOD, W. G.: "The Cranial Osteology of the Fishes of the Families *Mormyridae*, *Notopteridae* and *Hyodontidae*." Journ. Linn. Soc., vol. XXIX.

107. 1891. ROBINSON, A.: "On the Formation and Structure of the Optic Nerve, and its Relation to the Optic Stalk." Journ. Anat. and Phys., vol. X. N. S.

108. 1882. SANDERS, A.: "Contributions to the Anatomy of the Central Nervous System in Vertebrate Animals. I. On the Brain of the *Mormyridæ*." Phil. Trans. Roy. Soc., vol. 173.

109. 1905. SEDGWICK, A.: "Textbook of Zoology," vol. III. London.

110. 1898. SOBOTTA, J.: "Die morphologische Bedeutung der Kupffer'schen Blase." Verhandl. d. Physik.-med.-Gesellschaft zu Würzburg. Bd. XXXII.

111. 1893. STÖHR, P.: "Die Entwickelung von Leber und Pancreas der Forelle." Anat. Anz., Bd. VIII.

112. 1891. STOSS: "Zur Entwickelungsgeschichte des Pancreas." Anat. Anz., Bd. VI.

113. 1900. SUMNER, F. B.: "Kupffer's vesicle and its relation to Gastrulation and Concrescence." Mem. New York Acad. Sc., vol. II.

114. 1904. SWAEN, A. and BRACHET, A.: "Etude sur la formation des feuillets et des organes dans le bourgeon terminal et dans la queue des embryons des poissons téléostéens." Archives de Biologie, T. XX.

115. 1900. SWAEN, A. and BRACHET, A.: "Etude sur les premières phases du développement des organes dérivés du mesoblast chez les poissons téléostéens." Archives de Biologie, T. 16.

116. 1899. WALLACE, LOUISA B.: "The Germ ring in the egg of the Toad-fish (*Batrachus Tau*)." Journ. of Morphology, vol. XV.

117. 1820. WEBER, E. H.: "De aure et auditu Hominis et Animalium. Pars 1. De aure Animalium Aquatilium." Lipsiæ.

118. 1891. WILSON, V. HENRY: "The Embryology of the Sea Bass (*Serranus atrarius*)." Bulletin of the U. S. F. C., vol. IX.

VII. LIST OF LETTERINGS FOR PLATES XVI—XXI.

a.b.a.	Air bladder artery.	*m.b.*	Mid brain.
a.bl.	Air bladder.	*Md.*	Mandibular arch.
ab.l.l.	Left lobe of air bladder.	*mes.*	Mesoblast.
ab.l.r.	Right lobe of air bladder.	*mes.d.*	Mesonephric diverticulum.
ab.v.	Pulmonary vein.	*mes.bd.*	Mesonephric bud.
ao.	Axonost.	*mes.n.*	Mesonephros.
a.s.c.	Anterior ventral semicircular canal.	*Mx.*	Maxilla.
Au.	Auricle.	*n.*	Nerve rudiment.
au.v.	Auditory sac.	*n.c.*	Neural canal.
b.	Basibranchial cartilage; in fig. 57 bony sheath to notochord.	*nch.*	Notochord.
Bhy.	Basihyal.	*n.p.*	Neural plate.
bl.	Blastopore.	*o.l.*	Olfactory lobe.
br.st.	Branchiostegal ray.	*olf.*	Olfactory nerve.
bd.	Basidorsal.	*opl.*	Optic lobe.
bv.	Basiventral.	*opl'.*	Optic lobe.
b.o.	Baseost (cartilage).	*op.rec.*	Optic recess.
b.o.'	Cap of bone on Baseost.	*opt.*	Optic foramen.
C.A.	Anterior commissure.	*P.*	Paraphysis.
cbl.	Cerebellum.	*P.*	Roof of skull derived from paraphysial cartilage.
c.c.	Crura Cerebri.	*Pa.*	Pallium.
Ch.y.	Ceratohyal cartilage.	*P.C.*	Paraphysial cartilage.
cœ.mes.	Coeliaco-mesenteric artery.	*Pch.*	Parachordal.
c.s.	Corpus striatum.	*Per.*	Pericardium.
D.	Dentary.	*pn.pd.*	Pronephric duct.
d.a.	Dorsal aorta.	*Prm.*	Premaxilla.
d.b.	Diverticula on bulbus.	*Prn.*	Pronephros.
d.c.	Ductus Cuvieri.	*Pt.*	Pituitary body.
dk.	Deckplatte.	*Ptf.*	Pituitary foramen.
E.	Epiphysis.	*P.Pt.Q.*	Palato-pterygo-quadrate bar.
e.	Eye.	*Q.*	Quadrate end of pl-pt-quadrate bar.
ep.ep.	Covering layer of cells.	*R.*	Rostrum.
f.	Gill filament.	*r.*	Rib.
fb.	Forebrain.	*r.d.c.*	Right ductus Cuvieri.
f.r.	Fissura rhomboidalis.	*s.*	Stomach.
g.bl.	Gall bladder.	*s.au.*	Sinu-auricular opening.
g.c.	Germ cell.	*sp.*	Spleen.
g.l.l.	Ganglion on lateral line nerve.	*T.*	Temporal vacuity in the side of the skull.
H.	Heart.		
h.b.	Hind brain.	*t.a.*	Tuberculum acusticum.
h.s.c.	Horizontal semicircular canal.	*t.a.a.*	Tuberculum acusticum anterius.
h.v.	Hepatic sinus.	*t.a.p.*	Tuberculum acusticum posterius.
Hy.	Hypoblast.	*t.au.*	Tambour like diverticulum of the air bladder in connection with the auditory labyrinth.
Hy.a.	Hyoid arch.		
i.	Intestine.		
int.	Intestine.	*thy.*	Thyroid gland.
in.	Infundibulum.	*t.l.*	Torus longitudinalis.
i.j.	Inferior jugular vein.	*t.op.*	Tectum opticum.
i.j.	(in fig. 50) internal jugular foramen.	*u.bl.*	Urinary bladder.
k.	Keel of neural plate (infundibulum?)	*v.a.*	Vitelline afferent vessel (subintestinal).
l.	Liver.	*v.c.*	Valvula cerebelli.
l.d.c.	Left ductus Cuvieri.	*vent.*	Ventricle of the heart.
lep.	Lepidotrichia.	*v.t.*	Velum transversum.
l.i.	Lobus inferior.	*v.v.*	Vitelline vein.
Li.	Root of lateral line nerve.	*w.*	Wing of nerve plate which gives rise to cranial ganglion.
l.l.	Lateral line nerve.		
l.l.	(on Plate XIX) lobus lateralis.	*y.s.*	Yolk sac.
l.p.c.	Left posterior cardinal sinus.	*z.*	Zygapophysis.
ly.	Pseudo-lymphoidal tissue.	I.—X.	Cranial nerves, or their foramina in the skull.
m'm''.	Muscles of dorsal fin.		

EXPLANATION OF PLATE XVI.

Gymnarchus niloticus.

Fig. 1.   A surface view of the youngest stage of the collection.   First day.   The segmenting blastoderm occupies the centre of the flattened pole of the egg.   There is no sharp line between segmented and unsegmented parts of the egg.   *Vide* text-fig. 86, p. 295.   23. VII. 00.   × 5.

Fig. 2.   Another egg from the collection on the first day.   The segmentation is more advanced than in fig. 1, and the outer layer of epiblast is differentiated from the sub-lying cells and is continuous at its edge with the unsegmented yolk mass.   *Vide* text-fig. 87, p. 295.

23. VII. 00.   × 5.

Fig. 3.   Egg of the second day.   The blastoderm has expanded in all directions, and become reduced to a thin layer of cells in the centre, while the rim remains thicker all round and is especially thickened at one point, the embryonic knob.   *Vide* text-fig. 90, p. 303.   24. VII. 00.   × 5.

Fig. 4.   Surface view of the egg on the third day.   The central nervous system is clearly defined and lies directed towards the centre of the blastoderm.   No other organs can be detected in surface view as yet.   The neural plate has an evident groove along its anterior region.   The blastoderm has extended so as to cover about three-fifths of the surface of the egg.

25. VII. 00.   × 5.

Fig. 5.   The neural plate of an egg of the same age as that of fig. 4.   25. VII. 00.   × 10.

Fig. 6.   Another specimen of the third day which is more advanced and shows in addition to the neural plate several pairs of protovertebrae.   The neural plate is grooved.

25. VII. 00.   × 10.

Fig. 7.   Surface view of the posterior region of the egg of the fourth day.   The blastoderm has completely surrounded the yolk mass and closed as a linear streak.   26. VII. 00.   × 5.

Fig. 8.   A surface view of the whole "embryo" of the fourth day together with the closed blastoderm rim.   The central nervous system is now tubular throughout the greater part of its length.   The fore brain and optic vesicles and mid brain are seen in front; see also text-fig. 88, p. 298.   The hind brain and trunk are greatly attenuated.   There is no tail bud or fold.   The anterior part of the fore brain lies on its side.   26. VII. 00.   × 10.

Fig. 9.   The egg upon the fifth day.   The "embryo" lies upon its side greatly distorted and compressed by the pressure of the egg membrane.   All seven visceral arches are very prominent.   The pectoral fins, and vitelline afferent and efferent vessels are also visible.

27. VII. 00.   × 5.

Fig. 10.   The egg upon the sixth day.   The gill filaments first noticeable on the fifth day upon the first and second branchial arches are now easily seen on the third and fourth as well.

28. VII. 00.   × 5.

Fig. 11.   The larva of the seventh day.   The tough membrane has been ruptured, and the "embryo" has assumed its normal shape.   29. VII. 00.   × 5.

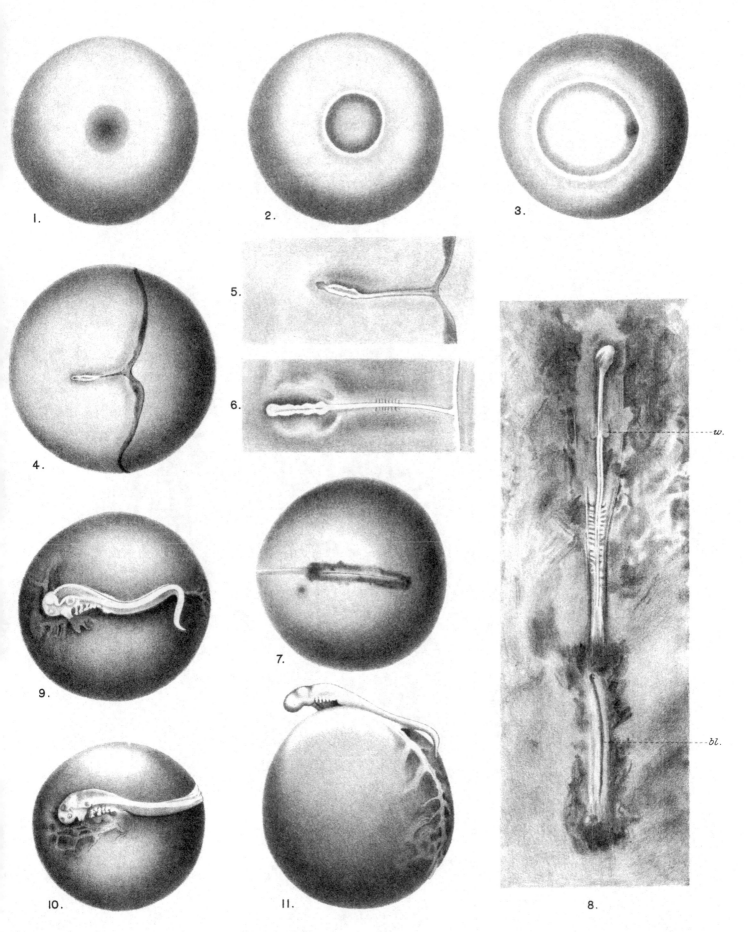

PLATE XVII.

EXPLANATION OF PLATE XVII.

Gymnarchus niloticus.

FIG. 12. Larva of the seventh day. The nasal pit and eye are visible. The yolk sac vessels are less prominent owing to the greater depth at which they now lie. The pectoral fin is spatulate.

29. VII. 00. × 5.

FIG. 13. Larva of the eighth day. The nasal pit is double. There is a continuous dorsal and caudal and ventral fin. The latter is broken by the anus. The yolk sac is much elongated. The operculum is growing back, tending to cover the external gill filaments.

30. VII. 00. × 5.

FIG. 14. Larva of the ninth day. 31. VII. 00. × 5.

FIG. 15. Larva of the tenth day. This is the greatest length which the yolk sac attains. The (epiblastic) external gill filaments are also at the height of their development. The constriction between embryo proper and yolk sac is about as narrow as it ever is. It may be noted that it is not much narrower than in the earlier stages. Outline figures in the text show the subsequent changes in general form and in condition of the median fin, p. 302. I. VIII. 00. × 5.

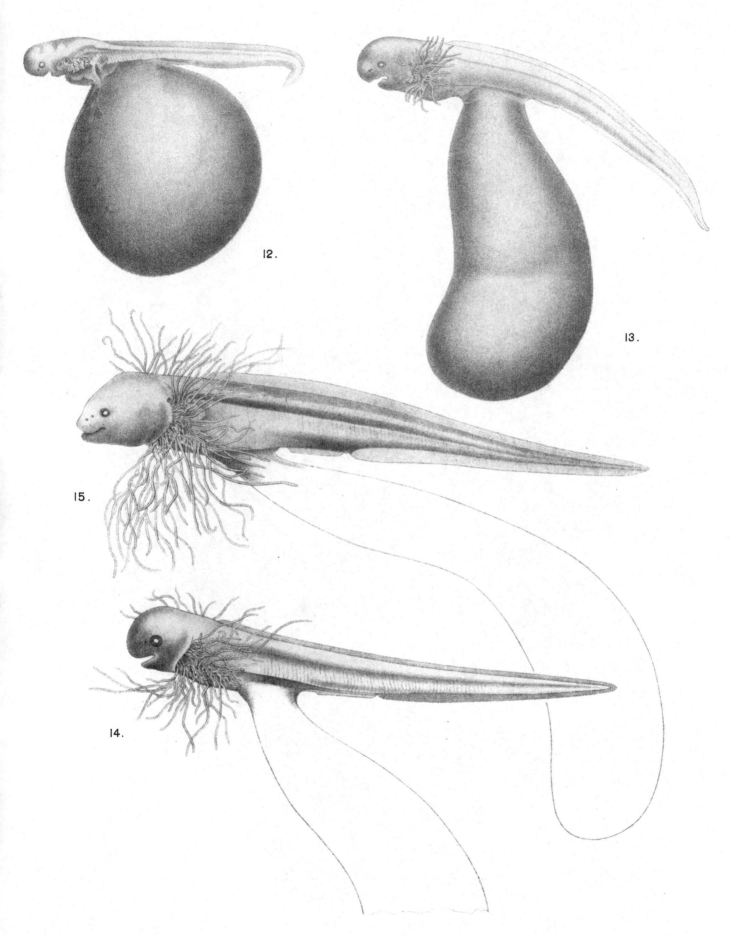

12.

13.

15.

14.

PLATE XVIII.

EXPLANATION OF PLATE XVIII.

Gymnarchus niloticus.

Fig. 16. Heart and right ductus Cuvieri and branchial afferents of larva of the forty-third day. × 10.

Fig. 17. Hind view of the same heart showing the opening of the left ductus into the left side of the auricle. The right ductus and hepatic veins open by a common opening more ventrally. Note the groove in the dorsal wall of the auricle, marking the line of a partial septum in the earlier stage. *Vide* text-fig. 129, p. 341. × 10.

Fig. 18. Heart of larva of about sixteen days. There is still a common opening for the two ductus Cuvieri and hepatic vein. The bulbus diverticula are small and rounded. × 10.

Fig. 19. Dorsal view of the same heart (sixteen days). The single opening into the auricle is seen upon the left side. There is a distinct groove along the dorsal wall of the auricle. × 10.

Fig. 20. Dorsal view of the heart upon the ninth day. × 10.

Fig. 21. Ventral view of the heart upon the ninth day. × 10.

Fig. 22. Dorsal view of the heart on the seventh day. × 10.

Fig. 23. Ventral view of the heart on the seventh day. A lateral view is seen in fig. 30. × 10.

Fig. 24. A view of a larva of the forty-third day dissected to show the liver (*l*) and yolk mass (*ys*) now totally within the ordinary contour of the fish. *au.*, auricle; *v.*, ventricle; *d.b.*, diverticula of bulbus; *d.c.*, ductus Cuvieri of left side; *v.v.*, vitelline vein. × 5.

Fig. 25. View of the abdominal viscera of the same specimen after removal of the yolk mass. *s.*, stomach; *l.*, liver; *i.*, intestine; *py.c.*, pyloric caeca; *sp.*, spleen; *g.bl.*, gall bladder. × 5.

Fig. 26. View of the abdominal viscera of the same specimen after removal of the liver and intestine, etc. and the heart and pericardium. *a.bl.*, air bladder; *ab.a.*, the "pulmonary" artery, being the third and fourth epibranchials of the left side; *ab.v.*, the pulmonary vein, being the left posterior cardinal and ductus Cuvieri which has a separate opening into the left side of the heart; *cœ.mes.*, coeliac mesentery; *d.a.*, dorsal aorta; *prn.*, convolutions of the pronephric duct in a state of degeneration; *mes.n.*, kidney; *u.bl.*, urinary bladder. 3. IX. 00. × 5.

Fig. 27. Dissection of a specimen of the thirteenth day. The abdominal cavity is open and the relation of the yolk sac to the liver is seen. *v.a.*, vitelline afferent vessel (sub-intestinal); other letters as in figs. 24 and 26. × 5. 3.

Fig. 28. Dorsal view of the alimentary canal and air bladder of a larva of the ninth day. The right lobe of the air bladder has a forward bilobed piece which has not yet reached the ear. The left lobe is growing backwards as the lung. See text-fig. 116, p. 324. × 5. 1.

Fig. 29. The alimentary canal and air bladder of a larva of the seventh day seen from the dorsal surface. The air bladder has already an anterior and posterior horn. × 10.

Fig. 30. A dissection from the side of the anterior end of a larva of the seventh day, to show the brain, the cranial nerve roots, the heart and parts of the alimentary canal. *per.*, pericardium; X, IX, VII, V, the tenth, etc., cranial nerve roots; *l.l.*, is the lateral line nerve, quite unconnected with the vagus nerve; *mb.*, mid brain; *inf.*, infundibulum; *br.af.*, branchial afferents cut across; *v.t.*, velum transversum. × 10.

Fig. 31. A dissection of the side of the head of a larva of the thirteenth day, to show the anterior bulb (*t.au.*) of the air bladder coming into position in connection with the sacculus of the auditory labyrinth. Note also the gill filaments (*f*) of the first branchial arch, flattened by the cushion-like inner surface of the opercular fold. *Br.st.*, branchiostegal rays.

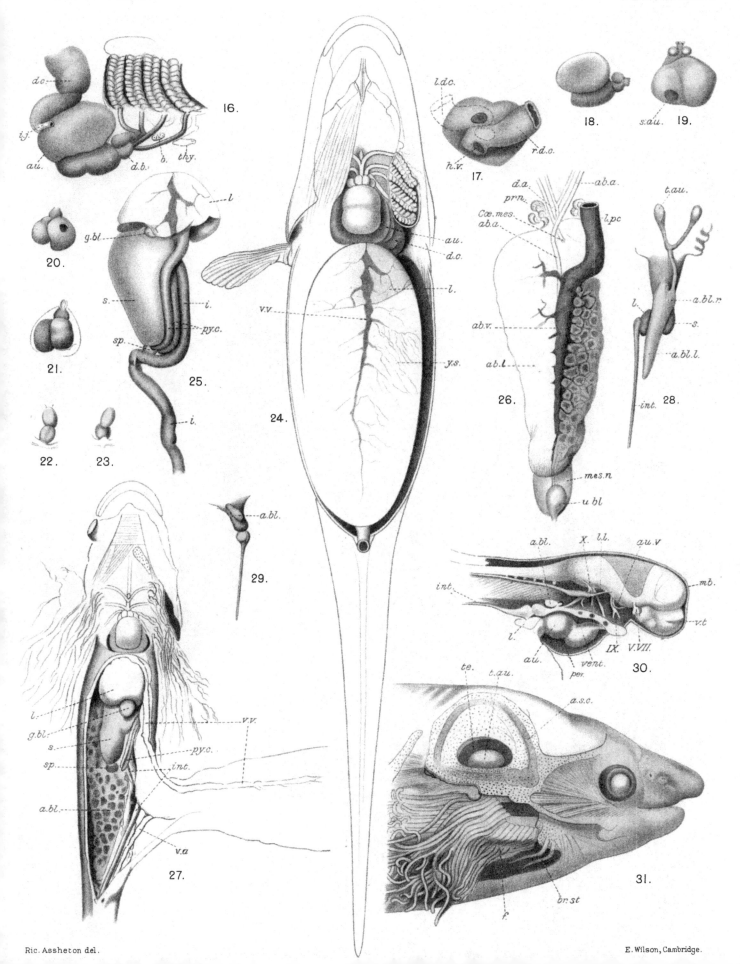

Ric. Assheton del.   E. Wilson, Cambridge.

PLATE XVII

PLATE XIX.

EXPLANATION OF PLATE XIX.

Gymnarchus niloticus.

FIG. 32. Dorsal view of the brain of a seventh day larva.   × 10.

FIG. 33. Dorsal view of the brain of a ninth day larva.   × 10.

FIG. 34. Dorsal view of the brain of a thirteenth day larva.   × 15.

FIG. 35. Dorsal view of the brain of a forty-third day larva.   × 15.

FIG. 36. Ventral view of the brain of a seventh day larva.   × 10.

FIG. 37. Ventral view of the brain of a ninth day larva.   × 10.

FIG. 38. Ventral view of the brain of a thirteenth day larva.   × 15.

FIG. 39. Ventral view of the brain of a forty-third day larva.   × 15.

FIG. 40. Lateral view of the brain of an embryo of the fourth day. Somewhat compressed.

× 10.

FIG. 41. View of the brain of an embryo of the fifth day. The embryo is now greatly distorted and the brain much flattened. It is seen partly from the side and partly from above.   × 10.

FIG. 42. Lateral view of the brain on the day of hatching, sixth day. It is not certain whether these furrows are divisions into neuromeres, or whether they are accidental.   × 10.

FIG. 43. Lateral view of the brain of a larva of the seventh day.   × 10.

FIG. 44. Lateral view of the brain of a larva of the ninth day.   × 10.

FIG. 45. Lateral view of the brain of a larva of the thirteenth day.   × 10.

FIG. 46. Lateral view of the brain of a larva of the forty-third day.   × 10.

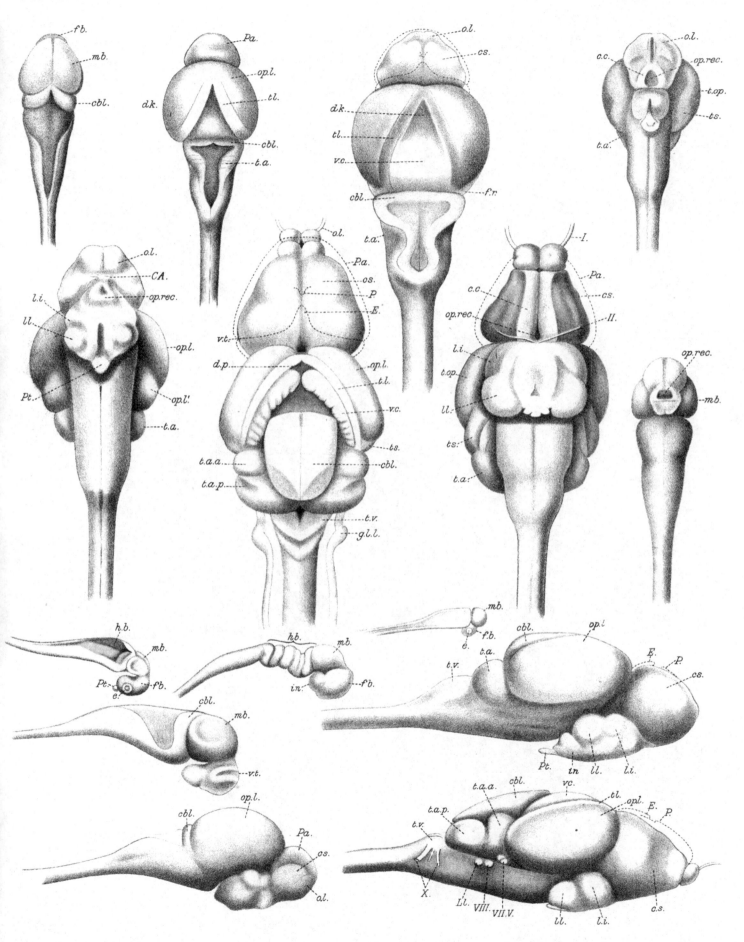

PLATE XX.

EXPLANATION OF PLATE XX.

Fig. 47. Dorsal view of the cartilaginous cranium of a ninth day larva.   × 20.

Fig. 48. Lateral view of the cartilaginous cranium of a ninth day larva.   × 20.

Fig. 49. Dorsal view of the cartilaginous cranium of a forty-third day larva.   × 20.

Fig. 50. Lateral view of the cartilaginous cranium of a forty-third day larva.   × 20.

Fig. 51. Transverse section through the neural plate and notochord on the third day in the region of the hind brain.

Fig. 52. Transverse section through the spinal cord region of the neural plate on the third day.

Fig. 53. Transverse section through the spinal cord region taken farther back than the preceding section.

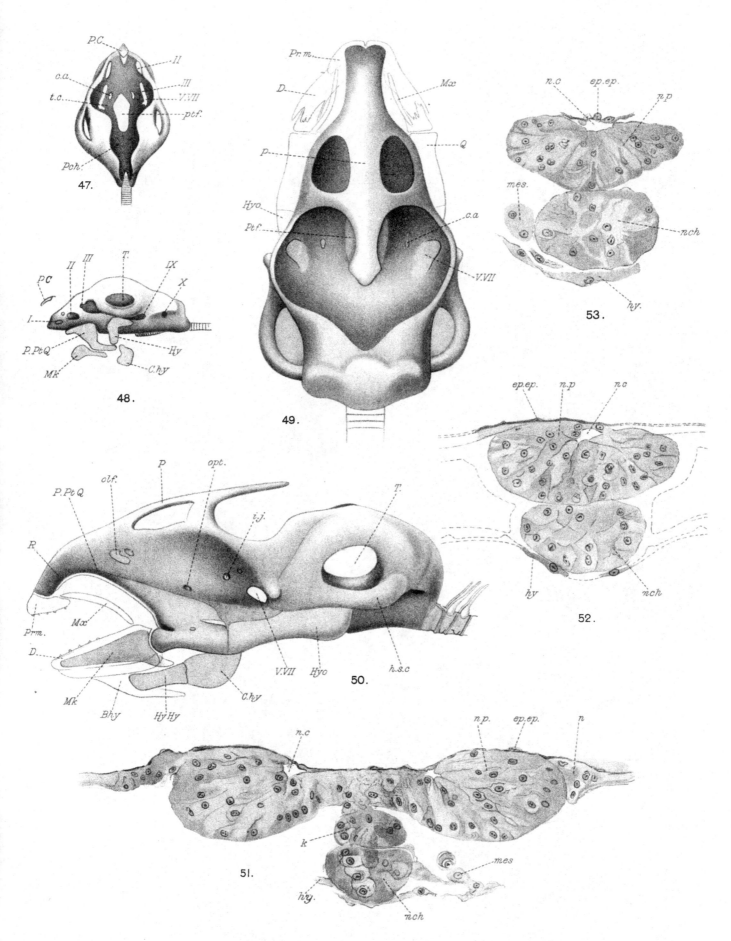

47.

48.

49.

50.

51.

52.

53.

PLATE XXI.

EXPLANATION OF PLATE XXI.

FIG. 54. A side view of the head of an embryo on the fifth day to show the vascular system and its relation to the visceral arches. The arterial system is coloured red, the venous blue. The embryo is greatly distorted so that both auditory vesicles are visible.

FIG. 55. Section through the corresponding region of the kidney on the seventh day. *ly.*, pseudo-lymphoidal cells still isolated. Other letters as above.

FIG. 56. Section through the kidney (mesonephros) of a tenth day larva to show the diverticula of the pronephric duct in connection with the development of the buds in the region in front of the permanent kidney tubules. *bl.*, blood vessels; *ly.*, pseudo-lymphoidal tissue; *Pn.*, pronephric duct; *b.pn.*, bud developing on walls of the duct; *div.pn.*, the rudimentary tubules developing as diverticula of the pronephric duct.

FIG. 57. A semi-diagrammatic transverse section through the trunk of a larva on the forty-third day, to show the cartilaginous and bony elements in the vertebral column. The cartilage is coloured purple.

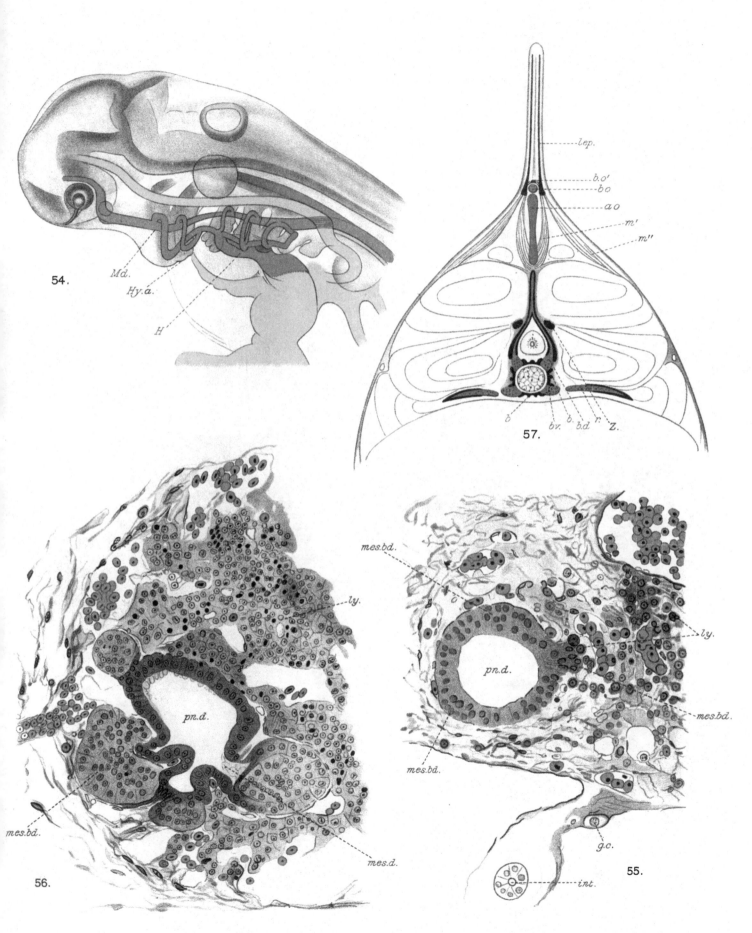

XV. REPORT UPON SUNDRY TELEOSTEAN EGGS AND LARVAE FROM THE GAMBIA RIVER.

BY RICHARD ASSHETON, M.A.,

Trinity College, Cambridge;

Lecturer on Biology in the Medical School of Guy's Hospital, in the University of London.

With text-figures 166—171.

HYPEROPISUS BEBE?

In addition to the series of *Gymnarchus* embryos there are eggs and larvae of several Teleostean fishes obtained by Budgett from the Gambia river. Amongst them are eggs and larvae, contained in twenty-seven tubes, of a species which Budgett considered to be possibly *Hyperopisus bebe* (*v.* this volume, p. 134). These specimens, probably selected from one nest of eggs, were preserved at intervals of a few hours on the 10th, 11th, and 12th July, 1899, and one lot each on the 13th, 14th and 15th. The youngest egg is a fairly early stage of segmentation, while the latest stage corresponds to about the eighth day of *Gymnarchus* which is still an early developmental stage and far too early to enable one to form any opinion upon the correctness of Budgett's diagnosis.

Hyperopisus is remarkable for the enormous development of the valvula cerebelli in the adult. The only trace of this in the oldest embryo is a slight depression outwards and backwards of the optic lobes, such as occurs in *Gymnarchus* where the valvula though well developed never becomes as prominent as in *Hyperopisus*.

In the youngest stage (3 p.m. X.) there is a cap of blastomeres which number about 114, many of which are in process of division, so that it probably represents a period nearing the end of the seventh generation of cell divisions. There are two well marked layers of segments, an outer one (text-fig. 166, *CL*), continuous round its margin with the protoplasm of the egg mass, which forms a thin layer extending over the whole surface of the egg, although nuclei occur only immediately round the cap of blastomeres, and an inner layer (*IM*) which lies between the outer layer and the yolk mass. Cells occasionally occur between these two layers, and between the lower layer and the yolk mass. There is little or no protoplasm on the yolk under the centre of the blastomere cap. At the end of segmentation a compact mass of cells is probably formed, though in some specimens spaces occur. At this stage the outer layer of cells may be distinguished from the bulk lying within by their lighter colour, larger size, larger nuclei, and greater vacuolation. They are directly continuous with the protoplasm

55

of the yolk mass (text-fig. 166, Diagram II, *CL*, *P*).   This layer of protoplasm is more evident now under the centre of the cap of cells, but nuclei occur only at the periphery.   Here they are larger than elsewhere and some already show signs of transformation into the yolk nuclei of later times.

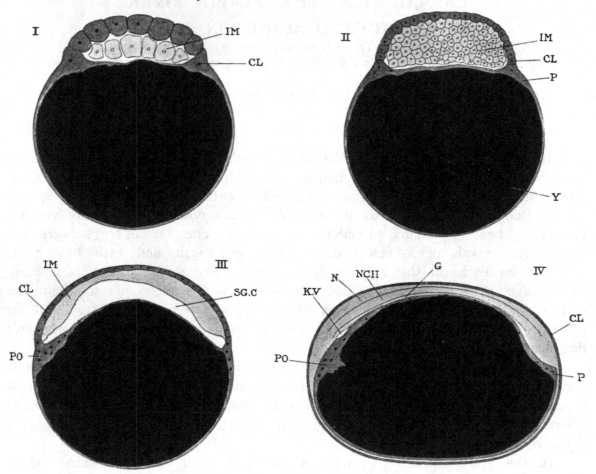

FIG. 166.   Diagrams I—IV, representing sagittal sections of stages in the development of a Teleostean supposed to be *Hyperopisus bebe*.   I.   Early segmentation stage.   An inner mass from which perhaps the whole embryo is developed is marked off from a surrounding protective and nutritive layer.   II.   A late stage of segmentation.   III.   Stage with subgerminal cavity.   IV.   Stage after envelopment of the yolk mass by the blastoderm. *CL*, covering layer protective and nutritive;   *G*, endoderm;   *IM*, inner mass from which probably the whole embryo is developed;   *K.V*, Kupffer's vesicle;   *N*, neural plate; *NCH*, notochord;   *P*, periblast (anterior end);   *PO*, periblast (posterior end);   *SG.C.*, subgerminal cavity.

In the more advanced specimens (12 noon XI.) the cap of cells is much flattened, and extends rather further over the yolk.   A cavity occurs between the cap of cells and the yolk which is probably a subgerminal cavity, that is to say, it may be regarded as archenteron, and is bounded below by the layer of protoplasm or syncytium containing many nuclei round the periphery and a few more centrally (text-fig. 166, Diagram III).

At one point, the future posterior border, the syncytial layer is especially thickened and contains many nuclei, the more internally placed of which are normal, those more externally placed are hypertrophied.

Text-fig. 167 is a semi-diagrammatic figure of a section through this part of the border.

In the next stage, said to be only one hour later (1 p.m. XI.), the head and tail ends are distinguishable.

I have been unable to follow the growth of a germ ring over the surface of the egg as occurs in other Teleosteans. The specimens are small and have been over seven years in spirit, and surface views are very difficult to observe. The process, if it occurs in the normal way, must be exceedingly rapid. It is possible that the extension of the cap of cells may occur chiefly by the spreading of a thin single layered edge, after the manner of the anterior edge of the blastoderm in Batrachus (Clapp (31[1])), so that an actual blastopore rim is developed in an even smaller degree than supposed by Hertwig in his explanation of this process on page 808 of his *Handbuch*, and perhaps the condition approaches closely that of the Amniota.

The subgerminal cavity has disappeared completely unless, as is probable, a small space lying between the syncytial layer and the cap, a rudimentary Kupffer's vesicle, is derived from it (text-fig. 166, Diagram IV, *K.V*). The segmented portion now lies flat down upon the yolk mass. The outer or covering layer is much attenuated.

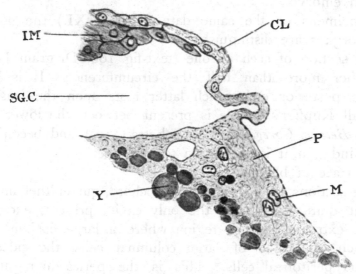

FIG. 167. *Hyperopisus bebe?* Section taken across the edge of the blastoderm showing the continuity of the outer layer with the periblast. *CL*, outer covering layer; *IM*, inner mass; *M*, hypertrophied yolk nucleus; *P*, periblast; *SG. C*, subgerminal cavity; *Y*, yolk grain.

---

[1] The numbers refer to the list of references in the preceding paper of this volume by the present writer on *Gymnarchus*.

The origin of this covering layer and its relation to the periblast is a matter of some interest. It is commonly regarded as an epidermic layer of epiblast, and is spoken of as usually extending beyond the limits of the general segmented part of the egg or blastoderm, and is said to rest upon the yolk mass. Sumner (113) has recently described for several species (e.g. *Salvelinus fontinalis*) what had previously been noticed by Kowalewski in *Carassius*, that this covering layer is thickened at one point, the posterior end, and produced inwards between the lower layer cells and the periblast. This he regards as an invagination and speaks of it as the prostomal thickening. In all the figures given, and in the description, this prostomal thickening of the covering layer is indicated as being separate from the periblast. In *Hyperopisus*—if such is the animal—there is certainly a greater mass of periblast at the future posterior end, and this is shown very clearly to be continuous with and to be part of the covering layer. I do not think however that this thickened periblast layer gives rise to the lining cells of the future gut. It forms the floor of the archenteron, including the Kupffer's vesicle, but I have no evidence to show that it forms gut epithelium. It may do, for certainly not all the nuclei are at this stage merocytic (text-fig. 167).

I think it more probable that all the future gut epithelium is derived from the inner cell mass.

In fact, the covering layer and the periblast are of common origin and are essentially a protective and yolk bearing embryonic envelope, and together bear precisely the same relation to the embryo proper that the trophoblast does to the mammalian embryo.

In other specimens of the same date (1 p.m. XI.) the notochord, neural plate, and other organs are distinguishable.

In a sagittal section of such an one (text-fig. 166, Diagram IV), the embryo extends over rather more than half the circumference. It is broader at the anterior than the posterior end, which latter rests upon the mass of thickened periblast. A small Kupffer's vesicle is present between the lower layer cells and the periblast as in *Belone*, *Coregonus*, etc. (Sobotta (110)), and becomes much larger later. I do not find that it is ever open to the exterior.

There is no trace of blastopore.

In transverse sections the neural tube, notochord, gut epithelium, and mesoblast can be made out quite easily, but the only cavity present, except the Kupffer vesicle, is in the extreme anterior region where a large coelomic space occurs, the floor of which is formed of large columnar cells, the sides and roof of ordinary squamous peritoneal cells. This is the pericardium, and in front of it a group of cells occur which are the earliest formed blood cells. In some of these specimens there are as many as seven protovertebrae.

Four hours later (5 p.m. XI.) there are about eighteen protovertebrae.

I find no trace of the renal system, which is peculiar, considering the advanced stage of the embryo.

There is now a head fold, so that the pericardial cavity has assumed its normal position. The optic vesicles are hollow, but not yet continuous with the cavity of the brain, which has appeared along the ventral region of the neural thickening.

Two pairs of hemispherical depressions occur on the head, just in front of, and, in general form, similar to the auditory vesicles. These are the mucous glands which secrete the strand of mucus by which the larva is suspended after leaving the egg shell.

A tail fold also is present.

With reference to the alimentary canal, it may be noted that an anterior pharyngeal and posterior gut region can be distinguished. The former is solid and compressed dorso-ventrally, the latter is tubular and hollow. There are no gill clefts. Solid pouches project outwards and come in contact with thickenings of the epiblast. I am not certain that I can detect more than one pair.

Spec. 11 p.m. XI. This is the earliest stage at which I have found the renal system. In this there is a solid strand of cells, the pronephric duct, which at its anterior end has a lumen and ends anteriorly in a groove, which I think is open to the general body cavity. There is not more than one opening, but a strip of thickened epithelium is continued forwards from this opening. The sections however are not good.

About this point the dorsal wall of the alimentary canal is thickened, the thickened area having very much the appearance of the isolated epithelial-like tissues which lie in a corresponding position at a later stage and give rise, probably, to the islands of Langerhans.

After hatching.

In spec. 4 p.m. XII. there is as yet no well formed pronephric chamber.

In 9 p.m. XII. specimens very considerable advances have been made. The pronephric chamber has the normal Teleostean character, lying beneath the dorsal aorta at the point where the two roots of the aorta join. The pronephric ducts pass back straight to the posterior end, where they open into the hind end of the gut.

The large mucous glands are at their greatest development. The two anterior ones are small and are placed on the anterior exterior extremity.

The posterior pair are large and are on the top of the head about the level of the optic lobes. The covering layer is not concerned in the formation of these organs and is non-existent at this point.

There are three pairs of aortic arches, though only the first pair is complete. The two hinder ones are visible only as outgrowths of the aortic roots. I cannot find any true gill cleft. The pharynx is a flattened solid sheet of hypoblast, double layered, and having pouches which run out towards the skin. I cannot say where the hypoblast and epiblast meet. In the region of the hind brain, at the point where the fifth nerve occurs, the first aortic arch passes round the pharynx, and is presumably the mandibular arch. Posterior to this, at the level of the auditory vesicle, the pharynx is split, and the cavity passes out towards a depression in the skin, but it is not continuous with it. I cannot be sure of the nature of this depression. It is no doubt the hyomandibular cleft, and I think it is probably a depression from without, that is to say, with epiplastic lining.

The brain and nervous system present no characters that call for any comment.

There is no paraphysis or epiphysis.

The remaining specimens of the thirteenth and fourteenth days do not show any very marked advance. In the oldest stage the heart lies in a large pericardium. The cardiac aorta divides into two branches which run backwards on the ventral wall of the pharynx. There are not more than three, perhaps only two complete aortic arches. The mandibular arch is by far the largest. The hyomandibular cleft alone is open ; this is widely open.

There is no ventral mesentery joining the yolk mass to the alimentary canal posterior to the liver.

The excretory system consists simply of pronephric ducts and chambers, and some large cells which probably are the forerunners of the pseudo-lymphatic tissue of other forms.

The skeletal system has not yet begun to be formed.

Heterotis niloticus.

There are seven stages of larva of *Heterotis* collected between the 22nd and 26th July, 1900, and another later stage collected on the 14th of the following month.

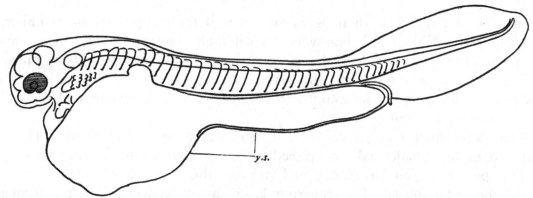

Fig. 168.   View of the youngest stage of *Heterotis niloticus*. The yolk sac extends along the whole length of the alimentary canal; *y.s.*, yolk sac.

The youngest larva is advanced, and already shows the external gill filaments which this species bears upon the visceral arches 3—6, as does *Gymnarchus*.

The yolk mass extends along the whole ventral wall of the gut as in the Amphibians, but it is more bulky anteriorly (text-fig. 168). There is a broad caudal fin without fin rays. The tissue immediately ventral to the most caudal region of the nerve tube and notochord is beginning to proliferate, and so causes an expansion of that region which results in the turning up of the ends of the notochord and spinal cord. This proliferating tissue eventually gives rise to the caudal fin rays of the adult homocercal tail. The broad larval caudal tail becomes much reduced as development proceeds, but later, on development of the fin rays, it recovers itself. The caudal fin of the adult is derived wholly from the ventral caudal fin of the larva as in other Teleosteans (Agassiz (1)).

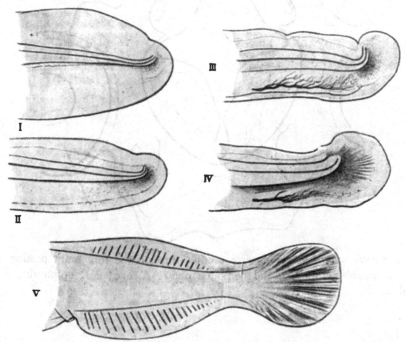

FIG. 169. *Heterotis niloticus.* Diagrams I—IV, drawn from transparent objects, show the caudal fin beginning as a general proliferation of tissue immediately ventral to the tip of the notochord. Diagram V, shows the recovery of ventral and dorsal fins, which had almost disappeared in stage IV.

The brain shows, as yet, no Teleostean characters. It is amphibian-like in appearance.

In a rather older specimen the diverticula on the roof of the third ventricle are interesting. I find nothing in front of the velum transversum comparable to the structure I have called the paraphysis in *Gymnarchus*. There is the structure which —one almost may say obviously—corresponds to the organ usually known as epiphysis (Trout, *Gymnarchus*, etc.), and in the posterior wall of this the habenular commissure runs.

Posterior to this, and between this commissure and the commissura posterior there is a ganglionic mass, and perhaps a diverticulum, which lies in the median line and projects slightly forwards and backwards.

In the older stages the preservation of the material is not satisfactory and I am unable to trace the later history.

One is inclined to wonder whether this is the true epiphysis, and whether the organ usually called the epiphysis in Teleosteans is a kind of false epiphysis produced by the velum transversum.   This organ described here (text-fig. 171) is far more like

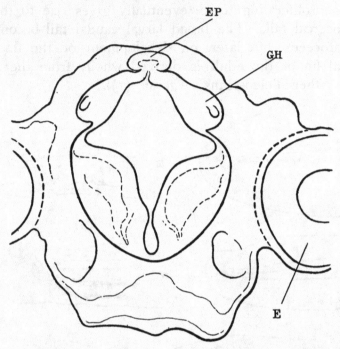

FIG. 170. *Heterotis niloticus.* Transverse section through the fore brain passing through the ganglia habenulae, epiphysis, and corpora striata. *E*, eye; *EP*, epiphysis; *GH*, ganglion habenulae.

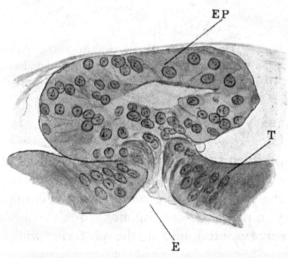

FIG. 171. *Heterotis niloticus.* Transverse section through the epiphysis more highly magnified. *E*, opening into the epiphysis; *EP*, epiphysis; *T*, roof of thalamencephalon.

the pineal organ of Amphibia than is the structure usually named epiphysis in Teleosteans—which resembles rather an everted choroid plexus.

The nerve fibres of the optic nerve intermingle very much where they cross, but I find no trace of a true chiasma.

Mucous Canals.

There are two special systems of mucous canals on the head which become embedded in bone as in many other Teleosts. These two large systems arise as depressions in the skin having tall columnar epithelial cells and occupy the position and have very much the same appearance as the two pairs of mucous glands which secrete the anchoring strand in *Hyperopisus*.

As in that larva, the two anterior ones are much smaller and are placed upon the face—so to speak—of the larva, while the two larger are placed on the top of the head. These glands, and the corresponding but very small ones of *Gymnarchus*, are probably all homologous, but are only very generally homologous with analogous glands in Amphibians and Ganoids.

Air Bladder.

This organ arises in the mid-dorsal line and grows backwards between the dorsal border of the mesentery and the body wall, expanding as it does so above the peritoneum. It is a single undivided chamber throughout. The wall is very vascular but there is no trace of air chambers.

Reproductive and Renal Organs.

The gonad in the youngest stage is single and lies beneath the right pronephric duct, but as the air bladder increases in size it is driven away from it and so all connection between the gonad and kidney tubules is rendered impossible.

The excretory organs show the usual Teleostean characters. In the youngest stage the pronephric chamber is on the wane. In the oldest the mesonephros is well established.

The first formed mesonephric tubules appear segmentally in the usual manner and are all dorsal to the pronephric ducts. Secondary ones are added later which arise like the primary ones as little vesicles separate from the pronephric duct, but may arise upon any side of it. They are almost, if not quite, confined to the pronephric duct itself. The Malpighian bodies are all near to the middle line, because the renal arteries come off as single median vessels from the dorsal aorta.

The mesonephric "buds" are difficult to trace, as they are not numerous. In the oldest specimens they have the appearance of lymphatic glands. When first formed the cells are like those of the walls of the pronephric duct, but they become smaller and acquire a much reduced and deeply staining nucleus. They can be quite easily distinguished from the pseudo-lymphatic tissue which has the usual distribution.

The Skeleton.

Of the skeleton the stages of cartilaginous formation are included within the limits of the series.

In one of the younger stages the cranial skeleton is represented by the trabeculae cranii which are present as two thin bars of cartilage, united in front anterior to the optic nerves, and continuous posteriorly with the parachordals which are separated by the notochord.

In the region of the horizontal semicircular canals the parachordals expand upwards so as to cover the sides of the medulla and auditory organ as far up as the horizontal canals.

There is no rostrum. There is no cartilage in connection with any part of the vertebral column. Cartilages are present in the visceral arches. The palato-pterygo-quadrate is a thin rod of cartilage quite unconnected with any other cartilages.

Later the anterior part of the brain is protected dorsally and laterally by the nasal capsules and rostral cartilage. This seems to be brought about by the continuation upwards of the fused trabeculae in front of the brain and the continuation backwards of the plate thus formed as three narrow strips. One dorsal and median strip extends backwards only as far as the epiphysial region.

The other two are lateral. Each passes back first as a fairly wide strip, then narrower along the dorsal border of the orbit to broaden out and become continuous with the sides of the cranium derived from the parachordals in the region of the cerebellum, where the hyomandibular now articulates with it.

The palato-pterygo-quadrate bar is now a stout V-shaped piece of cartilage, as in the Anura, but it is not fused with the cranium. It articulates firmly, however, in front with the trabeculae, but behind, although its hinder limb approaches closely to the cranium, it does not actually touch. The hyomandibular embraces the quadrate end of the bar, and clearly aids in the support of the lower jaw, which articulates, however, solely with the quadrate.

In the later stages the median dorsal band of cartilage of the rostral region extends backwards beyond the epiphysial region, and it meets and fuses with the roof of the occipital region of the cranium formed by the upgrowth of the parachordals.

The development of the vertebral column is on the usual Teleostean arco-centrous plan.

Sarcodaces, etc.

One tube containing six larvae, labelled as above, measuring from 11—14 mm. in length. They are all advanced larvae.

Another tube, unlabelled, clearly contains larvae of the same species, but a much younger stage, and is that described by Budgett (26).

A third tube contains three specimens of advanced larvae which resemble *Sarcodaces* in the arrangement of fins, but which are certainly not that animal.

In a fourth tube, unlabelled, are some advanced Teleostean larvae with forked homocercal tail showing externally no trace of the heterocercal condition. Internally, however, the notochord is found to be bent upwards passing to the base of the dorsal lobe of the caudal fin.

In *Sarcodaces*, and in the larvae of the third tube, the ventral fin is produced forwards along the mid-ventral line on to the abdomen between the pelvic fins which are present as simple longitudinal folds.

XVI. NOTES ON ANURAN DEVELOPMENT; PALUDICOLA, HEMISUS AND PHYLLOMEDUSA. BY EDWARD J. BLES, M.A., D.Sc., UNIVERSITY OF GLASGOW.

With Plates XXII—XXVII. and text-figures 172, 173.

The material for the descriptions contained in these notes was collected by Mr Budgett in the Paraguayan Chaco and on his second journey to the Gambia in 1900. There are also in my hands specimens of late stages in the development of *Xenopus calcaratus* Peters and of *Rana albilabris* Boul. and six adult specimens of *Hymenochirus boettgeri* Boul.; the examination of these I propose to defer until I can study them in conjunction with my own *Xenopus laevis* material. I am now proceeding with this and due acknowledgement will of course be made in any published results.

Budgett himself has made full use of his *Phyllomedusa* material in his paper on Frogs of the Paraguayan Chaco (this volume, p. 59) without having given much time to an examination of his *Paludicola* eggs and embryos. When this material was handed over to me I was much impressed by its interesting character and still further impressed by the novelty of the Engystomid embryos of *Hemisus* and, as the series of stages of the last two forms were not extensive, I determined to make an effort to obtain more material and obtained six months leave for a voyage to S. America. I spent almost the whole of the time available—May to August 1905—at San Bernardino on Lake Ipacaraÿ in Paraguay. Unfortunately the winter was most exceptionally cold and there were great floods over immense tracts in Brazil, Paraguay, and the Argentine Republic. Lake Ipacaraÿ had also flooded its shores and my main objects, to collect vertebrate embryological material, were completely defeated. The frogs did not breed during that winter in Paraguay, the only tadpoles found were late hibernating stages.

Budgett's observations in the field on *Paludicola* and *Phyllomedusa* are contained in his paper in this volume, pp. 61—63 and pp. 65—68. I will quote his observations on *Hemisus marmoratum* from his Gambia diary written in 1900.

"*July* 8. In the afternoon Sory called me to see a nest of the frog he had told me about. In a bank at the edge of a rice farm I 'saw one of the Engystoma-like frogs lying in a hole and covering with its belly a mass of egg-capsules containing developing embryos. The hearts were just beginning to beat. The frog, which was a female, did not attempt to move away when disturbed and had to be taken off. I noticed that the belly was very red with dilated capillaries.

July 9. A copious network of blood vessels now covers the yolk sac. The whole body apparently is covered with cilia. A current of fluid runs down the surface of the body tailwards and forwards to the head along the walls of the capsule.   Three distinct arterial trunks run along the outside of the three gill arches.

July 10.   Pigment begins to appear all over the upper surface.   The capsules are swollen.

July 11.   The operculum now covers the gill arches....At 10 o'clock the embryos began to hatch, but how they live henceforward I cannot make out.   They appear to hatch towards the centre of the mass of jelly and I think that the outer parts form a nest while the mother floods them with anal fluid" (i.e. from the bladder, which is almost pure water in frogs) "as she gave out quantities when taken off the eggs. The larvae certainly do not look fit to look after themselves yet.

July 12.   I have made a sketch of a larva this morning, there are neither external nor internal gills or lungs as far as I can see, neither, apparently, is the mouth functional. How they breathe I cannot make out as the tail though large is not specially well supplied with blood vessels.

July 13.   Had three more lots of the frog eggs brought with one female attending an advanced lot of larvae.   She lay perfectly still while the larvae swarmed up over her back and covered her[1].   In these advanced larvae the mouth was well formed and furnished with horny teeth in several rows and also many barbels.

Of the original lot the larvae have acquired a coiled intestine though much yolk remains.   I notice great variation among them especially in the size of the eye and in the amount of pigment.

July 14.   More of the frog with the gill-less larvae.   Males are hard to get. When the female is given too much water she leaves the larvae to swim in the pool she had previously been lying in.   The gills now appear to be functional as there is a stream of water to be seen coming out of the spiracles.

July 14.   There is the same difficulty with these frog eggs as with those of *Phyllomedusa hypochondrialis*: that they develop abnormally if by chance the egg mass has been placed upside down.   The yolk sinks to the lower half of the egg while the upper half becomes clear and empty so that the cells and nuclei could be readily seen in the transparent epiblast.   Segmentation and the formation of the yolk plug seem to be quite like *P. hypochondrialis.*

July 15.   The frog larvae appear to do equally well whether left in a moist mass or put into water.   In water they continually come to the surface and emit bubbles of air."

These observations show how much interest would be attached to further information on Engystomid development.   The only connected account I know of is by J. A. Ryder on a N. American *Engystoma*[2].   It is very brief and deals with a form

---

[1] This recalls Brauer's account of the Seychelles frog, *Arthroleptis sechellensis*, Bttgr.—which transports its tadpoles on its back. "Ein neuer Fall von Brut-Pflege bei Fröschen," *Zool. Jahrb. Syst* Bd. xi., pp. 89—94, 1898.

[2] "Notes on the Development of *Engystoma*," *Am. Nat.*, Vol. xxv., pp. 838—840, 1891.

very different in development to *Hemisus*. The tadpoles of two species of *Microhyla* have been figured by Flower[1].

Methods.

There is nothing of special interest to record on the methods used in this work. My simple prism reflector again proved itself useful[2]. I have taken advantage of the absence of pigment in the early stages of all three forms to examine them by transmitted light after clearing in clove oil; this gave better results than xylol. An ordinary Swift spot-lens was used for the illumination with dark background.

Mr A. K. Maxwell has prepared the drawings with his usual care and discrimination under my direction.

Paludicola fuscomaculata Steindachner.

The size of the egg in this species is given by Budgett (p. 62 of this volume) as 1 mm. diam. and he also describes the character of the spawn. The ovum is quite free from pigment. In the earliest stage examined (Pl. XXII. fig. 1) the blastopore is very small and its edge forms a closed curve. There is no yolk plug and the lip of the blastopore is swollen, but only to a very small extent. The medullary folds have not yet appeared.

The following stage (Pl. XXII. figs. 2 *S* and 2 *V*) is an embryo in the vitelline membrane removed from the surrounding froth. In spite of the, for an Anuran, small size of the egg the comparatively bulky embryo is curved round the yolk very much like the young embryo of a Pleuronectid Teleost. The head and the tail have rapidly risen above the yolk which is confined to the trunk as far as the main mass is concerned, and as the myotomes of the trunk are as much emancipated from the mass of yolk as the head or tail, the embryo may be described as having abdominal yolk. Of course the cells of all the tissues at this early stage contain a certain amount of yolk.

The shape of the abdominal yolk mass is almost spherical and only about 60° of its surface is not covered by embryo, including the part overhung by the tail (fig. 2 *S*). The dorsal flexure of the embryo is therefore very considerable. The rounded tail with only the merest trace of a fin-fold at the tip (fig. 2 *V*) is placed in the median plane and not yet bent laterally.

In the head the position of the optic vesicle is clear (figs. 2 *S* and 2 *V*) also that of the auditory pit (fig. 2 *S*). The stomodaeum is triangular, bounded by fore brain in front and mandibular arch on each side (fig. 2 *V*). At the tip of each of these arches is a faintly indicated rudiment of a cement organ. Behind the mandibles the outline of the pericardium is marked out on the anterior surface of the yolk. On the right side of the head in fig. 2 *V* beyond the eminence of the optic vesicle can be seen the

---

[1] S. S. Flower, "Notes on a Second Collection of Reptiles, made in the Malay Peninsula and Siam," *Proc. Zool. Soc.*, 1899, pp. 902, Pl. LX., figs. 1, 2.

[2] *Trans. Roy. Soc. Edin.*, Vol. XLVI., p. 793, 1905.

rounded branchial region with a shallow groove dividing hyoidean and first branchial arches.   The proctodaeum (fig. 2 *V*) is on the root of the tail, not on the surface of the yolk.

Practically the same stage is represented in three views in figs. 21 *S*, 21 *V* and 21 *D*, Pl. XXVI., drawn from an embryo cleared in clove oil.

The curvature of the embryo is even more striking in fig. 21 *S* than it is in surface view, it brings out so distinctly the relative smallness of the mass of yolk which dominates the attitude of the embryo.

The main outlines of the brain are fairly distinct (fig. 21 *S*), the fore brain very little dilated, the mid brain somewhat constricted and the hind brain much wider. Both the optic vesicle and the auditory sac are elongated in the longitudinal direction. The notochord can only be seen in the trunk region opposite the middle of the myotomes.   The wall of the pharynx, consisting of thick and opaque yolk-laden epithelium, stands out clearly except at the most anterior end where it lies deep as a median recess in close contact with the front end of the fore brain and bending upwards in front of the brain.   At this point it is seen in sagittal section to come into contact with the stomodaeum (see text-fig. 172).   The stomodaeum is not visible in this view as it is covered by the projecting end of the left mandibular arch which produces the swelling of the ventral contour behind the bulge of the fore brain.

The shape of the pharynx can be best realised by comparing figs. 21 *S*, 21 *V* and 21 *D* which show it in different aspects.   It is very wide, but very shallow from roof to floor.   The dark bands running obliquely across its sides (fig. 21 *S*) are the edges of the folds of endoderm lining the gill pouches.   The dark band running up almost vertically towards the anterior end of the ear vesicle represents the hyo-mandibular cleft, the paler band behind it the first branchial cleft, while the equally pale second branchial cleft if continued up dorsally would divide the ear vesicle into two equal parts.   Parallel with the second branchial cleft is the dark posterior wall of the pharynx, continued ventrally into the well defined anterior wall of the liver diverticulum seen bending round to the ventral contour of the yolk.   In the ventral view (fig. 21 *V*) the lumen of the liver is seen to lie over to the left side of the median plane.   In the space between the liver and the anterior part of the pharynx lies the pericardium.   The thick mesodermal lining runs parallel to the anterior surface of the liver and bends round under the pharynx between the lower ends of the first and second branchial clefts.   Running transversely across the Λ-shaped space is the rudiment of the heart.

Returning to the gill pouches, they are very distinct in both ventral and dorsal views.   The more massive hyoidean arch (fig. 21 *D* and *V*) separates the spiracular from the first and second branchial clefts which lie close together divided by the slender first branchial arch.   The anterior wall of the spiracle is continuous with the anterior wall of the pharynx (figs. 21 *V* and 21 *D*) while the posterior wall is continuous below with the vertically placed anterior part of the pharyngeal floor which in fig. 21 *V* shows as a pale band joining the posterior wall of the right with that of the left spiracle. In fig. 21 *V* the optic stalks can be made out and the infundibulum ventral to them with its floor covered by the little anterior dorsal pouch of the fore-gut.

Seen from the side (fig. 21 *S*) this embryo gives a remarkably clear view of the rudiment of the pronephros. The lower ends of myotomes I to VI are all continued into nephrotomes *A—F* which all terminate in the longitudinal archinephric duct seen running back between the lower ends of the myotomes and the yolk until it fades away opposite myotomes XIII and XIV. The most massive nephrotome at this stage is *B*, rather smaller are *D* and *C*. These three are the persistent pronephric nephrotomes, while the nephrotomes *A*, *E* and *F* are vestiges which disappear immediately without any further development. It will be observed that while *A* and *C* run back *B* to *F* run forwards and that while the whole lower part of the myotomes III and IV are turned forwards, myotome V shows a tendency, more marked in myotome VI, to run backwards at its lower end like myotome VII and the more posterior one. These observations have been confirmed by the examination of sections, but the yolk laden tissue of the pronephric rudiments is very difficult to trace and the whole preparations are much more instructive.

The embryos of the next stage (Pl. XXII. figs. 3 *S* and 3 *V*) show less flexure. The yolk in 2 *S* and 3 *S* has been kept in the same position, and when the positions of the head are compared, it will be seen that the anterior end of the embryo instead of being turned downwards is now turned forwards and consequently the trunk and head are brought more into alignment, while the tail is still curved round the posterior surface of the yolk mass. The stomodaeum is drawn away from the yolk. The groove running down from its median ventral angle still separates the rudiments of the right and left cement organs. The eyes are more prominent. The position of the visceral clefts shows in side view more distinctly and the third branchial cleft is just visible externally. The three pronephric tubules are just visible. The proctodaeum is (fig. 3 *V*) on the root of the tail, not on the surface of the yolk. The lengthened tail is folded over to one side and the ventral fin-fold is now continued round its tip to the dorsal surface.

A slightly younger stage than that just described is figured as seen by transmitted light in fig. 4, Pl. XXII. The raised position of the head is clearly seen to be due to the growth of the pharyngeal region. The bulk of the brain has hardly increased. The pineal body and the dorsal thickening[1] have formed and the infundibulum has grown down from the floor of the fore-brain. The lens thickening of the epidermis is beginning, while the ear is now a closed vesicle. Between the eye and the ear the rudiments of the fifth and seventh cranial nerves can be seen. The pharynx is now expanding dorso-ventrally and a fourth visceral cleft is appearing close behind the second branchial, appearing as a dark band close behind and parallel to the second branchial cleft. The liver

---

[1] See p. 448.

diverticulum is pushed forwards to such an extent that the anterior lip of its opening from the floor of the pharynx is anterior to the third branchial cleft. The anterior wall of the liver diverticulum can be seen in fig. 4 bending backwards and downwards into the anterior wall of the yolk, and the narrow chink of the lumen of the liver can be made out in the yolk. The roof of the alimentary canal opposite the opening into the hepatic cavity is depressed and shows in this view as a dark band passing upwards and forwards into the roof of the pharynx and upwards and backwards into the dorsal surface of the yolk. Behind the yolk the cloaca has formed with a widening caused by a bay in its dorsal wall where it leaves the yolk.

An early stage in the development of the pronephros can be made out in its main features in this preparation (fig. 4). The condition here visible closely resembles that figured by Field[1] ('91, Pl. IV. fig. 39) of the young pronephros in *R. sylvatica* Le Conte. Also the position corresponds exactly in relation to the somites, the three nephrostomes being connected respectively with somites II, III and IV, as in Field's case. The archinephric duct runs back just dorsal to the yolk and ends opposite the widened part of the cloaca into which it opens later in development.

The sagittal section of an embryo at the same stage (as fig. 4) is shown in text-figure 172. The striking feature of this section is the amount of cranial flexure. The bending of the fore and mid brain round the end of the notochord is of about the same extent as the bending at its maximum in *R. temporaria*, but in the common frog this amount of flexure occurs at a much earlier stage, when the blastopore is about to close, and the neural tube is very much less differentiated than it is in this *Paludicola* embryo. And in the common frog the reduction of the flexure has progressed considerably when the brain has differentiated to the extent seen in this *Paludicola* larva. In this section (text-fig. 172) the floor of the infundibulum is at right angles to the notochordal axis, whereas in the corresponding brain section of a *Rana* larva the floor of the infundibulum is inclined at about an angle of 45° to the notochordal axis. The pineal body and behind it a ridge projecting into the neural canal are seen on the dorsal wall of the brain. The solid transverse ridge is the "dorsaler Wulst" of v. Kupffer. He states that it occupies the position of the posterior commissure[2]. On the floor of the fore-brain the ridge of the optic chiasma, the recessus opticus, and the commissural portion of the lamina terminalis are in a very backward state of development. There is no trace of the neuropore remaining. The infundibulum is very long and the pituitary body does not reach it. This latter is not much more than an inwardly projecting knob of ectoderm. A depression

[1] H. H. Field, "The Development of the Pronephros and Segmental Duct in Amphibia." *Bull. Mus. Harvard*, Vol. XXI.

[2] O. Hertwig, *Handb. d. vergl. und exp. Entw.-Lehre der Wirbeltiere*, Bd. II. Teil III.; v. Kupffer *Die Morphogenie des Centralnervensystems*, pp. 189–90, figs. 204 and 205.

on the surface of the pituitary body is continuous with the stomodaeal depression which lies immediately behind and is very small. The bay of the endoderm of the fore-gut which is in contact with the ectoderm of the stomodaeum is clearly seen in this section (cf. Pl. XXVI. figs. 21 *S* and 21 *V*). The lumen of the liver diverticulum is shown in its full dimensions, it is very narrow from side to side. Opposite the liver the dorsal wall of the gut is thickened to form the dorsal pancreas rudiment. The main mass of the yolk is clearly abdominal and does not encroach on the head region. The intestine is straight and widely open, there is a bend where it passes into the cloaca, and the dorsal wall at

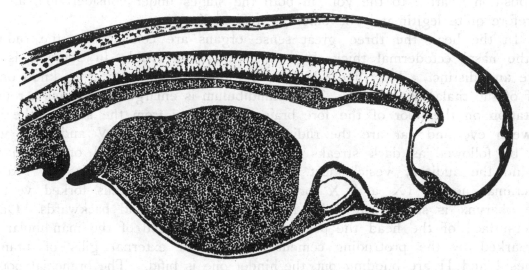

FIG. 172. Sagittal section of *Paludicola* embryo at the stage shown on Pl. XXII. fig. 4. The length of the scale is $\frac{1}{10}$ mm.

this point forms a bay into which the archinephric ducts open later in development. The cloacal opening is just forming; the lumen, visible in one section only (10μ), is choked with a hyaline mass apparently of protoplasm. The prominent lips of the opening are obviously thickened ectoderm, i.e. proctodaeal. In contact with the ventral surface of the notochord there is a well-marked sub-notochordal rod, slightly swollen at its anterior end, and quite separate from the endoderm. The anterior end of the notochord does not reach the infundibulum as it does in the corresponding stage of *R. temporaria*.

In the larva of the next stage available (Pl. XXVI. fig. 22) the tail has lengthened considerably and straightened. There is still no perceptible pigment present, and an instructive view of the internal structure can be obtained by clearing in clove oil and using spot-lens illumination. The abdominal mass of yolk-cells has become less protuberant and the head has elongated (compare Pl. XXII. fig. 4). The rectification of the cranial flexure has advanced a stage, and is obviously not entirely due to the outgrowth of the cerebral hemispheres now only just commencing and scarcely perceptible. By taking the angle between (1) the long axis of the body and (2) the diameter of the optic vesicle which passes through the choroid fissure in fig. 22, and comparing it with the angle similarly obtained in fig. 4, it will be seen that at the later stage this angle is decidedly less acute. This comparison can be used to show that cranial flexure is diminished by differential growth of dorsal and ventral parts of the brain and of the adjacent parts. The assumption is made that there is no rotation on each other of the optic vesicle and the fore brain, there being of course no evidence of twisting in these organs. The part of the axis used for line (1) in fig. 4 is that portion of the notochord opposite to the yolk, as this maintains its position relative to the yolk in both the stages under consideration, and can therefore quite legitimately be taken as a fixed line.

In the head the three great sense organs are conspicuous, the rudiment of the nasal ectodermal thickening which was not visible in stage 4 is now large and distinct. The pineal body still forms a small projecting knob on the roof of the thalamencephalon. The infundibulum is enlarged and forms a rounded dilatation on the floor of the fore brain. Emerging from the side of the brain between eye and ear are the rudiments of cranial nerves V and VII, which can be followed as dark streaks running down over the wall of the pharynx. Behind the auditory vesicle just dorsal to the external gills is the rudiment of cranial nerves IX and X forming a dark mass of cells forked ventrally. The pharynx is still much inclined and runs upwards and backwards. On the undersurface of the head the position of the lower end of the mandibular arch is marked by the protruding cement organ. The external gills of branchial arches I and II are budding out, the hinder one is bifid. The branchial pouches could not be made out. The simple twisted tubular heart is very distinct and opening into it behind three veins are seen. The left ductus Cuvieri runs down to the sinus venosus from the dilated portion of the left posterior cardinal vein which surrounds the left pronephros and forms a bay in the antero-dorsal surface of the yolk. The left posterior cardinal vein can be made out running from the root of the tail over the surface of the yolk towards the pronephric sinus. Above the pericardium is the dark anterior cardinal sinus expanding posteriorly into the pronephric sinus. Running over the lower surface of the yolk is a branched system of vitelline veins which appear to open into a hepatic sinus on the antero-ventral surface of the yolk and in front of the liver diverticulum.

The clear rod of the notochord and the darker neural tube can be easily distinguished in the trunk and tail regions. The ventral contour of the tail at its root bulges in consequence of the presence of a mass of post-abdominal yolk between which and the main mass of abdominal yolk the cloaca can be distinguished as a pale streak.

Several later stages are represented in Plate XXII. figs. 5, 6, 7 and 8. These show the growth of the fin-fold; it is very narrow in stage 5 and confined to the tail, at stage 7 the dorsal fin has extended forward over the back to the anterior end of the abdominal region. In fig. 8 S the dorsal fin-fold is much wider than the ventral. At stage 8 pigment has begun to appear in the skin over the brain, and along the dorsal edge of the myotomes, also over the pronephros and the dorsal surface of the yolk. By comparing the gills in figs. 5, 6, 7 and 8, where they have been drawn with great care, it will be seen that there is a good deal of irregularity in their development, and that they tend to fuse together at their bases. The three distinct external gills shown in fig. 6 S are in the slightly later stage of fig. 7 no longer distinguishable and have a common base. The palmate appearance of the branchial appendage thus formed is well seen in fig. 8 V. The opercular fold forms on the left hyoidean arch earlier than on the right (figs. 6 V and 8 V). The right and left cement organs do not fuse across the mid-ventral line, but remain quite separate during their development. The nasal pits are ventral at the stages of figs. 6 to 8, and shift to a dorsal position later (see text-fig. 173). The mouth opening is beginning to assume the shape of the horny jaws in fig. 8 V and the stomodaeum is perforate, while at the stage of fig. 6 V the stomodaeal membrane is still present.

The shape of the tadpole when the hind limbs are budding is drawn in text-fig. 173. The spiracle is single and on the left side. The tail is sharply pointed at the tip, its dorsal crest reaches forward over two-thirds of the length of the abdomen. This tadpole is the latest available stage but has not reached its full development as a tadpole, so that the shape of the tail, the mouth, etc. are not in their final state.

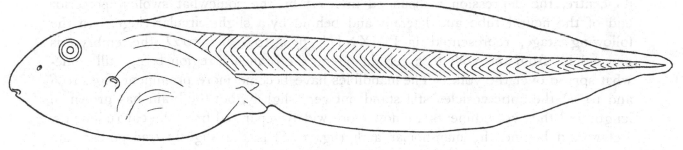

FIG. 173. Tadpole of *Paludicola fuscomaculata* Steind. × 8.

Hemisus marmoratum. Ptrs.[1]

The earliest stages which were preserved are late segmentation stages of which there is nothing special to note except that the eggs are quite without pigment; those preserved in formalin having a dense opaque pure white appearance which no doubt is not very different from the appearance of the living egg. It is extremely difficult in the absence of pigment to make out the cell outlines but there seem, from

---

[1] I rely on Budgett's identification of the species, the collection of embryos, etc. being labelled in his handwriting. He brought home living specimens of *H. marmoratum* Ptrs. from the Gambia expedition of 1900.

what could be seen, to be no features different from corresponding stages in the eggs of similar Anura.   The diameter of the segmenting egg is 2·4 mm.

The youngest embryos are at the stages shown in Pl. XXIII. figs. 9 *A*, 9 *S*, and 9 *P*.   The optic vesicles have only just begun to form, but the embryo has already risen above the surface of the yolk.   The mass of yolk compared with the size of the embryo is comparatively large, but the elevation of the embryo above the general surface of the yolk is as great as that of an older embryo of *Phyllomedusa* (compare Pl. XXVII. figs. 24 *A* and 24 *D*).   The length of the embryo covers only little more than 180° of the circumference of the egg (fig. 9 *S*).

In this stage of *Hemisus* the neural tube has just completely closed in, there is still a trace of the fused edges of the medullary folds at the extreme posterior end (fig. 9 *P*).   In figs. 9 *A* and 9 *S* the shape of the brain can be made out.   It is short and wide and the auditory rudiment is very near the anterior end.   Round the sides and in front of the brain are the mandibular arches meeting each other in front of the crescentic stomodaeum.   Surrounding the mandibular arches is the flattened-out wall of the branchial region, the right and left halves being continuous across the middle line below the mandibular symphysis.   Dorsally this *branchial plate*, as I propose to call it ("Sinnesplatte" of O. Schultze), is marked out by slight grooves into hyoidean and first branchial arches and a groove is just commencing to form dorsally between the second and third branchial arches.   These are best seen in the side view (fig. 9 *S*).   The myotomes could not be distinguished.   In the view from behind (fig. 9 *P*) the proctodaeum is seen as a slight depression with a small perforation in its centre, the depression is bounded in front by the somewhat swollen posterior end of the neural tube and laterally and behind by a slight circular ridge.   In the following stage, represented in Pl. XXIII. figs. 10 *S* and 10 *D*, the embryo is pinched off from the yolk to a certain extent, the branchial region being still somewhat spread over the yolk.   The mandibles have become more prominent (figs. 10 *S* and 10 *A*), the optic vesicles still stand out very slightly, but the brain has grown in length and the eye rudiments are now more widely separated from the ear rudiments. Below and behind the mandibular arch (fig. 10 *S*) is the slightly transparent and apparently darker wall of the pericardium, roughly triangular in shape.   The tail has grown outwards and downwards and the whole dorsal contour of the embryo forms a slightly convex curve.

In stage 11 (Pl. XXIII.) the head fold has risen above the yolk, the sides of the pharynx have almost assumed their final position, the visceral arches are clearly defined.   In the side view (fig. 11 *S*) the great prominence of the mandibular arch is obvious.   In this figure the long axis, it will be observed, is almost straight in an embryo freed from the vitelline membrane as this was. In the posterior view it can be seen that the proctodaeum is not on the root of the tail but on the surface of the yolk.

At stage 12 (Pl. XXIII.) the tail has elongated considerably and the head is sharply demarcated from the yolk by well-marked neck grooves; it has grown much more prominent.

Stage 13 (Pl. XXIV.) is slightly more advanced. It is drawn on a large scale in order to show the minute rudiment of the nasal pit immediately in front of the eye and almost touching it, and to show also the pronephros which will be referred to later. The lower end of the mandible is still very prominent, just below it is the slightly bulging pericardium.

An embryo of the stage shown in fig. 13 was cleared in clove oil and it is represented in two views in Pl. XXVI. figs. 20 S and 20 D. In the side view the olfactory pit, the eye and the ear stand out distinctly and behind the eye the short, deep infundibulum appears, while the regions of the brain, with the exception of the hind brain, are more distinct in the dorsal view. The trigeminal nerve crosses the infundibulum and divides into maxillary and mandibular branches, parallel to it run the facial nerve, likewise forked, and the glosso-pharyngeal which is faint. The root and branches of the vagus, however, are quite distinct : three branchial and the lateral line branch, the latter disappearing over the pronephros. The lower end of the mandibular arch forms a slight bulging of the ventral contour of the head, but otherwise the visceral clefts are only indicated in these two figures by the ectodermal grooves on the sides of the head in fig. 20 D, apart from the course of the branchial nerves. The distribution of the chromatophores over the brain, the sides of the head, and the pronephros is well marked. In the dorsal view the segmentation of the hind brain is the most interesting feature, there being nine "neuromeres," a larger number than has hitherto been observed in the hind brain. The pineal body is quite inconspicuous.

In all the early stages up to stage 13 the size of the embryo is completely dwarfed by the dimensions of the yolk, and this condition of things only begins to disappear in stage 15 (Pl. XXIV.), which appears to be the stage in the series at which the larva hatches. There is now a well-formed tail and the general appearance of the larva reminds one of a newly-hatched Teleostean fish with a bulky yolk. It is obvious now that external gills, as Budgett observed, are not developed in *Hemisus*, as there is up to this stage no indication of their appearance and the opercular fold is already commencing to grow back over the branchial region. In the dorsal view of this stage the regions of the brain show fairly clearly (fig. 15 D) and the pineal body is large and distinct.

Stages 15, 16 and 17 all have a peculiarly shaped tail with a kind of sigmoid flexure. This is only a passing phase and has disappeared at stage 18 (Pl. XXV.). Stage 16 (Pl. XXV.) is rapidly approaching the typical tadpole condition. The spiracle on the left side defines the limit of the head, behind it the yolk is still present in quantity, it persists in *Hemisus* to a remarkably late stage at which the first rudiment of the hind limb is just appearing.

Later stages of the young tadpoles are given in Pl. XXV. figs. 18 S, 18 D and 19 S. These are for a tadpole remarkably graceful as both abdomen and tail are rather elongated. The dorsal fin does not extend forwards over the abdomen.

The tadpole does not appear to need food until stage 19 (Pl. XXV. fig. 19 S) is reached.   This is the oldest in the series of specimens and a tube containing them is marked "Food" by Budgett.

The dorsal view of stage 18 (Pl. XXV. fig. 18 *D*) shows the pineal body covered by a small area of the integument free from chromatophores.   In both figs. 18 *D* and 18 *S* the lines of skin sense organs are indicated.

Some stages in the development of the pronephros as seen in embryos in spirit are drawn in Pl. XXIII. figs. 10 *K* and 12 *K*.   The portion of the preparation shown in fig. 10 *K* includes the left side of the embryo from the auditory vesicle to the tenth metotic myotome.   Just in front of the ear the facial nerve can be seen with its pre- and post-trematic branches and just behind the ear the glosso-pharyngeal nerve with its branches.   Horizontally behind the root of n. IX lies the ganglion of the vagus in contact with the small metotic myotome I.   Then follow three myotomes (II to IV) below which lie three nephrostomes, obviously segmental.   Below and parallel to the third nephrostome is the archinephric duct which runs obliquely upwards to the lower edge of the myotomes and then backwards as far as myotome X.   The pronephros of stage 12 (fig. 12 *K*) is not all visible, only the "collecting trunk" and the beginning of the duct could be made out, but it has been drawn to show the remarkable similarity of the arrangement of the coils to those visible in the pronephros of embryo 13 (Pl. XXIV. fig. 13).

The Development of the Face.

I have had a series of careful drawings prepared (Pl. XXIII. figs. 9 *A*, 10 *A* and 11 *A* ; Pl. XXIV. figs. 14, 14*, 15*A, 15* *S* and 16 *A* ; Pl. XXV. figs. 17 *A* and 19 *A*) to illustrate the development of the face.   Stage 9 has already been described.   The main differences between this and stage 10 (fig. 10 *A*) are the greater prominence of the fore brain and of the mandibular arches and the narrowing of the stomodaeum, while the optic vesicles now just appear at the sides of the fore brain and the branchial plate has risen from the surface of the yolk.   In fig. 11 *A* these characters have become still more prominent and the stomodaeum has become deeper.   Pl. XXIV. fig. 14 shows the forming maxillary processes, the notch between them and the notch at the symphysis of the mandibles give the mouth a <> shape.   The nasal pits are beginning to show distinctly at this time. On each side of the head the protruding visceral arches produce a curious effect. The tissues over the mandibles pass ventrally without a sharp boundary into the wall of the pericardium seen raised above the yolk below the head ; indeed the whole character of the face at this stage is best described by stating that there is a complete absence of marked features.

A slightly more advanced stage is shown in Pl. XXIV. fig. 14* where a groove has formed between the maxillary and mandibular regions and a slight swelling is just forming at the tip of the mandibles on the ventral edge of the stomodaeum ; this swelling is the rudiment of the lower lip, its length at present is only about twice the diameter of the nasal pit. The operculum is commencing to grow back over the branchial arches and it is quite clear from this figure that no trace of external gills exists, this being the stage at which they ordinarily reach their fullest development. The edge of the operculum stretches from below and behind the auditory vesicle downwards and forwards as far as the lateral wall of the pericardial region, where it ends. The operculum has only developed sufficiently to cover a part of the first of the four branchial arches.

A further stage, perhaps of some morphological interest, is shown in figs. 15S, 15*S and 15*A. Here the right and left opercular folds have met below the throat, in stage 15 before the branchial region is covered in, and later at stage 15* immediately after the operculum has reached its complete backward extension and just before its hinder free edges have fused with the body wall to enclose the branchial chamber. This condition, especially as it appears in fig. 15*S, recalls very vividly the gular fold of *Amblystoma* and many other Urodeles. It is easy to understand that in Urodeles, whose larvae are always characterised by long plumose external gills, the gular fold should remain in a rudimentary condition, whereas in the Anuran tadpole it has assumed a different function and forms the floor of the branchial chamber. Since the outer surface of this chamber becomes merged at the metamorphosis into the general surface of the under side of the head there can be no gular fold in the adult Anura equivalent morphologically to that of the Urodela.

The edge of the operculum at its upper end on the left side is seen to gape in fig. 15*A at the spot where the spiracular opening remains as the result of the edge of the operculum not fusing along this portion with the skin of the neck. The first state of the spiracle is shown in Pl. XXV. fig. 16S. Later it becomes more spout-like with the opening of the spout directed upwards (figs. 17S and 18S), and still later the spout disappears (fig. 19S), and the spiracle opens from below into a little triangular depression.

The small knob below the stomodaeum in Pl. XXIV. fig. 14* can be followed through its growth in fig. 15S and in fig. 15*A it appears as a low oval elevation with its horizontal long axis rather longer than the mouth opening. Until and at stage 15 the position of the mouth is almost terminal, at stage 16 it has become ventral as a result of the growth of the maxillary region between the nostrils and the edge of the upper lip. The change is evident from a comparison of Pl. XXIV. fig. 15S and Pl. XXV. fig. 16S. Seen from before and slightly from below the face of stage 16 has a remarkable and, for an Anuran larva, a somewhat unfamiliar appearance (Pl. XXIV. fig. 16A). This is due to the fact that at this particular stage all of the features, including the eyes, can be seen when the tadpole is looked at from the front, whereas at later stages when the eyes and the nostrils come to be placed laterally they are no longer visible. In fig. 16A the lips have begun to show their final disposition, namely, a flap-shaped upper lip and a cushion-like lower lip. The upper lip consists of a right and a left flap with a notch between them in the middle. Each flap is triangular and the apex of the triangle overhangs the angle of the mouth at each side. The thickened and enlarged knob of the lower lip is still smooth and is partially covered with small chromatophores. The swelling surrounding the lower lip is that of the mandible and below it, much fore-shortened, appears the pericardial swelling. Above and to the sides of the mouth are the nasal pits, they are raised on a slight eminence which is emphasized on the right side of the figure by the lighting of the object. Behind this nasal eminence and below the eye is a ridge formed by the prominence of the hyoidean arch. The eye stands out slightly above the surface of the head and remains prominent at later stages (see Pl. XXV. figs. 16D and 18D) but at the stages between 11 and 15 the eyes are sunk below the surface. The lens of the left eye stands out as an unpigmented patch. On the same side behind the eye is seen the bulging wall of the branchial chamber. The outer contour of this is continued downwards and inwards as a faint groove across the middle line behind the pericardium, this groove marks the attachment of the branchial operculum at the hinder end of the branchial chamber where the operculum fuses with the body wall. The sparse distribution of the chromatophores along the mid-dorsal line over the brain does not persist for long, they are especially densely clustered over this region at stage 18 (Pl. XXV. fig. 18D). The lines of skin sense organs begin to be conspicuous at this stage.

From now onwards the face in front view becomes more and more restricted in area, the head becomes more and more pointed in front and at the same time the eyes and nostrils come to lie at the sides and on the dorsal surface. At stage 17 the eyes lie well back on the sides of the head (Pl. XXV. fig. 17S) and hence they are omitted in fig. 17A. This figure shows the nasal pits and the mouth. The upper lip has become wider and its lateral angles more prominent. The lower lip is budding off from its ventral edge a pair of the long papillae seen in stage 19 (fig. 19A). The cushion-like thickening has grown out into two transverse ridges with a shallow groove between them. Each of these ridges is again divided by a shallower groove than the first so that four in all (see fig. 19A) are formed on the lower lip. Each of them is set with a row of horny teeth at stage 19, when the horny jaws still consist of imperfectly compacted rows of horny teeth. The upper jaw in fact at its outer ends is continued into a row of discrete horny teeth. The tooth formula at this stage, the

hind limbs being mere buds, is $\dfrac{\overset{1}{\overline{1-1}}}{\underset{3}{\overline{1-1}}}$, but this may be modified by the time the tadpole

has fully developed into the condition at which the formula is usually taken. The projecting outer angles of the upper lip are covered with short papillae (fig. 19A) and the edge of the lower lip bears two rows of papillae, an inner row of short ones and an outer row of longer ones. The latter apparently arise like the first two in pairs, a right and a left one; at this stage there are four pairs, the shortest and presumably the youngest pair in the middle. Seen in profile (figs. 18S and 19S) the only part of the mouth visible is the upper lip and its pendant outer angle.

Phyllomedusa hypochondrialis, Daud.

Two interesting early stages have been drawn from specimens cleared in clove oil.

An anterior and a posterior view is given of an early stage (Pl. XXVII. figs. 23 *A* and 23 *P*) where the medullary folds are closing in. At the extreme posterior end the folds have met (fig. 23 *P*) but at the upper part of the figure they are seen to be separate. In this view the proctodaeum is seen indenting the edge of a wide dark area with a narrow darker border. This area is that over which the endoderm roofing in the archenteron extends. In the anterior view of the same egg the medullary folds are just meeting at about the middle of their length, farther forwards the floor of the medullary groove can be seen with a pale streak running along its middle produced by the underlying notochord. The dark band on either side of the edge of the medullary folds shows the width of the medullary plate. Opposite the place where the notochord ends anteriorly the edges of the medullary folds diverge a little and then again become parallel while the edges of the medullary plate continue to diverge so that the widest part of the plate is at the anterior end. The small patch immediately anterior to the medullary plate and forming a slight indentation in the middle of its anterior edge is the stomodaeum. In certain lights this small patch looks most deceptively like a low knob projecting above the surface. One more feature in fig. 23 *A* should be noticed. This is the circular dark patch surrounding the cephalic part of the medullary plate. A comparison of fig. 23 *A* with fig. 24 *A* shows clearly that this dark area is the pharyngeal wall, the visceral arches, etc., flattened out on the yolk, it is the *branchial plate*.

The stage shown in the four figures 24 *A*, 24 *D*, 24 *P* and 24 is illustrated in some detail as it demonstrates with great clearness the remarkable extent of the spreading of the head organs on the surface of the large yolk. The fore brain and the optic vesicles are conspicuous in fig. 24 *A* and behind them the hinder brain vesicles in which three constrictions of the neural tube are distinctly seen in the region of the mid brain and hind brain. Spread out on either side of the brain, but not extending in front of it, is the flattened pharyngeal wall with its thickened folds along the lines of the future gill clefts. It is perfectly obvious that in such a condition as this the ventral ends of the branchial arches are widely separated and that this must apply equally to the mandibular arches. These cannot and do not meet below the stomodaeum as they do in *Hemisus* at a much earlier stage (Pl. XXIII. fig. 9 *A*) and as they do from their first appearance in other Anuran embryos with a smaller amount of yolk. The anterior border of the branchial plate is coincident with the anterior edge of the neural plate and the stomodaeum is anterior to these. The stomodaeal region and the end of the fore brain are shown in fig. 24. The mandibular arches then lie at the sides of the fore brain and their position is roughly defined as lying between the optic vesicle and the dark streak near it which is the hyomandibular gill pouch. The structures in the posterior part of the branchial plate can be more easily distinguished in the dorsal view of the same embryo (fig. 24D). Above is the fore brain with the flattened optic vesicles on either side, just behind these is the branchial plate which produces the slight bulgings behind the optic vesicles. The cavity of the pharynx shows as a clear space in this view, divided by dark bands running in a transverse direction and appearing to converge and meet at the edge. These dark bands are the folds of the pharyngeal wall representing the gill clefts. The band immediately behind the optic vesicles is the hyomandibular pouch, the band behind the first clear space is the first branchial pouch, and the less distinct band behind the second clear space is the second branchial pouch just appearing.

The whole extent of the neural tube behind the fore brain is seen in this view, with nine pairs of myotomes right and left of it. Posterior to the myotomes lies the unsegmented mesoderm from which the posterior myotomes are differentiated. This unsegmented myotomic mesoderm extends round the posterior end of the neural tube (see fig. 24P) and ends in front of the proctodaeum. On each side of the procto-daeum a band with a dark edge sweeps outwards and forwards parallel with the edge of the segmented mesoderm and this is the lateral mesoderm. In fig. 24D the proctodaeum is visible as a kind of short projecting spout.

PLATE XXII.

EXPLANATION OF PLATE XXII.

Paludicola.

The letters suffixed to the numbers of the figures indicate cases in which a side view (*S*) and a ventral view (*V*) of the same specimen has been drawn.

FIG. 1. Egg showing blastopore.   × 58.   (See p. 445.)

FIGS. 2*S* and 2*V*.   Embryo in vitelline membrane.   × 50.   (See p. 445.)

FIGS. 3*S* and 3*V*.   Older embryo in vitelline membrane.   × 50.   (See p. 447.)

FIG. 4.   Embryo intermediate between stage two and stage three seen by transmitted light in clearing reagent.   × 64.   (See pp. 447, 448.)

FIG. 5.   Larva.   × 40.   (See pp. 450, 451.)

FIG. 6*S*.   Larva, anterior end.   × 64.   (See pp. 450, 451.)

FIG. 6*V*.   Same larva.   × 35.   (See pp. 450, 451.)

FIG. 7.   Larva.   × 36.   (See pp. 450, 451.)

FIGS. 8*S* and 8*V*.   Larva.   × 33.   (See pp. 450, 451.)

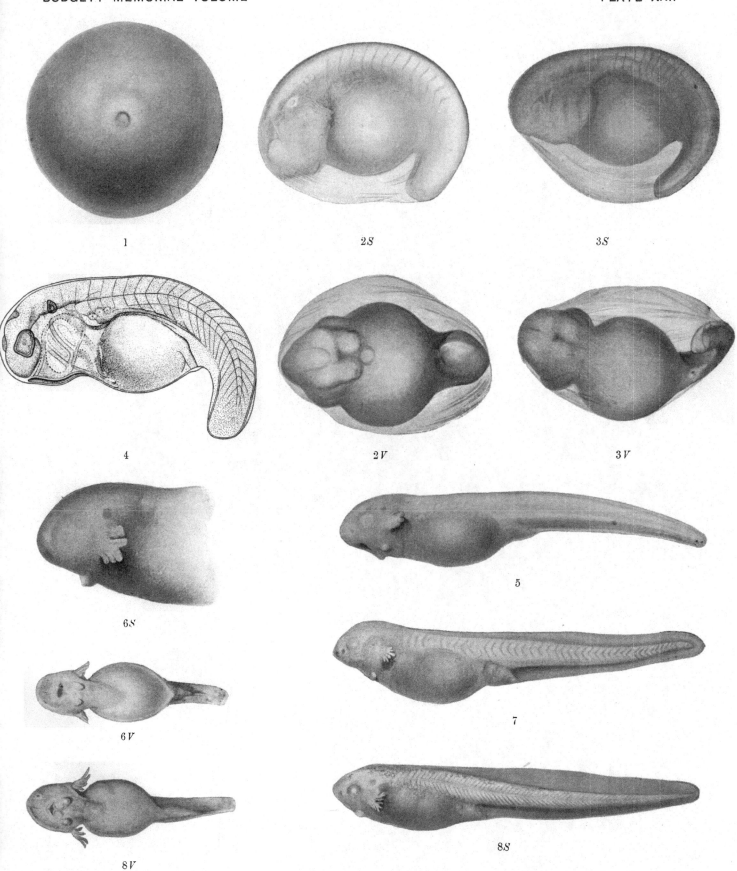

1

2S

3S

4

2V

3V

6S

5

6V

7

8V

8S

DEVELOPMENT OF PALUDICOLA

PLATE XXII. BURDETT MEMORIAL VOLUME

PLATE XXIII.

EXPLANATION OF PLATE XXIII.

Hemisus.

See explanation of Plate XXII. *D* indicates dorsal view, and *P* view from behind.

Figs. 9*A*, 9*S* and 9*P*. Egg removed from the vitelline membrane. × 17. (See p. 452.)

Fig. 10*S*. Embryo removed while living from the vitelline membrane. × 16. (See p. 452.)

Fig. 10*D*. The same embryo. × 13. (See p. 452.)

Fig. 10*A*. The same embryo, frontal view. × 19. (See p. 452.)

Fig. 10*K*. The left otic and pronephric region of the same embryo × 40, as seen in the spirit specimen. (See p. 454.)

Fig. 11*D*. An embryo removed from the envelopes while living. × 14. (See p. 452).

Fig. 11*A*. The same embryo, frontal view. × 18. (See p. 452.)

Figs. 11*S* and 11*P*. The same embryo. × 14. (See p. 452.)

Fig. 12*D*. Embryo preserved while still in the vitelline membrane. × 16. (See p. 452.)

Fig. 12*K*. The left pronephros of the same embryo × 42, as seen from the dorsal side in the spirit specimen. (See p. 454.)

PLATE XXIII

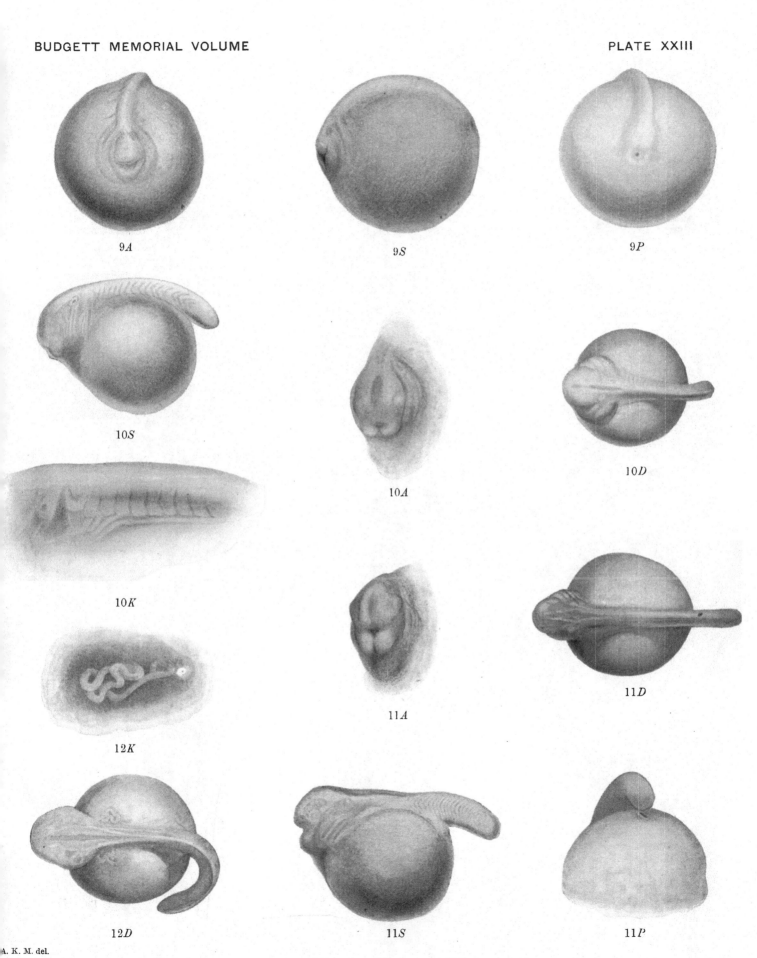

9A

9S

9P

10S

10A

10D

10K

11A

11D

12K

12D

11S

11P

A. K. M. del.

DEVELOPMENT OF HEMISUS

PLATE XXIV.

EXPLANATION OF PLATE XXIV.

Hemisus.

See explanation of Plate XXIII.

FIG. 13. Embryo about the same age as stage 12 removed from its envelopes after fixation. × 28. (See p. 453.)

FIG. 14. Similar embryo. × 21. (See p. 454.)

FIG. 14*. Anterior end of slightly older embryo. × 30. (See p. 454.)

FIGS. 15*S* and 15*D*. Still older embryo. × 10. (See pp. 453, 455.)

FIG. 15**A*. Slightly older embryo than fig. 15. × 16. (See p. 455.)

FIG. 15**S*. Slightly older embryo than fig. 15. × 20. (See p. 455.)

FIG. 16*A*. Anterior view of larva shown on Pl. XXV. fig. 16*S*. × 40. (See pp. 455, 456.)

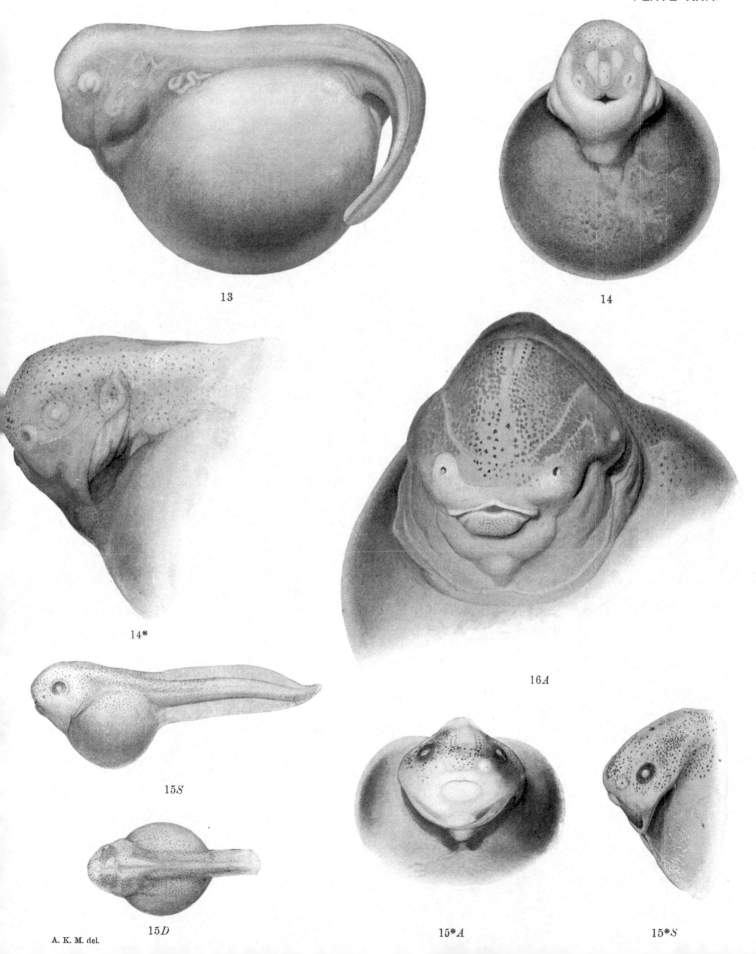

13

14

14*

16A

15S

15D

15*A

15*S

PLATE XXIV

PLATE XXV.

EXPLANATION OF PLATE XXV.

Hemisus.

FIGS. 16*S* and 16*D*. Larva. × 8·5. (See pp. 453, 455, 456.)

FIG. 17*S*. Outline of larva. × 10. (To show the stage of fig. 17*A*.) (See pp. 453, 456.)

FIG. 17*A*. Frontal view of the same larva. × 43. (See p. 456.)

FIGS. 18*S* and 18*D*. Larva. × 10. (See pp. 453, 454, 456.)

FIG. 19*S*. Oldest larva. × 10. (See pp. 453, 456.)

FIG. 19*A*. View of the same larva as seen in the direction of the arrow in fig. 19*S*. × about 50. (See pp. 453, 454, 456.)

16S

16D

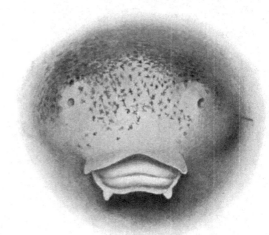

17A

17S

18D

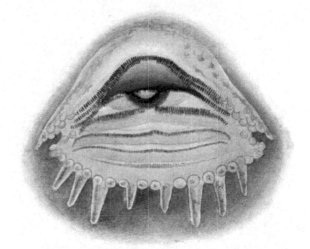

19A

18S

19S

A. K. M. del.

DEVELOPMENT OF HEMISUS

PLATE XXVI.

EXPLANATION OF PLATE XXVI.

Hemisus. Paludicola.

FIGS. 20*S* and 20*D*. An embryo of *Hemisus* of the stage shown in fig. 13, as seen in clove oil with spot-lens illumination. × 26. (See p. 453.)

FIGS. 21*S*, 21*D* and 21*V*. The same stage of *Paludicola* as is shown in fig. 2*S* and fig. 2*V* and seen as above. × 65. (See pp. 446, 447.)

FIG. 22. Larva of *Paludicola* slightly younger than that of fig. 5, Pl. XXII., cleared in clove oil and viewed with spot-lens illumination. × 45. (See pp. 449, 450.)

PLATE XXVI

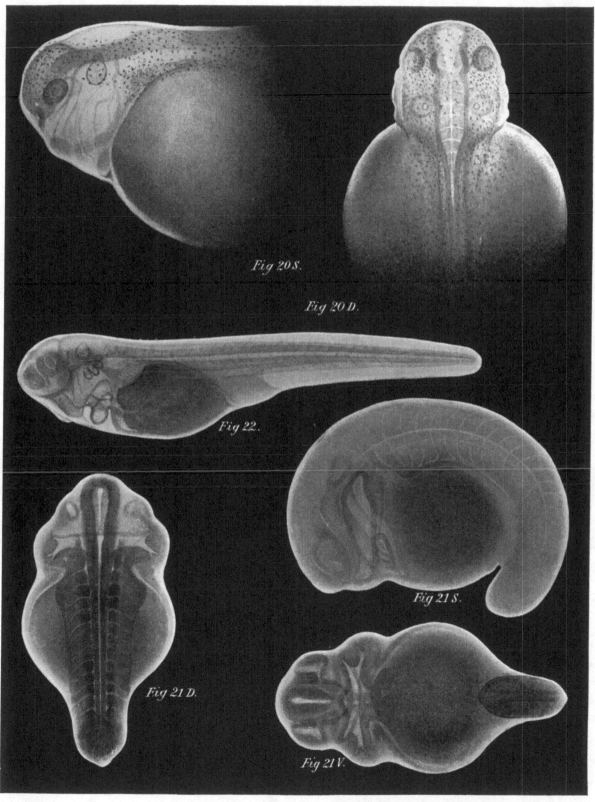

Fig 20 S.

Fig 20 D.

Fig 22.

Fig 21 S.

Fig 21 D.

Fig 21 V.

A. K. M. del.

DEVELOPMENT OF HEMISUS AND PALUDICOLA

PLATE XXVII.

EXPLANATION OF PLATE XXVII.

Phyllomedusa.

FIGS. 23*A* and 23*P*. Egg removed from vitelline membrane, seen in clove oil with spot-lens illumination. × 33. (See p. 457.)

FIGS. 24*A*, 24*D* and 24*P*. Older egg in the vitelline membrane, seen as above. × 30. (See pp. 457, 458.)

FIG. 24. The stomodaeal region of the same embryo. × 35. (See p. 457.)

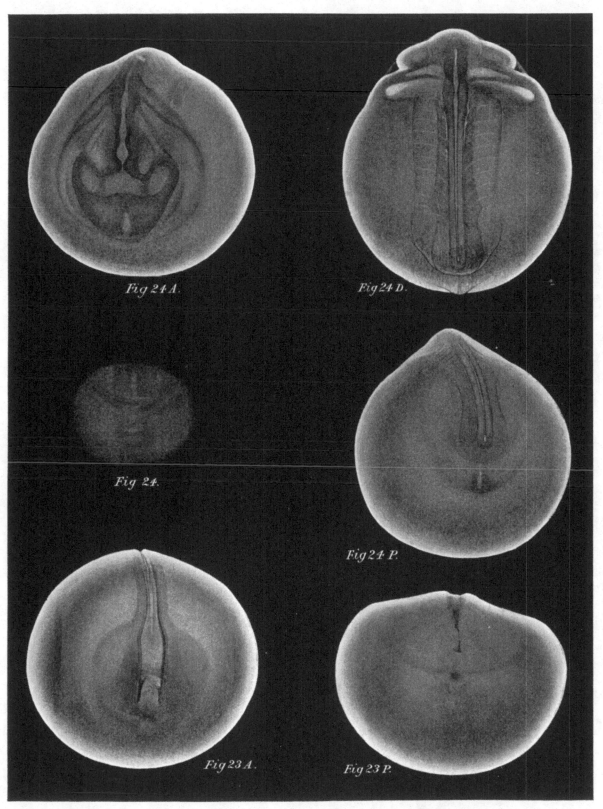

Fig 24 A.

Fig 24 D.

Fig 24.

Fig 24 P.

Fig 23 A.

Fig 23 P.

A. K. M. del.

DEVELOPMENT OF PHYLLOMEDUSA

XVII. ON THE FRESHWATER MEDUSA, LIMNOCNIDA TANGA-NICAE, DISCOVERED IN THE RIVER NIGER BY THE LATE J. S. BUDGETT.

By EDWARD T. BROWNE, B.A., Zoological Research Laboratory, University College, London.

With Plate XXVIII.

INTRODUCTION.

In the collection brought home by Mr J. S. Budgett from the river Niger there were five specimens of a medusa taken in a freshwater lagoon near Assay. The first two specimens came by post to me, but without a letter or label. As the box was stamped with the Cambridge postmark I thought, at the time, that the sender was Mr Stanley Gardiner, whose collection of medusae from the Maldives I was then examining. On looking at the specimens (without removing them from the bottle) I considered them to be Narcomedusae and placed the bottle upon the shelf where it remained for several weeks. At last I had to write to Mr Gardiner about some of the Maldive specimens and mentioned the arrival of the two medusae from Cambridge without a label. Mr Gardiner replied "I did not send them. On the morning that Budgett was taken ill I recommended him to send two practically freshwater medusae from the river Niger to you. I will make inquiries and write to you again." Mr Gardiner's inquiries established the fact that the medusae had been sent by Mr Budgett, who no doubt intended writing on the day the box was posted, but his sudden illness and death prevented that letter from ever coming.

Fortunately Mr Budgett kept a diary during his travels and Professor Graham Kerr, to whom Budgett bequeathed his collections, sent the following extracts from it in a letter to me :—" The jelly-fishes came apparently from a lagoon (freshwater) near Assay in the Niger delta. In Budgett's diary I find the position of the lagoon described as follows :—

"*Aug.* 22 (1903). Went down to nest of *Gymnarchus*. Opposite Bari entered creek, then a small stream very winding. After half a mile came to swamp. Then crossing entered another stream, all the way through dense forest carpeted with ferns; after quarter mile came to large lagoon free from sand....

"*Aug.* 25. Revisited lagoon. Tow-netted all morning. Many kinds of small fish and some jelly-fishes....

"*Sept.* 16. Trawled lagoon, but caught only jelly-fishes."

Later on I received three more specimens from the Budgett collection. These specimens look as if they had been badly mauled in a trawl net. They are little more than discs of jelly and would just prove the existence of a medusa in the Niger. The two specimens sent by Mr Budgett are in very good condition and had been preserved in formalin.

An earlier record of a Medusa in the Niger.

The occurrence of a medusa in the river Niger was, however, first noticed in 1888, by Dr Tautain, "a distinguished anthropologist and an excellent observer, who is well known on account of his travels in French Soudan, and in Bélédougou, and who was formerly medical officer to the Galliéni Expedition." Dr Tautain after leaving Africa read in *La Nature* a résumé of Günther's paper on the Medusa of Lake Tanganyika and he sent a letter to the editor of that journal, who forwarded it to M. Jules de Guerne to bring before the Société zoologique de France.

The following are the principal passages in Dr Tautain's communication, written from Taiohae (Nouka-Hiva Island), 30th September, 1893 :—

"In the issue of *La Nature* for June 24, 1893, I observe, under the signature of M. Jules de Guerne, an article on the subject of freshwater Medusae, with reference to the Medusa of Lake Tanganyika.

"It is stated by M. de Guerne that this species is the third freshwater Medusa that has been recorded. I believe that I am acquainted with another.

"In the month of January in the year 1888, at low water, I found in the Niger near Bamakou, in the still water at the edge of the river above the rocks of Sotuba, a Medusa which appears to me different from that of Lake Tanganyika. If my memory serves me, the diameter of this Medusa is from 20 to 25 millim.

"On the day when I noticed it I busied myself in collecting a certain number of individuals, and in a short time I had some fifty specimens of it in a bottle.

"On my return to Bamakou I endeavoured to preserve these Medusae, in order to bring them back to France; but the various methods which I employed, the only ones that I had at my disposal, miscarried, and after the elapse of a certain time, varying according to the different methods, I had nothing left. It was my fixed intention to return to Sotuba to make a fresh collection and attempt other systems or combinations of systems of preservation, but I was unable to do so.

"The distance between the habitat of the Medusa of the Niger and the sea is considerable, and it must be remarked that the number of rapids (besides those of Sotuba) between Timbuctu and Boussa renders communication with the ocean very difficult for a creature of the nature of a Medusa."

The medusa found by Budgett in the delta of the Niger I am sure is *Limnocnida tanganicae*, Böhm. After a careful examination of the specimens I have failed to find a character sufficiently distinct and likely to be constant, by which the Niger medusa could be clearly distinguished from those inhabiting Lake Tanganyika. The Niger medusa does not agree exactly with the description given of the Tanganyika species, but the points of difference are practically useless for the establishment of a new species or even of a new variety.

Description of Mr Budgett's specimens.

Umbrella. The umbrella is disc-shaped, with a flat or nearly flat top, about three times as broad as high. The central portion of the umbrella is very thick and projects into the cavity of the sub-umbrella. The peripheral portion is fairly thin and curves inwards. The velum is narrow.

Stomach and Mouth. The stomach is circular and its diameter is nearly that of the umbrella. The space which should be the central cavity of the stomach is occupied by a great mass of jelly, which almost blocks up the whole of the cavity of the sub-umbrella and forms the roof of the stomach. The lower or lateral wall of the stomach hangs down as a narrow band, encircling the periphery of this mass of jelly. The mouth is circular, measuring a little less than the diameter of the stomach and has a plain even margin.

A similar kind of stomach is found in *Mesonema pensile* (Browne, 1904), a large medusa belonging to the Aequoriidae. In *Mesonema* the stomach has become quite rudimentary and the mouth cannot be closed. In *Limnocnida* the abolition of the stomach is not carried quite so far, but to judge from the appearance of these specimens the mouth is incapable, owing to the shortness of the lower wall of the stomach, of completely closing up.

In the case of *Mesonema* I have suggested that the function of the stomach has been transferred to the canal system and that the animal lives upon organisms of microscopic dimensions, such as unicellular algae and protozoa, which are driven by ciliary currents into the radial canals. Such organisms may also form the food supply of *Limnocnida*.

Canal system. There are four radial canals, which are fairly broad, running from the periphery of the stomach to a broad marginal circular canal.

Gonads. The gonads are upon the lower or lateral wall of the stomach and form a continuous circular band extending in width from the base of the stomach almost to the mouth. The wall hangs in sinuous folds, like a curtain. Both specimens are females and the ova look immature. The position of the gonads is similar to that found in some of the Narcomedusae, for instance in the genus *Solmaris* which has the gonads forming a continuous circular band in the lower wall of the stomach, and it also has a circular mouth. In *Solmaris* the lower wall of the stomach is capable of extending across to the centre of the stomach and the mouth can be closed.

Tentacles. The tentacles are very numerous and are very closely packed together round the margin of the umbrella. They are arranged in definite series showing a well marked difference in size which is due to growth. The four perradial tentacles, opposite the radial canals, are the largest and these are closely followed in size by the interradial and adradial tentacles. Then follow about five series of tentacles, decreasing rapidly in size. The tentacles in each series arise between the tentacles of the previous series, so that beginning with the four primary perradial tentacles there come four interradial, eight adradial, and then 16, 32, 64 up to about 500 tentacles. The tentacles belonging to the last two series are so minute and so crowded together that it is practically impossible to trace their order or to count them. They seem to arise wherever there is room for them on the margin of the umbrella. This description of the tentacles is based upon the larger specimen, as the smaller one has the margin of the umbrella slightly damaged.

The basal portion of the tentacles is partly embedded in the jelly of the umbrella. The length of the portion which is embedded varies according to the size and age of the tentacle (fig. 2 B). In the primary series it extends about 3 millim. up the exumbrella, but in the seventh and eighth series of tentacles the attachment is not visible. The basal portion of the tentacle is hollow and its cavity is in direct communication with the marginal canal.

Near the point where the tentacle leaves the umbrella and becomes quite free and isolated a small portion of the umbrella projects out to form a support and a stronger attachment for the free hanging part of the tentacle. The size and shape of this projecting support (fig. 2) vary with the series of tentacles to which it belongs. In the primary series it is a large semi-globular swelling; hollowed out on the lower side and forming a kind of groove in which the tentacle is situated. It decreases in size and the groove becomes less marked as the tentacles become smaller in size, finally disappearing at about the fifth series where the tentacles do not require a support on account of their minuteness. The presence and shape of the support appear to me to indicate that the medusa generally carries its tentacles in a downward direction.

The nematocysts upon the tentacles are contained within small projecting knobs, which are arranged in nearly transverse bands in a contracted tentacle, but irregularly in an expanded tentacle. These clusters of nematocysts do not occur upon the basal portion of the tentacle.

Sense organs. Around the margin of the umbrella there is a conspicuous continuous band of ectoderm loaded with nematocysts (fig. 2 N). It lies over the marginal canal, just below the origin of the tentacles. It is in the lower part of this band that the sense organs are situated. This band appears to be arranged somewhat in radial folds producing a grouping of the sense organs, but I am not sure that the folding is not due to the contraction of the margin of the umbrella. The sense organs in the largest specimen are so closely packed together as to form a continuous single row round the margin of the umbrella. The number in the largest specimen is estimated at about 400. In the smaller specimen the sense organs are not quite so close together leaving here and there a slight gap. Their shape is either oval or roundish.

I cut off a small portion of the margin of the smaller specimen and made a series of transverse sections. The sections showed that the preservation was bad for histological work. I am, however, able to confirm the excellent anatomical description of *Limnocnida* given by Günther. The structure of the tentacles, sense organs and marginal canal corresponds exactly with his figures.

Size. 18 mm. in width and 5 mm. in height.

11 mm. in width and 4 mm. in height.

The largest specimen from the Niger does not quite agree with the description of the species given by Günther. It has more sense organs and more tentacles. Medusae which have a large number of sense organs and tentacles usually show a great variation in the number of these organs. Much depends upon the size of the umbrella and the abundance of food. The specimens of *Limnocnida* examined by Gravier (1903) from the Victoria Nyanza also have more sense organs and tentacles than the Tanganyika form. When more specimens have been examined from Tanganyika it will probably be found that their tentacles and sense organs greatly exceed the number stated in the original description which is based upon just a few specimens.

The History of Limnocnida.

Limnocnida was first discovered in Lake Tanganyika in 1883 by Böhm, but it was not until 1892 that specimens were sent to England. These were fully described and their anatomy investigated by Günther (1893). Since then Moore has made two visits to the lake for the purpose of investigating its fauna. His expeditions have increased our knowledge of the life-history of this medusa. He has come to the conclusion that the medusa has a direct development, omitting altogether the hydroid stage, but I am at present sceptical as to the non-existence of a hydroid and will state my reasons further on.

In August 1903 *Limnocnida tanganicae* was found by Mr Hobley in the Victoria Nyanza and the specimens were exhibited by Prof. Ray Lankester at a meeting of the Zoological Society of London in December.

In September 1903 M. Ch. Alluaud also found *Limnocnida* in the Victoria Nyanza and his specimens were described by Gravier (1903).

The occurrence of *Limnocnida tanganicae* in the river Niger is, I think, more interesting than the discovery of a new freshwater medusa, as it is found in a river which runs direct into the sea, and in localities far away from the great African lakes.

The Origin of Limnocnida.

According to Moore's theory the Tanganyika district formed the floor of a central African sea in prae-tertiary times. Later on an upheaval of land took place and enclosed a portion of the sea, which is now known as Lake Tanganyika. The changes in the configuration of the land resulted in the lake's receiving the rainfall of the surrounding mountains and it became gradually converted into a freshwater lake. The original inhabitants (the Halolimnic fauna) were marine forms, and some of them, including *Limnocnida*, have survived the gradual change in the salinity of the water and have adapted themselves to a new environment.

The occurrence of freshwater medusae in a river which has direct communication with the sea naturally suggests the idea that the medusae gradually migrated up the river and have changed their habitat from salt to fresh water, just as the hydroid *Cordylophora* has done. Even if the change of habitat originally took place in the Niger and not in Tanganyika we should still have to find the means of conveyance across the great African continent.

Mr Boulenger in his presidential address to the Zoological section of the British Association (1905) discussed very fully the distribution of the African freshwater fishes. From this address I have gained much information concerning the "Tanganyika Problem." Geological evidence points to the fact that Tanganyika is not a very ancient lake, its formation not dating back beyond Miocene times; and up to the present no traces of Jurassic or Cretaceous deposits have been found on the plateau of Central Africa. The absence of these rocks indicates the absence of a sea during Jurassic and Cretaceous times, and it was just during these periods that a sea was wanted over the Tanganyika district to account for the marine origin of a certain portion of the fauna of the lake. Mr Boulenger in the course of his address alluded to *Limnocnida* in the following words :—

"As regards the origin of this Medusa, recent palaeontological discoveries afford a much more rational explanation of the presence in Tanganyika of a Coelenterate of unquestionably marine derivation. The highly important finds of fossils between the Niger and Lake Chad by the English and French officers of the Boundary Commission, which have been reported upon by Prof. de Lapparent, Mr Bullen Newton, and Dr Bather, have conclusively established the existence of Middle Eocene marine deposits over the Western Soudan, and the Egyptian and Indian character of these fossils, as well as of others previously obtained in Cameroon and Somaliland, justifies the belief in a Lutetian (Middle Eocene) sea extending across the Soudan to India. In fact, as stated by Mr Newton, the palaeontological evidence seems to prove that the greater part of Africa above the equator was covered by sea during part of the Eocene period. On this sea retreating northwards, after the Lutetian period, Medusae became land-locked and gradually adapted themselves to fresh water: they had not far to travel to find themselves in what are now the Nile lakes, and later, through the changes which Mr Moore himself has shown to have taken place in the drainage of Lake Kivu, they were easily carried into the Tanganyika— probably at no very remote time—and maintained themselves to the present day. I understand that the Medusa reported from Bammaku, Upper Niger, in 1893, but still undescribed, has been re-discovered by Budgett, and is now being studied. Should it prove to be related to the Tanganyika species, it would also have to be regarded as a relic of the same Eocene sea, and it would add further support to the very simple explanation which I have ventured to offer of a case which seemed so tremendously puzzling in our previous state of ignorance of the geological conditions of Africa between the equator and the tropic of cancer."

This appears to me to be an explanation which intelligibly accounts for *Limnocnida* occurring in two localities so far apart. It removes the need to speculate about the medusa ascending the Niger from the Atlantic Ocean, and journeying across Africa. Its position, however, in the delta of the Niger is interesting for it may again become an inhabitant of another sea.

The Life-history of Limnocnida.

Our knowledge of the life-history of *Limnocnida* in Lake Tanganyika is due to Moore's observations, which extended from April to November. He says that "the jelly-fishes do not differ in the phase of their life cycle in different parts of the lake, and consequently it is really incorrect to say that their metamorphosis corresponds in any way directly with the seasons." At the end of the wet season in March, only a very few large medusae are to be found, and upon the wall of their stomach medusa-buds are in the process of development. The young medusae which are budded off also produce medusa-buds and the process of gemmation is repeated through many generations, "so that in June and July the lake swarms with medusae; vast shoals of them, as I found, extending for miles and miles, and containing individuals of all sizes, but nearly all of them presenting manubriums which were covered with hundreds of minute developing buds." After a time this budding of young medusae ceases, and the stomach grows a new wall in which the generative cells develop. The sexually mature individuals swarmed during September and October. The ova and spermatozoa were evacuated and later on Moore found "numbers of small planulae and small medusae, which were growing rapidly; but these showed no tendency to form buds during the autumn and had, without doubt, been formed from the fertilised ova of the sexual forms."

There is, unfortunately, a break in the record for the months of December, January and February. It appears to me that there should have been an extra large shoal of medusae during these months because a further increase in numbers must have taken place, owing to the breeding of the sexual adults, which had become so numerous after a series of asexual generations as to form shoals. Even allowing for an extra heavy death-rate during the wet season I fail to see why on a system of direct development there should have been such a great diminution early in the year, within a period of about three months.

It seems to me that the weakest part in the chain of evidence for direct development is the connection between the planulae and the young medusae. Granting that the planulae belonged to the medusa, and not to some other animal, there is no mention made of the very important stages between the planula stage and the young fully formed medusa. These are just the stages of which we require a full account, as they are likely to give a clue to the relationship of this peculiar medusa to other members of the group. The presence of young medusae when the sexual adults are breeding is a common occurrence among medusae in our seas. These young medusae are late arrivals, either direct from a hydroid, or from a medusa which reproduces asexually by gemmation for a definite period and then later on has gonads which finally develop into hydroids. When young and adult medusae occur together late in the year the young medusae usually die off during the winter months without reproducing. The adult medusae after shedding their gonads also die. Consequently medusae become very scarce in winter. It is not until the advent of spring when the hydroids begin to breed that medusae again become common objects in the sea.

Moore (1897) states that the medusae in Lake Tanganyika are "rather local in their distribution and are not always easy to find, but in some places they exist in countless numbers."

Cunnington (1906) says "I was struck by the irregularity in the appearance of the Tanganyika medusa, or rather the uncertainty of finding it at any particular time or place. It is doubtless, like all such forms, driven to and fro by the wind and currents, but it is curious that one may be a month or more on the lake without seeing a single specimen."

The fact that the medusa occurs in definite places and in varying quantities appears to me to indicate the existence of a hydroid stage living in definite localities. If the medusa reproduces by direct development only, one would expect to find it fairly uniformly distributed over the lake. The hydroid in all probability has a special habitat which accounts for it not yet having been found. In our seas where collecting has been carried on systematically for many years there are still many hydroids yet to be discovered. In several instances hydroids have been reared in bell-jars from ova of medusae, but these hydroids have not yet been found in the sea. The fact that Moore and others have failed to find a hydroid in Lake Tanganyika is but little evidence against the existence of a hydroid.

Until more is definitely known about the development of *Limnocnida* I think that the probability of the existence of a hydroid stage should not be lost sight of, and that future explorers of the lakes should still look out for it.

The Position of Limnocnida.

Limnocnida still remains outside any system of classification. It looks at first sight a Narcomedusa on account of the shape of the stomach and the position of the gonads, but I do not think that it has any connection with that division of the Hydromedusae. The structure of the tentacles of *Limnocnida* is not at all like that of a Narcomedusa, in fact the margin of the umbrella, the tentacles and sense organs have not a single character of a Narcomedusa but are very much like the Olindiadae.

The stomach of *Limnocnida* has become rudimentary and what should be the central cavity of the stomach is occupied by a large mass of jelly. Its ancestors probably had a hollow sac-shaped stomach and a mouth which could be closed, similar to the stomach of a normal Anthomedusa.

Mesonema pensile has a quite rudimentary stomach, whereas in its allied genus *Aequorea*, the stomach is very large and the mouth can be closed. A transverse section through the centre of the umbrella of *Mesonema* has nearly the same external shape as one through that of *Limnocnida*.

Goto (1903) has put forward a suggestion that *Limnocnida* (and also *Limnocodium*) is related to the Olindiadae, which are, according to Haeckel's classification, a sub-family of the Petasidae, a family belonging to the Trachomedusae. Goto also states that the Olindiadae should be placed in the Leptomedusae under the Eucopidae. His researches on the origin of the sense organs of *Olindioides formosa* has led him to believe that the sense organs are exclusively derived from ectodermal cells, hence the removal of the Olindiadae to the Leptomedusae.

Perkins (1903) in his account of the development of *Gonionema murbachii* (another genus of the Olindiadae), states that the sense organs must be regarded as modified tentacles. His figure clearly shows that the sense organ arises from an outgrowth of the endoderm of the circular canal. Perkins also reared the eggs of *Gonionema murbachii* and gives a good account of their development. He obtained free-swimming planulae, which fixed themselves and developed into small hydroids with four tentacles. These hydroids he kept alive for several months and saw them produce by fission buds which, after moving about, developed into hydroids similar to their parents. The hydroids unfortunately died before revealing the most important part of their life-history.

Perkins strongly favours the idea that the hydroid ultimately develops direct into a medusa and consequently regards the Olindiadae as Trachomedusae. I fail to find any evidence to support this theory. The hydroid is evidently a Gymnoblastic form, and it is more likely to liberate medusae by budding in the usual manner than to develop direct into a medusa.

All the Olindiadae are distinctly littoral medusae and are usually found amongst sea-weeds. Their habitat supports the idea that they are budded off from hydroids which live attached to sea-weeds and other objects not far from the place where the medusae are found.

Granting that *Limnocnida* has affinities with the Olindiadae we have still to settle the systematic position of the family.

References to Literature.

Bernard, F., 1898: "Note sur des Méduses rapportées par M. Foà du lac Tanganyika, et dénommées *Limnocnida tanganyicae*, Böhm." Bull. Mus. Hist. nat. Paris, p. 62.

Boulenger, G. A., 1905: "The Distribution of African Freshwater Fishes." Brit. Assoc. Report, 1905. Nature, vol. LXXII., pp. 413—421.

Browne, E. T., 1906: "On the Freshwater Medusa, *Limnocnida tanganicae*, and its Occurrence in the River Niger." Ann. Mag. Nat. Hist., Ser. 7, vol. XVII., pp. 304—306.

Cunnington, W. A., 1906: "The Third Tanganyika Expedition." Nature, vol. LXXIII., p. 310.

Goto, S., 1903: "The Craspedote Medusa *Olindias* and some of its Natural Allies." Mark Anniversary Volume, pp. 1—22, Pls. I.—III. New York.

GRAVIER, C., 1903: "Sur la Méduse du Victoria Nyanza et la faune des grands lacs africains." Bull. Mus. Hist. nat. Paris, Tom. IX., pp. 347—352.

GRAVIER, C., 1903: "Sur la Méduse du Victoria Nyanza." Comptes rendus Acad. Sci. Paris, Tom. CXXXVII., pp. 867—869.

GUERNE, JULES DE, 1893: "A propos d'une Méduse observée par le Dr Tautain dans le Niger, à Bamakou (Soudan français)." Bull. Soc. Zool. de France, Tom. XVIII., pp. 225—230. Translation, Ann. Mag. Nat. Hist., Ser. 6, vol. XIV., pp. 29—34 (1894).

GÜNTHER, R. T., 1893: "Preliminary Account of the Freshwater Medusa of Lake Tanganyika (*Limnocnida tanganyicae*)." Ann. Mag. Nat Hist., Ser. 6, vol. XI., pp. 269—275, Pls. XIII., XIV.

GÜNTHER, R. T., 1894: "A further Contribution to the Anatomy of *Limnocnida tanganyicae*." Quart. Journ. Micro. Sci., vol. XXXVI., pp. 271—293, Pls. XVIII., XIX.

LANKESTER, E. R., 1903: "Exhibition of Specimens of Medusae from the Victoria Nyanza." Proc. Zool. Soc., 1903, p. 340. Nature, 1903, p. 348.

MOORE, J. E. S., 1897: "The Freshwater Fauna of Lake Tanganyika." Nature, vol. LVI., pp. 198—200.

MOORE, J. E. S., 1899: "Exhibition of Specimens of the Freshwater Jelly-Fish, *Limnocnida tanganyicae*." Proc. Zool. Soc., 1899, pp. 291, 292.

MOORE, J. E. S., 1903: "The Tanganyika Problem." London.

MOORE, J. E. S., 1904: "The Victoria Nyanza Jelly-Fish." Nature, vol. LXIX., p. 365.

PERKINS, H. F., 1903: "The Development of *Gonionema murbachii*." Proc. Acad. Nat. Sci. Philadelphia, vol. LIV., pp. 750—790, Pls. XXXI.—XXXIV.

PLATE XXVIII.

EXPLANATION OF PLATE XXVIII.

Fig. 1. Lateral view of the medusa. × 7.

Fig. 2. Margin of the umbrella showing the position of the tentacles and sense organs. × 12. *T*, tentacle; *B*, basal portion of tentacle; *C*, projecting support of the tentacle; *N*, band of nematocysts; *S*, sense organ; *V*, velum.

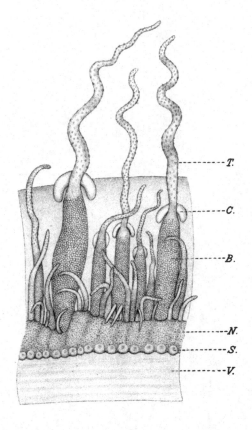

T.

C.

B.

N.

S.

V.

2.

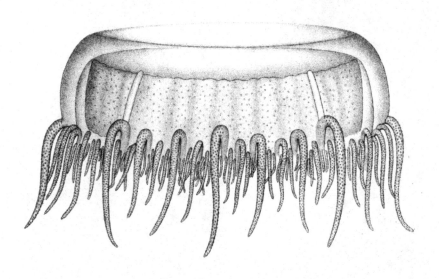

I.

E.Wilson, Cambridge.

INDEX.

NOTE: *Figures in black type refer to an illustration.*